Grasslands of the World

Diversity, Management and Conservation

Editors

Victor R. Squires

International Dryland Management Consultant
Formerly, University of Adelaide
Australia

Jürgen Dengler

Institute of Natural Resource Sciences
Zurich University of Applied Sciences
Wädenswil, Switzerland

Haiying Feng

Qinzhou University
Guangxi Region
People's Republic of China

Limin Hua

College of Grassland Science
Gansu Agricultural University
Lanzhou, Gansu Province
People's Republic China

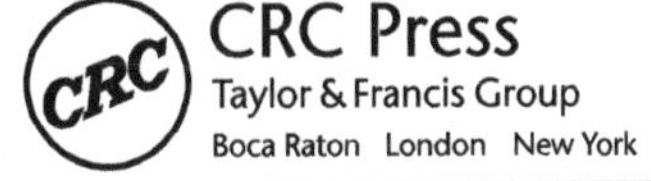

CRC Press is an imprint of the
Taylor & Francis Group, an **informa** business
A SCIENCE PUBLISHERS BOOK

CRC Press
Taylor & Francis Group
6000 Broken Sound Parkway NW, Suite 300
Boca Raton, FL 33487-2742

First issued in paperback 2021

Version Date: 20180619

ISBN 13: 978-0-367-78093-7 (pbk)
ISBN 13: 978-1-4987-9626-2 (hbk)

Visit the Taylor & Francis Web site at
http://www.taylorandfrancis.com

and the CRC Press Web site at
http://www.crcpress.com

Preface

There are many ways to define 'grasslands', but the central component of almost all of them is Gramineae (Poaceae) species and an upper limit of tree cover in the overstory (Gibson, 2009; Dixon et al., 2014; Wesche et al., 2016). Grasslands are present in most parts of the world and almost all kinds of climatic zones, except the north and south poles, the highest mountains and extreme arid zones. Because of that, grasslands are the most important ecosystems in the world after forests. The Earth's terrestrial surface (134 million km^2) comprises 30–31 per cent forest areas (FAO, 2010), 26 per cent grasslands, 10–11 per cent croplands and 6.8 per cent other land uses (Panunzi, 2008). Including the savannas, woodlands, shrublands and tundra to "non-woody grasslands", grasslands cover as much as 40 per cent of the globe.

Sub-Saharan Africa and Asia have the largest proportion of their land total area in grassland, with 14.5 and 8.9 million km^2, respectively. The largest grasslands (by surface area) on Earth are in Australia, the Russian Federation, China, United States of America, Canada, Brazil and Argentina. However, based on the fraction of total land area each nation used for grazing, Mongolia, Botswana, and Uruguay lead with 80 per cent, 76 per cent, and 76 per cent, respectively. Twenty-five of the 145 major watersheds of the world are made up of at least 50 per cent grassland (Squires et al., 2014). Sub-Saharan Africa has the most extensive grassland watersheds and Europe, the least. On the other hand, grasslands are found most commonly in semi-arid zones (28 percent of the world's grasslands), followed by humid (23 per cent), cold (21 per cent), and arid zones (19 per cent).

Natural grasslands on Earth occur under three major conditions (Wesche et al., 2016): *Savannas* found in the tropics and are never affected by frost. While the climate is favorable, fires, large herbivores and/or special soil conditions prevent the establishment of closed forests here. *Steppes* are extra-tropical grasslands in climates that are too dry for forest growth. And finally, *Arctic-alpine grasslands* are located where the climate is too cold for forests. Secondary grasslands, which have replaced natural forests or shrublands due to human land use, occupy substantial areas, particularly in Europe. If not overused (semi-natural grasslands), they can be of High Nature Value, but nowadays these have largely been converted to monotonous intensified grasslands.

Seeing the increasing risks grasslands face, we felt compelled to review the evidence and make an updated assessment of the extent and current status of grasslands worldwide. We assembled writing teams comprising more than 50 grassland experts from 27 different countries and added our own contributions based on decades of field research. The foci of the book are on grassland ecology and management as well to review the threat factors. However, biogeographic aspects were not ignored and we were at pains to provide the most up to date and comprehensive summary of the grassland ecosystems across the world and particularly in the Palaearctic region.

The multifunctional grasslands and rangeland contribute to the livelihoods of more than two billion people, including about 600 million in the arid zones alone, by providing forage for over 360 million cattle and 600 million sheep and goats (Huntsinger and Hopkinson, 1996). The area of pasture and fodder crops was 3.5 billion ha (35,000,000 km^2) in 2010, i.e., 26 per cent of the land surface of the world and 70 per cent of the world agricultural area. However, owing to several socio-political reasons the area under pasture and fodder crops is decreasing rapidly in many tropical countries. Despite the often-cited opinion that grassland systems have low productivity, it should be noted that the economies of several countries depend heavily on grasslands. Examples of highly productive grasslands include the vast savannas of the Serengeti with their spectacular assemblages of large fauna.

Managed grazing dominates the marginal bioclimatic and edaphic regions of drylands (Steinfeld et al., 2006; Zhaoli, 2004). Grazing occurs in the best bioclimatic areas of temperate forests and woodlands but is employed on marginal soils found throughout much of the humid tropics. Managed grazing systems are defined as any geographically extensive operation designed for the production of animals for consumption, including for meat, milk, and any major animal products. Recent work indicates that managed grazing systems have increased more than 600 per cent in geographic extent (from about 5.3 million km^2) during the past three centuries. More than 1.5 billion "animal units" (AU[1]) were present in managed grazing systems in 1990 (Asner et al., 2004). Livestock production is an important source of income and employment in the rural sector. More than 38 per cent of the global populations live in grasslands and a large proportion of the world's poorest are settled in them (Nalule, 2010). With respect to human populations, it is highest in the dry grasslands (arid, semi-arid, and dry sub-humid) of Sub-Saharan Africa (> 150 km^{-2}) followed by South and East Asia (India has 390 km^{-2}). Human populations are lowest in the dry grasslands of Oceania (< 2 km^{-2}).

In addition to food provisioning, other major ecosystem services of grasslands include protection of soil against erosion, wildlife habitat, carbon sequestration and water harvesting. Grasslands are also of major relevance for biodiversity conservation, not only for large ungulate grazers and the vascular plants. Also, many invertebrate groups are most diverse in grasslands. Moreover, grasslands hold a record number of vascular plant species at spatial scales below 100 m^2 (Wilson et al., 2012). Only from 100 m^2 upwards tropical rainforests are richer. Interestingly are all these extremely rich grasslands located in the temperate zone (Central Eastern and Northern Europe as well as Argentina) and are subjected to low-intensity livestock grazing or mowing.

Grasslands are under threat globally. There are hundreds of documented cases of increased woody plant cover in semiarid, subtropical rangelands of the world (Squires, 2015). In North and South America, Africa, Australia, and elsewhere, woody vegetation cover has increased significantly in grazing systems during the past few decades. Cited causes of woody encroachment include overgrazing of herbaceous cover that reduces competition for woody seedlings, fire suppression that enhances woody plant survival, atmospheric CO_2 enrichment that favors C_3 (woody) plant growth, and nitrogen pollution that likewise favors woody encroachment.

[1] An animal unit (AU) is defined as the number of cattle, buffalo, sheep, goats, horses, and camels weighted by their relative size and growth rates [AU = n (cows + buffalo) + 0.2 n (sheep + goats) + 1.2 n (horses + camels)].

For semi-natural grasslands as they are mainly known in the traditional cultural landscapes of Europe (and to a minor extent also East Asia), other threat factors are prevalent (Dengler et al., 2014; Janssen et al., 2016). Most importantly, the twin threats of agricultural intensification on productive land and abandonment on marginal sites, both can lead to a drastic loss in grassland biodiversity. On top of that comes eutrophication from agricultural fertilization, and also airborne nitrogen deposition from multiple sources as a third main driver of biodiversity impoverishment and homogenization.

References

Asner, G.P., A.J. Elmore, L.P. Olander, R.E. Martin and A.T. Harris. 2004. Grazing systems, ecosystem responses and global change. Ann. Rev. Environ. Res. 29: 261–299.

Dengler, J., M. Janišová, P. Török and C. Wellstein. 2014. Biodiversity of Palaearctic grasslands: A synthesis. Agric. Ecosyst. Environ. 182: 1–14.

Dixon, A.P., D. Faber-Langendoen, C. Josse, J. Morrison and C.J. Loucks. 2014. Distribution mapping of world grassland types. J. Biogeogr. 41: 2003–2019.

FAO. 2010. Global Forest Resources Assessment 2010—Main Report. FAO [Forestry Paper No. 163], Rome. Available at: http://www.fao.org/docrep/013/i1757e/i1757e00.htm.

Gibson, D.J. 2009. Grasses and Grassland Ecology. Oxford University Press, Oxford, 305 pp.

Huntsinger, L. and P. Hopkinson. 1996. Viewpoint: Sustaining rangeland landscapes: A social and ecological process. J. Range Manag. 49: 167–173.

Janssen, J.A.M., J.S. Rodwell, M. Garcia Criado, S. Gubbay, T. Haynes, A. Nieto, N. Sanders, F. Landucci, J. Loidi, (...) and M. Valachovič. 2016. European Red List of Habitats—Part 2. Terrestrial and Freshwater Habitats. European Union, Luxembourg, 38 pp.

Nalule, A.S. 2010. Social Management of Rangelands and Settlement in Karamoja Subregion. European Commission, Humanitarian Aid and FAO.

Panunzi, E. 2008. Are Grasslands under Threat? Brief Analysis of FAO Statistical Data on Pasture and Fodder Crops. Available at: http://www.fao.org/ag/agp/agpc/doc/grass_stats/grass-stats.htm.

Squires, V.R., H.M. Milner and K.A. Daniell. 2014. River Basin Management: Understanding People and Place. CRC Press. Boca Raton, 510 pp.

Squires,V.R. (ed.). 2015. Rangeland Ecology, Management and Conservation Benefits. NOVA Press, New York, 356 pp.

Steinfeld, H., P. Gerber, T. Wassenaar, V. Castel, M. Rosales and C. de Haan. 2006. Livestock's Long Shadows—Environmental Issues and Options. FAO, Rome, 390 pp.

Wesche, K., D. Ambarlı, J. Kamp, P. Török, J. Treiber and J. Dengler. 2016. The Palaearctic steppe biome: A new synthesis. Biodivers. Conserv. 25: 2197–2231.

Wilson, J.B., R.K. Peet, J. Dengler and M. Pärtel. 2012. Plant species richness: The world records. J. Veg. Sci. 23: 796–802.

Zhaoli, Y. 2004. Co-Management of Rangelands: An Approach for Enhanced Livelihoods and Conservation. ICIMOD [Newsletter No. 45], Kathmandu.

Contents

Scope and Purpose

Grasslands form an important land cover and have a world-wide distribution with grass-dominated communities from the equator to the polar tundra, from the sea level to the alpine zone, and occurring both naturally (such as tropical savannas and temperate steppes) and secondary (as in the cultural landscapes of Europe). *Grasslands of the World: Diversity, Management and Conservation*, edited by Victor Squires, Jürgen Dengler, Haiying Feng and Hua Limin, is a tribute to these important grass-dominated ecosystems. It was written by international teams of grassland experts, who compiled information from more than 90 countries, in the form of regional syntheses and case studies. Together, these chapters provide a fascinating glimpse into the various grasslands, their value for nature, culture and agriculture, and the threats they are facing today. The 18 chapters contain essential background information on topics such as the history of agriculture, grassland communities (including fauna) and the connection between grasslands and climate. The book also analyses the opportunities and risks of current policies to conserve these grasslands throughout the various grassland biomes against the background of global (including climate) change. It explores the exploitation of grasslands from the earliest time to the present and examines the impact of recent human interventions and global warming on their productivity, diversity and survival. The challenges of conserving biodiversity, maintaining livelihoods of land users while protecting the land and ensuring sustainable use are highlighted.

In almost every grassland ecosystem on Earth there are pressures from rising human and livestock populations, new demands for products from the grasslands as well as the ongoing expectation that grasslands will deliver ecosystem services under the changing climatic regimes. Climate change is likely to have major impacts on grasslands of several regions, and many of these will have serious implications for people whose livelihoods are absolutely dependent on the grasslands but also for people far away who depend on grasslands to provide water, clean air, and a steady stream of food, fiber and recreational opportunities. Environmental goods and services provided by grasslands are many and varied, and recent assessments of the monetary value of grasslands reveal that these ecosystems are severely underestimated by many. Non-monetary value, in terms of biodiversity conservation, put grasslands at the forefront of natural ecosystems worth conserving. Expanding cropland, wrought by converting grassland, is a serious threat in many regions. Often the encroachment onto grasslands is on marginal land (too dry, too steep, too rocky) where land abandonment quickly follows as the stored soil nutrients are exhausted or simply agriculture is not profitable anymore when agricultural products can be produced cheaper elsewhere. Abandoned pasture lands and hay meadows can cause accelerated land degradation in some regions, especially encroachment by undesirable plants, such as invasive species and woody plants. In other regions, succession following abandonment might lead back to a more natural state, which could be desirable in some cases.

This is what we believe will be a useful book for those who are interested in ecological, economic and policy issues in grassland ecosystem management, environmental protection and likely impacts of climate change. It will also be a valuable reference book for academics, researchers and policy makers in bi-lateral and multi-lateral donor organizations and various UN agencies. The book fills the gap in the literature admirably. Grassland ecologists, land use change specialists, policy makers and natural resource management agencies will find the book very useful. It is also a valuable reference book for university students in ecology, geography, environmental studies and development studies in general.

The book is in four PARTS

PART 1: Context and Setting

Squires and Feng open the book with a brief account of the extraordinary sequence of events that led to emergence of grasslands as major vegetation formation that now occupy some of the driest and hottest and the highest and coldest places on earth as well as vast steppes and prairies in more temperate climates. It is the story of grasses successfully competing with forests and woodlands, aided and abetted by grazing herbivores and by humans and their use of fire as a tool. It is the story of adaptation to changing climates and the changing biophysical environments.

PART 2: Grasslands of the Palaearctic Region

This PART is the first comprehensive study ever of both the natural and secondary grasslands across the whole Palaearctic biogeographic realm, which is the largest on Earth. It provides the reader with a new synthesis and updates data on many aspects where ambiguity prevailed. The series of treatises in this PART has been organized by the Eurasian Dry Grassland Group (EDGG), a large international organization devoted to ecology and conservation of all types of natural and semi-natural grasslands throughout the region. It comprises one synthesis chapter and seven regional chapters. Contributors were teams of grassland specialists drawn from 17 countries, ranging from Central Europe and the Mediterranean through the Middle East, Russia, Kazakhstan, Middle Asia, Mongolia and China to Japan.

Török and Dengler provide a synthetic view on the grasslands of the Palaearctic, a vast region that encompasses more than 70 countries in Europe, North Africa, West, North, Central and East Asia. Most importantly, they quantify, probably for the first time, areas of natural and secondary grasslands throughout the realm, broken down to major subtypes. Moreover, they provide a consistent assessment of the relative importance of various factors threatening grassland biodiversity in each of the comprised seven regions.

Dengler and Tischew describe the major grassland types in northern and western Europe including significant areas of secondary grasslands in place of former forests, but often of High Nature Value. They explain origin, extent, types and land use practices of grasslands in the region. A major focus is a comprehensive assessment of the threat factors that negatively affect the biodiversity of these grasslands. Of particular importance is the policy framework, specifically the Common Agricultural Policy of the European Union, which plays a key role in grassland abandonment and intensification, both of which negatively affect grassland biodiversity. Further the authors provide an overview on suitable management techniques to maintain or restore High Nature Value grasslands.

Török et al., enumerate the threats to grasslands throughout Eastern Europe, giving special attention to land abandonment, woody plant invasion and the biodiversity implications of the Common Agricultural Policy of the European Union. Cessation of traditional grassland management practices such as mowing, periodic burning and rotation grazing are identified as major drivers of change toward a reduction in biodiversity. Biodiversity in some 'hotspots' is critically endangered for some birds, mammals and invertebrates as is the loss of some important sites on major flyways (including grassland wetlands) for migrating birds. Particular attention is paid to grassland rehabilitation and to ecological restoration.

Ambarlı et al., provide an overview of the problems faced by grasslands and their users in the southern part of Europe and around the shores of the Mediterranean Sea including North Africa and the Middle East. Climate change impacts are discussed here but attention is also given to population pressure, land use and land cover change. The authors present the prospects for grassland communities and the peoples who depend on them in this vast region where political instability and mass migrations and re-settlement are occurring. The spatial extent, predominant land uses and biogeography of each of the countries in the region are discussed and analyzed. Problems are both on-going and emerging and relate to population pressure, land conversion, land abandonment and incursions by alien plants.

Reinecke et al., tackle the task of summarizing the issues occurring on the huge and diverse steppes and grasslands of Russia. Particular attention is paid to defining the management practices that maintain grassland ecosystem health and biodiversity. The status of nature reserves and other protected areas is updated. Climate change impacts are discussed. Lives and livelihoods of millions of people who depend on grasslands, pasturelands and hayfields are at risk in some places. Future prospects for grasslands and grassland users are discussed.

Bragina et al., address the problems and prospects confronting a major region of global importance. The Kazakhstan-Middle Asia region has some of the world's major alpine areas and some of the driest deserts. Grasslands (steppes) occupy huge tracts of land in Kazakhstan—many of them degraded by massive land conversion to cropland. Burgeoning populations, critical water shortages and expansion of soil salinity are problems that are addressed. Management of water resources and watersheds on which the rivers depend are key issues. Climate change impacts will impinge heavily on the millions whose livelihoods depend on using grasslands.

Pfeiffer et al., turn attention to the dry grasslands in Mongolia and in western China with special attention being given to the Tibetan plateau. Biodiversity of plants and animals (some of them endangered species) in the mountainous regions (in particular) receive attention here. An extremely high proportion of the human population in Mongolia and on the Qinghai-Tibet Plateau depends on grasslands for their livelihoods and as their spiritual home. Government policies aimed at arresting and reversing land degradation has led to large-scale resettlement of pastoralists and their herds, especially in critical watersheds that are the source of major rivers, some of them transnational. Climate variability and global warming have already begun to cause glaciers to melt and permafrost to thaw. Climate change over the next 50 years will present major challenges to achieving the balance between improving livelihoods, conserving biodiversity and land protection.

Ushimaru et al., present an overview of the status of Japan's semi-natural grasslands, focusing on the rapid reduction in area. Biodiversity has been severely compromised, and local extinctions of key species seem to be inevitable. Traditional grassland management approaches sustained grassland vigor and productivity. These practices are outlined and the role of local ecological knowledge in conserving biodiversity is highlighted. Climate variability impacts on land use and prospects of climate change are discussed.

PART 3: Grasslands of Other Biogeographic Regions: Problems and Prospects

Most biogeographic regions outside the Palaearctic also support extensive grassland vegetation. The six chapters in this PART deal with grassland and their management in India, East Africa, Southern Africa, North America, South America, and Australia.

Malaviya et al., present an overview of Indian grasslands from the tropics in the Southeast to the deserts in the Northwest. India supports the world's largest cattle inventory, and forage and fodder to support them are vital resources. Grasslands, either as grazing or as 'cut and carry' for stall-fed animals is important. India's human population is expected to peak at near 1.5 billion; so livestock, principally as draft animals and for milk, must be maintained and increased where possible.

Starrs et al., make an assessment of the scale, geographical distribution, and diversity of grasslands in North America (including Canada and Mexico). Discussion topics include: categorization of grassland according to species composition and productivity levels and the merits of making an indicative assessment of the carrying capacity of the different types of grassland including the seasonal variations in carrying capacity. They discuss invasive plants, fire frequency and biodiversity.

Maltitz reviews and describes the situation on grasslands in Southern Africa, that is, the countries south of 10° S, namely South Africa, Botswana, Namibia, Swaziland, Lesotho, Angola, Zambia, Zimbabwe, Malawi and Mozambique. A typology based primarily on functions of the grass layer, using a grass- and grazer-centric lens of analysis is proposed. The biogeography is outlined. The likely consequences of climate change receive attention.

Otieno and Kinyamario consider the grasslands of Eastern Africa. The landscape of Eastern Africa, which includes Ethiopia, parts of the Horn of Africa, Sudan, Kenya, Uganda, Rwanda, Burundi and Tanzania, receives attention here. The authors provide a review of Eastern African grasslands, their functions and the current status. Future climate-induced changes in land cover will occur concurrently with human-induced changes in land use, compounding their effects on grassland. The future of grasslands in the region will, therefore, depend on how they will respond to future warmer and wetter or drier climates and on how much national governments in the region will be able to curb human population increases and the mitigation measures against climate change.

Morales reviews the principal grassland types in South America, using Chile as a north-south transect stretching more than 4,600 km from the north (18° S) to the extreme tip of South America at 53° S latitude. This chapter provides a general overview of the rainfed grasslands in the vast South American continent. Key features of the biogeography of each of the major biomes are presented. Changing land use patterns wrought by the expansion of cropping agriculture and the decline in productivity of many grasslands are discussed. Special attention is given to the increased density of woody plants and the implications for ranching and livelihoods but also for the impact on carbon sequestration. The potential of grasslands to sequester carbon is discussed. Special attention is given here to the important role of grasslands in provision of environmental goods and services. Changes are occurring to the provision of ecosystem services by the main grasslands biomes in South America in response to accelerated land degradation that is now found in parts of every grassland biome.

Nie and Campbell describe the principal grassland types in Australia where upwards of 80 per cent of the land surface is (or has been) grassland, including sown pastures based on self-regenerating annuals. The focus is on the various livestock/crop systems that grasslands support. The current issues and challenges faced by the grassland industries and opportunities and strategies for future grassland research and development are briefly reviewed.

PART 4: Concluding Remarks and Summing Up

Hua et al., provide an analysis of the way in which climate change is already affecting lives and livelihoods of people who are entirely dependent on grasslands. The research (described here) on herders' responses to climate change was based on studies in Inner Mongolia, Ningxia, Gansu, Qinghai and Tibet. As background they give a brief overview of the main grasslands in China as a whole and present key elements of the climatic data associated with each major grassland type.

Starrs and Huntsinger review the different land tenure arrangements governing grassland use and the rights associated with each type in Mexico, USA and Canada. This is done in the context of rising environmental concerns about mining and mineral exploration, biodiversity conservation and the notion of 'rewilding' rangelands to better cater for needs of wildlife. The chapter also looks to the future as a new regulatory framework emerges. The implications for grasslands are discussed.

Finally, *Squires* et al., ask what future the world's grasslands have under global (not only climate) change? This concluding chapter examines a set of themes arising from the chapters that make up the bulk of this book. The following provide a focus for the text that follows in this chapter. The recent history of grasslands using examples drawn from many countries with a brief recap of current thinking and recent trends. Special attention is given to dry grasslands in the Palaearctic region. They offer an overview of the current status of grasslands and germplasm resources (biodiversity) in a range of grassland types from alpine to desert. They discuss the management systems that ensure sustainability and outline measures to recover degraded grasslands. The impacts of environmental problems such as future climate change and socio-economic issues like land abandonment and the impact of intensification on grasslands receive special attention.

PART 1
Context and Setting

1

Brief History of Grassland Utilization and Its Significance to Humans

Victor R. Squires[a,*] and *Haiying Feng*[b]

Origins and Spread of Grassland Ecosystems—A Brief Recap of Current Thinking

The rise of grass-dominated ecosystems is one of the most profound paleoecological changes in the Cenozoic[1] (Crepet and Feldman, 1991). According to Stromberg (2011) the evolution and subsequent ecological expansion of grasses (Poaceae) since the Late Cretaceous[2] have resulted in the establishment of one of Earth's dominant vegetation formations, the temperate and tropical grasslands, at the expense of forests. The development of grassland ecosystems on most continents was a multistage process involving the Paleogene[3] appearance of (C_3 and C_4) open-habitat grasses, the mid-late Cenozoic spread of C_3 grass-dominated habitats, and, finally, the Late Neogene[4] expansion of C_4 grasses at tropical-subtropical latitudes. Macrofossil records reveal that grasses were established in northern South America, northern Africa, and India in the latest Cretaceous and Paleocene (Anderson, 2006; Stromberg, 2011). According to Mix et al. (2013) changes in vegetation in western North America, for example, played a critical role in establishing the modern hydrologic regime. They attribute this to three primary reasons: (1) increased evaporation and transpiration fluxes in grassland regions affected the water balance, (2) shallower rooting depths of grasses led to the transpiration of soil water enriched in δ18O[5] due to evaporation, and (3) grasslands transpire δ18O rich waters during a shorter, more punctuated growing season. Mix et al. (2013) argue that the observed isotope signals are indicative of a feedback mechanism wherein grasslands not only respond to regional

[a] Institute of Desertification Studies, Beijing, China.
[b] Qinzhou University, Guangxi, China.
* Corresponding author: dryland1812@internode.on.net

[1] Cenozoic was about 70 MaBP [Ma is short for *mega annum* (million) and BP is before the present].
[2] Late Cretaceous about 60 MaBP.
[3] Paleogene was about 43 MaBP.
[4] Late Neogene lasted about 43 million years end about 23 MaBP.
[5] Oxygen isotope δ180 is widely used to help determine sequence stratigraphy.

and global climatic trends, but also act as drivers of hydrologic change. By enhancing seasonality and aridity, grasslands transmit hydrologic disturbances downstream, engineering climatic conditions favorable for their expansion. The Eocene saw a nearly worldwide distribution of grasses with Early Eocene records ranging from the British Isles and North America to Australia.

According to Stromberg (2011) the evolutionary history of grasslands can be broken down into several distinct stages: (a) the Paleogene appearance of C_3 open-habitat grasses; (b) the Paleogene appearance of C_4 open-habitat grasses; (c) the mid-late Cenozoic emergence of open, C_3 grass-dominated habitats; and finally, in some regions, (d) the late Neogene shift to C_4-dominated grass-dominated habitats. Grassland expansion has been proposed as a driver of global-scale paleo-environmental change via its roles in carbon storage, phosphorus and nitrogen fixation, albedo change, dehumidification, adaptability to wildfire and herbivore coevolution (Retallack, 2001). The evolution of herbivores adapted to grasslands did not necessarily coincide with the spread of open-habitat grasses (Sponheimer et al., 1999; Bobe et al., 2004; Bobe, 2006). Southern South America has long been thought to be the cradle of grassland evolution. This assumption is based on the observation that fossil South American herbivores had 'grazer'-grass-eating-tooth morphologies (hypsodonty) around 38 MaBP, while on other continents, this same tooth evolution didn't take place until 20 million years later. Hypsodonty is an adaptation to resist excessive tooth wear in animals as longer teeth significantly extend the life of the grass-eating mammal. Not everyone agrees with the theory that herbivores co-evolved with grasslands. Alternative explanations to account for hypsodonty have been propounded (Bobe, 2006; Stromberg et al., 2013).

The dynamics of the grassland area worldwide reflect the development of human society because grasslands are an integral part of the cultural landscape. The history of humans starts with the early hominids, our ancestors, *Homo habilis* and *Homo erectus* in Africa. It is presently believed that the hominid populations migrated out from eastern Africa. Adjacent areas in southern Asia and Europe were colonized circa 500,000 years ago (Barker, 1985). All of the assumed cradle sites for the evolution of the hominids can be classified as grasslands—savanna environment, although on a local scale they contain a complex environmental mosaic with availability to different habitats (Simmons, 1989). The evolution and subsequent ecological expansion of grasses (Poaceae) since the Late Cretaceous have resulted in the establishment of one of Earth's dominant vegetation formations, the temperate and tropical grasslands, at the expense of forests. The Cenozoic rise to dominance of grasses undoubtedly influenced climate systems (Mix et al., 2013) and was thought by some to be central to the evolution of grass-eating animals (Jacobs et al., 1999; Pagani et al., 2009). However, there are those who dispute the link (Stromberg, 2011, 2013). Understanding this ecological transformation provides the evolutionary context for today's grass-dominated ecosystems (Stromberg, 2011). Currently accepted pollen and macrofossil records reveal that grasses were established in northern South America, northern Africa and India in the Late Cretaceous and Paleocene (about 60–70 MaBP).

Origin of Grasslands in Europe

In terms of origin, grasslands in Europe can be classified into (i) natural grasslands, predetermined by environmental conditions and wild herbivores; (ii) seminatural grasslands, associated with long-term human activity from the beginning of agriculture during the Mesolithic–Neolithic transition; and (iii) improved (intensive) grasslands, a product of modern agriculture based on sown and highly productive forage grasses and legumes (Hejcman et al., 2013). A review paper on the origins of Central European grasslands explains more fully the stages in the development of grasslands from the Holocene (9500 BC) to recent times, using archaeobotanical (pollen and macro remains), archaeozoological (molluscs, dung beetles, animal bones) and archaeological evidence, together with written and iconographic resources and recent analogies (Hejcman et al., 2013).

Transition of Natural Communities into Seminatural Communities in Europe

Seminatural communities can be defined as plant and animal communities composed of species that are indigenous to the region, but where the development and maintenance of the communities require direct or indirect human activities.

The continuous grazing and browsing of livestock in combination with burning to maintain the grasslands led to a change from forested and half-open woodland ecosystems to open grass-dominated communities (Barker, 1985; Olsson, 1991; Behrensmeyer, 1992). Olsson (1991) gives an overview of the development of agriculture and agroecosystems in southern Scandinavia and the environmental and landscape changes related to this process. An overview of sources, evidence and hypotheses of prehistoric farming in Europe is provided in the books by Barker (1985), Behrensmeyer (1992) and Rackham (1980) and in the review by Hejcman et al. (2013). Where there was not a complete shift from forested ecosystems to grasslands, the grazing and browsing impact on the woodland led to biotic and edaphic changes, creating seminatural communities (Belsky, 1992; Simmons, 1989). Certainly the domestic herds were not the first big grazers affecting the woodland ecosystems. The existence of large wild herbivores, such as mastodons, mammoths, horses, antelopes, bison and so on during the Pleistocene, circa one million to 10,000 years ago, was of great significance for creating and maintaining grassland sites within the woodlands (Hejcman et al., 2013). Grazing pressure increased the possibilities for grassland species to exist within the forested ecosystem long before the introduction of domesticated ungulates (Harris and Hillman, 1989). The paleoecological, ethnographic and archeological evidence for exploitation of plant resources by hunter-gatherers, the domestication and spread of crop species and the introduction of agriculture in different parts of the world is elaborated in the book by Harris and Hillman (1989) and by Hejcman et al. (2013).

However, some authors argue from the standpoint that Neolithic humans met a fully forested landscape and that it was the human impact that created the seminatural habitats, such as grasslands and heathlands. The ecological effect of wild, large herbivores and of livestock grazing is related to a variety of factors: (i) herbivore species and breeds: cattle, camel, llama, horse, goat, sheep, pig, reindeer and so on, (ii) grazing by single species or mixture of herbivore species, (iii) grazing pressure (number of animals per area), (iv) time of the year and duration of the grazing periods (all year round; seasonal; rotation system) Mere grazing *per se* is also composed of sub-factors, such as the consumption of biomass, the possible effect of animal saliva on plant metabolism and growth, trampling (soil compaction, creation of vegetation free-patches, the effects on albedo and soil temperature), the deposition of dung, and so on (Belsky, 1986). The multiple interactions of all these factors in combination with environmental factors determine the ecosystem effect and the structure and composition of the communities created and maintained by human intervention (Dix and Beidleman, 1969; French, 1979). Such human impact includes forest clearing, burning, grazing by domestic animals, and/or mowing. The long-term persistence of these activities is essential; often the seminatural communities were already in existence during prehistoric time and their duration is thus measured in thousands of years (see chapters which follow for more detail and for scientific evidence to support the idea).

European examples of seminatural vegetation include most lowland heathlands (Gimingham and de Smidt, 1983) and grasslands (Emmanuelson, 2009a,b) and many woodlands and shrublands that have been affected by burning, grazing and coppicing (Rackham, 1980). By comparison, natural grasslands and heathlands in Europe are shaped mainly by climatic and edaphic factors which include alpine grasslands and heathlands and communities affected by periodic flooding along freshwater and marine shores (Rosa Garcia et al., 2013). Many European seminatural grasslands have their origins in early prehistoric times, circa 6,000 years ago or even earlier (Emmanuelson, 2009b). The Atlantic heathlands characteristic of coastal areas in Western Europe were developed from forested ecosystems during the early Iron Age, circa 2,000 years ago (Gimingham and de Smidt, 1983; Rosa Garcia et al., 2013). Seminatural ecosystems in temperate and alpine regions have had fundamental significance as the basis for the development of agriculture and the function of preindustrial agroecosystems (Nagy and Grabherr, 2009). These grassland types provided fodder resources (summer grazing and winter fodder—hay) from prehistoric time until the end of the pre-industrial era, which encompasses, in temperate Europe, about 6,000 years (Barker, 1985). Heathlands and grasslands are today threatened ecosystems that harbor a number of endangered plant and animal species and act as refuges of biodiversity. The threats are mainly changes in land use, such as cessation of grazing or mowing, or the onset of plowing or fertilizing. The number of species-rich seminatural grasslands in northern Europe has decreased significantly due to the abandonment of traditional land use practices (Dahms et al., 2010; Filho et al., 2017; Dengler and Tischew; and Torok and Dengler (this volume)).

In seminatural grasslands on limestone bedrock and calcareous soils, plant biodiversity reaches high levels. Among the highest species richness ever recorded of field-layer species (expressed as number of vascular plant species per unit area) is reported from seminatural meadows on limestone bedrock in South Estonia, 63 species per m^{-1} (Paal, 1998). The seminatural grasslands are definitely hot-spots of European biodiversity. In the absence of management, these seminatural grasslands and heathlands revert by natural succession to woodlands and forests and the species richness of vascular plants declines dramatically. The mowing and/or grazing regimes maintaining these communities might be seen as critical ecological disturbances, reducing the populations of dominating species and creating micro sites for colonization and completion of life cycles of shortlived species, thus positively contributing to high species diversity. Evidence of rapid evolution in the sense of evolutionary adjustment to specific environments is documented from experimental studies of annual plant species. Consequently, it is not surprising that the long-term existence of seminatural communities (6,000 years or more) has increased the grassland biodiversity due to the development of subspecies (and maybe even new species) with adaptations to the specific ecological conditions prevailing in a grazed or mowed grass-sward. Potter et al. (1993) in their book give an overview of biodiversity conservation efforts at species and genetic levels, based on a number of case studies from different ecosystems of the world, also including several cases of agroecosystems and domesticated species.

The alpine heathland species, *Pseudorchis straminea* and its sister species *Pseudorchis albida*, recorded only in seminatural grassland sites in Scandinavia are examples of this process. It is also plausible that some of the present grassland endemics confined to seminatural grasslands might have their origin in the Pleistocene grasslands—or for areas subjected to European colonization (e.g., the Americas)—in the pre-European grasslands.

A study of *Primula alcalina*, an endemic grassland plant species in the US, shows that this species might have benefited from domestic stock grazing (Hitchcock, 1959). This may be interpreted as an adaptation to ungulate grazing and where the present domestic herds are substitutes for the extinct native animal grazers. Incorporating a historical perspective, e.g., earlier land use and herbivore populations, is critical to interpret and to predict the dynamics of rare species and to assess relevant conservation measures.

North American Grasslands—Development since the Miocene

Although still poor, the pre-Miocene record of grasses reveals the broad patterns of Poaceae diversification. The expansion of grass-dominated ecosystems is one of the most prominent ecological changes of the Late Cenozoic. Paleobotanical, paleofaunal and stable isotope studies document the rapid rise of C_4 grasslands at the expense of forests in the Middle to Late Miocene. The post Miocene (20–25 MaBP) expansion of grasslands and savannas was associated with the adaptive radiation of large mammals conducive to grazing (Stebbins, 1981; Axelrod, 1985; McNaughton, 1993; Osterheld et al., 1999). Adaptive responses of grasses to herbivores that reflect a coevolutionary relationship between grazers and grasses includes the presence of silica in epidermal cells of grasses, perennating organs below ground and above ground production in excess of that which decomposes in a single year (Stebbins, 1981; Anderson, 1982, 1990). The widespread expansion of grassland is associated with the appearance of the C_4 photosynthetic pathway. The C_4 pathway provides an advantage over the more common C_3 pathway because it provides higher quantum yields for carbon dioxide uptake under conditions of high irradiance and temperature. Accelerated development of C_4 grasslands world-wide occurred during the Miocene-Pliocene transition (8–6 MaBP) when aridity increased world-wide due to the expansion of the Antarctic ice sheet when atmospheric carbon dioxide was below 550 ppmV. During this time, the area of forest and woodlands declined and there was an explosive evolution of grasses and forbs (Cerling et al., 1997; Ehleringer et al., 1997, 2002). Monsoon-like conditions were thought to prevail. It was warmer and wetter during this time (4–7 MaBP) and a warm, moist growing season resulted in high biomass production that was converted into combustible fuels by a pronounced dry season. The monsoon climate most likely would be accompanied by frequent lightning strikes at the end of the dry season. In the Keeley-Rundel (2005) model, fire would have been a major driver as it is today in the conversion of forest to grasslands and in the maintenance of grasslands.

Expansion of open grassland and savanna habitats was associated with increased fossilized silica which provides protection against grazing. Concomitantly, there was an increase in mammalian fossils with high-crowned teeth adapted to grazing (Stebbins, 1981; Axelrod, 1985; Fraser and Theodor, 2013) and evolution of animals that could run faster and jump higher (Retallick, 2001).

Grasslands and Fire

Natural fires are a part of the normal disturbance regime in tropical and temperate ecosystems. The human use of fires caused intentional or unintentional modification of the ecosystems and increased the natural disturbance. The human use and control of fire is dated back circa 1.4 million years (Balter, 1995). Where evidence of fire is found, it is usually in connection with settlements. It is assumed that fire was used for directing the game in the hunting process and for manipulation of vegetation, e.g., creation and maintenance of favorable patches for game, such as grasslands or semi-open forests.

It is believed that many tropical and subtropical grasslands are maintained by fire (Maltitz, this volume; Courtwright, 2011; Pyne, 1991). A significant component in this dynamic is human-induced fires that result in increased fire frequency. Because of the long history of the human use of fire, the human impact probably has contributed to the present ecosystem structure and function. The evolution of prairies in North America can be interpreted in part as seminatural ecosystems maintained by the grazing of wild animals and by fires set by native Americans (Courtwright, 2011). The absence of human-induced fires and the disappearance of big grazers led to significant shifts in the structure and functions of these ecosystems, as has been demonstrated in the prairie ecosystems of USA (Whiteny, 1994). Another example is the present savanna ecosystem of Cuba, which has been shown to be a product of the long-term human impact of cutting, burning and livestock grazing (Fiara and Herrara, 1988). Similarly in Australia, use of fire by the indigenous peoples gave rise to a new form of a primitive agriculture, called 'firestick farming' (Jones, 1969; Kimber, 1983). A major reinterpretation of the Australian ecosystems based on natural and human-induced fires was proposed by Pyne (1991). He demonstrated the ecosystem and landscape consequences of fire at the core of the Australian Aboriginals' culture.

The process of transformation of natural communities into seminatural communities had a significant influence on biological diversity at local and regional scales leading to landscape, ecosystem and community changes, in part from forested to nonforested, often grass-dominated ecosystems (Von Maydell, 1995). In alpine habitats, often the forest limit was lowered by anthropogenic activities. Biomass extractions in the form of fuel wood and fodder collection are the main source of anthropogenic pressure in the Himalayan forest (Baghar et al., 2014). Grace et al. (2002) explore the role of climate change in the altitudinal shifts in the treeline of alpine regions. Anthropogenic forces are at work there.

Agroforestry has contributed both to diminution of grassland and to its recovery (Huxley, 1999; Kidd and Pimental, 1992). Von Maydell (1995) gives a review of historical uses of different agroforestry systems, including wood pastures and silvopastoral practices and so on in Europe. Tropical areas are also sites for agroforestry and other management interventions to regulate the tree:grass ratio. In addition to a change in species composition of communities, transformations into seminatural communities have changed the genetic composition of populations over the last 10,000 years. This process led to evolution of new varieties and species (Huston, 1994).

Grasslands and Animals

For millennia, grasslands have been one of the foundations of human activities and civilizations by supporting production from grazing livestock. This is still the situation, particularly for developing countries where 68 per cent of grasslands are located. Of course, there are numerous animals, other than livestock, that depend on grassland for food and shelter. Wildlife, including birds, reptiles, molluscs and insects, are the most numerous users of grasslands.

From the perspective of animal scientists, the utilization of grasslands has historically focused on their use for livestock, particularly to produce meat and milk and to lesser extents, fiber and draft power. This has arguably been at the expense of many other current and potential functions of grasslands, and of many peoples who have historically derived their livelihoods and cultures from the same grasslands (DeFries and Rosenzweig, 2010; Ayantunde et al., 2011). However, perspectives and perceptions of the most appropriate roles and functions of grasslands have been changing in recent decades. There has been recognition that there are numerous regional, national and global issues with which utilization of grasslands are inextricably linked. These include the function of grasslands to provide social and cultural needs for many rural societies, their role in reducing greenhouse gas (GHG) emissions, as water catchments and the preservation of ecosystem biodiversity (De Fries and Rosenzweig, 2010). At the same time increased global demand for food must be met without unacceptable adverse effects (Food and Agriculture Organization (FAO), 2009; Godfray et al., 2010). Solutions to such issues are complicated by short-term and long-term needs of those whose livelihoods depend on grasslands. There are more than 800 million grassland-dwellers in the world with very low income and an additional 200 million in the more marginal arid and semi-arid areas, who are highly dependent on grasslands for their livelihoods. Extensive pastoral systems occupy the majority of global dry zone regions where agricultural production is generally marginal. Pastoral, and semi-natural and marginal areas, represent 47 per cent and 36 per cent, respectively, of total grasslands (Kruska et al., 2003). Many of these grasslands in Africa, Asia and Europe have been used for millennia.

Summary and Conclusion

Grasslands have been important to humans over a very long period of time. Northern Europe is in the forest zone, but wild megaherbivores and domesticated species have maintained grass-dominated vegetation since 1.8 million years ago. Thus, grasslands have existed continuously in temperate Europe for centuries, albeit with highly variable extent. This history of extension and contraction has shaped the biota of temperate European grasslands through speciation, extinction and exchange of species with other biomes, such as steppes, temperate forest, alpine grasslands, tundra and Mediterranean communities. A rich fauna of large mammalian herbivores was an integrated part of this grassland biome during the penultimate interglacial Eemian (circa 1,20,000 years ago). About 17 species of large herbivores were native to lowland north-western Europe (Svenning, 2002). Probably similar numbers have been present in all Quaternary interglacials except our own, the Holocene (Bradshaw et al., 2003). During periods of maximum forest cover, grasslands and their associated biota survived in refugia, often in places where edaphic and climatic conditions have been suboptimal to trees (Ellenberg, 1988). Similar to full glacial conditions, cold and dry places have prevented or retarded tree growth and regeneration and promoted grasslands. During interglacials, such conditions have mainly existed in alpine environments. In addition, warm and dry places, moderately unstable soils and floods have promoted grassland resistance to woody plant invasion. These conditions have been found in a number of topographical and geomorphological situations, perhaps most prominently on shallow soils, over solid rock or limestone, preventing penetration of tree roots, and, in addition, steep slopes, sites exposed to strong winds, sandy infertile soils and flood plains. To some extent, areas rich in such topographical features are, in present day relatively species-rich. Similarly, high-pH substrates tend to be more species-rich than low-pH substrates, due to the relative abundance of these substrate types in this biogeographical province during the evolution of grassland floras (Pärtel, 2002). Paleoecological evidence of a grassland element in the landscape-scale vegetation comes from, among others, chalk hills in England (Bush and Flenley, 1987; Bush, 1993; Waller and Hamilton, 2000) and sandy infertile soils in Jutland, Denmark (Odgaard, 1994). Pastoralism and arable farming were introduced to Central Europe in simultaneous waves; circa 5,000 BC (band-ceramic culture), circa 1,000 years later to Western Europe and circa 2,000–3,000 years later to the Baltic area and Scandinavia (Champion, 1984). When selecting suitable areas for settlement, the first farmers probably sought out regions where, among other things, open pastures were easy to maintain, in other words, places where edaphic and climatic conditions have been suboptimal to trees and natural grasslands were already present (Sammul et al., 2000).

The maintenance of grasslands at the expense of forests has been, to a large extent, due to the efforts of humans. Fire has been used as an important tool, but grazing animals, including ruminant livestock, have played an integral part. Continuity of the grassland biome through glacial-interglacial cycles and connection to steppe vegetation has resulted in evolution, immigration and survival of a large number of grassland species. During the last century, natural grasslands in Europe have faced a dramatic loss of area and increased isolation of the remaining fragments, cessation of proper management and increased load of nutrients (Filho et al., 2017). Redman (1999) reminds us that throughout history, agriculturalists have displaced pastoralists from lands that could be cultivated. However, the sustainability of crop production in areas of great climatic variability (Gaur and Squires, 2018) and rather infertile soils with a fragile soil structure is controversial (Morales, this volume).

In Europe, there has been intensification of agriculture with land amalgamation to achieve economy of scale. Small fragments of arable land are not commercially viable. Regular cultivation on some land has been abandoned and is reverting to woodland and forest (Filho et al., 2017). Grassland destruction and fragmentation have in many regions already gone so far that long-term sustainability of biodiversity is questionable. Grassland restoration may reduce the extinction debt and alleviate the effect of fragmentation in the long-term (Squires, 2016).

References and Further Readings

Anderson, R.C. 1982. An evolutionary model summarizing the roles of fire, climate and grazing animals in the origin and maintenance of grasslands: An end paper, p. 297–308. *In*: J. Estes, R. Tyrl and J. Brunken (eds.). Grasses and Grasslands: Systematics and Ecology. University of Oklahoma Press, Norman, OK.

Anderson, R.C. 1990. The historic role of fire in the North American Grassland. p. 8–18. *In*: L. Wallace and S. Collins (eds.). Fire in tallgrass prairie ecosystem. University of Oklahoma Press, Norman, OK.

Anderson, R.C. 2006. Evolution and origin of the Central Grassland of North America: Climate, fire and mammalian grazers. J. Torrey Botanical Society 133(4): 626–247.

Axelrod, D.I. 1985. Rise of the grassland biome, central North America. Bot. Rev. 51: 163–202.

Ayantunde, A.A., J. de Leeuw, M.D. Turner and M. Said. 2011. Challenges of assessing the sustainability of (agro)-pastoral systems. Livestock Science 139: 30–43.

Baghar, S., A. Chandra and J.S. Chandrashekar. 2014. Status and distribution of anthropogenic pressure in kedarnath wildlife sanctuary in western himalaya, India. Bulletin of Environmental and Scientific Research 3(2-3): 8–15.

Balter, M. 1995. Did *Homo erectus* tame fire first? Science 268(5217): 1570.

Barker, G. 1985. Prehistoric Farming in Europe. Cambridge Univ. Press, 327 pp.

Behrensmeyer, A.K., J.D. Damuth, W.A. DiMichele, R. Potts, H.-D. Sues and S.L. Wing (eds.). 1992. Terrestrial Ecosystems Through Time: Evolutionary Paleoecology of Terrestrial Plants and Animals. Chicago, IL, US: University of Chicago Press, 568 pp.

Belsky, A.J. 1986. Does herbivory benefit plants? A review of the evidence. The American Naturalist 127(6): 870–892.

Belsky, A.J. 1992. Effects of grazing, competition, disturbance and fire on species composition and diversity in grassland communities. Journal of Vegetation Science 3: 187–200.

Bobe, R., A.K. Behrenmeyer and R.E. Chapman. 2004. The expansion of grassland ecosystems in Africa in relation to mammalian evolution and the origin of the genus *Homo*. Paleogeography, Paleoclimatology Paleoecology 227: 394–420.

Bobe, R. 2006. The evolution of arid ecosystems in eastern Africa. J. Arid. Environ. 66: 564–584.

Bradshaw, R.H.W., G.E. Hannon and A.M. Lister. 2003. A long-term perspective on ungulate-vegetation interactions. Forest Ecology and Management 181: 267–280.

Bredenkamp, G.J., F. Spada and E. Kazmierczak. 2002. On the origin of northern and southern hemisphere grasslands. Plant Ecology 163: 209–229.

Bush, M.B. and J.R. Flenley. 1987. The Age of British chalk Grassland. Nature 329(1638): 434–436.

Bush, M.B. 1993. An 11400 year paleoecological history of a British chalk grassland. Journal of Vegetation Science 4: 47–66.

Cerling, T.E., J.M. Harris, B.J. MacFadden, M.G. Leakey, J. Quade et al. 1997. Global change through the Miocene/Pliocene boundary. Nature 389: 153–58.

Champion, T.C. 1984. Prehistoric Europe. London: Academic Press.

Courtwright, J. 2011. Prairie Fire: A Great Plains History, University Press of Kansas, 274 pp.

Crepet, W.L. and G.D. Feldman. 1991. The earliest remains of grasses in the fossil record. Am. J. Bot. 78: 1010–1014.

Dahms, H., L. Lenoir, R. Lindborg, V. Wolters and J. Dauber. 2010. Restoration of seminatural grasslands: what is the impact on ants? Restoration Ecology 18(3): 330–337.

DeFries, R. and C. Rosenzweig. 2010. Toward a whole-landscape approach for sustainable land use in the tropics. Proceedings of the National Academy of Sciences of the United States of America 107: 19627–19632.

Dengler, J. and S. Tischew. 2018. Grasslands of Western and Northern Europe—Between Intensification and Abandonment. pp. 27–63 (this volume).

Dix, R.L. and R.G. Beidleman (eds.). 1969. The grassland ecosystem: A preliminary synthesis. Proceedings of the Information Synthesis Project, Grassland Biome, U.S. International Biological Program. Sci. Sere 2, Range Science Department, Colorado State University, Fort Collins, 441 pp.

Ehleringer, J.R., T.E. Cerling and B.R. Helliker. 1997. C_4 photosynthesis, atmospheric CO_2, and climate. Oecologica 112: 285–299.

Ehleringer, J.A., T.E. Cerling and M.D. Dearing. 2002. Atmospheric CO_2 as global change driver influencing plant-animal interactions. Integr. Compr. Biol. 42: 424–430.

Ellenberg, H. 1988. Vegetation Ecology of Central Europe. Cambridge: Cambridge University Press.

Emanuelsson, U. 2009a. Semi-natural Grasslands in Europe Today. CABI, Wallingford.

Emanuelsson, J. 2009b. The Rural Landscapes of Europe—How Man has Shaped Europe's Nature, Formas.

FAO. 2009. The State of Food and Agriculture. FAO, Rome, 180 p.

Fiara, K. and R. Herrara. 1988. Living and dead belowground biomass and its distribution in some savanna communities in Cuba. Folia Geobot. Phytotax., Praha 23: 225–237.

Filho, W.L., M. Mandel, A. Quasem Al-Amin, A. Feher and C.J.C. Jabbour. 2017. An assessment of the causes and consequences of agricultural land abandonment in Europe. International Journal of Sustainable Development & World Ecology 24(6): 554–560.

Fraser, D. and J.M. Theodor. 2013. Ungulate diets reveal patterns of grassland evolution in North America. Palaeogeography, Palaeoclimatology, Palaeoecology 369: 409–421.

French, N.R. (ed.). 1979. Perspectives in Grassland Ecology. Springer, New York, 204 pp.

French, N.R. 2012. Perspectives in Grassland Ecology: Results and Applications of the US/IBP Grassland Biome Study. Springer Science & Business Media. Science, 204 pp.

Gaur, M.K. and V.R. Squires. 2018. Climate variability impacts on land use and livelihoods in drylands. Springer, N.Y. 348 p.

Gimingham, C.H. and J.T. de Smidt. 1983. Heaths as natural and semi-natural vegetation. pp. 185–199. *In*: W. Holzner, M.J.A. Werger and I. Ikusima (eds.). Man's Impact on Vegetation. The Hague, Dr. W. Junk Publishers.

Godfray, H.C.J., I.R. Crute, L. Haddad, D. Lawrence, J.F. Muir, N. Nisbett, J. Pretty, S. Robinson, C. Toulmin and R. Whitely. 2010. The future of the global food system. Philos. Trans. R Soc. Lond. B Biol. Sci. 365(1554): 2769–2777.

Gomez-Pompa, A. and A. Kaus. 1992. Taming the wilderness Myth. Bioscience 42: 271–279.

Grace, J., F. Berninger and L. Nagy. 2002. Impacts of climate change on the tree line. Annals of Botany 90(4): 537–544.

Harris, D.R. and G.C. Hillman (eds.). 1989. Foraging and Farming. The Evolution of Plant Exploitation. London: Unwin Hyman, 733 pp.

Hejcman, M., P. Hejcmanov, V. Pavlu and J. Bene. 2013. Origin and history of grasslands in Central Europe—A review. Grass and Forage Science 68(3): 345–363.

Hitchcock, C.L. 1959. *Primula*. *In:* C.L. Hitchcock, A. Cronquist, M. Ownbey and J.W. Thompson. Vascular Plants of the Pacific Northwest. University of Washington Press, Seattle 4: 50–53.

Huston, M.A. 1994. Biological Diversity. The Coexistence of Species in Changing Landscapes. Cambridge University Press, 681 pp.

Huxley, P. 1999. Tropical Agroforestry. Oxford, UK: Blackwell Science Ltd, 371 pp.

Jacobs, B.F., J.D. Kingston and L.L. Jacobs. 1999. The origin of grass-dominated ecosystems. Ann. Missouri Bot. Garden 86: 590–643.

Jones, R. 1969. Fire-stick farming. Aust. Nat. Hist. 16: 224–228.

Keeley, J.E. and P.W. Rundel. 2005. Fire and the Miocene expansion of C_4 grasslands. Ecol. Lett. 8: 685–690.

Kidd, C.V. and D. Pimentel (eds.). 1992. Integrated Resource Management. Agroforestry for Development. San Diego: Academic Press, 223 pp.

Kimber, R.G. 1983. Black lightning: Aborigines and fire in central Australia and the western desert. Archaeol. Oceania 18: 38–45.

Kruska, R.L., R.S. Reid, P.K. Thornton, N. Henninger and P.M. Kristjanson. 2003. Mapping livestock-oriented agricultural production systems for the developing world. Agricultural Systems 77: 39–63.

Maltitz, G.P. von. 2018. Southern African Grassland in an Era of Global Change. pp. xx this volume.

McNaughton, S.J. 1993. Grasses and grazers, science and management. Ecol. Appl. 3: 17–20.

Mix, H.T., M.J. Winnick, A. Mulch and C.P. Chamberlain. 2013. Grassland expansion as an instrument of hydrologic change in Neogene western North America. Earth and Planetary Science Letters 377-378: 73–83.

Nagy, L. and G. Grabherr. 2009. The Biology of Alpine Habitats. Oxford University Press.

Odgaard, B.V. 1994. The Holocene vegetation history of northern West Jutland, Denmark. Opera Botanica 123: 3–171.

Oesterheld, M., J. Loreti, M. Semmartin and J.M. Paruelo. 1999. Grazing, fire, and climate effects on primary productivity of grasslands and savanna. pp. 287–306. *In*: L.R. Walker (ed.). Ecosystems of the World: 16 Ecosystems of Disturbed Ground. Elsevier, New York, NY.

Olsson, E.G.A. 1991. Agro-ecosystems from Neolithic Time to the Present. The Cultural Landscape During 6000 Years. Ecological Bulletins 41. B.E. Berglund (ed.)., 293–314.

Olson, D.M., E. Dinerstein, E.D. Wikranamayake et al. 2001. Terrestrial ecoregions of the world: A new map of life on earth. Bioscience 51: 933–938.

Paal, J. 1998. Rare and threatened plant communities of Estonia. Biodiversity & Conservation 7(8): 1027–1049.

Pagani, M., K. Caldeira, R. Berner and D.J. Beerling. 2009. The role of terrestrial plants in limiting atmospheric CO_2 decline over the past 24 million years. Nature 460: 85–88.

Pärtel, M. 2002. Local plant diversity patterns and evolutionary history at the regional scale. Ecology 83: 2361–2366.

Pignatti, S. 1978. Evolutionary trends in Mediterranean flora and vegetation. Vegetation 37: 175–185.

Potter, C.S., J.I. Cohen and D. Janczewski (eds.). 1993. Perspectives on Biodiversity: Case Studies of Genetic Resource Conservation and Development. Washington: AAAS Publications, 254 pp.

Pyne, S. 1991. Burning Bush. A Fire History of Australia. New York: Henry Holt and Company, 520 pp.

Rackham, O. 1980. Ancient Woodland: Its History, Vegetation and Uses in England. London: Edward Arnold.

Raven, P.H. 1973. The evolution of Mediterranean floras. *In*: F. Di Castri and H.A. Mooney (eds.). Mediterranean-type Ecosystems. Origin and Structure. Ecological Studies 5: 213–224.

Redman, C. 1999. Human Impacts on Ancient Environments. University of Arizona Press.

Retallack, G.J. 2001. Cenozoic expansion of grasslands and climatic cooling. J. Geol. 109: 407–426.

Rosa García, R., M.D. Fraser, R. Celaya et al. 2013. Grazing land management and biodiversity in the Atlantic European heathlands: A review. Agroforest. Syst. 87: 19.

Sammul, M., K. Kull and T. Kukk. 2000. Natural grasslands in Estonia: evolution, environmental and economic roles. pp. 20–27. *In*: R. Viiralt, R. Lillak and M. Michelson (eds.). Conventional and Ecological Grassland Management. Tartu: Estonian Grassland Society.

Simmons, I.G. 1989. Changing the Face of the Earth. Culture, Environment, History. Oxford: Blackwell, 487 pp.

Sponheimer, M., K.E. Reed and J.A. LeeThorp. 1999. Combining isotopic and eco morphological data to refine bovid paleodietary reconstruction: A Case Study from Makapansgat hominin locality. J. Hum. Evol. 36: 705–718.

Squires, V.R. 2016. Restoration of wildlands including wilderness, conservation reserves and rangelands. pp. 131–150. *In*: Victor R. Squires (ed.). Ecological Restoration: Global Challenges Social Aspects and Environmental Benefits. Nova Science Publishers, N.Y.

Stebbins, G.L. 1981. Coevolution of grasses and herbivores. Ann. Mo. Bot. Gard. 68: 75–86.

Stromberg, C.A.E. 2011. Evolution of grasses and grassland ecosystems. Annu. Rev. Earth Planet. Sci. 39: 517–44.

Stro¨mberg, C.A.E. and F.A. McInerney. 2011. The Neogene transition from C3 to C4 grasslands in North America: Assemblage analysis of fossil phytoliths. Paleobiology 37: 50–71.

Stromberg, C.A.E., R. Dunn, R.H. Madden and A.A. Carlini. 2013. Decoupling the spread of grasslands from the evolution of grazer-type herbivores in South America. Nature Commun 4: 1478 doi: 10.1038/ncomms2508.

Svenning, J.-C. 2002. A review of natural vegetation openness in Northwestern Europe. Biological Conservation 104(2): 133–148.

Torok, P. and J. Dengler. 2018. Palaearctic Grasslands in Transition: Overarching Patterns and Future Prospects. pp. 15–26 (this volume).

Von Maydell, H.J. 1995. Agroforestry in central, northern, and eastern Europe. Agroforestry Systems 31: 133–142.

Waller, M.P. and S. Hamilton. 2000. Vegetation history of the English chalklands: A mid-Holocene pollen sequence from the Caburn, East Sussex. Journal of Quaternary Science 15: 253–272.

Whiteny, G.G. 1994. From Coastal Wilderness to Fruited Plain. A History of Environmental Change in Temperate North America from 1500 to the Present. Cambridge University Press, 451 pp.

PART 2
Grasslands of the Palaearctic Region

2

Palaearctic Grasslands in Transition: Overarching Patterns and Future Prospects

Péter Török[1,2] and *Jürgen Dengler*[3,4,5,*]

Introduction

The Palaearctic biogeographic realm covers about 45 million km² (Table 2.1), which corresponds to 35 per cent of the terrestrial ice-free surface of the Earth and thus it is the largest out of the eight biogeographic realms (Olson et al., 2001). In geographic terms, this means Europe, Africa north of the Sahara and the Mediterranean, temperate, boreal and arctic zones of Asia (Fig. 2.1). The realm currently comprises circa 9.7 million km² grasslands, which correspond to 22 per cent of its total area (Table 2.1) and thus the largest amount and likely also the biggest share of grasslands among all biogeographic realms. These grasslands are partly natural, partly secondary, that is, anthropogenic. In any case, they are of high ecological and economic importance, but at the same time subject to various severe threats.

Here we introduce the Palaearctic section of *Grasslands of the World*. Apart from this introductory and synthesis chapter, the section consists of seven regional treatises, roughly arranged from the west to the east (Fig. 2.1): Western and Northern Europe (Dengler and Tischew, 2018), Eastern Europe (in the socioeconomic sense) (Török et al., 2018), the Mediterranean Basin and the Middle East (Ambarlı et al., 2018), Russia (Reinecke et al., 2018), Kazakhstan and Middle Asia (Bragina et al., 2018), China and Mongolia (Pfeiffer et al., 2018) and, last but not least, Japan (Ushimaru et al., 2018). These chapters have been organised by the *Eurasian Dry Grassland Group* (EDGG), an international scientific network, which deals with ecology, biodiversity, conservation and management of all types of natural and semi-natural grasslands throughout the Palaearctic biogeographic realm (Box 2.1).

[1] MTA-DE Lendület Functional and Restoration Ecology Research Group, 4032 Debrecen, Egyetem sqr. 1, Hungary.
[2] Department of Ecology, University of Debrecen, Egyetem sqr. 1, 4032 Debrecen, Hungary.
[3] Vegetation Ecology Group, Institute of Natural Resource Sciences (IUNR), Zurich University of Applied Sciences (ZHAW), Grüentalstr. 14, Postfach, 8820 Wädenswil, Switzerland.
[4] Plant Ecology, Bayreuth Center of Ecology and Environmental Research (BayCEER), University of Bayreuth, Universitätsstr. 30, 95447 Bayreuth, Germany.
[5] German Centre for Integrative Biodiversity Research (iDiv), Deutscher Platz 5e, 04103 Leipzig, Germany. Email: molinia@gmail.com
* Corresponding author: juergen.dengler@uni-bayreuth.de

Box 2.1 The *Eurasian Dry Grassland Group* (EDGG).

The *Eurasian Dry Grassland Group* (EDGG; http://www.edgg.org; Vrahnakis et al., 2013; Venn et al., 2016) was founded in 2008 under the name *European Dry Grassland Group*, resulting from an internationalization of the German *Arbeitsgruppe Trockenrasen*. Notwithstanding that EDGG deals with both animals and plants, it became an official working group of the *International Association for Vegetation Science* (IAVS; http://www.iavs.org) in 2009, which gave the group access to financial and organizational support of a global scientific organization. EDGG is also a member of the *European Forum on Nature Conservation and Pastoralism* (EFNCP; http://www.efncp.org/). With EDGG's name change from 'European' to 'Eurasian', the group in parallel also widened its scope even beyond what the new name suggests, which had been chosen to be able to retain the well-known acronym. According to its bylaws, EDGG now deals with biodiversity, ecology and conservation of all natural and semi-natural grasslands of the Palaearctic biogeographic realm. This means that instead of the former focus on dry grasslands, EDGG now deals with grasslands, whether they are wet, mesic or dry, base-rich, acidic or saline, and from the coastline to the alpine zone. In autumn 2017, the EDGG had about 1,250 members from nearly 70 countries, including both scientists and conservation practitioners. Membership is free of charge. EDGG is governed by a seven-head Executive Committee, elected by the members for a two-year term.

EDGG coordinates scientific and policy-related actions in grassland research, conservation and restoration. It facilitates international communication between researchers, site managers, policy- and decision-makers, using its mailing list and the quarterly published open-access electronic journal, *Bulletin of the Eurasian Dry Grassland Group* (now: *Palaearctic Grasslands*), available from the EDGG website. The main recurrent activity of the EDGG is its annual scientific conference in varying locations. In summer 2017, the 14th *Eurasian Grassland Conference* (EGC) was jointly organized by Latvian and Lithuanian EDGG members in Riga, with excursions to various grasslands in both the countries, scientific talks and posters as well as practical workshops. Slightly younger are the *EDGG Field Workshops* (formerly known as *EDGG Research Expeditions*, e.g., Aćić et al., 2017) during which interested EDGG members of all levels join for one to 1.5 weeks to collect high-quality grassland diversity data (originally vascular plants, bryophytes and lichens, now increasingly also including animal taxa) in less well-studied regions of the Palaearctic. They use a standardized methodology, involving multi-scale sampling, which allows for many different analyses (Dengler et al., 2016b). The 9th such event in early summer 2017 took place in central Italy, as usual with participants from many different countries and a wide range of academic levels. The data of these sampling events are then used for joint publications on patterns and drivers of grassland biodiversity (Kuzemko et al., 2016; Polyakova et al., 2016) or grassland classification (e.g., Dengler et al., 2013), and contribute to collaborative vegetation-plot databases that allow drawing further academic benefits. EDGG was strongly involved in establishing comprehensive national grassland databases in various regions of Europe (e.g., Vassilev et al., 2012, 2018; Dengler et al., 2017), and has recently (re-) started the *Database of Scale-Dependent Phytodiversity Patterns in Palaearctic Grasslands* (GrassPlot; http://bit.ly/2qKTQt2; Janišová et al., 2017). This database combines all the data from the EDGG field workshops (Dengler et al., 2016a) plus many comparable datasets from other projects and aims at using these for multiple broad-scale vegetation ecological and macroecological studies.

Last but not least, EDGG has organized numerous special features and special issues on grassland-related topics in international journals. Since 2005, it has an annual special feature in *Tuexenia*

Box 2.1 contd. ...

...Box 2.1 contd.

focused on grassland vegetation in Central Europe (e.g., Deák et al., 2017). Reflecting the now taxonomically and geographically wider scope, since 2014 this series has been complemented by EDGG Special Issues in *Hacquetia*, about every 1.5 years (e.g., Valkó et al., 2016). Beyond that, EDGG has also organized special issues in other international journals, focusing on specific topics, namely *Conservation of dry grasslands* (in *Plant Biosystems*; Janišová et al., 2011), *European grassland ecosystems* (in *Biodiversity and Conservation*; Habel et al., 2013), *Biodiversity of Palaearctic grasslands* (in *Agriculture, Ecosystems and Environment*; Dengler et al., 2014) and *Palaearctic steppes* (in *Biodiversity and Conservation*; Török et al., 2016). EDGG also has edited two special issues aimed at advancing the consistent broad-scale classification of Palaearctic grassland vegetation, in *Applied Vegetation Science* (together with the *European Vegetation Survey*, another IAVS working group: Dengler et al., 2013) and in *Phytocoenologia* (Janišová et al., 2016). Some of the contributions in these EDGG special issues/features became much-cited reference works because they reviewed and synthesized the knowledge on certain grassland-related topics, most importantly, perhaps *Biodiversity of Palaearctic grasslands* (focused on secondary grasslands; Dengler et al., 2014) and *The Palaearctic steppe biome* (exclusively dealing with the natural, zonal grasslands; Wesche et al., 2016). The eight chapters in this book go a step further by providing seven consistent regional reviews covering nearly the complete Palaearctic biogeographic realm and both natural and secondary grasslands, complemented by this synthesis of syntheses at hand.

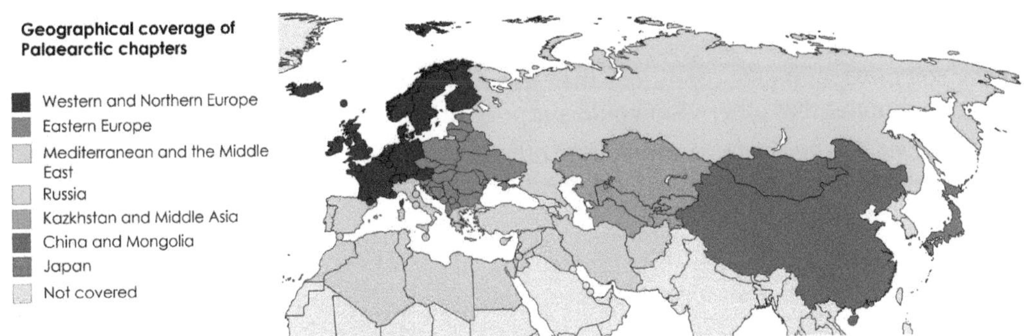

Fig. 2.1 Chapter division of the Palaearctic realm. There are no chapters dealing with the Caucasus countries (Armenia, Azerbaijan and Georgia) nor with North and South Korea. Note that, deviating from this simplified map, the chapter 'Western and Northern Europe' excludes Mediterranean France (treated in the chapter 'Mediterranean Basin and the Middle East') but includes the Italian Alps (instead of the chapter 'Mediterranean Basin and the Middle East'), and from China only the Palaearctic northern part is considered.

The chapters are arranged according to biogeographic and socioeconomic criteria because both can have a strong impact on the current state of grasslands, their diversity, management and threats. Since many statistics are only available on a per-country basis and socioeconomic drivers usually act on a country level, we normally included complete countries in a chapter even if parts of a country territory show stronger biogeographic relationships to the region of another chapter. The only exceptions are France and Italy, which have been divided between Western and Northern Europe (Dengler and Tischew, 2018) and the Mediterranean Basin and the Middle East (Ambarlı et al., 2018), as well as China, whose subtropical parts (belonging to the Indo-Malayan biogeographic realm) are not considered. Overall, we managed to cover the whole Palaearctic biogeographic realm, with the only exception of two smaller regions, namely the Caucasus countries (Armenia,

Azerbaijan and Georgia) and the Korean Peninsula (North and South Korea). To facilitate comparisons between the regions, the chapters use a similar structure and terminology. They have been written by a team of 28 experts from the EDGG, resident in 17 different countries.

Grasslands in the Palaearctic

To get a hang of the topic, one first needs to define what grasslands are, which is not easy as there are many different definitions from ecological, physiognomic, agronomic or remote-sensing points-of-view (Gibson, 2009; Dixon et al., 2014; Wesche et al., 2016). Here we adopt the definition of Janišová et al. (2011) modified by Dengler et al. (2014): *Grasslands are herbaceous vegetation types that are mostly dominated by grasses (Poaceae) or other graminoids (Cyperaceae, Juncaceae) and have a relative dense vegetation cover (usually > 25 per cent).* On top of that, we only need to exclude artificial grasslands that are re-seeded every year, such as cereal fields. Based on this definition, we can find four main types of grasslands in the Palaearctic (Dengler et al., 2014):

(1a) *Steppes* (climatogenic grasslands in climates that are too dry to sustain forests and are affected by frost).

(1b) *Arctic-alpine grasslands* (climatogenic grasslands in climates that are too cold to sustain forests).

(1c) *Azonal and extrazonal grasslands* (pedogenic or topogenic grasslands under special soil or topographic conditions that, at small spatial scales, allow grassland to exist in climates that otherwise would support forests, shrublands or deserts).

(2) *Secondary grasslands* (resulting from other natural vegetation, mainly forests, but also wetlands, through human land use, like mowing, grazing, burning or abandoning arable fields).

Types (1a) to (1c) together form the natural grasslands, and there are many transitions between these, for example alpine steppes, which are both too cold and too dry for forests (Wesche et al., 2016).

With these concepts in mind, we tried to compile overall statistics on the grasslands of the Palaearctic (Table 2.1). Already getting an overall extent of grasslands in the realm was a challenge as there are many statistics that often strongly deviate from each other. The widely recognised 'world grassland types' of Dixon et al. (2014) give a value of 10.1 million km^2 of 'International Vegetation Classification Divisions with Dominant Grassland Types' for the Palaearctic. Deviating from what the title of this paper suggests, this value is not meant to provide the actual distribution of grasslands and their areas, but the area of ecozones that are assumed to have a natural vegetation dominated by grasslands. Thus on the one hand, all the secondary grasslands, are excluded but on the other, grassland biomes that have been converted to arable fields or other land-cover types are included. In consequence, Dixon et al. (2014) do not even provide a well-founded estimate of the area of grassland biomes because, while including various units dominated by forests and shrublands, they excluded other units that are clearly grassland-dominated (see Wesche et al., 2016). The latter authors therefore re-evaluated the same basic units (ecoregions) from the Terrestrial Ecosystems of the World (Olson et al., 2001) and concluded that the original extent of steppes (including alpine steppes) was circa 8.9 million km^2. While this number is

Table 2.1 Grassland areas and fractions in the Palaearctic biogeographic realm and its seven main regions according to the chapter division of this book. (Note that the majority of figures are based on expert estimates by the chapter authors and the authors of this synthesis; this is particularly true for the subordinate grassland categories. NA = not available.)

Region	Western and Northern Europe[1]	Eastern Europe	Mediterranean Basin and the Middle East[1]	Russia	Kazakhstan and Middle Asia	China and Mongolia[2]	Japan	Total[3]
Chapter	**Dengler and Tischew (2018)**	**Török et al. (2018)**	**Ambarlı et al. (2018)**	**Reinnecke et al. (2018)**	**Bragina et al. (2018)**	**Pfeiffer et al. (2018)**	**Ushimaru et al. (2018)**	
Number of countries included[1]	19	19	21	1	5	2	1	71
Total area included [km²]	2,714,355	2,187,878	10,744,442	17,125,000	4,008,139	7,900,000	377,972	45,057,786
Total extant grasslands [km²]	470,000	300,000	1,810,000	1,790,000	1,480,000	3,800,000	20,000	9,710,000
- Fraction of territory	17%	14%	17%	10%	37%	48%	5%	22%
- Proportion of natural grasslands	21%	7%	60%	79%	87%	95%	30%	78%
- Proportion of HNV grasslands[4]	36%	69%	NA	91%	76%	83%	NA	ca. 80%[7]
(1) Natural grasslands[5] (extant) [km²]	100,000	20,000	1,080,000	1,420,000	1,290,000[8]	3,610,000	6,000	7,550,000
- as fraction of their original extent	98%	7%	80%	51%	63%[8]	95%	80%[9]	72%
(i) Steppes [km²]	0	11,000	830,000	500,000	1,120,000[8]	1,900,000	0	4,380,000
(ii) Arctic-alpine grasslands [km²]	80,000	5,000	210,000	820,000[0]	100,000[8]	1,610,000	500	2,830,000
(iii) Azonal + extrazonal grasslands [km²]	20,000	4,000	40,000	100,000	70,000[8]	100,000	5,500	340,000
(a) In good state [km²]	95,000	13,000	NA	1,280,000	9,50,000	3,040,000	NA	ca. 84%[7]
(b) Degraded [km²]	5,000	7,000	NA	140,000	340,000	570,000	NA	ca. 16%[7]
(2) Secondary grasslands[6] (extant) [km²]	370,000	280,000	730,000	370,000	190,000	190,000	14,000	2,160,000
- as fraction of their maximum extent (in the past)	60%	50%	80%	NA	NA	100%	30%	ca. 72%[7]
(a) Semi-natural grasslands [km²]	75,000	195,000	NA	345,000	170,000	95,000	3,500	ca. 62%[7]
(b) Intensified grasslands [km²]	295,000	85,000	NA	25,000	20,000	95,000	10,500	ca. 38%[7]

[1]France and Italy, partly treated under 'Western and Northern Europe', partly under 'Mediterranean Basin and the Middle East', appear twice. [2]Only the Palaearctic part of China is considered under 'China and Mongolia'. [3]Total also includes the Caucasus countries Armenia, Azerbaijan and Georgia as well as North and South Korea, for which there are no regional chapters. We thus assumed that the situation in the Caucasus is roughly similar to that in Russia and that of the Korean Peninsula to that in Japan. [4]HNV (High Nature Value) grasslands are here defined as those of categories 1a and 2a. [5]Under 'Natural grasslands' we include those secondary grasslands that grow in the place of natural grasslands (e.g., overgrazed natural grasslands or recovering grasslands after temporary use as arable fields). [6]Under 'Secondary grasslands' we include here only those grasslands that grow in places that naturally would be occupied by non-grassland formations, e.g., forests and shrublands. [7]Percentage calculated on the basis of those regions with data. [8]The values of original coverage of natural grasslands in the region are based on the estimates in Wesche et al. (2016: Online Resource 3) for the included ecoregions. We roughly assume that circa 60 per cent of the original steppes (as given by Wesche et al. [2016: Online Resource 7] for Kazakhstan) remained unconverted, but circa 75 per cent of the original azonal/extrazonal grasslands and 90 per cent of the original arctic-alpine grasslands. [9]No published values seem to exist. However, under forest climates as in Japan not much more than the currently 6,000 km² of natural grasslands are to be expected (compare the situation in Western and Northern Europe). [10]This number includes 6,70,000 km² of a total of 3,352,000 km² reindeer pastures, assuming that circa 20 per cent of these can roughly be ascribed to grasslands in the wide sense.

useful, it also does not quantify how much of these steppes are still extant, nor how much area needs to be added for arctic-alpine, azonal or extrazonal and secondary grasslands.

Table 2.1 for the first time attempts to provide such an overview of grassland areas and their types for the Palaearctic biogeographic realm as a whole and its main regions corresponding to our chapters. To compile this table, we used the expertise of the seven regional author teams as well as additional sources. Still in many cases the values are to be considered as rather rough estimates. The total areas of all extant grasslands were mostly derived from statistics of the Food and Agriculture Organisation of the United Nations (FAO), based on land use statistics. Unfortunately, even in Western European countries, there are considerable unexplained discrepancies between grassland areas provided by FAO and two European statistical sources (see Dengler and Tischew, 2018), while in Iran, an estimate based on national sources gave a more than three-fold larger grassland area than the FAO value (907,000 vs. 295,000 km^2; A. Naqinezhad, pers. comm.). We also tried to estimate the amount of High Nature Value (HNV) grasslands in the regions because this is a term that becomes more widely applied in discussions on biodiversity conservation (Veen et al., 2009; Oppermann et al., 2012). While originally this term was only applied to grasslands in low-input farming systems that host a high biodiversity or high concentrations of species with particular conservation interest (Paracchini et al., 2009), we extended it here to match its intuitive meaning, that is, to include both natural and secondary grasslands that contribute significantly to biodiversity conservation. Thus, we subdivided natural grasslands (in the sense of all grasslands that grow in places naturally covered by grasslands) into (a) those of good state and (b) degraded ones and secondary grasslands into (a) semi-natural ones and (b) intensified ones (Table 2.1). The natural grasslands in good state and the semi-natural grasslands together would then constitute the HNV grasslands. Evidently, in both cases there is a gradual transition between (a) and (b), and we are not aware of any previous clear definition. Thus, we consider, for the purpose of this synthesis, HNV grasslands roughly as those that still host 50 per cent or more of their 'original' diversity and whose floristic composition and structure are still so similar to the 'original' state that they would conventionally be considered the same vegetation type (phytosociological association or at least alliance), albeit possibly a different subtype. This was the rule of thumb with which the regional author teams were asked to 'classify' their grasslands, acknowledging all imprecisions that come with that.

As a result of our exercise, we can now state with reasonable confidence that currently there are about 9.7 million km^2 of grasslands in the Palaearctic, of which 78 per cent (7.6 million km^2) are natural and 22 per cent (2.2 million km^2) are secondary (Table 2.1). The extant natural grasslands are the remains of originally about 10.4 million km^2 (i.e., 78 per cent). Their biggest share are steppes (58 per cent), followed by arctic-alpine grasslands (37 per cent), while azonal and extrazonal grasslands are of subordinate importance only (5 per cent). Secondary grasslands have also lost about one-fourth of their maximum extent. The HNV fraction among the remaining natural grasslands (84 per cent) is higher than that among the secondary grasslands (62 per cent). Of particular interest are the regional differences revealed by Table 2.1—the fraction of grasslands among the current land cover types ranges from only 5 per cent in Japan to nearly 50 per cent in the Palaearctic parts of China and Mongolia. The extant grasslands are strongly dominated by natural types (more than 75 per cent) in Russia, Kazakhstan and Middle Asia as well as China and Mongolia, while these contribute an intermediate share in the Mediterranean Basin and the Middle East (60 per cent) and a relatively small fraction in Western and Northern Europe as well as Japan (20–30 per cent). A special case is eastern Europe, which once had extensive natural steppes (mainly in Ukraine), which were almost completely destroyed (Korotchenko and Peregrym, 2012) so that nowadays secondary grasslands are strongly prevailing (only 7 per cent for all three groups of natural grasslands).

Drivers of Biodiversity Loss in Palaearctic Grasslands

Palaearctic grasslands are in very intense transition; in many regions grassland biodiversity is facing many threats, which are strongly linked to changes in human activities (Dengler et al., 2014; Wesche et al., 2016). We summarized the most important threats and their relative importance by region in Table 2.2. The table was composed based on the information provided in the seven Palaearctic chapters and refined and supplemented by the author teams of these regions. Inspired by the reference works of World Resource Institute (2005), Salafsky et al. (2008) and Janssen et al. (2016), we distinguished 14 threat categories arranged into six main groups. While the drivers of biodiversity loss vary from region to region, some general patterns are obvious nevertheless.

It is evident that overall and in most regions grassland abandonment or underuse can be considered as the most important threat to grassland biodiversity (Table 2.2: relevance score 17). Other generally influential threat factors (relevance scores ≥ 10) are overgrazing and other types of intensification of grassland use and various types of grassland losses due to conversions to arable land, forest or built-up areas. Alterations of site conditions, climate change, invasive species and direct human impacts are considered of lesser impact across the Palaearctic grasslands. Beyond these general patterns, there are also striking regional differences. Most importantly, abandonment/underuse, as one of the prevailing threats in the other five regions, was considered of low importance in Kazakhstan and Middle Asia, as well as China and Mongolia. By contrast, in these two regions as well as in Russia, conversion of grasslands to arable land and overgrazing are still the most important threats. The conversion of grasslands to arable fields also has a long history in Western, Central and Northern Europe, with a peak probably in the early decades of the 20th century, but it hardly accounts for biodiversity losses in recent decades. By contrast, the countries of the former Soviet Union, particularly Ukraine, Russia and Kazakhstan with their large share of the steppe biome, experienced the most intense conversion ('Virgin Land Campaign') during the communist period, but continue on an alarming scale until today. Finally, eutrophication is considered one of the two main threat factors in Western and Northern Europe and of some importance in other parts of Europe, while it is hardly seen as relevant in the Asian regions.

Our new assessment based on expert knowledge from the individual regions reveals some marked differences to previous seminal works on relevance of threat factors to biodiversity. Sala et al. (2000) suggested that for extratropical grassland land use (meaning both land use change and conversion to other land-cover types) is the most important group of threats, while each of their four other factor groups (climate, N deposition, biotic exchange, increase in atmospheric CO_2) are on a similar and lower level, approximately at one-third. According to our assessment, 'land use' would be even more influential, while the negative impacts of climate change and biotic exchange for Palaearctic grasslands are probably lower and those of elevated CO_2 currently probably negligible. With a slightly different categorization of threats, the Millennium Ecosystem Assessment (World Resources Institute, 2005) considered for temperate grasslands habitat change and eutrophication as the two categories of very high importance for past biodiversity loss, invasive species as moderately relevant while climate change and overexploitation so far had low importance. While our new assessment agrees with the ranking of the last two and of land use, rating

Table 2.2 Relative importance of recent threats (< past 30 years) for grassland biodiversity reviewed in the regional chapters of the Palaearctic biogeographic realm. (Relative importance of a respective threat: –: absent or very rare, +: rare/local/low impact, ++ moderately widespread/moderate impact, +++: widespread/high impact. The overall relevance score, intended to provide an idea of the relative importance of threat factors, was pragmatically defined as the number of +'s per threat, not taking into account the different sizes of the regions.)

	Region							
Threats/causes of Biodiversity Loss	**Western and Northern Europe**	**Eastern Europe**	**Mediterranean Basin and the Middle East**	**Russia**	**Kazakhstan and Middle Asia**	**China and Mongolia**	**Japan**	**Relevance Score**
Habitat loss of grasslands								
- Conversion to arable land	+	++[1]	+	+++[2]	+++[3]	++[4]	–	12
- Afforestation	+	++	+	++[5]	+	+	++[6]	10
- Mining and energy production	+	+	+	+	+	++	–	7
- Urbanisation, transport and touristic infrastructure	+	+	++	+	+	++	++	10
Changes in grassland use								
- Abandonment and/or underuse	+++	+++	+++	+++	+	+[7]	+++	17
- Overgrazing	+	+	++	++[8]	++[9]	++[10]	+	11
- Other types of agricultural intensification	++	+	+	+	+	+	+++[11]	10
Alteration of site conditions								
- Eutrophication	+++	++[12]	++	+	+	–	–	9
- Altered water regime	++	++	+	+	++	–	–	9
Climate change	+	+	+	+	++	+	+[13]	8
Invasive species	+	++	++	+	+	–	+	8
Direct human impact								
- Military and armed conflicts	–[14]	+[15]	+[16]	–	+	+	–	4
- Recreational activities	+	+	+	+	+	–	+	6
- Collecting wild plants/hunting wild animals	–	+	++	++	++	++	–	9

[1]Lower rate in mountainous regions and countries are typical. [2]Heterogeneous; main threat in Stavropol and Krasnodar, but low impact in other provinces. [3]Massive re-use of fallows (secondary grasslands), but not primary steppes. [4]Heterogeneous in the region; in China higher rate of conversion is typical. [5]Heterogeneous; important threat in Belgorod and a few other provinces, but minor impact in others; [6]mostly conifer plantations, but no data on extent. [7]Some abandonment in Mongolia. [8]Insignificant in central Russia and the Volga region. [9]Overgrazing is an important issue in Kazakhstan around settlements. [10]Heterogeneous; lower rates are typical in Mongolia. [11]Conversion to sown pastures and paddy consolidation caused rapid biodiversity loss. [12]In the Baltic countries and lowland regions this is a typical threat. [13]Affects mostly alpine grasslands in the region. [14]In Western and Northern Europe military is not only not negative, but in general quite positive for maintaining grassland diversity as many of the most valuable grasslands are located inside military training areas, where they are kept open, while no eutrophication occurs. [15]While in some countries of the region military acts protective, in Ukraine it is an important threat for biodiversity. [16]Armed conflicts in north Africa and Syria became a major driver of land cover changes.

eutrophication as very high in the past across all temperate grasslands seems to be a result of the biased view or researchers resident in highly industrialized regions that clearly suffer from this factor, while large areas in the inner part of Eurasia do not. Lastly, our three European regions show a good correspondence to the recent assessment of threat factors of grasslands in the European Union and in neighboring countries (Janssen et al., 2016).

The impact of military and armed conflicts is difficult to assess, but was pointed out to be important by several author teams. While in Western and Northern Europe, eastern Europe and Japan military help to protect and maintain grasslands (i.e., low accessibility of military training areas to agriculture, tourism, etc.), in some other regions and countries (e.g., Ukraine, North Africa and the Middle East or Kazakhstan and Middle Asia) armed conflicts and regular military training are an emerging threat to grassland biodiversity.

Regardless of region, we can point out that lowlands and mountainous areas generally differ with respect to threats. In mountain areas, the recent historical rate of conversion to arable land, the effects of eutrophication and altered water regime are considered to be much lower.

Conclusions and Future Prospects

Building on previous syntheses of the EDGG, in particular Wesche et al. (2016) for the Palaearctic steppes, we provided here for the first time a comprehensive and regionalized assessment of grassland areas and types as well as the relative importance of threat factors considering all types of grasslands across the whole Palaearctic biogeographic realm. We based Tables 2.1 and 2.2 on the aggregation of many different sources, mostly at country level, with varying and often low reliability. Assuming that there was no systematic bias, but over- and underestimation were equally frequent, we are confident that the overall picture reflects the reality. Moreover, some of the parameters were more easy to assess than others: Generally the importance of threats in a region can be estimated easily than their overall impact on grassland biodiversity. Moreover, in some regions, such as the Mediterranean Basin and the Middle East (Ambarlı et al., 2018), the separation of natural and secondary grasslands was challenging because landscape-modifying human impact here dates back more than 10,000 years to the onset of the Neolithic Revolution (see Wesche et al., 2016). Therefore in the countries of this region and some others, national statistics do not distinguish natural and secondary grasslands at all or, as in Spain, use the term 'steppes' (according to common definitions defined as climatogenic grasslands, see above) for widespread dry grasslands in the Mediterranean forest biome (M. Pulido Fernández, pers. comm.; compare with Bohn et al., 2004).

Biodiversity of Palaearctic grasslands is shaped by a complex of interacting abiotic, biotic and human-mediated socio-economic factors. To halt biodiversity loss, caused mostly by changes in land use type and intensity, it is vital to develop a realm-scale inventory and database of grasslands and to identify and conserve the key areas by establishing ecological networks. The very first step would be to agree on joint definitions of grassland types according to origin, use, conservation value, ecology and biogeography, clear delimitations against other formations, etc. This would establish better and more consistent statistics across the Palaearctic realm. It is necessary to develop and coordinate targeted research for sustainable-management practices fine-tuned to regional and local biodiversity patterns. It is also vital to evaluate the natural capital of grassland habitats and calculate realistic values of their ecosystem services.

We can conclude that for sustainable use and biodiversity conservation in Palaearctic grasslands an integrative view and holistic thinking are inevitable. This implies that effective policy tools acting at a transnational level should be implemented. In particular this means (i) to develop a more effective international-level policy tools and actions for grassland conservation and restoration, (ii) to initiate a transnational knowledge transfer and networking for enhanced food security and sustainable use of grasslands, and (iii) to develop a platform integrating the opinions of key stakeholders and policy-makers in tuning decisions related to sustainable grassland management and conservation.

Acknowledgements

The authors are thankful to the 28 authors of the Palaearctic chapters for their invaluable contribution to the book and to this introductory chapter. P. Török was supported by NKFIH 119 225 grant during manuscript preparation.

Abbreviations

EDGG = Eurasian Dry Grassland Group; FAO = Food and Agriculture Organisation of the United Nations; HNV = High Nature Value

References

Aćić, S., J. Dengler, I. Biurrun, T. Becker, U. Becker, A. Berastegi, S. Boch, I. Dembicz, I. García-Mijangos, (...) and Z. Dajić Stevanović. 2017. Biodiversity patterns of dry grasslands at the meeting point of Central Europe and the Balkans: Impressions and first results from the 9th EDGG Field Workshop in Serbia. Bull. Eurasian Dry Grassl. Group 34: 19–31.

Ambarlı, D., M. Vrahnakis, S. Burrascano, A. Naqinezhad and M. Pulido Fernández. 2018. Grasslands of the Mediterranean Basin and the Middle East and their management (pp. 89–112 this volume).

Bohn, U., G. Gollub, C. Hettwer, Z. Neuhäuslová, T. Raus, H. Schlüter, H. Weber and S. Hennekens (eds.). 2004. Map of the natural vegetation of Europe. Scale 1 : 2,500,000. Interactive CD-ROM: explanatory text, legend, maps. Bundesamt für Naturschutz, Bonn: CD-ROM + 19 pp.

Bragina, T.M., A. Nowak, K.A. Vanselow and V. Wagner. 2018. Grasslands of Kazakhstan and Middle Asia: The ecology, conservation and use of a vast and globally important area (pp. 141–169 this volume).

Déak, B., V. Wagner, A. Csecserits and T. Becker. 2017. Vegetation and conservation of Central-European grasslands—Editorial to the 12th EDGG Special Feature. Tuexenia 37: 375–378.

Dengler, J., E. Bergmeier, W. Willner and M. Chytrý. 2013. Towards a consistent classification of European grasslands. Appl. Veg. Sci. 16: 518–520.

Dengler, J., M. Janišová, P. Török and C. Wellstein. 2014. Biodiversity of Palaearctic grasslands: A synthesis. Agric. Ecosyst. Environ. 182: 1–14.

Dengler, J., I. Biurrun, I. Apostolova, E. Baumann, T. Becker, A. Berastegi, S. Boch, I. Dembicz, C. Dolnik, (...) and F. Weiser. 2016a. Scale-dependent plant diversity in Palaearctic grasslands: A comparative overview. Bull. Eurasian Dry Grassl. Group 31: 12–26.

Dengler, J., S. Boch, G. Filibeck, A. Chiarucci, I. Dembicz, R. Guarino, B. Henneberg, M. Janišová, C. Marcenò, (...) and I. Biurrun. 2016b. Assessing plant diversity and composition in grasslands across spatial scales: the standardised EDGG sampling methodology. Bull. Eurasian Dry Grassl. Group 32: 13–30.

Dengler, J., T. Becker, T. Conradi, C. Dolnik, B. Heindl-Tenhunen, K. Jensen, J. Kaufmann, M. Klotz, C. Kurzböck, (...) and J. Went. 2017. GrassVeg.DE – die neue kollaborative Vegetationsdatenbank für alle Offenlandhabitate Deutschlands. Tuexenia 37: 447–455.

Dengler, J. and S. Tischew. 2018. Grasslands of Western and Northern Europe—Between intensification and abandonment (pp. 27–63 this volume).

Dixon, A.P., D. Faber-Langendoen, C. Josse, J. Morrison and C.J. Loucks. 2014. Distribution mapping of world grassland types. J. Biogeogr. 41: 2003–2019.

Gibson, D.J. 2009. Grasses and Grassland Ecology. Oxford University Press, Oxford, 305 pp.

Habel, J.C., J. Dengler, M. Janišová, P. Török, C. Wellstein and M. Wiezik. 2013. European grassland ecosystems: Threatened hotspots of biodiversity. Biodivers. Conserv. 22: 2131–2138.

Janišová, M., S. Bartha, K. Kiehl and J. Dengler. 2011. Advances in the conservation of dry grasslands—Introduction to contributions from the 7th European Dry Grassland Meeting. Plant Biosyst. 145: 507–513.

Janišová, M., J. Dengler and W. Willner. 2016. Classification of Palaearctic grasslands. Phytocoenologia 46: 233–239.

Janišová, M., J. Dengler and I. Biurrun. 2017. GrassPlot—the new database of multi-scale plant diversity of Palaearctic grasslands. IAVS Bull. 2017(2): 18–21.

Janssen, J.A.M., J.S. Rodwell, M. Garcia Criado, S. Gubbay, T. Haynes, A. Nieto, N. Sanders, F. Landucci, J. Loidi, (...) and M. Valachovič. 2016. European Red List of Habitats—Part 2. Terrestrial and Freshwater Habitats. European Union, Luxembourg, 38 pp.

Korotchenko, I. and M. Peregrym. 2012. Ukrainian steppes in the past, at present and in the future. pp. 173–196. *In:* M.J.A. Werger and M.A. van Staalduinen (eds.). Eurasian Steppes. Ecological Problems and Livelihoods in a Changing World. Springer, Dordrecht.

Kuzemko, A.A., M.J. Steinbauer, T. Becker, Y.P. Didukh, C. Dolnik, M. Jeschke, A. Naqinezhad, E. Ugurlu, K. Vassilev and J. Dengler. 2016. Patterns and drivers of phytodiversity of steppe grasslands of Central Podolia (Ukraine). Biodivers. Conserv. 25: 2233–2250.

Olson, D.M., E. Dinerstein, E.D. Wikramanayake, N.D. Burgess, G.V.N. Powell, E.C. Underwood, J.A. D'Amico, I. Itoua, H.E. Strand, (...) and K.R. Kassem. 2001. Terrestrial ecoregions of the world: A new map of life on Earth. BioScience 51: 933–938.

Oppermann, R., G. Beaufoy and G. Jones (eds.). 2012. High Nature Value Farming in Europe: 35 European Countries—Experiences and Perspectives. verlag regionalkultur, Ubstadt-Weiher, 544 pp.

Paracchini, M.L., C. Bamps, J.-E. Petersen, Y. Hoogeveen, I. Burfield and C. van Swaay. 2009. Identification of high nature value farmland at the EU27 level on the basis of land cover and biodiversity data. pp. 53–56. *In:* P. Veen, R. Jefferson, J. de Smidt and J. van der Straaten (eds.). Grasslands in Europe of High Nature Value. KNNV Publishing, Zeist.

Pfeiffer, M., C. Dulamsuren, Y. Jäschke and K. Wesche. 2018. Grasslands of China and Mongolia: Spatial extent, land use and conservation (pp. 170–198 this volume).

Polyakova, M.A., I. Dembicz, T. Becker, U. Becker, O.N. Demina, N. Ermakov, G. Filibeck, R. Guarino, M. Janišová, (...) and J. Dengler. 2016. Scale- and taxon-dependent patterns of plant diversity in steppes of Khakassia, South Siberia (Russia). Biodivers. Conserv. 25: 2251–2273.

Reinecke, J.S.F., I.E. Smelansky, E.I. Troeva, I.A. Trofimov and L.S. Trofimova. 2018. Land use of natural and secondary grasslands in Russia (pp. 113–138 this volume).

Sala, O.E., F.S. Chapin III, J.J. Armesto, E. Berlow, J. Bloomfield, R. Dirzo, E. Huber-Sanwald, L.F. Huenneke, R.B. Jackson, (...) and D.H. Wall. 2000. Global biodiversity scenarios for the year 2100. Science 287: 1770–1774.

Salafsky, N., D. Salzer, A.J. Stattersfield, C. Hilton-Taylor, R. Neugarten, S.H.M. Butchart, B.E.N. Collen, N. Cox, L.L. Master and D. Wilkie. 2008. A standard lexicon for biodiversity conservation: unified classifications of threats and actions. Conserv. Biol. 22: 897–911.

Török, P., K. Wesche, D. Ambarli, J. Kamp and J. Dengler. 2016. Step(pe) up! Raising the profile of the Palaearctic natural grasslands. Biodivers. Conserv. 25: 2187–2195.

Török, P., M. Janišová, A. Kuzemko, S. Rūsiņa and Z. Dajić Stevanović. 2018. Grasslands, their threats and management in eastern Europe (pp. 64–88 this volume).

Ushimaru, A., K. Uchida and T. Suka. 2018. Grassland biodiversity in Japan: threats, management and conservation (pp. 197–218 this volume).

Valkó, O., M. Zmihorski, I. Biurrun, J. Loos, R. Labadessa and S. Venn. 2016. Ecology and conservation of steppes and semi-natural grasslands. Hacquetia 15: 5–14.

Vassilev, K., Z. Dajić, R. Cušterevska, E. Bergmeier and I. Apostolova. 2012. Balkan Dry Grasslands Database. Biodivers. Ecol. 4: 330–330.

Vassilev, K., E. Ruprecht, V. Alexiu, T. Becker, M. Beldean, C. Biţă-Nicolae, A.M. Csergő, I. Dzhovanova, E. Filipova, J.P. Frink, (...) and J. Dengler. 2018. The Romanian Grassland Database (RGD): historical background, current status and future perspectives. Phytocoenologia 48: 91–100.

Veen, P., R. Jefferson, J. de Smidt and J. van der Straaten (eds.). 2009. Grasslands in Europe of High Nature Value. KNNV Publishing, Zeist, 320 pp.

Venn, S., D. Ambarlı, I. Biurrun, J. Dengler, M. Janišová, A. Kuzemko, P. Török and M. Vrahnakis. 2016. The Eurasian Dry Grassland Group (EDGG) in 2015–2016. Haquetia 15: 15–19.

Vrahnakis, M.S., M. Janišová, S. Rūsiņa, P. Török, S. Venn and J. Dengler. 2013. The European Dry Grassland Group (EDGG): stewarding Europe's most diverse habitat type. pp. 417–434. *In:* H. Baumbach and S. Pfützenreuter (eds.). Steppenlebensräume Europas—Gefährdung, Erhaltungsmaßnahmen und Schutz: Thüringer Ministerium für Landwirtschaft, Forsten, Umwelt und Naturschutz, Erfurt.

Wesche, K., D. Ambarlı, J. Kamp, P. Török, J. Treiber and J. Dengler. 2016. The Palaearctic steppe biome: A new synthesis. Biodivers. Conserv. 25: 2197–2231.

World Resources Institute. 2005. Ecosystem and Human Well-being: Biodiversity Synthesis—A Report of the Millennium Ecosystem Assessment. Island Press, Washington, D.C., 86 pp.

3

Grasslands of Western and Northern Europe—Between Intensification and Abandonment

Jürgen Dengler[1,2,3,*] and *Sabine Tischew*[4]

Introduction

The Western and Northern European region as delimited here corresponds to the two subregions of the same name as defined by the United Nations Statistical Division. Compared to other subdivision schemes, we also included the western part of Central Europe (Germany, Austria and Switzerland) while eastern Central Europe, based on socioeconomic similarities, is treated in the Eastern European chapter (Török et al., 2018). That way, the current chapter comprises those countries of Europe that were subject to capitalist economy after World War II from the temperate to the arctic zone, while the countries of the Mediterranean zone are treated in a separate chapter (Ambarlı et al., 2018). For pragmatic reasons, we include eastern Germany (the former German Democratic Republic, GDR) in the present chapter. From those countries that partly lie in the temperate and partly in the Mediterranean zone, we consider the northern part of France and the Italian Alps, while Mediterranean France, the major part of Italy and the whole of Spain are covered by Ambarlı et al., 2018. This chapter thus deals with 15 countries and several dependent territories with a total of about 2.7 million km² (Table 3.1).

The region extends from 43° to 81° northern latitude, from 25° western to 17° eastern longitude and from sea level to 4,807 m a.s.l. According to the Köppen climate classification, the types Cfb (temperate, with warm summer and no drought), Dfb, Dfc (cold with cold or warm summers, but no drought) and ET (polar: tundra) are represented.

[1] Vegetation Ecology Group, Institute of Natural Resource Sciences (IUNR), Zurich University of Applied Sciences (ZHAW), Grüentalstr. 14, Postfach, 8820 Wädenswil, Switzerland.

[2] Plant Ecology, Bayreuth Center of Ecology and Environmental Research (BayCEER), University of Bayreuth, Universitätsstr. 30, 95447 Bayreuth, Germany.

[3] German Centre for Integrative Biodiversity Research (iDiv), Deutscher Platz 5e, 04103 Leipzig, Germany.

[4] Department for Nature Conservation and Landscape Planning, Anhalt University of Applied Sciences, Strenzfelder Allee 28, 06406 Bernburg, Germany.
Email: sabine.tischew@hs-anhalt.de

* Corresponding author: juergen.dengler@uni-bayreuth.de

Table 3.1 Basic statistics on the Western and Northern European region and its grasslands from different sources.

Country	Total Area (km^2)	Population (million)	Population Density (persons/km^2)	Grassland Area (km^2) [FAO 2015]	Grassland Fraction of Total Area (%)			Grassland Fraction of Agriculturally Used Area (%)		
					FAO 2015	Eurostat 2015	Smit et al. (2008)	FAO 2015	Eurostat 2015	Smit et al. (2008)
Belgium*	30,530	11.250	369	5,527	18.1	31.0	20.2	28.4	52.1	44.5
Channel Islands	190	0.163	840	19	10.0	–	–	16.4	–	–
France (without Mediterranean part)*	477,000	58.196	122	90,212	18.9	26.7	23.1	28.5	48.0	42.9
Ireland*	70,280	4.762	68	47,759	68.0	56.3	56.1	91.5	90.7	89.8
Isle of Man	570	0.084	147	381	66.8	–	–	85.9	–	–
Luxembourg*	2,590	0.576	223	767	29.6	28.9	29.9	50.3	55.4	60.0
Netherlands*	41,540	17.100	424	12,526	30.2	36.3	24.9	50.2	60.0	51.3
United Kingdom*	243,610	65.648	269	108,530	44.6	36.2	45.8	61.2	64.8	67.9
Subtotal Western Europe	**866,310**	**157.779**	**182**	**264,722**	**30.7**	**–**	**–**	**44.0**	**–**	**–**
Austria*	83,879	8.783	105	15,372	18.3	24.7	23.5	47.0	61.8	58.3
Germany*	367,380	82.176	230	55,416	15.5	21.9	14.9	26.4	40.4	30.9
Italy (only Alps)*	47,000	9.452	201	22,725	8.7	21.7	16.0	15.0	46.4	31.3
Liechtenstein	160	0.037	231	39	24.6	–	–	74.6	–	–
Switzerland	41,290	8.401	203	12,838	31.1	–	18.0	70.3	–	68.8
Subtotal Central Europe (western part)	**529,709**	**108.849**	**205**	**106,391**	**20.1**	**–**	**–**	**32.1**	**–**	**–**
Denmark*	42,922	5.749	133	3,880	9.0	17.5	9.7	11.6	25.7	14.8
Faroe Islands	1,396	0.050	36	1,013	72.5	–	–	97.6	–	–
Finland*	338,420	5.506	16	8,009	2.4	4.4	2.0	23.6	42.7	30.0
Iceland	103,000	0.333	3	36,000	35.0	–	22.7	100.0	–	99.7
Norway (with Svalbard + Jan Mayen)	385,178	5.267	17	30,853	8.0	–	3.7	72.5	–	93.1
Sweden*	447,420	10.043	23	19,691	4.4	5.4	3.2	35.4	56.3	45.1
Subtotal Northern Europe	**1,318,336**	**26.948**	**20**	**99,447**	**7.5**	**–**	**–**	**47.6**	**–**	**–**
Total	**2,714,355**	**293.576**	**108**	**471,560**	**17.4**	**–**	**–**	**43.4**	**–**	**–**

FAO 2015: http://www.fao.org/faostat/en/#data/LC (accessed 12 August 2017), data for 2015; Eurostat 2015: http://appsso.eurostat.ec.europa.eu/nui/show.do (accessed 13 August 2017, online data code: lan_lcv_ovw), data for 2015; Smit et al. (2008), data averaged over 1995–2004. Note that the differences between the sources only partly are due to different reference years because data for older years in FAO (2017) also deviated from the respective years in the two other sources.

* = Member country of the European Union. Areas for 'France (without Mediterranean part)' and 'Italy (only Alps)' are estimates, and statistics in these lines assume that human population and grasslands are similarly distributed over the whole country.

From the 13 Environmental Zones of Europe of Metzger et al. (2005), ATN (Atlantic North), ATC (Atlantic Central) and ALN (Alpine North) are completely and BOR (Boreal) largely covered, while from NEM ('Nemoral', actually meaning hemi-boreal), CON (Continental) and ALS (Alpine South) the western parts and from LUS (Lusitanian) the northern parts are included. During the Pleistocene, large parts of the region were repeatedly covered by the Nordic and Alpine glaciers, which, during their maximum advances, left only SW England, France, Belgium and the central part of Germany ice-free, but as an open tundra with permafrost. Thus the ecosystems of the region are geologically of recent origin. Today, Western and Northern Europe are home to about 290 million people, but with population densities varying as widely as from about 3 persons/km^2 in Iceland to more than 400 persons/km^2 in the Netherlands (Table 3.1). Apart from Norway, Iceland, Switzerland and Liechtenstein, all countries are part of the European Union (EU). According to their per capita incomes all of them range among the 30 most prosperous countries on Earth. These factors, in particular geological history, current climate, human population and wealth, jointly shape the extent and type of grasslands that occur as well as their threats and future prospects.

Grasslands of the Region

Origin of the Grasslands

Western and Northern Europe have a humid climate throughout so that forest growth is nowhere precluded by climatic drought. Therefore, different types of forests form the zonal (climax) vegetation in most of the region (Bohn et al., 2004). In consequence, three major types of grasslands can be distinguished in the region according to their origin (Ellenberg and Leuschner, 2010; Hejcman et al., 2013; Dengler et al., 2014):

- **Arctic-alpine grasslands** together with dwarf-shrub heaths, cryptogam-dominated and sparsely vegetated types form the zonal vegetation under climates that are too cold to allow tree growth. This happens in the highest elevations of the Alps, the Pyrenees, the Scottish Highlands and the Scandes as well as in the lowlands of the northernmost tip of Fennoscandia, in Iceland and Svalbard (Bohn et al., 2004). In the Alps, for example, natural (zonal) grasslands dominate the alpine zone, ranging from the timberline at 1,500–2,400 m a.s.l. up to the subnival zone ending around 3,000 m a.s.l. (Reisigl and Keller, 1987).
- **Azonal and extrazonal grasslands** are natural grasslands that occur under climates favorable to forest growth, but where edaphic peculiarities or natural disturbance regimes prevent the succession towards woodlands. Soils could be too wet, too dry, too shallow/rocky, too saline or rich in heavy metals or too instable to allow trees to establish (Ellenberg and Leuschner, 2010; Klötzli et al., 2010). Typically such conditions hostile to trees are found only at a very small spatial extent, e.g., a few square meters on and around rocky outcrops. The only cases where such natural grasslands can extend over square kilometers and more are coastal salt marshes and coastal dunes as well as some graminoid-dominated mire-types (Bohn et al., 2004). Likewise disturbances, such as wild fire, wind throw, erosion, avalanches or intensive browsing by wild herbivores can locally kill trees and shrubs and thus allow the establishment of grasslands.

- **Secondary grasslands** nowadays constitute the biggest part of grasslands in the region. They originated through cutting and burning of forests by humans or livestock grazing in the woodlands and were maintained as open habitat by continuous pasturing, haymaking or a combination of these. Locally, secondary grasslands might exist due to fire or military training or, as a successional stage, following on abandoned arable fields. Among secondary grasslands one can distinguish **semi-natural grasslands** from so-called **'improved' or 'intensive' grasslands** (anthropogenic grasslands in the narrow sense), where humans strongly modify the site conditions and the community composition through fertilization, high cutting/grazing intensity and/or repeated re-seeding with grass cultivars (Dierschke and Briemle, 2002; Stevens et al., 2010).

Types of Grasslands

According to phytosociological classification (modified after Mucina et al., 2016), the grasslands of the region can be placed in 10 main vegetation classes (Table 3.2). Among these are five with more or less arctic-alpine distribution (classes 1–5) and five with mainly temperate distribution (classes 6–10).

Among the arctic-alpine grasslands the classes *Juncetea trifidi* (Fig. 3.1B) and *Elyno-Seslerietea* can be considered as zonal types of acidic and base-rich bedrocks, respectively. By contrast, the other three classes occur more locally under specific edaphic and/or disturbance conditions.

Most widespread among the temperate grasslands is currently the class *Molinio-Arrhenatheretea* (Figs. 3.1C–D), which includes various semi-natural and all intensive grasslands on nutrient-rich, moist to wet sites (see also Dierschke and Briemle, 2002). The class *Nardetea strictae* also comprises only secondary grasslands, but on nutrient-poor, acidic sites. The two dry grassland classes *Festuco-Brometea* (Figs. 3.1E) and *Koelerio-Corynephoretea* (Fig. 3.1F) occur as zonal vegetation in the steppe zone of Eastern Europe and Russia as well as extrazonal natural grasslands on coastal grey dunes and around rocky outcrops, but in the region largely occupy (as secondary grasslands) sites originally covered by forests. Finally, the class *Juncetea maritimi* (Fig. 3.1A) comprises grasslands adapted to saline soils and occurs naturally in coastal salt marshes and very locally at inland salt springs. Though essentially natural, the structure and composition of the stands have in many cases been altered through livestock grazing.

Apart from the 10 mentioned classes of grasslands in the strict sense, there are some further units that could be considered as grasslands in the wider sense (classification according to Mucina et al., 2016): The *Trifolio-Geranietea* are tall-forb communities that typically grow at the transition of dry grasslands to woodlands, while the *Artemisietea vulgaris* comprise one order of semi-ruderal grasslands (*Agropyretalia intermedio-repentis*). The classes *Ammophiletea* (vegetation of fore-dunes and white dunes), *Phragmito-Magno-Caricetea* (open vegetation of eutrophic mires and swamps) and *Scheuchzerio-Caricetea* (open vegetation of mesotrophic mires) and *Salicetea herbaceae* (alpine snow-bed vegetation) comprise natural vegetation types that are partly dominated by grasses and other graminoids.

Table 3.2 Phytosociological classes and orders of grassland vegetation in Western and Northern Europe.

No.	Class	Brief Description	Origin/Type
1	*Saxifrago cernuae-Cochlearietea groenlandicae*	Open grassy tundra on disturbed sites in Svalbard	arctic
2	*Carici ruprestris-Kobresietea bellardii*	Arctic-alpine wind-exposed short grasslands on base-rich substrata	arctic-alpine
3	*Juncetea trifidi*	Arctic-alpine acidophilous grasslands	arctic-alpine
4	*Elyno-Seslerietea*	Arctic-alpine limestone grasslands	arctic-alpine
5	*Mulgedio-Aconitetea*	Tall-grass and tall-herb vegetation of moist sites in the montane and subalpine belts	arctic-alpine
6	*Juncetea maritimi* (incl. *Saginetea maritimae*)	Grass- and herb-dominated salt marshes of the European coasts	azonal and semi-natural
7	*Koelerio-Corynephoretea* (incl. *Sedo-Scleranthetea* and *Helichryso-Crucianelletea*)	Dry grasslands of sandy and shallow skeletal soil	extrazonal and semi-natural
7.a	- O. *Artemisio-Koelerietalia*	Atlantic grey dunes	
7.b	- O. *Thero-Airetalia*	Submeridional-subatlantic sand grasslands rich in annuals	
7.c	- O. *Corynephoretalia canescentis*	Pioneer sand grasslands of temperate subatlantic Europe	
7.d	- O. *Trifolio arvensis-Festucetalia ovinae*	Meso-xeric sand grasslands of temperate Europe	
7.e	- O. *Sedo acris-Festucetalia*	Xeric sand grassland of subcontinental and continental Europe	
7.f	- O. *Sedo-Scleranthetalia*	Succulent- and cryptogam-rich communities on acidic skeletal soil	
7.g	- O. *Alysso alyssoidis-Sedetalia*	Succulent- and cryptogam-rich communities on base-rich skeletal soil	
8	*Festuco-Brometea*	Dry grasslands of deep base-rich soil	extrazonal and semi-natural
8.a	- O. *Brachypodietalia pinnati*	Meso-xeric calcareous grasslands of temperate Europe	
8.b	- O. *Festucetalia valesiacae*	Xeric continental grasslands (steppes) on base-rich, deep soil	
8.c	- O. *Stipo pulcherrimae-Festucetalia pallentis*	Dealpine and circum-Pannonian rocky grasslands	
8.d	- O. *Artemisio albae-Brometalia erecti*	Submeridional-subatlantic xeric calcareous grasslands	
9	*Nardetea strictae*	Secondary mat-grass swards on nutrient-poor soil	semi-natural
9.a	- O. *Nardetalia strictae*		
10	*Molinio-Arrhenatheretea*	Secondary mesic and wet grasslands of nutrient-rich soil in the temperate and boreal zone	semi-natural and intensive
10.a	- O. *Arrhenatheretalia elatioris*	Mesic grasslands of low- to mid-elevations	
10.b	- O. *Poo alpinae-Trisetetalia*	Mesic grasslands of the mountain ranges	
10.c	- O. *Molinietalia caeruleae*	Wet grasslands	

Classification is generally based on Mucina et al. (2016), but modified for the class *Koelerio-Corynephoretea* s.l. according to Dengler (2003). Orders (O.) are given only for the four widespread classes of semi-natural grasslands.

Fig. 3.1 Main types of grasslands in the region and some typical species. (A) Salt marsh, class *Juncetea maritimi*, Wadden Sea National Park, Schleswig-Holstein, Germany—natural azonal grassland, but now managed as sheep pasture. (B) Natural alpine grassland on acidic bedrock, class *Juncetea trifidi*, Gran Paradiso National

Fig. 3.1 contd. ...

Extent of Grasslands in the Region

Estimating the current extent of grassland in the region is challenging because values from different sources vary considerably, probably due to deviating grassland definitions (Table 3.1). In total, Western and Northern Europe comprise about 4,70,000 km² grasslands, which corresponds to 17.4 per cent of the territory or 43.4 per cent of the agriculturally used area, respectively. Generally, the fraction of grasslands in the landscape decreases from Western through Central to Northern Europe, with a particularly high share of more than 50 per cent found in Ireland, Isle of Man and Faroe Islands (Table 3.1). Within countries, typically the mountainous regions, such as the Highlands of Scotland and Wales in the UK or the Massif Central in France, contain more grasslands than the surrounding lowlands. Likewise, the grassland fraction is higher in the regions adjacent to the North Sea with their young and often poorly drained soils (http://ec.europa.eu/eurostat/documents/205002/274769/1204EN.pdf).

Breaking the total extent of grasslands down to individual types according to origin, as attempted in the Palaearctic synthesis by Török and Dengler (2018), is even more tricky. Since there are no such statistics available, at least not across the whole area, we tried to subdivide the known grassland area per country (FAO statistics in Table 3.1) into the four distinguished types based on several sources of knowledge and expert judgment. The map of the natural vegetation of Europe (Bohn et al., 2004) made us conclude that we are clearly outside the range of natural steppes and arctic-alpine grasslands have a significant share of the country area in Iceland, Norway, Austria and Switzerland, but, of course, not everything mapped as arctic-alpine vegetation (unit B) is grassland since this unit also includes heathlands and sparsely vegetated areas. The second-largest fraction among the natural grasslands are probably certain mire types and natural Atlantic heathlands that are grass-dominated, possibly partly in consequence of low-intensity grazing, e.g., on the British Isles and on the Faroe Islands (circa 15,000 km²). The third largest fraction are the coastal dunes and salt marshes, which at least in certain coastal areas are so extensive that they show up in European scale maps (as units P1 and P2, respectively, in Bohn et al. [2004]). Delbaere (1998) estimated the total area of coastal dunes in Europe at more than 5,300 km². If we assume that the extent of coastal salt marshes is similar and about 40 per cent of Europe's coastlines are in our region, we conclude that circa 5,000 km² naturally open coastal habitats are located in our area, resulting in perhaps 3,000–4,000 km² grasslands in the widest sense. The other types of azonal/extrazonal grasslands (e.g., dry grasslands around rock outcrops, grasslands on unstable substrata, such as cliffs, as well as heavy metal and inland salt vegetation) cover only small areas of not more than 1,000 km² in total.

...Fig. 3.1 contd.

Park, Italian Alps. (C) Semi-natural wet grassland, alliance *Calthion palustris*, order *Molinietalia caeruleae*, class *Molinio-Arrhenatheretea*, Eastern Ore Mountains, Saxony, Germany. (D) Semi-natural mesic grassland, alliance *Arrhenatherion elatioris*, order *Arrhenatheretalia elatioris*, class *Molinio-Arrhenatheretea*. (E) Semi-natural, meso-xeric basiphilous grassland (low-intensity pasture), alliance *Bromion erecti*, order *Brachypodietalia pinnati*, class Festuco-Brometea, low-intensity pasture with abundant *Orchis purpurea*, Saxony-Anhalt, Germany. (F) Semi-natural, xeric acidic sand pioneer grassland with *Corynephorus canescens* (grass) and *Cladonia arbuscula* (lichen), alliance *Corynephorion canescentis*, order *Corynephoretalia canescentis*, class *Koelerio-Corynephoretea*, NE Brandenburg, Germany. (G) *Conocephalus fuscus*, a typical grasshopper species of wet grasslands, NE Brandenburg, Germany. (H) *Lysandra coridon*, a typical butterfly species of dry grasslands, NE Brandenburg, Germany. Photos by S. Tischew (C, E) and J. Dengler (all others).

Based on these considerations, we assume that there are about 100,000 km² extant natural grasslands in the region (which are often also subject to human land use), including circa 80,000 arctic-alpine and 20,000 azonal/extrazonal ones. All the remaining circa 370,000 km² of Table 3.1 accordingly can be attributed to secondary grasslands.

Agronomic Use of Grasslands

Historical Development of Secondary Grasslands

Secondary (anthropogenic) grasslands first emerged when Neolithic settlers arrived in the region (Pott, 1995; Poschlod et al., 2009; Ellenberg and Leuschner, 2010; Hejcman et al., 2013), which was around 6000 BC in southern Central Europe and parts of France, but only about 4000 BC on the British Isles, in Denmark and southern Fennoscandia (Poschlod, 2015). These people brought with them the main domesticated livestock, i.e., goats, sheep, cattle and pigs (Poschlod, 2015). However, in these times there was no specific grassland management, but clear-cutting and burning of forests aimed at establishment of arable fields, while livestock grazed in the woodlands or on arable fields after harvest. Still, there are indications that certain present-day plant community types of nutrient-poor anthropogenic pastures occurred for the first time in the Neolithic period, namely base-rich dry grasslands (*Brachypodietalia pinnati*; class: *Festuco-Brometea*), sandy dry grasslands (*Koelerion glaucae* and *Armerion elongatae*; class: *Koelerio-Corynephoretea*) and acidic grasslands (*Violion caninae*; class: *Nardetea strictae*; see Table 3.2) (Poschlod, 2015). There is circumstantial evidence that during post-glacial reforestation of the region, woodlands never got so closed that they excluded light-demanding steppe species that existed in the steppe-tundra of the non-glaciated areas of the Last Glacial Maximum (e.g., Bush and Flenley, 1987).

The early anthropogenic grasslands thus drew their species composition from surviving fragments of steppic grasslands in the surroundings as well as from open woodlands, fens and coastal dunes. However, the species composition was continuously enriched by non-native species (archaeophytes) that came with the settlers from southeastern and southern regions—this addition being particularly prominent during the Roman Empire. Whilst the scythe was invented during the Iron Age, the first hay meadows should have occurred then, as haymaking became more relevant on a landscape scale only during the Roman period (Poschlod, 2015). However, the main current types of hay meadows in the region (e.g., alliances *Polygono-Trisetion* [belonging to unit 10.b of Table 3.2] and *Alopecurion pratensis* [to 10.c]) occurred only in modern times, i.e., around AD 1700 (Poschlod et al., 2009). The most-widespread two-cut hay meadow type of Western and Central Europe, the alliance *Arrhenatherion elatioris* (in 10.a of Table 3.2) is even dominated and characterized by *Arrhenatherum elatius* var. *elatius* (French Ray grass)—a grass taxon that apparently is not native to most parts of the region, but has been derived from its wild relative *Arrhenatherum elatius* var. *bulbosum* of SW Europe. It was intentionally sown in the region from the 17th century onwards to 'improve' the hay meadows (Poschlod et al., 2009). Of even younger origin is the majority of wet litter meadows (*Molinion caeruleae*; in 10.c of Table 3.2), which were created by draining wetlands or by transformation of former hay meadows, e.g., through late cutting and additional seeding of suitable species. Their biomass mainly consists of purple moor grass (*Molinia caerulea* agg.) and *Cyperaceae*, which are of very low nutritional value, but good as litter in the stables. This practice flourished mainly in the foothills of the Alps during the 19th century, when litter had become more limiting for animal husbandry than fodder (Poschlod et al., 2009; Poschlod, 2015), but today these grassland communities are among the most threatened ones (see unit E3.5 of Table 3.5).

Current Grassland Use

Agriculture in Europe nowadays strongly depends on subsidies and this is particularly true for grassland use and for the rich regions in Western and Northern Europe (Hopkins and Holz, 2006; Klimek et al., 2008; Buchgraber et al., 2011). The EU spends a total of 42.6 billion EUR (49.4 billion US$) on 1.75 million km² of agricultural area and Switzerland 2.8 billion CHF (2.8 billion US$) on 10,719 km² of agricultural area. This means average subsidies of 244 EUR (283 US$) and 2,621 CHF (2,623 US$) per hectare per year, respectively. Money is spent on elaborate and complex schemes, partly for doing agriculture, partly for amount of food produced and partly for doing agriculture in a more environment-friendly manner or for providing certain ecosystem services.

Market forces and these systems of subsidies led to an enormous change in grassland use after World War II. While in Germany numbers of cattle slightly and those of pigs strongly increased from 1800 to 2010 and that of sheep and goats from around 1870 and 1920, but afterwards dropped to perhaps 5 per cent of their maxima (Poschlod, 2015). In the same period, the average milk production per cow increased from less than 1,000 L to more than 8,000 L per year, requiring fodder of much higher nutritional value than that delivered by traditional semi-natural hay meadows or pastures. Since the nutritional value (proteins) is higher when the grasslands are fertilized with nitrogen and cut earlier, the requirement of higher milk yields led to intensive fertilization, more and earlier cuts or grazing periods and a shift from hay use to silage of grasslands (Fuller, 1987; Isselstein et al., 2005; Kirkham et al., 2008).

Moreover, intensified grasslands are now re-seeded at relatively short intervals with mixtures of a few highly productive species in specially bred cultivars (Dierschke and Briemle, 2002; Smit et al., 2008). Nowadays, cattle are kept far more often just in the stable, without any grazing outside. The high number of cattle and pigs, to a significant part fed with fodder imported from other continents, also leads to the usage of grasslands in these areas for 'deposition' of animal excrements in excessive amounts (see below in *Alterations of Site Conditions*; Scotton et al., 2014).

The sharp decline in sheep and goats led to the discontinuation of use of low-productive, dry grasslands. The former transhumance with sheep hardly exists any more in the region (Poschlod, 2015). Also, other traditionally grassland management types, such as high mountain pastures (German: *Almen*) connected with alpine transhumance (Jurt et al., 2015), *Streuobstwiesen* (low-intensity orchards of tall-stem fruit trees planted in hay meadows), formerly widespread in southern Central Europe, or the wooded meadows of southern Sweden (Mitlacher et al., 2002) can only survive under present-day socioeconomic conditions due to specialized subsidy programs or, at small scales, due to activities of NGOs.

Value of Grasslands

Direct Economic Value

Western and Northern Europe comprise regions with the highest productivity of grasslands in Europe, with averages of 7.4 and 7.0 t ha^{-1} yr^{-1} in the Atlantic North and Atlantic Central

Environmental Zone, respectively, and reach 11 t ha^{-1} yr^{-1} in the Netherlands (Smit et al., 2008). The grasslands of the region are the main fodder source for 61.1 million 'bovine animals' (cattle and a tiny fraction of buffaloes), 37.1 million sheep and 2.0 million goats in the region's countries belonging to the EU (counting France in total, but not Italy; here and in the following statistics the non-EU countries Iceland, Norway and Switzerland are excluded but would likely add less than 5 per cent; data for 2015 from European Union, 2016). This corresponds to 68.6 per cent, 43.4 per cent and 16.0 per cent of the total number of livestock species in the EU, respectively. In terms of annual meat production, the region contributes even more to European overall production, with 5,261,000 t bovine meat (69.3 per cent), 491,000 t sheep meat (67.7 per cent) and 9,000 t goat meat (20.1 per cent) per year. The other main agricultural product resulting from grasslands is milk, with 110 million t cow milk (72.8 per cent of the EU) and circa 1 million t sheep and goat milk (29.4 per cent) produced in 2016 (European Union, 2016). This milk mainly goes into the production of cheese (36 per cent), followed by butter (30 per cent), cream (13 per cent) and drinking milk (11 per cent) (values for the whole EU from European Union, 2016). Based on the EU totals given in European Union (2016), one can estimate the value of the products ultimately derived from grasslands of the region in 2015 as 37.5 billion EUR (43.7 billion US$) for milk, 23.4 billion EUR (27.2 billion US$) for cattle meet, 6.1 billion EUR for sheep and goat meat (7.1 billion US$) and perhaps 1 billion EUR (1.2 billion US$) for all other animal products, such as wool and leather. These total 71.0 billion EUR (82.7 billion US$) correspond to 0.7 per cent of the Gross Domestic Product (GDP) of the EU member countries of the region.

Importance for Biodiversity

Despite grasslands currently cover 'only' about one-sixth of the area and would cover even less than 4 per cent in the potential natural vegetation (see Section *Extent of Grasslands in the Region*), they have a disproportional importance for biodiversity of nearly any major taxonomic group (except strictly aquatic organisms). For example, among the 436 butterfly species of Europe whose habitat preferences are known, 88 per cent occur on grasslands and 43 per cent are grasslands specialists (WallisDeVries and van Swaay, 2009; only information on continental scale available, but the situation in the region should be similar; Fig. 3.1H). Grasshoppers are another group of invertebrates, whose majority of species is bound to grasslands. Among the 70 grasshopper species of Baden-Württemberg (SW Germany), 48 alone can be found in semi-dry basiphilous grasslands, but there are additional specialized species in wet (Fig. 3.1G), mesic and sandy dry grasslands (Detzel, 1998).

Among vertebrates, grasslands are particularly important for reptiles and birds. Being ectothermous and thus requiring warm places, all reptile species of Western and Northern Europe, with few exceptions, use different grassland types as (part of their) habitat (Laufer et al., 2007). Among the European bird species, 152 (29 per cent) are associated with one or more grassland types (Nagy, 2009). While typical steppe birds are rare in the region due to the climatic conditions (e.g., very small remaining population of *Otis tarda* [Great Bustard] in eastern Germany), birds of wet grasslands, both in flood plains and in coastal salt marshes, are abundant and rich in species. They include numerous waders (e.g., *Haematopus ostralegus*, *Limosa limosa*, *Vanellus vanellus*), ducks and geese (e.g., *Anas acuta*, *Anser anser*), passerines (e.g., *Anthus pratensis*, *Motacilla flava*, *Saxicola rubetra*), birds of prey (e.g., *Circus cyaneus*) and the enigmatic White Stork (*Ciconia ciconia*). Coastal salt marshes of the regions also serve as crucial wintering places of thousands of geese and waders

Table 3.3 Maximum plant species richness values known from grasslands in Western and Northern Europe as delimited in the chapter based on unpublished data from the GrassPlot database (vers. 88; Dengler et al., 2016; 2018). Global and European records are based on Wilson et al. (2012) and Dengler et al. (2016). S_{max} is the highest known richness value for the respective grain size, S_{total} is for all terricolous vascular plants, bryophytes and lichens, $S_{vasc.\ pl.}$ only for vascular plants.

Area [m²]	$S_{max,\ total}$ (Region)	Location	$S_{max,\ vasc.\ pl.}$ (Region)	Location	$S_{max,\ vasc.\ pl.}$ (Europe)	$S_{max,\ vasc.\ pl.}$ (Global)
0.0001	7	DE (S Bavaria)	7	DE (Franconia)	9	9
0.001 or 0.0009	12*	DE (S Bavaria)	13	DE (Franconia)	19	19
0.01	27	DK	23	DK	25	25
0.1 or 0.09	43	SE (Öland)	34	DE (S Bavaria)	43	43
1	63	SE (Öland)	53	DE (S Bavaria)	82	89
10 or 9	81*	SE (Öland)	86	CH	98	98
100	111	DE (S Bavaria)	76*	DE (S Bavaria)	133	233

* Indicates values that are lower than on a smaller area or total richness values that are smaller than vascular plant species richness values on the same scale which is due to the fact that not in all regions all grain sizes and all taxonomic groups were sampled. CH = Switzerland, DE = Germany, DK = Denmark, SE = Sweden.

that breed in the Siberian tundra. While bird abundances are most impressive in coastal and inland marshes, also mesic and dry grasslands have their specific bird communities, particularly if they are rich in additional structures like thorn shrubs or stone heaps, e.g., *Saxicola rubicola, Lanius collurio, Sylvia* spp. or *Perdix perdix*.

Among the circa 6,200 vascular plant taxa endemic to Europe, 18.1 per cent are bound to grasslands, making this the second-most important habitat for range-restricted species after rocks and screes (Hobohm and Bruchmann, 2009). Another 6.3 per cent endemics inhabit coastal habitats (which are also largely grasslands *sensu lato*) while forests only host 10.7 per cent. Grassland endemics (without coastal habitats) are particularly rich in France (498 species), Austria (403), Switzerland (355) and Germany (315) (Hobohm and Bruchmann, 2009). In Germany, dry grasslands host 488 of the 2,997 evaluated vascular plant species and thus more than any other of the 23 formations (Korneck et al., 1998). All 'formations' roughly correspond to our grassland definition account for 45 per cent of all assignments (including 1,120 double assignments; Korneck et al., 1998). Semi-natural grasslands of Europe have been highlighted as the global record holders in small-scale vascular plant diversity for areas < 100 m², with several examples from Sweden and Germany (Wilson et al., 2012). While generally the most diverse grasslands of Eastern Europe are richer, some types of the our region, mainly on the Swedish Island of Öland, in southern Germany and Switzerland, are not too far below the European and global maxima (Table 3.3). Some grassland types of the region, such as the *alvar* grasslands of Sweden (belonging to type 7.c of Table 3.2) can also be extraordinarily rich in non-vascular plants with up to 40 bryophyte and 24 lichen taxa on 4 m² and a share of up to 75 per cent of total richness of the vegetation in one association (Löbel and Dengler, 2008).

Other Ecosystem Services

An assessment of ecosystem services of grasslands from the Czech Republic, thus under relatively similar conditions, found that the agronomic benefits account for only circa 20

A
B
C
D
E
F
G
H
Maestranza da prada 2015
Wiesenmeisterschaft 2015
WIESEN
MEISTERSCHAFT
SCHWEIZ
Nebenstehend sind die angemeldeten Wiesen hervorgehoben. Der rote Punkt kennzeichnet ihren Standort.
Die Prämierung der Wiesen findet am Freitag, 10. Juli 2015 ab 16 h in Tschlin statt.
Bitte betreten Sie die Wiesen nicht, sondern betrachten Sie deren Schönheit vom Wegrand aus.

Fig. 3.2 contd. ...

per cent of the monetary value of the ecosystem services evaluated (Hönigová et al., 2012). They suggest that the most valuable non-agronomic service of grasslands is the water flow regulation, accounting for two-thirds to four-fifth of the total value, followed by erosion control, nitrogen retention and carbon sequestration. Soussana et al. (2007) found for nine European grassland sites (six intensive, three semi-natural, among them eight in the region) that grasslands are indeed a net carbon sink, but, taking into account the off-site emissions of climate-relevant gasses (mainly methane from cattle), the net greenhouse gas balance was not significantly different from zero. While aesthetic and touristic value was rated least among the ecosystem services in Czech grasslands (Hönigová et al., 2012), this is different in touristic regions whose positive perception largely depends on a semi-open diverse agricultural landscape and which certainly would lose much of the attractiveness if returned to closed forest cover or, in other cases, to croplands. Even within grasslands, ordinary people in their majority prefer rather species-rich over monotonous types (Lindemann-Matthies and Bose, 2006; Fig. 3.2G).

Threats to Grasslands

Habitat Loss of Grasslands

The amount of grasslands converted to other land uses is not easy to quantify. For natural grasslands the rate should generally be low; we estimate it at overall less than 5 per cent. While among arctic-alpine grasslands only a negligible area has been converted (roads, touristic infrastructure, hydroelectric power dams), certainly a higher fraction of coastal grasslands (dunes and salt marshes) has been completely destroyed because the coasts of Western and Northern Europe are among the most densely populated areas and thus suffer from urban sprawl, industrial and touristic infrastructure. However, in some regions, salt marshes expanded in area due to active land reclamation. For example, the North Sea salt marshes of Germany gained about 25 per cent area from 1932 to 1997, after losses in the preceding centuries (Riecken et al., 2006).

...Fig. 3.2 contd.

Fig. 3.2 Management and conservation of High Nature Value Grasslands. (A) Year-round grazing with Konik horses in the nature reserve 'Tote Täler', Saxony-Anhalt, Germany. (B) Restoration of severely shrub-encroached dry grasslands with low-intensity rotational grazing with goats, nature reserve 'Tote Täler', Saxony-Anhalt, Germany. (C) Year-round grazing with Heck cattle and Konik horses in the sandy dry grasslands was able to push back the 'native invasive' *Calamagrostis epigejos* (left side, outside the grazing fence), nature reserve 'Oranienbaumer Heide', Saxony-Anhalt, Germany. (D) Open soil patches as an important feature of structural heterogeneity created by low-intensity, year-round multi-species grazing, nature reserve 'Oranienbaumer Heide', Saxony-Anhalt, Germany. (E) Collecting seed-rich thrashing material for restoration purposes in the nature reserve 'Alperstedter Ried', Saxony-Anhalt, Germany. (F) Application of collected thrashing material on a prepared grassland restoration site, Wulfen, Saxony-Anhalt, Germany. (G) Announcement of the 'meadow championship' in which the farmer who maintained the most biodiverse and beautiful meadow is awarded, Grisons, Switzerland. (H) What looks like a particularly beautiful *Bromion erecti* stand (order *Brachypodietalia pinnati*, class *Festuco-Brometea*), with plenty of *Orchis morio*, is in fact an extensive green roof, established on top of the water treatment plant in Zürich-Wollishofen, Switzerland, back in 1914. Photos by A. Zehm (A), J. Dengler (G), S. Brenneisen (H) and S. Tischew (all others).

For secondary grasslands, the decrease in area compared to their maximum extent is higher. Poschlod (2015: p. 83) provided an estimate of how the fractions of the three main land cover types 'forest', 'grassland + heathland' and 'cropland' changed in Germany throughout the last 2,000 years, based on soil erosion data and settlement history. Accordingly, 'grassland + heathland' declined from a maximum of 47 per cent around AD 1300 to about 29 per cent at present, which would be a decrease by 37 per cent. We need to take these figures with a grain of salt as they are rough reconstructions and evidently 'grassland + heathland' not only contain grassland, but also other open habitats, like heathland and settlements (compare the grassland fraction of Germany in Table 3.1). Moreover, assuming a mean reduction in area of secondary grasslands compared to their maximum extent across the region of about 40 per cent (Török and Dengler, 2018) hides the fact that there are extreme differences between regions and grassland types. Conversion to land use other than grassland is particularly strong for grassland types of low productive sites, e.g., basiphilous semi-dry grasslands (8.a of Table 3.2). Here Riecken et al. (2006) reported a habitat loss for the lower mountain ranges of the central-western parts of Germany of 97 per cent from 1895 to 1998, but 'only' 18 per cent for southern Germany. Conversion mainly goes in the direction of forest (or shrubland) via afforestation or spontaneous succession, to smaller parts to settlements, touristic infrastructure or croplands. More productive grassland types, such as lowland hay meadows (10.a of Table 3.2) and wet grasslands (10.c), also experienced significant losses, but more towards croplands (Riecken et al., 2006; Wesche et al., 2009; Krause et al., 2011; Poschlod, 2015). Zooming in to recent developments in Germany, the area of permanent grassland decreased from 1991 to 2013 by about 14 per cent, mainly due to the demand for bioenergy crops, such as maize, but meanwhile has slightly increased again (UBA, 2017).

Beyond pure reduction in area, fragmentation can negatively impact biodiversity of grassland habitats. In a multi-taxon study of Austrian dry grassland patches, Zulka et al. (2014) found that the area of the grassland patch positively affected the small-scale richness within the patch for most invertebrate taxa, but not for vascular plants. Moreover, the patch area 50 years back was a much better predictor than current area, pointing to delayed extinction, the so-called 'extinction debt' (Cousins, 2009).

Changes in Grassland Use

Those grasslands that remained as grasslands additionally experienced often strong modifications in grassland use and profound changes in site conditions. In the more productive and better accessible sites, mainly in lowlands, there was a strong trend towards intensification (Poschlod, 2015; Stevens et al., 2010). On the other hand, the usage of grasslands in low-productive situations and remote areas has ceased at large scale, leading to natural succession, if forests were not actively planted (Wesche et al., 2009; Poschlod, 2015).

Intensification

For the lowlands of Wales, where total grassland area had remained more or less stable, the fraction of 'improved grasslands' increased from less than 5 per cent of the territory in the 1930s to 90 per cent in the 1980s/1990s, while only 10 per cent of the former semi-natural grasslands remained (Stevens et al., 2010). While their definition of 'semi-natural' probably was narrower than that of Török and Dengler (2018), this still demonstrates a dramatic change. 'Improved' or 'intensified' grasslands (Dierschke and Briemle, 2002; Wesche et al., 2009; Stevens et al., 2010; Poschlod, 2015) refer to types that are species-poor and very similar across the chapter region.

Intensification thus involves, apart from increased fertilization and drainage, which are treated in the section *Alteration of Site Conditions*, re-seeding, higher cutting frequencies in meadows and higher livestock densities in pastures. Re-seeding means the replacement of existing grassland habitat with a system that contains only few productive species of high fodder value (namely *Lolium perenne* as grass and *Trifolium repens* and *T. pratense* as legumes) and these even in a limited number of standardized cultivars (Dierschke and Briemle, 2002; Smit et al., 2008). Instead of one to two relatively late usages (cuttings or grazing periods) as formerly in semi-natural grasslands, intensified grasslands nowadays are subject to three to six usages per year (Dierschke and Briemle, 2002), leading to lower stand height, reduced structural diversity and preventing most of the grassland plants to flower and set seeds. Combining the different components of land use (mowing frequency, grazing intensity, fertilization) into an overall index of land-use intensity (LUI), Gossner et al. (2016) showed that α-diversity of most above-ground taxonomic groups decreased with increasing LUI, while that of below-ground groups generally increased and most groups showed a strong decline in β-diversity, i.e., a biotic homogenization.

Another component of intensification are land consolidation programs (German: *Flurbereinigung*) that have been carried out to improve agricultural efficiency by combining small fields into larger ones and removing small structures, like stone walls, outcrops, field margins or uneven terrain. As a consequence, grasslands are now mown synchronously over larger areas, being particularly detrimental for less mobile animals who cannot escape to another grassland patch nearby for fodder and shelter.

Abandonment

Abandonment particularly affects mountain hay meadows, dry and semi-dry as well as wet grasslands. Already before such grasslands cease to be grasslands, their species composition clearly changes. Jandt et al. (2011) found for semi-dry basiphilous grasslands (Type 8.a in Table 3.2) in Germany that typical, low-growing and less competitive dry grassland species have decreased over the past decades while species of mesic grassland and forest edges as well as woody species have increased, pointing to an 'underuse' of the grasslands. As a consequence of denser and taller stands as well as litter accumulation, both mean Ellenberg values for nutrients and for soil moisture had increased. Dupré and Diekmann (2001) reported that abandonment of formerly extensively grazed grasslands in South Sweden led to a decrease in species richness, particularly at small spatial scales, and to a shift towards more monocots and fewer legumes and more geophytes and fewer therophytes. Not only previously agriculturally used grasslands were subject to abandonment, but also many highly biodiverse grassland habitats in former military training areas became superfluous after the end of the Cold War (see von Oheimb et al., 2006; Johst and Reiter, 2017).

Alterations of Site Conditions

Eutrophication from Agricultural Fertilization and Air-borne Depositions

Next to Abandonment (see *Changes in Grassland Use*), we consider eutrophication as the second mega-threat to grassland biodiversity in the region. Eutrophication mainly refers to nitrogen and phosphorus. Nitrogen emissions (which ultimately become inputs) in the EU have increased in the last century until about 1980 and moderately decreased since then. The four main sources of nitrogen emissions to European ecosystems are artificial fertilizer (55 per cent), industry and traffic (combustion processes: 20 per cent), imported feed and food from other continents (17 per cent) and N_2 fixation by legumes (8 per cent) (Sutton et al., 2011). Agricultural input of nitrogen (mineral fertilizer + manure) on agricultural lands (croplands and grasslands) reaches its highest value within Europe of more than 170 kg ha^{-1} yr^{-1} in part of the study region, namely the British Isles except Scotland, Northern France, Benelux, Germany and Denmark (Grizzetti et al., 2007). On top of that comes the atmospheric deposition, which in semi-natural systems of the region (except Iceland, Fennoscandia and Scotland) is typically above 10 kg ha^{-1} yr^{-1}, but in northern France, Benelux and north-west Germany can exceed 30 kg ha^{-1} yr^{-1} (Sutton et al., 2011). Both together sum up to a surplus of nitrogen at the landscape scale, i.e., more deposition than the growing plants take up, nearly everywhere in the region, but with peaks of more than 100 kg ha^{-1} yr^{-1} in the previously named regions plus Denmark and Bavaria in Germany (Grizzetti et al., 2007).

For phosphorus the situation is slightly different as this element is not transported via the atmosphere and thus only results from direct agricultural application (manure + inorganic fertilizer). Accordingly, the phosphorus balance at landscape scale is in most parts of the region only slightly positive (< +5 kg ha^{-1} yr^{-1}) or even negative, while in several regions with extremely intensive animal production (including pigs and poultry), namely north-west France, parts of Benelux and north-west Germany, the phosphorus surplus exceeds 25 kg ha^{-1} yr^{-1} (Grizzetti et al., 2007).

The negative effects of high nutrient inputs on grassland diversity have been extensively demonstrated. Stevens et al. (2004) reported for semi-natural acidic grasslands in the United Kingdom that their vascular plant species richness is strongly affected by the atmospheric nitrogen deposition, with a loss of about one species (on 4 m^2) per each additional 2.5 kg ha^{-1} yr^{-1}. Dupré et al. (2010) showed that species loss of both vascular plants and bryophytes in acidic grasslands in several countries of the region negatively depends on cumulative nitrogen deposition and with decreasing richness in parallel, the cover of grasses increases. By contrast, in semi-dry basiphilous grasslands, nitrogen deposition in north-west Germany did not lead to a change in richness, but to a clear change in species composition away from typical dry grassland species that are small, light-demanding and/or with scleromorphic leaves (Diekmann et al., 2014). While in the past, nitrogen had been considered as the main limiting factor of terrestrial ecosystems and thus in turn the major driver of community change and species loss, recent studies highlighted that phosphorus in European grassland systems might have an even more negative effect. Ceulemans et al. (2013) found this for mesic to dry acidic grasslands and Wassen et al. (2005) for wet grasslands. Ceulemans et al. (2014) then showed that for a wide range of High Nature Value grasslands in Western and Northern Europe their diversity is strongly constrained by available phosphorus in the soil, with no lower threshold. Phosphorus might also be more problematic because it persists longer in the soil, though there is no atmogenic deposition; thus unused grasslands are not affected. By contrast, negative effects of nitrogen start at certain deposition levels, the so-called critical loads, which range from 5 to 25 kg ha^{-1} yr^{-1}, depending on the grassland type (Rihm and Achermann, 2016). Critical loads are exceeded in most parts of Western and Northern Europe, except middle and northern Fennoscandia, but the situation is slowly improving since the 1990s (Sutton et al., 2011). While in Switzerland, for example, critical loads exceeded in 81 per cent of all dry grasslands and 91 per cent of all fens (including wet grasslands), these numbers declined to 49 per cent and 76 per cent, respectively in 2010 (Rihm and Achermann, 2016). Generally, the negative effects of eutrophication on biodiversity are rather indirect via increased productivity and thus competitive than direct (Socher et al., 2012).

Alterations in Water Regime

Changes in the water regime are also an important threat factor, particularly large-scale drainages that affected all types of wet grasslands, mainly in the period after World War II. Wesche et al. (2009) demonstrated that in floodplain meadows of northern Germany, mainly in consequence of drainages, plot-scale species richness of wet grasslands significantly decreased from the 1950s/1960s until 2008. In parallel, species composition strongly shifted, with steep decline in formerly frequent wet meadow species (e.g., *Silene flos-cuculi, Caltha palustris*), while some mesophilic (e.g., *Arrhenatherum elatius, Lolium perenne*) and ruderal species (e.g., *Cirsium arvense, Urtica dioica*) became more frequent.

Climate Change and Increased CO_2 Concentrations

Early projections based on species distribution models (SDMs) suggested that regions in Europe would experience a strong loss of vascular plant species by end of the century due to anthropogenic climate change, with rates of loss being relatively low in Western and Northern Europe, but still in the range of 25–31 per cent (Thuillier et al., 2005). However, studies like this made several unrealistic assumptions, that is, (i) that species distributions are exclusively driven by climate and (ii) they are in equilibrium with climate. Taking into account the soil and land use data, Pompe et al.'s (2008) simulations suggest that depending on the land use scenario and the modeling technique, vascular plant species richness in German grid cells on an average will remain more or less the same if full dispersal is assumed. One also must take into account that long-lived species, say, of perennial grasslands, are able to persist under suboptimal climate for quite long periods (Lenoir and Svenning, 2015). Anyway, a more serious problem than global warming itself might be that niches of several species that depend on each other shift differently. For example, Schweiger et al. (2012) demonstrated that the climatic niches of the wet grassland forb *Bistorta officinalis* and the butterfly *Boloria titania*, whose caterpillars are monophagous on this plant, may shift to different regions in Europe and in the future may show little overlap.

Time-series data of permanent plots in the region of this chapter could not confirm the assumed negative climate change effects on species richness. On mountain tops of the Alps, the Scandes and the Scottish Highlands, which theoretically should be among the most sensitive habitats towards climate change, species richness consistently increased from 2001 to 2008. Similarly, in 362 1-m² quadrats throughout the alpine zone from 2,900 to 3,450 m a.s.l. on Mt. Schankogel, Austria, mean vascular plant species richness had increased from 11.4 to 12.7, with 43 per cent of all species increasing in frequency, only 6 per cent decreasing and none being lost (Pauli et al., 2007). In the alpine zone of the Scandes (mainly alpine heathlands and grasslands), the richness of vascular plants remained more or less stable over a 15-year period, whereas lichens became poorer and bryophytes richer (Vanneste et al., 2017). In systematic permanent plots across all elevational belts and vegetation types of Switzerland, cold-adapted vascular plant species kept their frequency, while warm-adapted and intermediate species became more frequent in recent years (BAFU, 2017). Overall global warming seems to have relatively small, but rather positive direct effects on vascular plant species richness in grasslands of the region. There might, however, be an indirect negative effect mediated through sea-level rise that could cause destruction of coastal dunes and salt marshes (Janssen et al., 2016). By contrast, there is currently no evidence that changing CO_2 levels, as suggested by Sala et al. (2000), would have any direct negative impact on grassland diversity of the region. This might be due to the fact that C_4 species are quite rare in the European flora (compared to North America for instance, see Wesche et al., 2016), thus no 'biome shift' is likely to happen even if C_4 species should gain competitive advantages in the future.

Invasive Species

In general, grasslands do not contain a big proportion of neophytes compared to other habitat types in Europe (Chytrý et al., 2008). While the most-invaded habitat types among 33 habitats, arable land and coniferous woodland, have on average about 8 per cent neophytes in their species composition, the most-invaded grassland types, wet grasslands and saline habitats, are only ranked 11th and 14th with about 1.5 per cent neophytes in their species composition. With 0.3–0.8 per cent neophyte species, mesic and dry grasslands as well as subalpine tall forb stands follow on ranks 19 to 22, while alpine grasslands with about 0.5 per cent are among the least invaded habitats. Generally, both fraction and cover of neophytes strongly decrease with elevation in all grassland types, becoming insignificant above circa 600 m a.s.l. (Chytrý et al., 2009).

While the overall impact of neophytes on grassland diversity is low, certain invasive neophytes still can change grassland ecosystems locally in a drastic manner. Presumably the tall forbs *Solidago canadensis* and *S. gigantea* from North America as well as *Impatiens glandulifera*, *Fallopia sachalinensis*, *F. japonica* and *F.* ×*bohemica* from temperate Asia are currently the most widespread neophytes that tend to form dominance stands in former grasslands. This typically happens in abandoned mesic to moist grasslands, particularly in river valley and along streams. In mono-dominant stands of these alien tall forbs, vascular plant species richness on 16 m² can be drastically reduced compared to neighboring non-invaded plots. *Fallopia* species have the strongest negative effect on diversity (–66 to –86 per cent), followed by *Solidago gigantea* (–26 per cent) and *Impatiens glandulifera* (–12 per cent) (Hejda et al., 2009). However, invasion typically takes place in grasslands and other habitats that already, without the respective neophyte, have been species poor (10–17 species on 16 m²; Hejda et al., 2009). In dominance stands of *Fallopia* spp., not only plant species richness was reduced but also richness, abundance and biomass of herbivorous and carnivorous invertebrates, while that of detritovorous species remained unchanged as compared to adjacent grassland communities in European alluvial plains (Gerber et al., 2008). *Lupinus polyphyllus* invasions are more localised in certain low mountain ranges with humid climate and acidic bedrock, but since they affect species-rich mountain hay meadows, they might be more problematic (Otte and Maul, 2005; Hejda et al., 2009; Thiele et al., 2010). There are also some invasive woody aliens, namely the North American tree *Robinia pseudoacacia* in some continental grasslands and the East Asian shrub *Rosa rugosa* in many coastal dunes, where it can cause a strong reduction in species richness of yellow and grey dunes (Thiele et al., 2010). While neophytes among bryophytes are rare, the South African *Campylopus introflexus* poses a serious problem to coastal and inland dune grasslands, but it apparently can replace native cryptogam species when nitrogen pollution is high (Sparrius and Kooijman, 2011) and sand blowing is stopped (Ketner-Oostra et al., 2012).

More relevant for loss of grassland diversity in Western (and Northern) Europe than non-native invasives are probably 'native invasives', namely the grasses *Calamagrostis epigejos* (Rebele and Lehmann, 2001; Schuhmacher and Dengler, 2013; Fig. 3.2C) and *Brachypodium pinnatum* (Bobbink and Willems, 1987) as well as the fern *Pteridum aquilinum* (Pakeman and Marrs, 1992). All are tall, competitive species and have rhizomes, which serve as below-ground storage organs that allow internal nutrient cycling and enable fast and effective vegetative spread. All three species are native to the region but more in forest edges and disturbed sites, and have greatly expanded into grasslands during recent decades, mainly due to abandonment and eutrophication, leading to species-poor mono-dominant stand.

Direct Human Impact

In general, we consider direct human impacts as of much lower relevance than the five groups of (more indirect) human impacts on grassland biodiversity discussed before (Table 3.4). First, military training areas might have negative impacts locally, but overall they make a considerable positive contribution to the maintenance of biodiverse grasslands

Table 3.4 Relative importance of threat factors (= drivers of biodiversity loss) in grasslands of Western and Northern Europe according to our assessment and that of Janssen et al. (2016) for the same habitats in EU28+. In our assessment, the following coding applies: –: absent or very rare, +: rare/local/low impact, ++: moderately widespread/moderate impact, +++: widespread/high impact. For Janssen et al. (2016), we counted how often a certain threat factor was mentioned in the 'fact sheets' of the 37 grassland habitats that are found in Western and Northern Europe (see Table 3.4), expressed as percentage.

Threats/Causes of Biodiversity Loss	This Chapter	Janssen et al., 2016
Habitat loss of grasslands		
- Conversion to arable land	+	19%
- Afforestation	+	24%
- Mining and energy production	+	8%
- Urbanization, transport and touristic infrastructure	+	49%
Changes in grassland use		
- Abandonment and/or underuse	+++	65%
- Overgrazing	+	35%
- Other types of agricultural intensification	++	22%
Alteration of site conditions		
- Eutrophication	+++	62%
- Altered water regime	++	19%
Climate change	+	19%
Invasive species	+	14%
Direct human impacts		
- Military and armed conflicts	–	0%
- Recreational activities	+	27%
- Collecting wild plants/hunting wild animals	–	0%

Note that the correspondence of threat factors in both assessments is only approximate. Temporal horizon for this chapter have been the past 30 years, whereas Janssen et al. (2016) used a more complex assessment, also looking into the future.

in the region as they maintain open landscapes (i.e., prevent succession), while fertilization, agricultural intensification and site alterations are generally not applied within such territories (Ellwanger et al., 2016). Second, recreational activities, including trampling by tourists, motocross driving or eutrophication by dogs, can have some negative impact, but this is largely restricted to grasslands in certain coastal areas and those in the vicinity of large urbanizations (Janssen et al., 2016 for Europe). Even there the perception of tourists by conservationists might be more negative than their actual impact is, since grasslands in the region generally depend on some degree of disturbance. Lastly, collection of grassland plants and mushrooms or hunting of grassland animals seems to be of low importance in the region (Korneck et al., 1998 for Germany).

Relative Importance of Threat Factors

The preceding individual assessments are summarized in an overall ranking of threat factors (Table 3.4). Accordingly, we consider abandonment/underuse on the one hand and eutrophication on the other as the two most important drivers of biodiversity loss in grasslands of the region during recent decades, which is in full agreement with independent assessment in Janssen et al. (2016; see last column of Table 3.4). Also quite negative and acting in many grasslands were other types of agricultural intensification (mainly meaning: high frequency of cutting and frequent re-seeding) and altered water regime (mainly referring to drainage of wet grasslands). In comparison with that, all other threat factors seem to be of low or local importance. Our ranking for Western and Northern Europe generally matches with that of World Resources Institute (2005) for temperate grasslands and that of Janssen et al. (2016) for grasslands in Europe, except that the latter source considers grassland conversion to built-up areas as the third most relevant driver (which is partly due to the fact that they distinguished many grassland types in coastal areas, which are indeed affected by this).

Overall Threat Situation of Grassland Habitats and Species

According to the recent *European Red List of Habitats* (Janssen et al., 2016), which only considers natural and semi-natural habitats (i.e., excluding segetal and ruderal communities as well as intensified grasslands), there are 233 terrestrial and freshwater habitats in the 28 member countries of the European Union plus Iceland, Norway, Switzerland and the Balkan countries (EU28+). Of these, 65 can count as grassland habitats (those of category E = Grassland and the grassland-dominated types of category B = Coastal habitats, excluding some other grass-dominated open habitats that can be found in the categories C = Freshwater habitats, D = Mires and bogs and H = Sparsely vegetated habitats), of which 37 occur in the area of this chapter (Table 3.5). Among the natural and semi-natural grassland habitats of Western and Northern Europe, 23 (62 per cent) are threatened and five more (14 per cent) near-threatened (Table 3.6). Thus the degree of threat is twice as high on an average for European habitats (32 per cent) and even more than twice as high for the two highest categories, CR (Critically Endangered) and EN (Endangered). Actually half of all European CR habitats are grasslands occurring in Western and Northern Europe, namely 'Pannonian and Pontic sandy steppe' (E1.1a) and 'Hemiboreal and boreal wooded pasture and meadow' (E7.2). Moreover, several grassland habitats that have a relatively low Red List status for Europe as a whole would have to be rated higher if only Western

Table 3.5 Red List status of grassland habitat types occurring in Western and Northern Europe based on Janssen et al., 2016. Apart from all habitats listed in category E = Grassland, also the grassland types from category B = Coastal habitats are considered. The corresponding vegetation type refers to the phytosociological unit in Table 3.2. RL gives the Red List status for EU28+ (i.e., the 28 member states of the European Union + Iceland, Norway, Switzerland and the Balkan countries): CR = Critically Endangered, EN = Endangered, VU = Vulnerable, NT = Near Threatened, LC = Least Concern. Main criteria give the prevailing reason for assignment to the specific Red List status: A1 = Present decline (over the last 50 years), A2a = Future decline (over the next 50 years), A3 = Historic decline, B2 = Restricted Area of Occupancy (AOO), CD1 = Reduction in abiotic/biotic quality over the last 50 years, CD2 = Reduction in abiotic/biotic quality in the future or in a period including present and future.

Code	Name in Red List	Corresponding Vegetation Type	RL	Main Criteria
Coastal habitats (B)				
A2.5a	Arctic coastal salt marsh	6	NT	A2a, B2
A2.5b	Baltic coastal meadow	6	EN	A1
A2.5c	Atlantic coastal salt marsh	6	VU	CD1
B1.3a	Atlantic and Baltic shifting coastal dune	other	NT	A3, CD1
B1.4a	Atlantic and Baltic coastal dune grassland (grey dune)	7.a, 7.c, 7.d, 7.e	VU	A1, CD1
B1.9	Machair	?	LC	–
B3.4a	Atlantic and Baltic soft sea cliff	various	LC	–
Grassland (E)				
E1.1a	Pannonian and Pontic sandy steppe	7.e	CR	A3, CD1
E1.1b	Cryptogam- and annual-dominated vegetation on siliceous rock outcrops	7.f	VU	A1
E1.1d	Cryptogam- and annual-dominated vegetation on calcareous and ultramafic rock outcrops	7.g	VU	A1
E1.1g	Perennial rocky grassland of Central Europe and the Carpathians	8.c	LC	–
E1.1i	Perennial rocky calcareous grassland of subatlantic-submediterranean Europe	8.d	VU	A1
E1.2a	Semi-dry perennial calcareous grassland	8.a	VU	A1, A3
E1.2b	Continental dry steppe	8.b	NT	A1
E1.7	Lowland to submontane, dry to mesic *Nardus* grassland	9.a	VU	A1
E1.9a	Oceanic to subcontinental inland sand grassland on dry acidic and neutral soils	7.d	EN	A1, A3
E1.9b	Inland sanddrift and dune with siliceous grassland	7.c	EN	A1, A3
E1.A	Mediterranean to Atlantic open, dry, acidic and neutral grassland	7.b	NT	CD1
E1.B	Heavy-metal grassland in Western and Central Europe	various	EN	A1
E2.1a	Mesic permanent pasture of lowlands and mountains	10.a	VU	A1
E2.2	Low and medium altitude hay meadow	10.a	VU	A1, A3, CD1
E2.3	Mountain hay meadow	10.b	VU	A1, A3
E3.4a	Moist or wet mesotrophic to eutrophic hay meadow	10.c	EN	A1
E3.4b	Moist or wet mesotrophic to eutrophic pasture	10.c	EN	A1

Table 3.5 contd. ...

...Table 3.5 contd.

Code	Name in Red List	Corresponding Vegetation Type	RL	Main Criteria
E3.5	Temperate and boreal moist or wet oligotrophic grassland	10.c	EN	A1
E4.1	Vegetated snow patch	other	VU	CD2
E4.3a	Boreal and arctic acidophilous alpine grassland	3	LC	–
E4.3b	Temperate acidophilous alpine grassland	3	LC	–
E4.4a	Arctic-alpine calcareous grassland	1, 2, 4	LC	–
E5.2a	Thermophilous woodland fringe of base-rich soils	other	NT	A1
E5.2b	Thermophilous woodland fringe of acidic soils	other	LC	–
E5.3	*Pteridium aquilinum* stand	other	LC	–
E5.4	Lowland moist or wet tall-herb and fern fringe	other	VU	A1
E5.5	Subalpine moist or wet tall-herb and fern fringe	5	LC	–
E6.3	Temperate inland salt marsh	6	EN	A1
E7.1	Temperate wooded pasture and meadow	various	VU	A1, CD1
E7.2	Hemiboreal and boreal wooded pasture and meadow	various	CR	A1, CD1

Detailed Red List assessments, broken down to individual countries and parameters, for each individual habitat can be found in the 'fact sheets' provided online at https://forum.eionet.europa.eu/european-red-list-habitats/library/terrestrial-habitats. Distribution maps and list of diagnostic species underlying the refined EUNIS habitat classification used in the Red List assessment have been published for the grassland types by Schaminée et al. (2016).

Table 3.6 Comparison of the Red List status at European scale (EU28+) of the 37 grassland habitats that occur in Western and Northern Europe with all the 233 terrestrial and freshwater habitats that are distinguished in EU28+.

Red List Status	Grasslands (of W and N Europe)		All Habitats (EU 28+)	
	Number	Fraction	Number	Fraction
Threatened categories				
CR = Critically Endangered	2	5%	4	2%
EN = Endangered	8	22%	23	10%
VU = Vulnerable	13	35%	46	20%
Other categories				
NT = Near Threatened	5	14%	29	12%
LC = Least Concern	9	24%	116	50%
DD = Data Deficient	0	0%	15	6%
Total	**37**		**233**	

and Northern Europe were considered, e.g., 'Continental dry steppe' (E1.2b) or 'Perennial rocky grassland of Central Europe and the Carpathians' (E1.1g), which only reach the most-continental parts of the territory.

Also, species of grasslands are particularly threatened—in Germany, dry grasslands and wet grasslands are ranked third and sixth, respectively, among 24 main vegetation types, concerning the threat to their typical vascular plant species, while dry grasslands alone contain already one-fourth of all threatened species and thus more than any of the other 23 types (Korneck et al., 1998). Similarly, in Switzerland, dry grasslands (of lower elevations) and mires (including the wet grasslands) are the two out of 12 habitat types with the highest number of threatened and near-threatened vascular plant species (Bornand et al., 2016). Moreover, in Switzerland the Red List Index for vascular plants showed the strongest deterioration from 2002 to 2016 for exactly these two habitat types while others even improved their situation (Bornand et al., 2016). Among animals, only a few threat assessments on habitat level exist. For butterflies in Europe as a whole, WallisDeVries and van Swaay (2009) found that grassland specialist species are the group that shows the strongest decrease, and among these are the species bound to humid grasslands and tall herb communities. For carabid beetles in Germany, Schmidt et al. (2016) report that the proportion of threatened species in dry open habitats (dry grasslands *sensu lato*) with 84 per cent is the highest among all habitat types. For birds, the signals are a bit mixed—in Germany among the critically endangered taxa with very negative population trend in recent years, grassland species are overrepresented (e.g., *Otis tarda*, *Philomachus pugnax*, *Limosa limosa*, *Anthus campestris*) (Pauly et al., 2009; Poschlod, 2015), the Swiss Bird Index (SBI) suggests that birds of the open cultural landscape on an average have stable populations (Sattler et al., 2015).

One tendency not covered by the usual Red Lists is the similarity of species assemblages within landscapes. The Swiss Biodiversity Monitoring (BDM) recorded a clear homogenization of vascular plant assemblages in grassland communities of low elevations from 2001 to 2007 (Koordinationsstelle Biodiversitätsmonitoring Schweiz, 2009).

Legal and Policy Instruments for Conservation of Grasslands

Western and Northern Europe have numerous protected areas that aim at conserving both natural and semi-natural habitats. Some of the coastal and alpine National Parks conserve larger tracts of natural grasslands, while semi-natural grasslands are protected in smaller nature reserves and larger biosphere reserves, the latter specifically focusing on maintenance of traditional cultural landscapes. Overall, nationally protected areas (of varying strictness) account for about 17 per cent of the territory of EU countries in the region, which is slightly less than the EU average (EEA, 2012). Across the EU, protected areas comprise only about 9 per cent grassland ecosystems (EEA, 2012; no country-specific data is available, but this should reflect the situation in the region), while the share of grasslands of the region's territory is about twice as large (Table 3.1), pointing to an underrepresentation of grasslands in the reserve network.

The EU has established the Habitats Directive that requires from the member states to maintain certain habitat types (European Commission, 2013) and species of 'community importance' in good conservation status. To this end, the member states have established Sites of Community Importance (SCIs), so-called Natura 2000 sites, partly within nationally protected areas, and partly in addition. In the Habitats Directive, grassland species and habitats are well-covered, with 31 grassland habitat types in the region alone, including 15 of priority (Table 3.7). The Natura 2000 system has certainly expanded the formal protection of grasslands. Likewise, Switzerland has national inventories of four habitat types of national importance, covering about 4 per cent of the country (mostly outside reserves), where there is an obligation to maintain the status unchanged, including dry

Table 3.7 Grassland habitats of Community importance included in the Habitats Directive (European Commission, 2013) in Western and Northern Europe (* denotes priority habitats). The last two columns provide an approximate cross-walk to the units of Tables 3.2 and 3.5 ('–' indicates that the unit is not listed in the respective table).

Number	Name	Table 3.2	Table 3.5
1230	Vegetated sea cliffs of the Atlantic and Baltic coasts	various	E3.4a
1320	*Spartina* swards (*Spartinion maritimae*)	–	A2.5c
1330	Atlantic salt meadows (*Glauco-Puccinellietalia maritimae*)	6	A2.5c
1340	* Inland salt meadows	6	C5.4
1630	* Boreal baltic coastal meadow	6	A2.5b
2120	Shifting dunes along the shoreline with *Ammophila arenaria* (white dunes)	–	B1.3a
2130	* Fixed coastal dunes with herbaceous vegetation (grey dunes)	7.a–e	B1.4a
21A0	Machairs (* in Ireland)	?	B1.9
2330	Inland dunes with open *Corynephorus* and *Agrostis* grassland	7.c+d	E1.9a+b
2340	* Pannonic inland dunes	7.e	E1.1a
6110	* Rupicolous calcareous or basophilic grasslands of the *Alysso-Sedion albi*	7.g	E1.1d
6120	* Xeric sand calcareous grasslands	7.e	E1.1a or E1.9a
6130	Calaminarian grasslands of the *Violetalia calaminariae*	various	E1.B
6150	Siliceous alpine and boreal grassland	3	E4.3a+b
6170	Alpine and subalpine calcareous grasslands	2 + 4	E4.4a
6190	Rupicolous pannonic grasslands (*Stipo-Festucetalia pallentis*)	8.c	E1.1g
6210	Semi-natural dry grasslands and scrubland facies on calcareous substrates (*Festuco-Brometalia*) (* important orchid sites)	8.a, (8.b), 8.d	E1.2a, (E1.2b), E1.1i
6230	* Species-rich *Nardus* grasslands, on siliceous substrates in mountain areas (and submountain areas, in Continental Europe)	9.a	E1.7
6240	* Sub-pannonic steppic grasslands	8.b	E1.2b
6250	* Pannonic sand steppes	7.e	E1.1a
6270	* Fennoscandian lowland species-rich dry to mesic grasslands	7.d, 8.a, 10.a	E1.2a, E1.9a, E2.1a, E2.2
6280	* Nordic alvar and precambrian calcareous flatrocks	7.g	E1.1b
6410	*Molinia* meadows on calcareous, peaty or clayey-silt-ladden soils (*Molinion caeruleae*)	10.c	E3.5
6430	Hydrophilous tall herb fringe communities of plains and of the montane to alpine levels	10.c, 5, etc.	E5.4
6440	Alluvial meadows of river valley of the *Cnidion dubii*	10.c	E3.4a, E3.4b
6450	Northern boreal alluvial meadows	10.c	E3.4a
6510	Lowland hay meadows (*Alopecurus pratensis*, *Sanguisorba officinalis*)	10.a	E2.2
6520	Mountain hay meadows	10.b	E2.3
6530	* Fennoscandian wooded meadows	8.a, 10.a	E1.2a, E2.2
8230	Siliceous rock with pioneer vegetation of the *Sedo-Scleranthion* or of the *Sedo albi-Veronicion dillenii*	7.f	E1.1b
8240	* Limestone pavements	7.g	E1.1d

grasslands as well as mires (with wet grasslands). However, designating conservation areas alone does not help for semi-natural grasslands that require agricultural use.

Since many years, the EU is supporting regional and national conservation actions that aim at improving the status of species and habitats of community importance within the so-called LIFE and LIFE+ programs. From 1992 to 2006, nearly 40 per cent of the LIFE co-funded conservation projects dealt directly or indirectly with grasslands, often involving purchasing land and re-establishing biodiversity-friendly management systems, many of them in the chapter region (Silva et al., 2008). More effect on grassland diversity than such conservation programs certainly have agricultural policies as they spend much more money and cover much larger areas. In 2017, the EU spent 42,613 million EUR (49,374 million US$) subsidies on agriculture but only 494 million EUR (572 million US$) on environmental issues and conservation (including LIFE+). With recent CAP reforms, the EU in 2005 largely decoupled subsidies from production (reducing the incentives for non-biodiversity-friendly farming) and in 2014 introduced 'greening' in the subsidy scheme, with the aim of supporting measures that are beneficial to biodiversity and climate change mitigation. Agri-environmental schemes (AESs) that can be financed by member states from CAP money are at least partly successful in maintaining/increasing grassland biodiversity (Klejn et al., 2006). However, overall, scientists have concluded failure of the CAP on its conservation aims because subsidized measures are not clearly directed at biodiversity conservation or counteracted by ongoing other subsidies with negative impact (Pe'er et al., 2014). Overall, Switzerland appears to be more efficient as it spends 14 per cent of its 2,809 million CHF (2,810 million US$) annual agricultural subsidies directly into precisely described measures, most of them in grasslands, that aim at biodiversity conservation, while other significant amounts are designated for maintenance of cultural landscapes (18 per cent) and landscape quality (5 per cent).

Management and Restoration of High Nature Value Grasslands

Grassland Management Methods for Conservation

To decide on the best and long-term sustainable management system for a certain locality, comprehensive background information on biotic and abiotic features, main targets in grassland conservation as well as the consideration of the current socio-economic situation are necessary (Török et al., 2016). Supporting the high diversity of traditional land use practices (see sections *Origin of the Grasslands* and *Historical Development of Secondary Grasslands*) should be the first choice for maintaining the biodiversity of grassland in the region. However, due to changes in the socio-economic situation, these traditional systems have often been replaced by intensive, more uniform and less adaptive management systems or are abandoned (Poschlod and WallisDeVries, 2002; Scotton et al., 2014), making their reintroduction often impossible (e.g., Isselstein et al., 2005; Köhler et al., 2016). For such situations, Vadász et al. (2016) suggest constructing management systems that fulfil the environmental criteria of sustainability by applying management attributes (e.g., intensity, timing and type of management) with known positive effects on grassland biodiversity.

In general, rotational or staggered management at local or landscape scale allows late-flowering plants to reproduce and animals to find refuge or nectar and pollen sources when most of the meadows are cut or paddocks are grazed. Another precondition for the conservation of grassland biodiversity is the application of site-adapted fertilization regimes that range from no fertilization at all in marginal grasslands with grazing management (e.g., dry and sandy grasslands) to moderate fertilization in grasslands with traditional two-cutting systems (hay meadows; Klimek et al., 2007). If the choice is between grazing or mowing, the meta-analysis of Tälle et al. (2016) suggests that in Central and Northern European grasslands grazing tends to be slightly more beneficial for biodiversity, but with huge differences between grassland types as well as the timing of usage.

In the following, the potential of traditional and alternative management practices is explored regarding their socio-economic sustainability as well as suitability for the preservation and/or enhancement of conservation values in grassland communities of the region. For more detailed information, we recommend the handbook of Blakesley and Buckley (2016) as well as the series of management guidelines for Natura 2000 habitats (e.g., Calaciura and Spinelli, 2008; Eriksson, 2008; Galvánek and Janák, 2008; Šeffer et al., 2008).

Grazing Practices

The management of open habitats by grazing requires knowledge of breed-specific grazing effects on species composition and diversity (Bullock et al., 2001; Isselstein et al., 2004). Grazing efficiently supports the creation of a structurally heterogeneous sward, particularly as a result of dietary choices (Rook et al., 2004; Rosenthal et al., 2012), bare soil patches due to trampling as well as improved ground-light conditions (Borer et al., 2014). Due to selective grazing, high-competitive grasses can be diminished (Pykälä, 2003; Schwabe et al., 2013; Henning et al., 2017).

An appropriate adjustment of stocking rates and grazing periods to fodder quantity and quality as well as to the requirements of grassland species is crucial for maintaining the species diversity in High Nature Value grasslands. Too low stocking rates foster the expansion of tall grass species (e.g., *Arrhenatherum elatius, Brachypodium pinnatum, Bromus erectus*), increasing litter deposition as well as shrub encroachment and consequently the loss of valuable grassland species and communities (Bobbink and Willems, 1987). This is particularly true for low-productive and unprofitable sites, which harbor most of the biodiversity in Western and Northern European grasslands and which were traditionally managed by seasonal husbandry (Calaciura and Spinelli, 2008; Eriksson et al., 2008; Galvánek and Janák, 2008).

In principle, migratory herding is preferable, as it enables zoochorous long-distance seed dispersal (e.g., Poschlod et al., 2009; Bentjen et al., 2016), provided an adequate grazing pressure prevents grass and shrub encroachment. In regions, where grazing by small ruminants is no longer economically feasible, paddock grazing by goats or sheep can be a successful tool to maintain species-rich, particularly dry, grasslands. On already encroached dry grasslands, goat paddock grazing with comparatively high-stocking rates (0.6–0.8 LU ha^{-1} yr^{-1}) and an early start in spring can be an efficient method for diminishing shrub encroachment during the restoration phase (Elias and Tischew, 2016). In this study the browsing goats decreased woody coverage within the pastures from 70 per cent to 37 per cent over seven years, which was positively related to the frequency of typical and endangered dry grassland species (Fig. 3.2B). The pasture regime must be realized without supplementary feeding (except for minerals) to avoid eutrophication.

Year-round large herbivore grazing may be also an economical alternative where traditional sheep or goat grazing is no longer profitable. Contrary to seasonal (summer) husbandry, the costs of year-round husbandry with robust breeds are minor, because management facilities, such as barns, supply of winter forage and transportation are not necessary. Supplementary feeding is only necessary in strong winters; therefore, additional nutrient input to the habitat can be largely avoided. Several studies confirmed that year-round large herbivore grazing can maintain and even improve the typical floristic species composition and vegetation structure of grassland communities (sandy grassland: Schwabe et al., 2013 and Henning et al., 2017; dry calcareous grassland: Köhler et al., 2016; Figs. 3.2A, 3.2C and 3.2D).

Traditionally grazed wooded pastures contribute greatly to the maintenance of local plant and animal diversity. However, they face a tremendous threat of abandonment due to excessive maintenance costs for individual farmers (Bergmeier et al., 2010). These pastures thus need to be better included into the CAP subsidies to secure their survival. In general, basic CAP rules for the eligibility of grazed areas should be modified to aim at restoration or maintenance of farmland biodiversity and heterogeneity instead of only focusing on 'good practice of intensive farming'. This means that the whole spectrum of ecotones from woodland to open habitats, including sparsely vegetated or even bare soil patches (Fig. 3.2D) or grazed margins of ponds must be eligible, if they support farmland biodiversity (Török et al., 2016).

Prompted by the fact that traditional grazing systems become less and less profitable, 'semi-open pasture landscapes' have turned out to be a promising instrument in the conservation of High Nature Value grasslands and other open habitats, and are successfully applied to various larger conservation areas, mainly in the Benelux countries and Germany (Redecker et al., 2002; Finck et al., 2004; von Oheimb et al., 2006; Bunzel-Drüke et al., 2015). This whole concept has been largely inspired by the 'megaherbivore theory', suggesting that the region would be much more open than some textbooks tell because the expansion of humans in the Holocene led to the extinction of various previously widespread megaherbivores, such as European bison (*Bos bonasus*), Aurochs (*Bos primigenius*) and Wild horse (*Equus ferus*), which are thought to have opened the landscape at least in certain places similar to the African savanna (Bunzel-Drüke et al., 2008). This together with the experiences gained suggests that semi-open pasture landscapes can be both viable and successful in keeping the landscape open and retain its species diversity when there is year-round grazing at low stocking rates, preferentially with multi-species herds. These include robust livestock breeds, de-domesticated horses and cattle (physiognomically similar to the European wild horse and the aurochs, respectively) and re-introduced formerly native species, such as European bison and Elk (*Alces alces*).

Mowing

Traditional hay production is currently restricted to regions where suckler cow husbandry and sheep or goat farming requires the production of hay for feeding the livestock during the winter period. Hay produced on species-rich grassland becomes less and less important for feeding dairy cows. By contrast, hay production for horse keeping is gaining importance in some regions (Donath et al., 2004).

Both, earlier mowing (Hansson and Fogelfors, 2000; Köhler et al., 2005) and increased mowing frequency (Zechmeister et al., 2003) negatively affect plant species diversity. Some authors, however, note that overly restrictive management guidelines with a late first cut inflict a considerable loss in forage quality and thus contribute to the abandonment of species-rich grassland (Tallowin and Jefferson, 1999; Šeffer et al., 2008). Therefore, Šeffer et al. recommend more flexible management schemes to support the use of hay from species-rich meadows. For alluvial meadows, Leyer (2002) showed that the majority of target plant species are able to tolerate a first cut in June. Only species lacking a vegetative regeneration strategy and the ability to produce seeds after a first cut depend from time to time on a late first cut to reproduce successfully from seeds (Kirkham and Tallowin, 1995; Šeffer et al., 2008).

More recently, grazing has proved as an alternative when hay making on former meadows is abandoned (Voß, 2001). However, decisions should be made on a site-by-site basis to prevent the loss of characteristic grassland species (Fischer and Wipf, 2002) and should consider other aspects, such as ornithological values. Anyway, a combination of mowing and pasturing has been traditionally employed on semi-natural grasslands in large parts of Europe. Such a combination is highly recommended because mowing as a non-selective method of biomass removal promotes different species than selective grazing (Galvánek and Janák, 2008).

Mulching

Several studies investigated the suitability of mulching for grassland maintenance when hay making or grazing is no longer profitable. While (two-time) mulching can maintain the typical grassland diversity under moderate nutrient conditions to a certain extent (Moog et al., 2002), in nutrient-richer sites, mulching will lead to dense litter layers with a subsequent loss of the characteristic species. Unfortunately, many grasslands currently are only mulched to maintain the 'good agricultural status' to ensure payments of CAP subsidies. Based on this motive, these grasslands are predominantly mulched only once a year and too late, and that leads to a rapid decrease in species richness in future. In Switzerland, by contrast, biodiversity subsidies in grasslands are only paid if the cut material is removed from the site.

Controlled Burning

Controlled burning was historically practiced in grasslands where a litter layer or standing overaged biomass impeded the generative and vegetative regrowth. After burning in autumn or winter, these grasslands were grazed or mown in the next season. Nowadays burning is prohibited by law in most countries of the region (Valkó et al., 2014).

Based on a literature review, Valkó et al., 2014 conclude that annual burning is usually not an appropriate option for conservation of species-rich European grasslands. Burning as the sole management can lead to loss in richness of the typical grassland species, compared to mowing or grazing (Kahmen et al., 2002; Köhler et al., 2005; Hansson and Fogelfors, 2000), but better results can be achieved by combined burning and grazing (Vogels, 2009) as it was traditionally practiced. Bensettiti et al. (2005) emphasize that burning may lead to changes in the species composition and, if applied regularly, promote the spreading of 'native invasives' like *Pteridium aquillinum* or *Molinia caerulea*, particularly in *Nardus* grasslands.

Restoring and Creating High Nature Value Grasslands from other Land Uses

Restoration from Intensified Grasslands, Croplands, Woodlands and Open-cast Mines

The restoration of species-poor grasslands as well as creation of new grasslands has gained high importance during the last two decades (Scotton et al., 2012; Harnisch et al., 2014; Blakesley and Buckley, 2016). Most published studies focused on establishment of species-rich grasslands on former arable land (e.g., Pakemann et al., 2002; Lawson et al., 2004; Lepš et al., 2007), on marginal land (e.g., mining areas: Baasch et al., 2012; Kirmer et al., 2012a) or on road-verges (Rydgren et al., 2010; Auestad et al., 2015). Moreover, studies using species introduction techniques for enhancing the diversity of species-poor grasslands have increasingly been conducted during the last decade (Edwards et al., 2007; Pywell et al., 2007; Hellström et al., 2009; Baasch et al., 2016). Unfortunately, in restoration practise the establishment of the target species often failed due to incorrect application of the introduction techniques (Conrad and Tischew, 2011), calling for a better knowledge transfer into practice and consequent application of well-developed species introduction methods, such as the transfer of hay or seed-rich thrashing material from species-rich grassland (Figs. 3.2E and 3.2F) or seeding of regionally propagated seed mixtures (Kiehl et al., 2010; Török et al., 2011; Scotton et al., 2012). In addition, adequate site preparation schemes must be applied to counteract the reasons for the previous decline of grassland species, including improving the water balance conditions (e.g., Klimkowska et al., 2007) and reduction of overly high soil-nutrient contents (Hölzel and Otte, 2003). In marginal lands, such as former open-cast mines, improvement of the harsh conditions of pristine substrate by applying a mulch layer might be advisable (Baasch et al., 2012). In grasslands that developed a dense grass cover, the successful establishment of target species will be successful after disturbing the grass sward by rotovating (Harnisch et al., 2014; Baasch et al., 2016).

The natural recolonization of target species into existing species-poor grasslands is only successful if populations of these species are available in the surroundings of the grasslands (Rosenthal, 2006). Many grassland species show severe dispersal limitation, or the remaining (small) population cannot secure a successful recolonization (Bischoff, 2002; Pywell et al., 2002). Thus, beside the implementation of an appropriate improvement in the site conditions and management, the active introduction of desired plant species is an indispensable measure to overcome the limitation of seed availability and to restore species diversity and composition (Klimkowska et al., 2007; Kiehl et al., 2010). Surviving species-rich grasslands in the surroundings harbor the regionally adapted species pool that is indispensible for a successful restoration or re-creation of grasslands. Therefore, such grasslands are increasingly integrated into donor site registers, while supporting activities for maintenance of their species richness (Krautzer and Pötsch, 2009; Kirmer et al., 2012b).

New Grassland Habitats on Green Roofs

Following early activities in the 1970s in Germany, modern 'extensive green roofs' were adopted in many countries of the temperate zone, particularly in Central Europe, France and the British Isles. Traditional grass roofs, where turf was laid over bark, were in use in Scandinavia for millennia (Thuring and Grant, 2016). Modern grass roofs combine the benefits of isolation of the building and reducing water run-off with biodiversity conservation. Depending on the soil-depth applied, their vegetation typically resembles either succulent-rich low-grown communities of the *Alysso-Sedetalia* (7.g of Table 3.2) or semi-dry grasslands of *Brachypodietalia pinnati* (8.a), i.e., threatened grassland types of high conservation priority. Provided they are sown or planted with native species, green roofs can make an important contribution to grassland conservation in urban areas and comprise beyond ubiquitous taxa many rare plants (Landolt, 2001; Fig. 3.2H) and animals (Kadas, 2006; Madre et al., 2013).

Résumé and Future Prospects

Semi-natural and natural grasslands together account for about one-sixth of the area of Western and Northern Europe, but they contribute a much higher share to biodiversity of the region. While natural grasslands, except in some coastal areas, are relatively slightly affected by anthropogenic threats, HNV semi-natural grasslands, which once originated due to low-intensity human land use, have experienced a sharp decline in area and quality in recent decades. These semi-natural grasslands are an integral part of the traditional cultural landscapes of Europe, with their intricate networks of croplands, semi-natural elements, such as grasslands, heathlands, hedgerows and so on, and remaining forests and wetlands that are a globally unique feature. While East Asia, with its very old human cultures, has some semi-natural grasslands, their fractional extent is very small as compared to Europe (Ushimaru et al., 2018). On the other hand, agriculturally used grasslands in the Americas and Australasia are of very recent origin and essentially consist of introduced species from Europe.

During the last decades, a large fraction of the former semi-natural grasslands has been intensified towards species-poor, frequently-used intensive grasslands ('grass fields'), while others have been converted to arable fields, built-up areas or forests. The remaining semi-natural grasslands face the main threats of succession following abandonment and eutrophication which, via high airborne nitrogen inputs, negatively impact grasslands that are not commercially used. While there are numerous local and regional good examples of grassland conservation and restoration, the negative trend of grassland biodiversity continues because the inclusion of grasslands in conservation networks alone does not help when major threats are not addressed. The CAP of the EU, like similar instruments in the non-EU member states of the region, having implemented some 'greening' recently with the majority of their instruments still support further intensification and eutrophication or at least is not effective in counteracting these trends. It is clear that maintenance of grassland biodiversity in the region can only succeed if the CAP is further restructured by questioning old paradigms and clearly transforming it into a 'green' program (Pe'er et al., 2014; BiodivERsA, 2017). We believe that a society that is willing to give so much money to a tiny economic sector has the right to request that subsidies are restricted to those farmers who produce in a way that safeguards clean water and air as well as high biodiversity and aesthetic cultural landscapes. On the other hand, conservationists need to be more open-minded. Under present socio-economic conditions with their associated far-distance ecological impacts, traditional conservation strategies, like establishing protected areas and mimicking historical land use, will not be sufficient on their own. Instead, one needs to be open to other solutions, including controlled burning, semi-open pasture landscapes or extensive green roofs, which might function better under certain conditions, but often are not tested and implemented because they do not fit in the traditional views of conservationists or even are prohibited by law.

Acknowledgements

We thank Aletta Bonn, Riccardo Guarino, Chiara Catelano and Stephan Brenneisen for providing relevant information and references; the latter also for a photo and Idoia Biurrun for constantly updating the GrassPlot database. Péter Török and Victor Squires kindly reviewed former drafts of this chapter and thus helped to improve it.

Abbreviations

CAP = Common Agricultural Policy (of the European Union); EU = European Union; EU28+ = 28 Countries of the European Union + Iceland, Norway, Switzerland and the Balkan Countries; HNV = High Nature Value; LU = Livestock Unit; NGO = Non-governmental Organization

References

Ambarlı, D., M. Vrahnakis, S. Burrascano, A. Naqinezhad and M. Pulido Fernández. 2018. Grasslands of the Mediterranean Basin and the Middle East and their Management (pp. 89–112 this volume).

Auestad, I., K. Rydgren and I. Austad. 2015. Near-natural methods promote restoration of species-rich grassland vegetation—Revisiting a road verge trial after nine years. Restor. Ecol. 24: 381–389.

Baasch, A., A. Kirmer and S. Tischew. 2012. Nine years of vegetation development in a postmining site: effects of spontaneous and assisted site recovery. J. Appl. Ecol. 49: 251–260.

Baasch, A., K. Engst, R. Schmiede, K. May and S. Tischew. 2016. Enhancing success in grassland restoration by adding regionally propagated target species. Ecol. Eng. 94: 583–591.

[BAFU] Bundesamt für Umwelt. 2017. Biodiversität in der Schweiz: Zustand und Entwicklung. Ergebnisse des Überwachungssystems im Bereich Biodiversität, Stand 2016. BAFU [Umwelt-Zustand No. 1630], Bern, 60 pp.

Bensettiti, F., V. Boullet, C. Chavaudret-Laborie and J. Deniaud (eds.). 2005. Cahiers d'habitats Natura 2000. Connaissance et gestion des habitats et des espèces d'intérêt communautaire. Tome 4 – Habitats agropastoraux. URL: http://natura2000.environnement.gouv.fr/habitats/cahiers.html.

Bentjen, O., J. Bober, J. Castens and C. Stolter. 2016. Seed dispersal capacity of sheep and goats in a near-coastal dry grassland habitat. Basic Appl. Ecol. 17: 508–515.

Bergmeier, E., J. Petermann and E. Schröder. 2010. Geobotanical survey of wood-pasture habitats in Europe: Diversity, threats and conservation. Biodivers. Conserv. 19: 2995–3014.

BiodivERsA. 2017. Policy Brief. The Common Agricultural Policy Can Strengthen Biodiversity and Ecosystem Services by Diversifying Agricultural Landscapes. URL: http://www.biodiversa.org/policybriefs.

Bischoff, A. 2002. Dispersal and establishment of floodplain grassland species as limiting factors in restoration. Biol. Conserv. 104: 25–33.

Blakesley, D. and P. Buckley. 2016. Grassland Restoration and Management. Pelagic Publishing, Exeter, 277 pp.

Bobbink, R. and J.H. Willems. 1987. Increasing dominance of *Brachypodium pinnatum* (L.) Beauv. in chalk grasslands: A threat to a species-rich ecosystem. Biol. Conserv. 40: 301–314.

Bohn, U., G. Gollub, C. Hettwer, Z. Neuhäuslová, T. Raus, H. Schlüter, H. Weber and S. Hennekens (eds.). 2004. Map of the natural vegetation of Europe. Scale 1 : 2,500,000. Interactive CD-ROM: Explanatory text, legend, maps. Bundesamt für Naturschutz, Bonn: CD-ROM + 19 pp.

Borer, E.T., E.W. Seabloom, D.S. Gruner, W.S. Harpole, H. Hillebrand, E.M. Lind, P.B. Adler, J. Alberti, M. Anderson, (...) and L.H. Yang. 2014. Herbivores and nutrients control grassland plant diversity via light limitation. Nature 508: 517–520.

Bornand, C., A. Gygax, P. Juillerat, M. Jutzi, A. Möhl, S. Rometsch, L. Sager, H. Santiago and S. Eggenberg. 2016. Rote Liste Gefässpflanzen. Gefährdete Arten der Schweiz. Bundesamt für Umwelt [Umwelt-Vollzug No. 1621], Bern, 178 pp.

Buchgraber, K., A. Schaumberger, E.M. Pötsch, B. Krautzer and A. Hopkins. 2011. Grassland farming in Austria—Status quo and future prospective. pp. 13–24. *In:* E.M. Pötsch, B. Krautzer and A. Hopkins (eds.). Grassland Farming and Land Management Systems in Mountainous Regions. Book of Abstracts. Walllig Ennstaler Druckerei und Verlag, Gröbming.

Bullock, J.M., J. Franklin, M.J. Stevenson, J. Silvertown, S.J. Coulson, S.J. Gregory and R. Tofts. 2001. A plant traits analysis of responses to grazing in a long-term experiment. J. Appl. Ecol. 38: 253–267.

Bunzel-Drüke, M., C. Böhm, G. Finck, R. Luick, E. Reisinger, U. Riecken, J. Riedl, M. Scharf and O. Zimball. 2008. 'Wilde Weiden'—Praxisleitfaden für Ganzjahresbeweidung in Naturschutz und Landschaftsentwicklung. Arbeitsgemeinschaft Biologische Umweltschutz im Kreis Soest (ABU), Bad Sassendorf-Lohne, 215 pp.

Bunzel-Drüke, M., C. Böhm, G. Ellwanger, P. Finck, H. Grell, L. Hauswirth, A. Herrmann, E. Jedicke, R. Joest, (…) and O. Zimball. 2015. Naturnahe Beweidung und NATURA 2000. Ganzjahresbeweidung im Management von Lebensraumtypen und Arten im europäischen Schutzgebietssystem NATURA 2000. Heinz Sielmann Stiftung, Duderstadt.

Bush, M.B. and J.R. Flenley. 1987. The age of the British chalk grassland. Nature 329: 434–436.

Calaciura, B. and O. Spinelli. 2008. Management of Natura 2000 Habitats. 6210 Seminatural dry grasslands and scrubland facies on calcareous substrates (*Festuco-Brometalia*) (*important orchid sites). European Commission, Brussels, 38 pp.

Ceulemans, T., R. Merckx, M. Hens and O. Honnay. 2013. Plant species loss from European semi-natural grasslands following nutrient enrichment—Is it nitrogen or is it phosphorus? Global Change Biol. 22: 73–82.

Ceulemans, T., C.J. Stevens, L. Duchateau, H. Jacquemyn, D.J.G. Gowing, R. Merckx, H. Wallace, N. van Rooijen, T. Goethem, (...) and O. Honnay. 2014. Soil phosphorus constrains biodiversity across European grasslands. Global Change Biol. 20: 3814–3822.

Chytrý, M., L.C. Maskell, J. Pino, P. Pyšek, M. Vilà, X. Font and S.M. Smart. 2008. Habitat invasions by alien plants: A quantitative comparison among Mediterranean, subcontinental and oceanic regions of Europe. J. Appl. Ecol. 45: 448–458.

Chytrý, M., J. Wild, P. Pyšek, L. Tichý, J. Danihelka and I. Knollová. 2009. Maps of the invasion level of the Czech Republic by alien plants. Preslia 81: 187–207.

Conrad, M. and S. Tischew. 2011 Grassland restoration in practice: Do we achieve the targets? A case study from Saxony-Anhalt/Germany. Ecol. Eng. 37: 1149–1157.

Cousins, S.A.O. 2009. Extinction debt in fragmented grasslands: paid or not? J. Veg. Sci. 20: 3–7.

Delbaere, B.C.W. (ed.). 1998. Facts and figures on Europe's Biodiversity: State and Trends 1998–1999. European Centre for Nature Conservation, Tilburg.

Dengler, J. 2003. Entwicklung und Bewertung neuer Ansätze in der Pflanzensoziologie unter besonderer Berücksichtigung der Vegetationsklassifikation. Arch. Naturwiss. Diss. 14: 1–297.

Dengler, J., M. Janišová, P. Török and C. Wellstein. 2014. Biodiversity of Palaearctic grasslands: A synthesis. Agric. Ecosyst. Environ. 182: 1–14.

Dengler, J., I. Biurrun, I. Apostolova, E. Baumann, T. Becker, A. Berastegi, S. Boch, I. Dembicz, C. Dolnik, (...) and F. Weiser. 2016. Scale-dependent plant diversity in Palaearctic grasslands: A comparative overview. Bull. Eurasian Dry Grassl. Group 31: 12–26.

Dengler, J., V. Wagner, I. Dembicz, I. García-Mijangos, A. Naqinezhad, S. Boch, A. Chiarucci, T. Conradi, G. Filibeck, (…) and I. Biurrun. 2018. GrassPlot—a database of multi-scale plant diversity in Palaearctic grasslands. Phytocoenologia. DOI: 10.1127/phyto/2018/0267.

Detzel, P. 1998. Die Heuschrecken Baden-Württembergs. Ulmer, Stuttgart, 580 pp.

Diekmann, M., U. Jandt, D. Alard, A. Bleeker, E. Corcket, D.J.G. Gowin, C.J. Steven and C. Dupré. 2014. Long-term changes in calcareous grassland vegetation in North-western Germany—No decline in species richness, but a shift in species composition. Biol. Conserv. 172: 170–179.

Dierschke, H. and G. Briemle. 2002. Kulturgrasland—Wiesen, Weiden und verwandte Staudenfluren. Ulmer, Stuttgart, 239 pp.

Donath, T.W., N. Hölzel, S. Bissels and A. Otte. 2004. Perspectives for incorporating biomass from non-intensively managed temperate flood meadows into farming systems. Agric. Ecosys. Environ. 104: 439–451.

Dupré, C. and M. Diekmann. 2001. Differences in species richness and life-history traits between grazed and abandoned grasslands in southern Sweden. Ecography 24: 275–286.

Dupré, C., C.J. Stevens, T. Ranke, A. Bleeker, C. Peppler-Lisbach, D.J.G. Gowing, N.B. Dise, E. Dorland, R. Bobbink and M. Diekmann. 2010. Changes in species richness and composition in European acidic grasslands over the past 70 years: The contribution of cumulative atmospheric nitrogen deposition. Global Change Biol. 16: 344–357.

Edwards, A.R., S.R. Mortimer, C.S. Lawson, D.B. Westbury, S.J. Harris, B.A. Woodcock and V.K. Brown. 2007. Hay strewing, brush harvesting of seed and soil disturbance as tools for the enhancement of botanical diversity in grasslands. Biol. Conserv. 134: 372–382.

[EEA] European Environment Agency. 2012. Protected Areas in Europe—An Overview. Publication Office of the European Union [EEA Report No. 5/2012], Luxembourg, 130 pp.

Elias, D. and S. Tischew. 2016. Goat pasturing—A biological solution to counteract shrub encroachment on abandoned dry grasslands in Central Europe? Agric. Ecosyst. Environ. 234: 98–106.

Ellenberg, H. and C. Leuschner. 2010. Vegetation Mitteleuropas mit den Alpen in ökologischer, dynamischer und historischer Sicht. 6th ed. Ulmer, Stuttgart, 1333 pp.

Ellwanger, G., C. Müller, A. Ssymank, M. Vischer-Leopold and C. Paulsch (eds.). 2016. Management of Natura 2000 sites on military training areas. Naturschutz Biol. Vielfalt 152: 1–200.

Eriksson, M.O.G. 2008. Management of Natura 2000 Habitats. 6450 Northern Boreal Alluvial Meadows. European Commission, Brussels, 14 pp.

European Commission. 2013. Interpretation Manual of European Union Habitats – EUR 28, April 2013. DG Environment, European Commission, Brussels, 144 pp.

European Union. 2016. Agriculture, Forestry and Fishery Statistics – 2016 Edition. Publication Office of the European Union, Luxembourg, 230 pp.

Finck, P., W. Härdtle, B. Redecker and U. Riecken (eds.). 2004. Weidelandschaften und Wildnisgebiete – Vom Experiment zur Praxis. Schriftenr. Landschaftspflege Naturschutz 78: 1–539.

Fischer, M. and S. Wipf. 2002. Effect of low-intensity grazing on the species-rich vegetation of traditionally mown subalpine meadows. Biol. Conserv. 104: 1–11.

Fuller, R.M. 1987. The changing extent and conservation interest of lowland grasslands in England and Wales: A review of grassland surveys 1930–1984. Biol. Conserv. 40: 281–300.

Galvánek, D. and M. Janák. 2008. Management of Natura 2000 habitats. 6230 *Species-rich *Nardus* grasslands. European Commission, Brussels, 20 pp.

Gerber, E., C. Krebs, C. Murrell, M. Moretti, R. Rocklin and U. Schaffner. 2008. Exotic invasive knotweeds (*Fallopia* spp.) negatively affect native plant and invertebrate assemblages in European riparian habitats. Biol. Conserv. 141: 646–654.

Gossner, M.M., M.L. Lewinsohn, T. Kahl, F. Grassein, S. Boch, D. Prati, K. Birkhofer, S.C. Renner, J. Sikorski, (...) and E. Allen. 2016. Land-use intensification causes multitrophic homogenization of grassland communities. Nature 540: 266–269.

Grizzetti, B., F. Bouraoui and A. Aloe. 2007. Spatialised European Nutrient Balance. Office for Official Publications of the European Communities, Luxembourg, 98 pp.

Hansson, M. and H. Fogelfors. 2000. Management of a semi-natural grassland: results from a 15-year-old experiment in southern Sweden. J. Veg. Sci. 11: 31–38.

Harnisch, M., A. Otte, R. Schmiede and T.W. Donath. 2014. Verwendung von Mahdgut zur Renaturierung von Auengrünland. Eugen Ulmer, Stuttgart, 150 pp.

Hejcman, M., P. Hejcmanová, V. Pavlů and J. Beneš. 2013. Origin and history of grasslands in Central Europe—A review. Grass Forage Sci. 68: 345–363.

Hejda, M., P. Pyšek and V. Jarosík. 2009. Impact of invasive plants on the species richness, diversity and composition of invaded communities. J. Ecol. 97: 393–403.

Hellström, K., A.-P. Huhta, P. Rautio and J. Tuomi. 2009. Seed introduction and gap creation facilitate restoration of meadow species richness. J. Nat. Conserv. 17: 236–244.

Henning, K., A. Lorenz, G. von Oheimb, W. Härdtle and S. Tischew. 2017. Year-round cattle and horse grazing supports the restoration of abandoned, dry sandy grassland and heathland communities by supressing *Calamagrostis epigejos* and enhancing species richness. J. Nat. Conserv. 40: 120–130.

Hobohm, C. and I. Bruchmann. 2009. Endemische Gefäßpflanzen und ihre Habitate in Europa – Plädoyer für den Schutz der Grasland-Ökosysteme. Ber. Reinhold-Tüxen-Ges. 21: 142–161.

Hölzel, N. and A. Otte. 2003. Restoration of a species-rich flood meadow by topsoil removal and diaspore transfer with plant material. Appl. Veg. Sci. 6: 131–140.

Hönigová, I., D. Vačkář, E. Lorencová, J. Melichar, M. Götzl, G. Sonderegger, V. Oušková, M. Hošek and K. Chobot. 2012. Survey on Grassland Ecosystem Services – Report of the European Topic Centre on Biological Diversity. Nature Conservation Agency of the Czech Republic, Prague, 78 pp.

Hopkins, A. and B. Holz. 2006. Grassland for agriculture and nature conservation: production, quality and multi-functionality. Agron. Res. 4: 3–20.

Isselstein, J., B.A. Griffith, P. Pradel and S. Venerus. 2004. Effects of livestock breed and stocking rate on sustainable grazing systems: 3. agronomic potential. Grassl. Sci. Eur. 9: 620–622.

Isselstein, J., B. Jeangros and V. Pavlu. 2005. Agronomic aspects of biodiversity targeted management of temperate grasslands in Europe—A review. Agron. Res. 3: 139–151.

Jandt, U., H. von Wehrden and H. Bruelheide. 2011. Exploring large vegetation databases to detect temporal trends in species occurrences. J. Veg. Sci. 22: 957–972.

Janssen, J.A.M., J.S. Rodwell, M. Garcia Criado, S. Gubbay, T. Haynes, A. Nieto, N. Sanders, F. Landucci, J. Loidi, (...) and M. Valachovič. 2016. European Red List of Habitats—Part 2. Terrestrial and Freshwater Habitats. European Union, Luxembourg, 38 pp.

Johst, A. and K. Reiter. 2017. Das Nationale Naturerbe: Naturschätze für Deutschland. BMUB, Berlin, 35 pp.

Jurt, C., I. Häberli and R. Rossier. 2015. Transhumance farming in Swiss mountains: Adaptation to a changing environment. Mt. Res. Dev. 35: 57–65.

Kadas, G. 2006. Rare invertebrates colonizing green roofs in London. Urban Habitats 4: 66–73.

Kahmen, S., P. Poschlod and K.-F. Schreiber. 2002. Conservation management of calcareous grasslands. Changes in plant species composition and response of functional traits during 25 years. Biol. Conserv. 104: 319–324.

Ketner-Oostra, R., A. Aptroot, P.D. Jungerius and K.V. Sýkora. 2012. Vegetation succession and habitat restoration in Dutch lichen-rich inland drift sands. Tuexenia 32: 245–268.

Kiehl, K., A. Kirmer, T. Donath, L. Rasran and N. Hölzel. 2010. Species introduction in restoration projects—evaluation of different techniques for the establishment of semi-natural grasslands in Central and Northwestern Europe. Basic Appl. Ecol. 11: 285–299.

Kirkham, F.W. and J.R.B. Tallowin. 1995. The influence of cutting date and previous fertilizer treatment on the productivity and botanical composition of species-rich hay meadows on the Somerset Levels. Grass Forage Sci. 50: 365–377.

Kirkham, F.W., J.R.B. Tallowin, R.A. Sanderson, A. Bhogal, B.J. Chambers and D.P. Stevens. 2008. The impact of organic and inorganic fertilizers and lime on the species-richness and plant functional characteristics of hay meadow communities. Biol. Conserv. 141: 1411–1427.

Kirmer, A., A. Baasch and S. Tischew. 2012a. Sowing of low and high diversity seed mixtures in ecological restoration of surface mined-land. Appl. Veg. Sci. 15: 198–207.

Kirmer, A., B. Krautzer, M. Scotton and S. Tischew. 2012b. Praxishandbuch zur Samengewinnung und Renaturierung von artenreichem Grünland. Eigenverlag Lehr- und Forschungszentrum Raumberg-Gumpenstein, Irdning, 221 pp.

Kleijn, D., R.A. Baquero, Y. Clough, M. Díaz, J. de Esteban, F. Fernández, D. Gabriel, F. Herzog, A. Holzschuh, (...) and J.L. Yela. 2006. Mixed biodiversity benefits of agri-environmental schemes in five European countries. Ecol. Lett. 9: 243–254.

Klimek, S., M. Hofmann and J. Isselstein. 2007. Plant species richness and composition in managed grasslands: the relative importance of field management and environmental factors. Biol. Conserv. 134: 559–570.

Klimek, S., H.H. Steinmann, J. Freese and J. Isselstein. 2008. Rewarding farmers for delivering vascular plant diversity in managed grasslands: A transdisciplinary case-study approach. Biol. Conserv. 141: 2888–2897.

Klimkowska, A., R. van Diggelen, J.P. Bakker and A.P. Grootjans. 2007. Wet meadow restoration in Western Europe: a quantitative assessment of the effectiveness of several techniques. Biol. Conserv. 140: 318–328.

Klötzli, F., W. Dietl, K. Marti, C. Schubiger-Bossard and C.-R. Walther. 2010. Vegetation Europas. Das Offenland im vegetationskundlich-ökologischen Überblick unter besonderer Berücksichtigung der Schweiz. Ott, Bern, 1190 pp.

Köhler, B., A. Gigon, P.J. Edwards, B. Krüsi, R. Langenauer, A. Lüscher and P. Ryser. 2005. Changes in the species composition and conservation value of limestone grasslands in northern Switzerland after 22 years of contrasting managements. Perspect. Plant Ecol. Evol. Syst. 7: 51–67.

Köhler, M., G. Hiller and S. Tischew. 2016. Year-round horse grazing supports typical vascular plant species, orchids and rare bird communities in a dry calcareous grassland. Agric. Ecosyst. Environ. 234: 48–57.

Koordinationsstelle Biodiversitäts-Monitoring Schweiz. 2009. Zustand der Biodiversität in der Schweiz – Ergebnisse des Biodiversitätsmonitorings Schweiz (BDM) im Überblick. Stand: Mai 2009. Bundesamt für Umwelt [Umwelt-Zustand 09/11], Bern, 112 pp.

Korneck, D., M. Schnittler, F. Klingenstein, G. Ludwig, M. Takla, U. Bohn and R. May. 1998. Warum verarmt unsere Flora? – Auswertung der Roten Liste der Farn- und Blütenpflanzen Deutschlands. Schriftenr. Vegetationskd. 29: 299–444.

Krause, B., H. Culmsee, K. Wesche, E. Bergmeier and C. Leuschner. 2011. Habitat loss of floodplain meadows in north Germany since the 1950s. Biodivers. Conserv. 20: 2347–2364.

Krautzer, B. and E.M. Pötsch. 2009. The use of semi-natural grassland as donor sites for the restoration of high nature value areas. pp. 478–492. *In:* B. Cagaš, R. Macháč and J. Nedělník (eds.). Alternative Functions of Grassland. Proceedings of the 15th European Grassland Federation Symposium, Brno, Czech Republic, 7–9 September 2009.

Landolt, E. 2001. Orchideen-Wiesen in Wolllishofen (Zürich) – ein erstaunliches Relikt aus dem Anfang des 20. Jahrhunderts. Vierteljahresschr. Naturforsch. Ges. Zürich 146(2-3): 41–51.

Laufer, H., K. Fritz and P. Sowig (eds.). 2007. Die Amphibien und Reptilien Baden-Württembergs. Ulmer, Stuttgart, 807 pp.

Lawson, C.S., M.A. Ford and J. Mitchley. 2004. The influence of seed addition and cutting regime on the success of grassland restoration on former arable land. Appl. Veg. Sci. 7: 259–266.

Lenoir, J. and J.-C. Svenning. 2015. Climate-related range shifts—A global multidimensional synthesis and new research directions. Ecography 38: 15–28.

Lepš, J., J. Doležal, T.M. Bezemer, V.K. Brown, K. Hedlund, M. Igual Arroyo, H.B. Jörgensen, C.S. Lawson, S.R. Mortimer, (...) and W.H. van der Putten. 2007. Long-term effectiveness of sowing high and low diversity seed mixtures to enhance plant community development on ex-arable fields. Appl. Veg. Sci. 10: 97–110.

Leyer, I. 2002. Auengrünland der Mittelelbe-Niederung: Vegetationskundliche und – ökologische Untersuchungen in der rezenten Aue, der Altaue und am Auenrand der Elbe. Diss. Bot. 363: 1–193.

Lindemann-Matthies, P. and E. Bose. 2006. Species richness, structural diversity and species composition in meadows created by visitors of a botanical garden in Switzerland. Landsc. Urban Planning 79: 298–307.

Löbel, S. and J. Dengler. 2008 [2007]. Dry grassland communities on southern Öland: Phytosociology, ecology, and diversity. Acta Phytogeogr. Suec. 88: 13–31.

Madre, F., A. Vergnes, N. Machon and P. Clergeau. 2013. A comparison of 3 types of green roof as habitats for arthropods. Ecol. Eng. 57: 109–117.

Metzger, M.J., R.G.H. Bunce, R.H.G. Jongman, C.A. Mücher and J.W. Watkins. 2005. A climatic stratification of the environment of Europe. Global Ecol. Biogeogr. 14: 549–563.

Mitlacher, K., P. Poschlod, E. Rosén and J.P. Bakker. 2002. Restoration of wooded meadows—A comparative analysis along a chronosequence on Öland (Sweden). Appl. Veg. Sci. 5: 63–73.

Moog, D., P. Poschlod, S. Kahmen and K.F. Schreiber. 2002. Comparison of species composition between different grassland management treatments after 25 years. Appl. Veg. Sci. 5: 99–106.

Mucina, L., H. Bültmann, K. Dierßen, J.-P. Theurillat, T. Raus, A. Čarni, K. Šumberová, W. Willner, J. Dengler, (...) and L. Tichý. 2016. Vegetation of Europe: Hierarchical floristic classification system of vascular plant, bryophyte, lichen, and algal communities. Appl Veg. Sci. 19, Suppl. 1: 3–264.

Nagy, S. 2009. Grasslands as a bird habitat. pp. 35–41. *In:* P. Veen, R. Jefferson, J. de Smidt and J. van der Straaten (eds.). Grasslands in Europe of High Nature Value. KNNV Publishing, Zeist.

Otte, A. and P. Maul. 2005. Verbreitungsschwerpunkte und strukturelle Einnischung der Stauden-Lupine (*Lupinus polyphyllus* Lindl.) in Bergwiesen der Rhön. Tuexenia 25: 151–182.

Pakeman, R.J. and R.H. Marrs. 1992. The conservation value of bracken *Pteridium aquilinum* (L.) Kuhn dominated communities in the UK and an assessment of the ecological impact of bracken expansion or its removal. Biol. Conserv. 62: 101–114.

Pakeman, R.J., R.F. Pywell and T.C.E. Wells. 2002. Species spread and persistence: implications for experimental design and habitat re-creation. Appl. Veg. Sci. 5: 75–86.

Pauli, H., M. Gottfried, K. Reiter, C. Klettner and G. Grabherr. 2007. Signals of range expansions and contractions of vascular plants in the high Alps: observations (1994–2004) at the GLORA master site Schrankogel, Tyrol, Austria. Global Change Biol. 13: 147–156.

Pauly, A., G. Ludwig, H. Haupt, H. Gruttke, R. May and C. Otto. 2009. Auswertungen zu den Roten Listen dieses Bandes. Naturschutz Biol. Vielfalt 70(1): 321–337.

Pe'er, G., L.V. Dicks, P. Visconti, R. Arlettaz, A. Báldi, T.G. Benton, S. Colllins, M. Dieterich, R.D. Gregory, (...) and A.V. Scott. 2014. EU agricultural reform fails on biodiversity. Science 344: 1090–1092.

Pompe, S., J. Hanspach, F. Badeck, S. Klotz, W. Thuiller and I. Kühn. 2008. Climate and land use change impacts on plant distributions in Germany. Biol. Lett. 4: 564–567.

Poschlod, P. and M. WallisDeVries. 2002. The historical and socioeconomic perspective of calcareous grasslands—Lessons from the distant and recent past. Biol. Conserv. 104: 361–376.

Poschlod, P., A. Baumann and P. Karlik. 2009. Origin and development of grasslands in Central Europe. pp. 15–26. *In*: P. Veen, R. Jefferson, J. de Smidt and J. van der Straaten (eds.). Grasslands in Europe of High Nature Value. KNNV Publishing, Zeist.

Poschlod, P. 2015. Geschichte der Kulturlandschaft: Entstehungsursachen und Steuerungsfaktoren der Entwicklung der Kulturlandschaft, Lebensraum- und Artenvielfalt in Mitteleuropa. Ulmer, Stuttgart, 320 pp.

Pott, R. 1995. The origin of grassland plant species and grassland communities in Central Europe. Fitosociologia 29: 7–32.

Pykälä, J. 2003. Effects of restoration with cattle grazing on plant species composition and richness of semi-natural grasslands. Biodivers. Conserv. 12: 2211–2226.

Pywell, R.F., J.M. Bullock, A. Hopkins, K.J. Walker, T.H. Sparks, M.J.W. Burke and S. Peel. 2002. Restoration of species-rich grassland on arable land: assessing the limiting processes using a multi-site experiment. J. Appl. Ecol. 39: 294–309.

Pywell, R.F., J.M. Bullock, J.B. Tallowin, K.J. Walker, E.A. Warman and G. Masters. 2007. Enhancing diversity of species-poor grasslands: an experimental assessment of multiple constraints. J. Appl. Ecol. 44: 81–94.

Rebele, F. and C. Lehmann. 2001. Biological flora of Central Europe: *Calamagrostis epigejos* (L.) Roth. Flora 196: 325–344.

Redecker, B., P. Finck, W. Härdtle, U. Riecken and E. Schröder (eds.). 2002. Pasture Landscapes and Nature Conservation. Springer, Berlin, 435 pp.

Reisigl, H. and R. Keller. 1987. Alpenpflanzen im Lebensraum—Alpine Rasen, Schuttund Felsvegetation. Fischer, Stuttgart, 148 pp.

Riecken, U., P. Finck, U. Raths, E. Schröder and A. Ssymank. 2006. Rote Liste der gefährdeten Biotoptypen Deutschlands – zweite fortgeschriebene Fassung 2006. Naturschutz Biol. Vielfalt 34: 1–318.

Rihm, B. and B. Achermann. 2016. Critical Loads of Nitrogen and their Exeedances. Swiss Contribution to the Effects-oriented Work under the Convention on Long-range Transboundary Air Pollution (UNECE). BAFU [Environmental Studies No. 1642], Bern, 78 pp.
Rook, A.J., B. Dumont, J. Isselstein, K. Osoro, M.F. WallisDeVries, G. Parente and J. Mills. 2004. Matching type of livestock to desired biodiversity outcomes in pastures—A review. Biol. Conserv. 119: 137–150.
Rosenthal, G. 2006. Restoration of wet grasslands—Effects of seed dispersal, persistence and abundance on plant species recruitment. Basic Appl. Ecol. 7: 409–421.
Rosenthal, G., J. Schrautzer and C. Eichberg. 2012. Low-intensity grazing with domestic herbivores: A tool for maintaining and restoring plant diversity in temperate Europe. Tuexenia 32: 167–205.
Rydgren, K., J.-F. Nordbakken, I. Austad, I. Auestad and E. Heegaard. 2010. Recreating semi-natural grasslands: A comparison of four methods. Ecol. Eng. 36: 1672–1679.
Sala, O.E., F.S. Chapin III, J.J. Armesto, E. Berlow, J. Bloomfield, R. Dirzo, E. Huber-Sanwald, L.F. Huenneke, R.B. Jackson, (...) and D.H. Wall. 2000. Global biodiversity scenarios for the year 2100. Science 287: 1770–1774.
Sattler, T., V. Keller, P. Knaus, H. Schmid and B. Volet. 2015. Zustand der Vogelwelt in der Schweiz: Bericht 2015. Schweizerische Vogelwarte, Sempach, 36 pp.
Schaminée, J.H.J., M. Chytrý, J. Dengler, S.M. Hennekens, J.A.M. Janssen, B. Jiménez-Alfaro, I. Knollová, F. Landucci, C. Marcenò, J.S. Rodwell and L. Tichý. 2016. Development of Distribution Maps of Grassland Habitats of EUNIS Habitat Classification. European Environment Agency [Report EEA/NSS/16/005], Copenhagen, 171 pp.
Schmidt, J., J. Trautner and G. Müller-Motzfeld. 2016. Rote Liste und Gesamtartenliste der Laufkäfer (*Coleoptera: Carabidae*) Deutschlands. 3. Fassung, Stand April 2015. Naturschutz Biol. Vielfalt 70(4): 139–204.
Schuhmacher, O. and J. Dengler. 2013. Das Land-Reitgras als Problemart auf Trockenrasen. Handlungsempfehlungen zur Reduktion von *Calamagrostis epigejos*. Ergebnisse aus einem Praxisversuch. NABU Hamburg, Hamburg, 16 pp.
Schwabe, A., K. Süss and C. Storm. 2013. What are the long-term effects of livestock grazing in steppic sandy grassland with high conservation value? Results from a 12-year field study. Tuexenia 33: 189–212.
Schweiger, O., R.K. Heikkinen, A. Harpke, T. Hickler, S. Klotz, O. Kudrna, I. Kühn, J. Pöyry and J. Settele. 2012. Increasing range mismatching of ineracting species under global change is related to their ecological characteristics. Global Ecol. Biogeogr. 21: 88–99.
Scotton, M., A. Kirmer and B. Krautzer (eds.). 2012. Practical Handbook for Seed Harvest and Ecological Restoration of Species-rich Grasslands. Cooperativa Libraria Eitrice Università di Padova, Padova, 124 pp.
Scotton, M., L. Sicher and A. Kasal. 2014. Semi-natural grasslands of the Non Valley (Eastern Italian Alps): Agronomic and environmental value of traditional and new Alpine hay-meadow types. Agric. Ecosyst. Environ. 197: 243–254.
Šeffer, J., M. Janak and V. Šefferová. 2008. Management models for habitats in Natura 2000 Sites. 6440 Alluvial meadows of river valleys of the *Cnidion dubii*. European Commission, Brussels, 24 pp.
Silva, J.P., J. Toland, W. Jones, J. Eldrige, E. Thorpe and E. O'Hara. 2008. LIFE and Europe's Grassland—Restoring a Forgotten Habitat. European Communities, Luxembourg, 53 pp.
Smit, H.J., M.J. Metzger and F. Ewert. 2008. Spatial distribution of grassland productivity and land use in Europe. Agric. Syst. 98: 208–219.
Socher, S.A., D. Prati, S. Boch, J. Müller, V.H. Klaus, N. Hölzel and M. Fischer. 2012. Direct and productivity-mediated indirect effects of fertilization, mowing and grazing on grassland species richness. J. Ecol. 100: 1391–1399.
Soussana, J.F., V. Allard, K. Pilegaard, P. Ambus, C. Amman, C. Campbell, E. Ceschia, J. Clifton-Brown, S. Czobel, (...) and R. Valentini. 2007. Full accounting of the greenhouse gas (CO_2, N_2O, CH_4) budget of nine European grassland sites. Agric. Ecosyst. Environ. 121: 121–134.
Sparrius, L.B. and A.M. Kooijman. 2011. Invaisveness of *Campylopus introflexus* in drift sands depends on nitrogen deposition and soil organic matter. Appl. Veg. Sci. 14: 221–229.
Stevens, C.J., N.B. Dise, J.O. Mountford and D.J. Gowing. 2004. Impact of nitrogen deposition on the species richness of grasslands. Science 303: 1876–1879.
Stevens, D.P., S.L.N. Smith, T.H. Blackstock, S.D.S. Bosanquet and J.P. Stevens. 2010. Grasslands of Wales—A Survey of Lowland Species-rich Grasslands 1987–2004. University of Wales Press, Cardiff, 387 pp.
Sutton, M.A., C.M. Howard, J.W. Erisman, G. Billen, A. Bleeker, P. Grennfelt, H. van Grinsven and B. Grizzetti (eds.). 2011. The European Nitrogen Assessment: Sources, Effects and Policy Perspectives. Cambridge University Press, Cambridge, 664 pp.
Tälle, M., B. Deák, P. Poschlod, O. Valkó, L. Westerberg and P. Milberg. 2016. Grasing vs. mowing: A meta-analysis of biodiversity benefits for grassland management. Agric. Ecosyst. Environ. 222: 200–212.

Tallowin, J.R.B. and R.G. Jefferson. 1999. Hay production from lowland semi-natural grasslands: a review of implications for ruminant livestock systems. Grass Forage Sci. 54: 99–115.

Thiele, J., M. Isermann, A. Otte and J. Kollmann. 2010. Competitive displacement or biotic resistance? Disentangling relationships between community diversity and invation success of tall herbs and shrubs. J. Veg. Sci. 21: 213–220.

Thuiller, W., S. Lavorel, M.B. Araújo, M.T. Sykes and I.C. Prentice. 2005. Climate change threats to plant diversity in Europe. Proc. Natl. Acad. Sci. USA 102: 8245–8250.

Thuring, C. and G. Grant. 2016. The biodiversity of temperate extensive green roofs—A review of research and practice. Isr. J. Ecol. Evol. 62: 44–57.

Török, P. and J. Dengler. 2018. Palaearctic grasslands in transition: overarching patterns and future prospects (pp. 15–25 this volume).

Török, P., E. Vida, B. Deák, S. Lengyel and B. Tóthmérész. 2011. Grassland restoration on former croplands in Europe: An assessment of applicability of techniques and costs. Biodivers. Conserv. 20: 2311–2332.

Török, P., N. Hölzel, R. van Diggelen and S. Tischew. 2016. Grazing in European open landscapes: How to reconcile sustainable land management and biodiversity conservation? Agric. Ecosyst. Environ. 234: 1–4.

Török, P., M. Janišová, A. Kuzemko, S. Rūsiņa and Z. Dajić Stevanović. 2018. Grasslands, their threats and management in Eastern Europe (pp. 64–88 this volume).

[UBA] Umweltbundesamt. 2017. Daten zur Umwelt 2017 – Indikatorenbericht. Umweltbundesamt, Dessau, 150 pp.

Ushimaru, A., K. Uchida and T. Suka. 2018. Grassland biodiversity in Japan: threats, management and conservation (pp. 199–220 this volume).

Vadász, C., A. Máté, R. Kun and V. Vadász-Besnyői. 2016. Quantifying the diversifying potential of conservation management systems: An evidence-based conceptual model for managing species-rich grasslands. Agric. Ecosyst. Environ. 234: 134–141.

Valkó, O., P. Török, B. Deák and B. Tóthmérész. 2014. Prospects and limitations of prescribed burning as a management tool in European grasslands. Basic Appl. Ecol. 15: 26–33.

Vanneste, T., O. Michelsen, B.J. Graae, M.O. Kyrkjeeide, H. Holien, K. Hassel, S. Lindmo, R.E. Kapás and P. De Frenne. 2017. Impact of climate change on alpine vegetation of mountain summits in Norway. Ecol. Res. 32: 579–593.

Vogels, J. 2009. Fire as a restoration tool in the Netherlands—First results from Dutch dune areas indicate potential pitfalls and possibilities. Int. Forest Fire News 38: 23–35.

von Oheimb, G., I. Eischeid, P. Finck, H. Grell, W. Härdtle, U. Mierwald, U. Riecken and J. Sandkühler. 2006. Halboffene Weidelandschaft Höltigbaum–Perspektiven für den Erhalt und die umweltverträgliche Nutzung von Offenlebensräumen. Naturschutz Biol. Vielfalt 36: 1–280.

Voß, K. 2001. Die Bedeutung extensiv beweideten Feuchtund Überschwemmungsgründlandes in Schleswig-Holstein für den Naturschutz. Mitt. Arbeitsgem. Geobot. Schleswig-Holstein Hamb. 61: 1–185.

WallisDeVries, M.F. and C.A.M. van Swaay. 2009. Grasslands as habitats for butterflies in Europe. pp. 27–34. *In*: P. Veen, R. Jefferson, J. de Smidt and J. van der Straaten (eds.). Grasslands in Europe of High Nature Value. KNNV Publishing, Zeist.

Wassen, M.J., H. Olde Venterink, E.D. Lapshina and F. Tanneberger. 2005. Endagered plants persist under phosphorus limitation. Nature 437: 547–550.

Wesche, K., B. Krause, H. Culmsee and C. Leuschner. 2009. Veränderungen in der Flächen-Ausdehnung und Artenzusammensetzung des Feuchtgrünlandes in Norddeutschland seit den 1950er Jahren. Ber. Reinhold-Tüxen-Ges. 21: 196–210.

Wesche, K., D. Ambarlı, J. Kamp, P. Török, J. Treiber and J. Dengler. 2016. The Palaearctic steppe biome: a new synthesis. Biodivers. Conserv. 25: 2197−2231.

Wilson, J.B., R.K. Peet, J. Dengler and M. Pärtel. 2012. Plant species richness: the world records. J. Veg. Sci. 23: 796–802.

World Resources Institute. 2005. Ecosystem and Human Well-being: Biodiversity Synthesis—A Report of the Millennium Ecosystem Assessment. Island Press, Washington, DC., 86 pp.

Zechmeister, H.G., I. Schmitzberger, B. Steurer, J. Peterseil and T. Wrbka. 2003. The influence of land-use practices and economics on plant species richness in meadows. Biol. Conserv. 114: 165–177.

Zulka, K.P., M. Abensperg-Traun, N. Milasowszky, G. Bieringer, B.-A. Gereben-Krenn, W. Holzinger, G. Hölzler, W. Rabitsch, A. Reischütz, (...) and H. Zechmeister. 2014. Species richness in dry grassland patches in eastern Austria: A multi-taxon study on the role of local, landscape and habitat quality variables. Agric. Ecosys. Environ. 182: 25–36.

4

Grasslands, their Threats and Management in Eastern Europe

Péter Török,[1,*] *Monika Janišová,*[2] *Anna Kuzemko,*[3,4] *Solvita Rūsiņa*[5] and *Zora Dajić Stevanović*[6]

Introduction

The Eastern European region covers the post-socialist countries of central and eastern Europe (excluding East Germany and the European part of Russia) and the Balkan countries (excluding Greece and Turkey) (Fig. 4.1). The total area of the region is 2,154,005 km^2, characterized mostly by extensive lowland regions to the north and north-east and with considerable mountainous regions in the central (Carpathians) and the southern (Balkan mountains, Crimean mountains) parts of the region. The region experiences a cool continental climate with increasing Mediterranean influence to the south (Peel et al., 2007). Based on the European Environmental Stratification system provided by Metzger et al. (2005), most of the Eastern European plains and lowlands and the uplands and low mountains of the Balkan Peninsula are situated in the Continental Environmental Zone (CON), naturally dominated by deciduous, mixed and coniferous forests. In the lowland regions, grasslands were formed on fine or coarse-grained alluvial and fluvial deposits and are characterized by the high influence of large rivers and their tributaries. The

[1] MTA-DE Lendület Functional and Restoration Ecology Research Group, Egyetem sqr. 1, 4032 Debrecen, Hungary.
[2] Institute of Botany, Plant Science and Biodiversity Center, Slovak Academy of Sciences, Ďumbierska 1, 974 11 Banská Bystrica, Slovakia.
[3] M.G. Kholodny Institute of Botany, National Academy of Sciences of Ukraine, Tereshchenkivska, 2, 01601, Kyiv, Ukraine.
[4] Ukrainian Nature Conservation Group, Kamianyy Lane, 4-64, Uman, 20300, Ukraine.
[5] Academic Center for Natural Sciences, Faculty of Geography and Earth Sciences, University of Latvia, 1 Jelgavas Street, Riga, 1004, Latvia.
[6] Department of Botany, Faculty of Agriculture, University of Belgrade, Nemanjina 6, 11080 Belgrade, Republic of Serbia.
Emails: monika.janisova@gmail.com; anyameadow.ak@gmail.com; rusina@lu.lv; dajic@agrif.bg.ac.rs
* Corresponding author: molinia@gmail.com

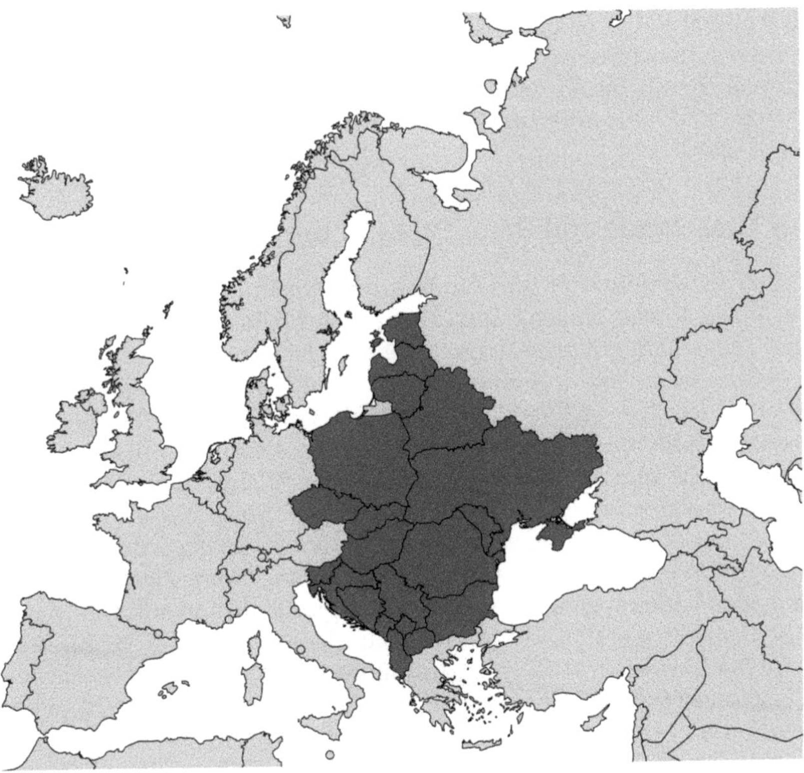

Fig. 4.1 Delimitation of the Eastern European socio-economic region as used in this chapter. The map was created by using MapChart (https://mapchart.net/).

northern part of the Baltic countries in the Boreal zone (BOR) is covered with coniferous forests (taiga). Most parts of the Baltic countries, some regions of Poland, Ukraine and Belarus falls into the Nemoral zone (NEM) with primary deciduous and mixed forests, wetlands and bog mosaics. The lowland and foothill regions of the Carpathian basin, the Middle and Lower-Danube Plains and the Black-Sea Lowland is within the Pannonian-Pontic environmental zone (PAN) and characterized by natural forest-steppe and steppe vegetation. The highest altitudes of the Carpathian and the Balkan mountains are in the Alpine South Environmental Zone (ALS) and home to heathland and alpine grassland vegetation. The low and medium mountains of the northern Balkans with an increased Mediterranean influence form the Mediterranean Mountains Environmental Zone (MDM), where the potential vegetation is Mediterranean evergreen forests and beech forests, but which are now mostly covered with overgrazed pastures and grasslands.

The region harbors a high proportion of grassland habitats; the permanent grassland area in the region based on the available literature and statistics, is higher than 300,000 km^2 out of which at least 10–30 per cent are High Nature Value natural or semi natural grasslands (see Appendix). The marked difference in the grasslands cover between the Western and Eastern European regions is that although the proportion of highly valuable grasslands is quite similar, in most countries of Eastern Europe there are large areas covered with partly degraded grasslands (fallows, semi-improved grassland, abandoned grasslands), which can be turned with appropriate restoration and conservation measures to diverse semi-natural grasslands. Most Western European countries do not have such a resource; instead, they have a high proportion of very intensively managed grasslands. In Eastern Europe, there is the western border of Palaearctic steppe zone in Europe, with high cover of steppe and steppe-like grasslands in Bulgaria, Hungary, Moldova and Ukraine (Wesche et al., 2016).

Origin of Grasslands and their Types in the Region

The historical development of grasslands in Eastern Europe follows several pathways, which are linked to the biogeographical division of the area and origin of grassland ecosystems. The majority of grasslands of the boreal and nemoral zonobiome (Walter and Breckle, 1991) are secondary or semi-natural grasslands of anthropozoogenic origin. When at the end of the Ice Age (14,400–12,000 BP) the glaciers retreated, the landscape remained open for several millennia and enabled long distance dispersal of plant species, which had survived in the more southern regions. In the Atlantic (8,500–6,800 BP), woodland returned and suppressed open grassland vegetation (Ložek, 2008). However, at the same time, the human population increased and due to its activities (this so-called Neolithic Revolution included deforestation and import of various domesticated plants and animal species) the open landscape was maintained and gradually spread in the region. The first Neolithic settlements were build in 8,500 BP in Macedonia and Romania, 7,700–7,600 BP in Transdanubia (Poschlod, 2015), and during the next 2,000 years the lifestyle of settled communities spread from these parts of the Eastern Europe further to northeast (eastern part of Romania, Ukraine) and northwest (Pannonia, Carpathian and Hercynian mountains). There are notes on the first human settlements in the Balkans dating from about 6,000 BC, known as 'Vinča culture', also known as the oldest European copper metallurgy and technologically the most advanced pre-historical world civilization, primarily focusing on livestock and crop production, such as wheat, lens, barley and flax (Barker, 1985).

So, between 8,500 and 6,500 BP the first semi-natural and anthropogenic grasslands might have been created. However, the vast majority of grasslands were established much later, during the Middle Ages and reached their largest spatial extensions during the last two centuries (Ružičková and Kalivoda, 2007). Pasture ecosystems are generally older than meadows, especially in the boreal part of the region where the scythe appeared only in 3rd–4th century AD (Anon, 1974; Rabinovič et al., 1985), while farming and livestock herding appeared 6,000 years ago. The continuously increasing age of grasslands toward the south is due to a longer-lasting period of climatic conditions favorable for grassland development and a longer history of agriculture. The time of farming establishment, as the main source of food, could be attributed to semi-natural grassland age, generally dated back to 3,000–6,500 BP (Melluma, 1994; Price, 2000).

The extraordinary variability of European grasslands is reflected in the huge number of distinguished phytosociological classes and alliances. Rodwell et al. (2002) listed 19 grassland classes with 326 alliances, while Mucina et al. (2016) recently proposed 27 classes with 365 alliances. The primary or natural grasslands of Eastern Europe can be grouped into three major types: (1) steppes (in areas too dry for forests); (2) alpine grasslands (in areas too cold for forests); (3) azonal and extrazonal grasslands (where hydrology, soil conditions, relief or natural disturbances within the forest biomes prevent tree growth locally). Some of these grasslands need human intervention by grazing and mowing to maintain their continuity and prevent the forest regeneration or reed bed development (Emanuelsson, 2009). Primary grasslands of climatogenic origin belonging to Palaearctic steppe biome cover large areas in south-eastern part of Eastern Europe and in natural conditions are maintained by drought, wildfire and wild herbivores (Wesche et al., 2016). Alpine grasslands of the region are distributed above the tree line (about 1,800 m a.s.l. in the Carpathians and generally above 1,950–2,150 m a.s.l. on the Balkan Peninsula). Semi-natural grasslands of secondary origin (4) were created mostly by tree cutting and are maintained by extensive management of mowing and/or grazing. These grasslands, ranging from semi-dry to wet conditions, are situated from lowlands to mountainous regions, in which in lack of management the shrub and tree encroachment is typical (see types 4a–4c below). The most important grassland types and subtypes of Eastern Europe are as follows (nomenclature of syntaxa follows Mucina et al., 2016):

1. Steppe grasslands (*Festuco-Brometea*: *Festucetalia valesiacae*) are primary grasslands in the Eastern European region associated with the steppe and forest steppe zones typically distributed in lowlands and at the foothills. In the Eastern European region, at least fragments of such vegetation are present in Romania, Ukraine, Poland, Moldova, Hungary, Slovakia, Slovenia, Czech Republic, Croatia, Bosnia and Herzegovina, Montenegro, Albania, Serbia and Bulgaria. Steppe grasslands are characterized by the dominance of *Festuca* and *Stipa* species and are rich in forbs, including multiple genera (among the most typical genera are *Astragalus*, *Artemisia*, *Aster*, *Salvia* and *Linum*).

2. Alpine grasslands are predominantly natural species-rich grasslands, which may be formed both on base-rich (*Elyno-Seslerietea*) and siliceous (*Caricetea curvulae*, *Carici rupestris-Kobresietea*, *Juncetea trifidi*, *Nardetea strictae*) bedrocks, occurring in the subalpine to subnival belts of the European boreal and nemoral mountain ranges in Slovakia, the Czech Republic, Romania, Ukraine, Poland and all Balkan countries. They are mostly dominated by tussock-forming graminoids of the genera *Festuca*, *Calamagrostis*, *Sesleria*, *Carex* and *Juncus* (Fig. 4.2).

3a. Rocky grasslands (*Sedo-Scleranthetea; Festuco-Brometea: Stipo pulcherrimae-Festucetalia pallentis*) include pioneer vegetation and xeric open steppic grasslands on shallow skeletal soils on rocky calcareous and siliceous substrates. Although they are often primary, their spread was supported in the past by intensive human deforestation activities and grazing. Some of them represent relic vegetation of Pleistocene periglacial steppes. These grasslands occur in all countries of the region, having larger distribution in Ukraine, Czech Republic, Slovakia, Poland, Hungary, Romania, Bulgaria, Moldova, Balkan countries and being rare in the Baltic countries. Quite often the dominants are succulents (*Sedum* spp., *Sempervivum* spp., *Jovibarba* spp.), therophytes (*Spergula* spp., *Cerastium* spp., *Veronica* spp.) or tussock-forming grasses (*Festuca* spp., *Stipa* spp., *Poa* spp.), while cryptogams (mosses and lichens) are also abundant (Fig. 4.2-3a).

3b. Sandy grasslands (*Koelerio-Corynephoretea*) are tussock grasslands and sandy steppes on acidic to alkaline sandy soils on inland sand dunes and plains. They are most common in the boreal zone on acidic sands of glaciofluvial deposits and weekly acidic to neutral sands of coastal dunes (calcium-rich sands with a local supply of calcium from crushed shells) and alluvial sands in floodplains (Latvia, Lithuania, Belarus, northern Poland, the Czech Republic and Ukraine) as well as on base-rich to alkaline sands of alluvial deposits in the Pannonian (Hungary, Slovakia, Serbia, Slovenia, Croatia) and Pontic (southern Ukraine) regions. In these communities, tussock grasses, such as closely related *Festuca* species (*F. psammophila*, *F. polesica*, *F. vaginata*, *F. beckeri*), *Corynephorus canescens*, *Koeleria glauca* and *Stipa borysthenica*, as well as mosses and lichens play a significant role (Fig. 4.2-3b).

3c. Coastal and inland halophytic grasslands (*Festuco-Puccinellietea; Juncetea maritimi*) are azonal and intrazonal grasslands occurring on soils with moderate to high salt content and generally astatic or semi-static water regime in the lowlands. Most typical stands of inland halophytic grasslands occur in Hungary and in Ukraine, but fragments are present also in Slovakia, Serbia, Bulgaria, and Macedonia. Estonia and Latvia possess large areas of coastal grasslands in the geolittoral zone of the Baltic Sea where soil salinity is lower and semi-halophytic vegetation develops under the periodic flooding with brackish sea water. This type of vegetation is dominated by stress-tolerant graminoids (e.g., *Festuca pseudovina, F. regeliana, Puccinellia* spp., *Juncus* spp.), *Plantago* spp. and several other halophytic forbs of the genera *Salicornia, Suaeda, Aster, Podospermum, Artemisia, Salsola, Spergularia* or *Limonium* (Fig. 4.2-3c).

4a. Dry and semi-dry semi-natural grasslands (*Festuco-Brometea: Brachypodietalia pinnati; Molinio-Arrhenatheretea: Galietalia veri*) are meso-xerophytic secondary grasslands occurring predominantly on moderate or deeper calcareous soils. They are distributed from lowlands to the mountain belt throughout the region; in the Czech Republic, Slovakia, Ukraine, Moldova, Poland, Hungary, Romania Serbia, Bulgaria, Montenegro, Bosnia and Herzegovina, Macedonia, Latvia, Lithuania, Poland, as well as on *alvars* (species-rich grasslands on shallow soils over flat limestone bedrock) along the eastern coast of the Baltic Sea in Estonia. In Latvia and Estonia, some of the most-species rich wooded grasslands occur in dry and semi-dry conditions. Many of these grasslands harbor steppe elements and are extraordinarily species-rich in both vascular plants and cryptogams including many rare and endangered taxa (Fig. 4.2-4a).

4b. Mesic and moist semi-natural grasslands (*Molinio-Arrhenatheretea: Arrhenatheretalia; Molinietalia*) include anthropogenic managed pastures, meadows and secondary mat-grass swards on well-drained mineral fertile deep soil or nutrient-poor soil. These grasslands represent the most widespread type of semi-natural grasslands distributed from lowlands to the mountain and rarely to subalpine belts occurring in all countries in the region. Dominants are mainly the loose tussock-forming and rhizomatous grasses (e.g., *Festuca pratensis, F. rubra, Poa pratensis, P. trivialis, Phleum pratense, Arrhenatherum elatius, Trisetum flavescens, Agrostis tenuis, Alopecurus pratensis, Cynosurus cristatus,* and *Anthoxanthum odoratum*) and representatives of the *Fabaceae* (*Trifolium* spp., and *Medicago* spp.), Cyperaceae, and Juncaceae. Various species of the genera *Plantago, Veronica, Ranunculus* and *Rhinanthus* are common as well (Fig. 4.2-4b).

4c. Wet (semi-) natural grasslands (*Phragmito-Magnocaricetea; Scheuchzerio-Caricetea fuscae*) include herb-rich temporarily wet meadows, sedge-bed marsh vegetation and sedge-moss vegetation on mineral and peaty temporarily wet, heavy soil, on oligo- to eutrophic organic sediments, calcareous and extremely mineral-rich brown-moss fens or moderate to low calcium-rich slightly acidic fens at low altitudes of temperate and boreal regions as well as the sub-Mediterranean precipitation-rich regions of the Balkan. This type of vegetation is common in all countries in the region, mostly in lowland regions. Typical dominants are tall sedges (e.g., *Carex acuta, C. acutiformis, C. elata,* and *C. cespitosa*) and/or grasses (e.g., *Phalaris arundinacea, Glyceria* spp.) or tall forbs (e.g., *Lysimachia vulgaris, Lythrum salicaria, Filipendula ulmaria,* and *Cirsium* spp.; Fig. 4.2-4c).

Fig. 4.2 Grassland types of Eastern Europe. 1. Steppe grassland (Askania-Nova, Ukraine), 2. Alpine grassland (Hoverla Mt., Ukraine), 3a. Rocky dry grassland in the Považský Inovec Mountains (Lúka nad Váhom, Slovakia), 3b. Sandy grassland (Fülöpháza, Hungary), 3c. Inland halophytic grassland (Oril River valley, Ukraine), 4a. Semi-dry semi-natural grassland (Synytsia River valley, Ukraine), 4b. Mesic semi-natural grassland in the Chywchyny Mountains (Sarata, Ukraine), 4c. Wet grassland (South Bug River valley, Ukraine). Photos by A. Kuzemko (1, 2, 3c, 4a, 4b and 4c), M. Janišová (3a and 4b) and P. Török (3b).

Trends of Agronomic Use of Grasslands

Most of Europe in ancient times was covered by forests but from Renaissance era onwards, a high proportion of forests were cut and the lands were transformed to extensively-managed agricultural areas and secondary habitats, like extensively-managed grasslands. In the area of the steppe biome (i.e., Ukraine, Moldova, some parts of Hungary, Croatia and Serbia) the natural and semi-natural grasslands reached their maximum extension before crop cultivation expanded, starting from the beginning of the 19th century (Wesche et al., 2016). In the northernmost countries of the region, this happened at the end of the 19th and the beginning of 20th century up to the 1920s. The 19th century could be a turning point in the history of grassland management throughout Eastern Europe. Intensification became necessary for feeding a growing urban population with increased demands for food quality and security (Hopkins and Holz, 2006). The main driver of changes in lowland natural grasslands (i.e., steppes) was the high demand of arable fields at the expense of grasslands. Decreasing areas of pastures provoked overgrazing especially because animal (draft) power was demanded for crop production. Thus, the countries in the region experienced rapid decline in grassland biodiversity because of conversion of grasslands to arable land and overexploitation of residual grassland areas. These grassland transformations resulted in massive soil erosion and habitat degradation. Humus loss, damage of secondary soil structure and compaction are interdependent factors of soil degradation (Leah, 2016).

While a high level of agricultural industrialization occurred in Western European countries from the first half of the 20th century onwards and resulted in a massive decrease in the area of extensively managed land, fragmentation and decline in biodiversity, in most parts of the Eastern Europe these negative trends were not so marked until the switch to communist economy (Pullin et al., 2009). After the First World War, the socio-economic settings in the eastern part of the region were influenced by the Soviet Union. Ukraine and Belarus became members of the Soviet Union in 1922, Moldova in 1924 (as part of Ukrainian SSR and from 1940 as Moldavian SSR) and the Baltic countries were annexed in 1940. After the Second World War, the Soviet communist influence became strong in the other central European and Balkan countries of the region. This meant forced collectivization in agriculture and, adaptation to socialist centrally-planned economy (Bogovin, 2006) in the industry.

In the last few decades two simultaneous processes either intensification or marginalization of agriculture (Vanwambeke et al., 2012; Jepsen et al., 2015) were seen. These processes were common for all Eastern European countries but with different rates of change. While intensification (collectivization) was a common process in the whole of Eastern Europe from the 1950s through 1970s to 1990s, but starting about 2000, countries diverged in land-management regimes. The northernmost countries of the region experienced two simultaneous processes. Industrialization (larger farms and fields, specialization in production) occurred in the agriculturally most-productive regions. Abandonment of agricultural lands was common throughout these countries. The dominant process in nemoral and continental countries was de-intensification (Jepsen et al., 2015). After the collapse of socialist economy in all the countries, most state-owned land became privatized and/or returned to the former owners of advanced age. Because of a lack of resources and funding, most of these lands were abandoned. With the access to various constructions of support in EU agri-environmental schemes, re-utilization of a high proportion of formerly abandoned land was enabled in some countries.

Ecosystem Services

Natural and semi-natural grasslands are key contributors to several ecosystem services, like food, genetic resources, pollination, invasion resistance and many cultural services. Potential provisioning of several services are still poorly known and not evaluated; for example, the provisioning service of natural medicines and regulating services of seed dispersal and disease regulation (Harrison et al., 2010). Natural and semi-natural extensively-managed grasslands provide more diverse and much higher quality ecosystem services than sown and intensively-managed grasslands. They are better CO_2 sinks, provide more effective water infiltration and storage; extensive management ensures less pollution, and provide extensive cultural and intangible services (Benayas et al., 2009; Bullock et al., 2011). Nevertheless, human use of semi-natural grassland services has been mostly subsided in recent decades in Eastern Europe because of high levels of decline in semi-natural grassland area. The monetary value of ecosystem services of semi-natural grasslands has been calculated only in a few countries of Eastern Europe. The best example is the Czech Republic (Hönigová et al., 2012) with the calculated amount of 11,000 to 103,000 EUR (13,000 to 120,000 US$) per hectare depending on the habitat type.

Potential of semi-natural grasslands for biogas and biofuel production has been evaluated in the Baltic countries (Heinsoo et al., 2010; Hensgen et al., 2007; Melts, 2014; Strazdiņa et al., 2015). In Latvia, the energetic potential of biomass from permanent grassland was estimated as 4,407–6,661 kWh ha^{-1} yr^{-1}, the methane potential from grassland biomass as 441–666 normal $m^3 ha^{-1} yr^{-1}$ and the economic potential of biomass resources calculated as income from biogas production as 139–220 EUR (161–256 US$) ha^{-1} yr^{-1} (Strazdiņa et al., 2015). Energy production from semi-natural grassland is most profitable in alluvial grasslands, followed by dry to mesic meadows. Methane production yield is highest in grasses and sedges/rushes and lowest in forbs. Energy yield through combustion is higher than from methane production. The energy yield from semi-natural grasslands can be comparable with that of energy crops in the boreal region (Melts, 2014).

Only a few attempts have been made to evaluate the cultural services of semi-natural grasslands in Eastern Europe. A contingent valuation study was carried out for Estonian semi-natural grasslands to evaluate them as a non-market environmental good. Based on 1,061 respondents, the total annual demand for semi-natural grasslands was evaluated to be 17.9 million EUR (20.8 million US$; Lepasaar and Ehrlich, 2015). In Slovakia, local residents prioritized provisioning and regulating services, and did not evaluate grasslands as important providers of cultural services (Bezák and Bezáková, 2014). In Hungary, aesthetics and social values were more appreciated by organic farmers, while the conventional farmers stressed the economical values (Kelemen et al., 2013). There are some indications that Eastern European farmers are less aware of biodiversity values and more sceptic to conservation policy if compared to Scandinavian and central European countries. Comparison of Finnish and Estonian farmers showed that Estonian farmers were less sceptic to undesirable effects of intensification to farmland wildlife. Hungarian farmers were more sceptic to nature conservation than French and Italian farmers (Kelemen et al., 2013). The possible reason is a long history of top-down nature conservation policy in Eastern Europe but without a tradition to involve the general public in environmental decision making (Young et al., 2007).

Grassland Biodiversity

Temperate and hemi-boreal grasslands are known for their high and, in some cases, extraordinary, small-scale diversity of vascular plants (Wilson et al., 2012) as well as bryophytes and lichens (Löbel et al., 2006). Comparative studies of species richness of different grassland types, carried out in Eastern Europe, showed that semi-dry basophilous grasslands are characterized by the greatest richness of vascular and non-vascular plants (Dengler et al., 2016). The extraordinary plant species richness was revealed for semi-dry grasslands of White Carpathians, Czech Republic and Slovakia (Chytrý et al., 2015), foothills of the Eastern Carpathians, Ukraine (Roleček et al., 2014), and Transylvania, Romania (Turtureanu et al., 2014) (Table 4.1).

Along with high phytodiversity, grassland ecosystems provide refuge to a large number of rare and endangered animal and plant species and they can be considered as one of the global biodiversity hotspots (Habel et al., 2013). Mesic and wet grasslands of Eastern Europe are habitats of many species of *Orchidaceae* (*Orchis, Anacamptis, Dactylorhiza,*

Table 4.1 Total plant and vascular plant species richness for some grasslands in Eastern Europe. BG = Bulgaria, CZ = Czech Republic, EE = Estonia, LV = Latvia, RO = Romania, SK = Slovakia, UA = Ukraine.

Country	Study Area	Grassland Type	Total Plant Richness (max.)			Vascular Plants Richness (max.)			Source
			1 m^2	10 m^2	100 m^2	1 m^2	10 m^2	100 m^2	
BG	NW Bulgarian Mountains	dry	41	62	89	36	60	87	Dengler et al. (2016)
CZ	White Carpathians	semi-dry	65	88	117	58	79	105	Dengler et al. (2016)
CZ	White Carpathians	semi-dry	–	–	133	82	–	119	Chytrý et al. (2015)
CZ	Bošovice (S Moravia)	semi-dry	–	–	–	57	–	107	Chytrý et al. (2015)
EE	Saaremaa	semi-dry	49	72	100	35	49	70	Dengler et al. (2016)
LV	Northern Latvia, Gauja River Valley	semi-dry	51	–	–	50	–	–	Rūsiņa (2008)
LV	Western Latvia, Sventaja River Valley	moist calcareous (*Molinion*)	–	–	–	47	–	–	S. Rūsiņa (unpubl.)
RO	Transylvania	dry	82	101	134	79	98	127	Dengler et al. (2016)
SK	Strážovské Vrchy Mts	semi-dry	–	–	–	–	–	97	Chytrý et al. (2015)
UA	Central Podolia	dry	48	67	108	42	64	86	Dengler et al. (2016)
UA	Foothills of the Eastern Carpathians, Dziurkac	semi-dry	–	–	–	–	92.8*	–	Roleček et al. (2014)

*standardized to 10 m^2.

Ophrys, Traunsteinera, etc.) as well as *Liliaceae* (*Lilium, Fritillaria*), *Iridaceae* (*Iris, Gladiolus*) and some other rare forbs as well as *Cyperaceae* and *Juncaceae*. All these taxa are particularly vulnerable to changes in management regime. However, rare and endangered species occur in the highest number in dry grasslands. Moreover, their rarity is driven by habitat destruction and fragmentation. For example, in Ukraine steppe ecosystems occupying only about one per cent of the territory are habitats for almost 30 per cent of all species of flora and fauna listed in the *Red Book of Ukraine* (Burkovsky et al., 2013). A similar situation was reported from Latvia—semi-natural grasslands cover 0.7 per cent of the area of the country, but they host 30 per cent of the total number of red-listed vascular plant species (Gavrilova, 2003).

Many representatives of the grassland flora are endemic (narrow-ranged) species or relict species. There are particularly many narrow-ranged species among the steppe and forest steppe flora: *Colchicum fominii, Hyacinthella pallasiana, Ornithogalum amphibolum, Elytrigia stipifolia, Stipa syreistschikowii, Rumia crithmifolia, Artemisia hololeuca, Carlina onopordifolia, Gymnospermium odessanum, Crambe aspera, Cerastium biebersteinii, Dianthus pseudoserotinus, Eremogone cephalotes, Euphorbia volhynica, Astracantha arnacantha, Calophaca wolgarica, Chamaecytisus graniticus, Erodium beketowii, Hyssopus cretaceus, Cymbochasma borysthenica, Androsace koso-poljanskii, Pulsatilla taurica* and *Viola oreades*.

However, there are also narrow-ranged species in mesic and wet grasslands (*Nigritella carpatica, Pinguicula bicolor*) as well as in saline (*Allium regelianum, Phlomis scythica*) and sandy grasslands (*Allium savranicum, Centaurea breviceps, Alyssum borzaeanum, Astragalus tanaiticus, Goniolimon graminifolium*). Although the majority of grasslands in the region are semi-natural, they serve as refugia for some relict species. The primary steppe habitats are the richest in relicts: *Allium obliquum, Sternbergia colchiciflora, Carex pediformis, Psathyrostachys juncea, Schivereckia podolica, Globularia trichosantha, Dracocephalum austriacum, Thalictrum foetidum*, etc. (Didukh et al., 2009).

Natural and semi-natural grasslands are the main nesting habitat for several tens of bird species. From 200 bird species that regularly nest in Latvia, one-fourth nest in grasslands on a regular basis, while for 15 of them the grassland is the only or almost the only nesting habitat in Latvia. Coastal grasslands of the Baltic Sea are directly related to the critically-endangered Baltic subspecies of the Dunlin—*Calidris alpina schinzii*. Three of six globally endangered bird species—the Aquatic warbler (*Acrocephalus paludicola*; 'vulnerable' status according to IUCN criteria), the Great snipe (*Gallinago media*) and the Black-tailed godwit (*Limosa limosa*; 'near threatened' status for both) depend on wet floodplain grasslands. Another two globally endangered bird species are the Eurasian curlew (*Numenius arquata*) and the European roller (*Coracias garrulus*; 'near threatened' status for both). The Corn crake (*Crex crex*) also had this status until recently, but thanks to the species protection and grassland habitat restoration measures in recent decades, especially in Western Europe, its population has increased and its status has been changed (Rūsiņa and Auniņš, 2017).

Conservation of Grassland Biodiversity

The most valuable grasslands have traditionally been preserved in protected areas, mainly in nature reserves and national parks. For example, almost all large areas of watershed steppes in Ukraine that survived until now are part of protected areas, such as the Biosphere Reserve 'Askania Nova', Ukrainian steppe reserve branches, some nature reserves and national parks, with a total area of over 700 km^2. In the Carpathian region, a whole network of protected areas was established, including those created in the framework of international cooperation, such as the 'Eastern Carpathians' trans-boundary Biosphere Reserve, which includes parts of Poland, Slovakia and Ukraine, or the bilateral Polish-Slovak 'Tatra' National Park, which protects the most valuable areas of mountain grasslands, including natural alpine grasslands.

Conservation policy has changed in post-Soviet countries substantially after the breakdown of the Soviet regime. In boreal countries, conservation of semi-natural grasslands was not given due consideration until the late 20th century. In general, the active protection of semi-natural grasslands only began in the late 20th century when the approach of nature conservation changed from absolute non-intervention to active nature conservation. Until then, the emphasis was mainly placed on species conservation, sometimes not even considering or misunderstanding habitat ecology and the requirements of the species. Entire nature conservation was mainly based on the reserve principles, described as absolute *zapovednost* (protection status) by Boreiko et al. (2013). For example, Decision No. 421 by the Latvian SSR Council of Ministers of 1977 mandated that hay must not be harvested during the entire year in ornithological reserves with substantial grassland areas. Such grassland management bans in ornithological reserves resulted in reduction of bird species for which these bans were established. These practices contributed to a significant reduction in semi-natural grassland area in protected nature areas (Kaltenborn et al., 2002; Klein, (ed.) 2008; Rūsiņa, (ed.) 2017). The approach of absolute 'zapovednost' in Ukraine is still popular and even reflected in some of the laws that prohibit regulatory measures in reserves and protected areas of national natural parks; this prevent implementation of proper protection of grasslands in these areas.

Until the late 20th century, due to the prevailing non-intervention nature conservation approach, there were very few grasslands in the protected nature areas, many of which formed in the Soviet era. In Latvia, only half of the 153 Natura 2000[1] areas containing protected grassland habitats had been established before 1990. From 1999 to 2004, new Natura 2000 areas for the conservation of protected grasslands were established. These were mostly for floodplain bird habitats and EU habitat, '6450 Floodplain grasslands'. Other protected grassland habitats mostly occur in the mosaic of agricultural land and forests and are heavily fragmented; therefore, it is administratively complicated to establish protected areas for them. Thus, only half of the total area of protected grassland habitats are situated inside the Natura 2000 network in several countries (see Appendix).

The main legislative instrument that regulates protection of ecosystems in Europe, including grasslands, is the Convention on the Conservation of European Wildlife and Natural Habitats (Bern Convention). It was adopted in Bern, Switzerland in 1979, and came into force in 1982. Signatories of the Bern Convention include, among others, the central and eastern European countries and the EU Member States. The principal aims of the Convention are to ensure conservation and protection of wild plant and animal species and their natural habitats (listed in Resolutions 4 and 6 of the Convention). This Convention provides the basis for development of the Emerald network of areas of special conservation interest (ASCIs). For EU Member States, Emerald network sites are those of the Natura 2000 network. Natura 2000 is based on the 1979 Birds Directive and the 1992 Habitats Directive. Now there are more than 4,100 sites that comprise certain types

[1] The Natura 2000 network is designated to protect core breeding and resting sites for rare and threatened species, and some rare natural habitat types in the European Union. The aim of the network is to ensure long-term survival of the most valuable and threatened species and habitats in Europe.

of grasslands (Table 4.2). Also in the post-Soviet countries, the process of establishing the Emerald network is ongoing. Today, the network has 821 sites that include certain types of grasslands (Table 4.3).

Another legal instrument for grasslands protection in some post-Soviet countries was the publication of so-called *Green Books*, which list plant communities that need protection. The first *Green Book* was published in Ukraine in 1987. The *Green Book of Ukraine* is a public document in accordance with the *Regulations on the Green Book of Ukraine*, approved by the Cabinet of Ministers of Ukraine in 2002. The current edition of the *Green Book of Ukraine* includes 24 types of herbaceous and shrub steppe communities, eight types of herbaceous and shrub communities of xeric type on outcrops and sands and six types of meadow communities (Didukh, 2009). The *Lithuanian Red Data Book* includes several endangered plant communities of grassland vegetation (Balevičiene et al., 2000). A list of rare and threatened plant communities of Estonia has been published in 1998 (Paal, 1998). Latvia does not have a published list of threatened plant communities.

While in the EU, national laws should be harmonized with the EU regulations for habitat protection, outside the EU, the protection of grasslands is exclusively regulated by national laws. For example there are laws *On Environmental Protection* (Belarus, Moldova, Ukraine), *The Law on the National Ecological Network* (Moldova, Ukraine), or *The Law on Plant World* (Ukraine). In Belarus, in 2012 a draft of a normative legal act says: "Compensation

Table 4.2 Number of Natura 2000 sites, comprising a habitat type of the group '6. Natural and semi-natural grassland formations' from the Habitats Directive Annex I and its subtypes (Source: http://www.eea.europa.eu/data-and-maps/data/natura-7). BG = Bulgaria, CZ = Czech Republic, EE = Estonia, HR = Croatia, HU = Hungary, LV = Latvia, LT = Lithuania, PL = Poland, RO = Romania, SI = Slovenia, SK = Slovakia.

Habitat Type	BG	CZ	EE	HR	HU	LV	LT	PL	RO	SI	SK
61 Natural grasslands (6110, 6120, 6150, 6170, 6190)	125	58	0	21	78	24	20	136	52	21	136
62 Semi-natural dry grasslands and scrubland facies (6210, 6220, 6230, 6240, 6250, 6260, 6270, 6280, 62A0, 62C0, 62D0)	167	194	250	92	430	105	45	44	95	29	183
64 Semi-natural tall-herb humid meadows (6410, 6420, 6430, 6440, 6450)	8	81	191	4	376	56	54	59	124	27	49
65 Mesophile grasslands (6510, 6520, 6530, 6540)	23	108	163	13	157	17	10	31	85	23	174

Table 4.3 Number of Emerald sites, comprising a habitat type of the group 'E Grasslands and lands dominated by forbs, mosses or lichens' from the Resolution 4 of the Bern Convention and its subtypes (Source: http://www.coe.int/en/web/bern-convention/ecological-networks-meetings-2016).

Habitat Type	Belarus	Moldova	Ukraine
E1 Dry grasslands (incl. E1.11, E1.12, E1.13, E1.2, E1.3, E1.71, E1.9)	19	2	255
E2 Mesic grasslands (E2.2, E2.3)	6	11	102
E3 Seasonally wet and wet grasslands (E3.4, E3.5)	24	6	176
E4 Alpine and subalpine grasslands (E4.11, E4.12, E4.3, E4.4)	0	0	16
E5 Woodland fringes and clearings and tall forb stands (E5.4, E5.5)	26	0	135
E6 Inland salt steppes (E6.2)	0	0	43

system for users of land plots and (or) water bodies for the introduction of restrictions on economic and other activities in natural areas under special protection (habitats of wild animals and plants species included in the *Red Book of the Republic of Belarus* passed under the protection of users of land plots and (or) bodies of water)" has been developed and submitted to the Ministry of Environment. Article 82 of the Law *On Environmental Protection* provides economic incentives for environmental protection by establishing (for legal and physical entities) tax and other benefits in respect of the protection and use of regime of protected areas, areas subject to special protection and rational (sustainable) use of their natural resources in the transition zones of biosphere reserves (CBD National Report of Belarus, 2014).

Threats

Land use change (land conversion, intensification and abandonment of management), eutrophication caused by industry and nutrient runoff from neighboring agricultural systems and climate change are the main direct drivers of ecosystem change listed in *Millennium Ecosystem Assessment* (World Resources Institute, 2005). The influence of the mentioned direct drivers of biodiversity in semi-natural and natural grasslands in Eastern Europe are accelerated by demographic, economic and socio-political changes.

Land conversion into arable land, forest (through encroachment following abandonment or active forest planting) and to a lesser degree also into urban areas was the main driving force leading to decrease in semi-natural and natural grassland area in the region in the last century and is still continuing at a high rate. In Latvia, only 28–44 per cent of the area of rare grassland habitat type (predominantly hard management conditions, e.g., wet, steep slopes) and 60 per cent of the more common habitat types (predominantly with easy management conditions) were still managed in 2007–2013. Moreover, 1.8 per cent of the total area was destroyed in this period by turning it into arable land (Rūsiņa, 2016). In Belarus, the area of grasslands has decreased by 1,219 km^2 or 3.86 per cent in recent years (Bogovin, 2006; CBD National Report of Belarus, 2014). In Poland, during 2009–2012 the total amount of farmland—most importantly, pastures and grasslands—decreased by 1,600 km^2. This decrease was caused by the conversion of farmland to non-agricultural uses and changing its classification. Many farms, especially small ones, abandoned production in the recent years (CBD National Report of Poland, 2014). It is predicted that depopulation and severe ageing will continue in Eastern Europe (Gavrilova and Gavrilov, 2009; Davoudi et al., 2010), leading to more empty rural areas and polarization of the landscape. Still, there are also reverse trends in grassland area dynamics. Thus, for example, in Moldova in the last 25 years, the area of grasslands has increased at the expense of arable land left fallow or abandoned. The area of pastures and hay meadows is growing while the area of intensively used arable land and cropland decreases. This increase of grasslands resulted from a failure of the agrarian reform after 1990 (Leah, 2016). In Estonia, 80 km^2 of semi-natural grasslands have been restored in the last decade and more than 30 km^2 are planned to be restored in the ongoing restoration projects (Helm et al., 2016).

Grassland abandonment and cessation of former extensive management by mowing or grazing was identified as one of the most crucial drivers of grassland biodiversity, especially in the mountain areas of Europe (Valkó et al., 2012). The increasing rate of abandonment was in parallel with the decrease of livestock in the region, typical for most countries in Eastern Europe (see Appendix). This resulted, in most cases, in shrub and tree encroachment and the decrease of grassland biodiversity.

Climate change has been identified as one of the major drivers of grassland biodiversity in the near future. It is forecast for Eastern European region that (i) the temperature will rise by 1–3°C with considerable sub-regional differences until the middle of the century. The highest increase is projected for the summer, while a lower increase in temperature is expected for the winter. For most sub-regions, the projection is, however highly uncertain (Anders et al., 2014). (ii) There will be complicated changes in precipitation with marked sub-regional differences, but likely there will be a precipitation shift from summer to winter. (iii) The frequency of extreme climatic events and the likeliness of summer arsons (in line with the decreased precipitation) will also increase (Anders et al., 2014; Wesche et al., 2016). In line with these changes, a high species turnover is expected: the cover of drought-tolerant species and the proportion of Mediterranean species are supposed to increase, especially in the Carpathian Basin (Thuiller et al., 2005). Further, the decreased precipitation and increased temperature (with increased rate of arsons) will suppress forest vegetation in many places and increase the area of open habitats, including drought-tolerant grasslands communities (IPCC, 2014).

In addition to the three main drivers listed above, the spread of invasive species forms a fourth threat for grassland biodiversity. In general, grassland habitats are characterized by intermediate levels of invasion and low invasion risk (Pyšek et al., 2010). However, considering grassland types separately, we can see that there are some grassland types of low invasibility (i.e., saline and dry grassland types, rocky grasslands), while others can be characterized by a high risk of invasion (sand grasslands) (Botta-Dukát, 2008). High-intensity management or other forms of disturbance, which cause the degradation of grasslands, can also enhance the risk of invasion. The most dangerous invasive species that can completely change the composition and structure of grasslands are invasive woody species, such as *Robinia pseudoacacia*, *Ailanthus altissima*, *Elaeagnus angustifolia*, *Hippophaë rhamnoides*, *Amorpha fruticosa* and *Acer negundo*. Among herbaceous plants, most dangerous for grasslands are *Asclepias syriaca*, *Heracleum sosnowskyi* (incl. *H. mantegazzianum*), *Phalacroloma annuum*, *Solidago canadensis*, *Conyza canadensis*, *Ambrosia artemisiifolia*, *Grindelia squarrosa*, *Impatiens glandulifera* and *Centaurea diffusa* (Protopopova et al., 2006; GISD, 2017).

Fifth, eutrophication caused by (i) the deposition of aerial nitrogen or (ii) the increase of nutrients by cropland run-off strongly affects the diversity and biomass production in semi-natural grasslands. Nutrient enrichment favors generally the dominant graminoids and increases their cover and biomass production, leading to the decrease of biodiversity and suppression to subordinated species (Bobbink, 1991). The nutrient enrichment also reduces the positive effects of grassland management on biodiversity, especially in nutrient poor grassland types (Habel et al., 2013).

Grassland Management and Restoration

Most grasslands in the Eastern European socio-economic region, similarly to other regions of Europe, were created and/or their biodiversity is maintained by an extensive form of management (Fischer and Wipf, 2002; Dengler et al., 2014). This entails, in most cases, grazing or mowing management. As discussed above, because of intensive agriculture a high proportion of grassland areas in the lowland regions has been converted to croplands; thus, the remaining grasslands have become fragmented and were degraded by the generally intensified use. By contrast, in mountain and foothill areas grasslands with low accessibility or productivity were subject to abandonment, which resulted in a strong shrub and tree encroachment. To conserve grassland biodiversity, it is crucial to maintain extensive management regimes (best represented by a traditional agricultural regimes) to avoid both abandonment and too high land-use intensity. In case of already degraded grassland stands, the change of management intensity is also suggested, but in case of completely destroyed grasslands, recovery by spontaneous succession or technical reclamation methods is recommended.

Eastern Europe belongs to the European regions with the best preserved remnants of the traditional rural culture based on traditional agricultural practices (Oppermann et al., 2012). The positive effects of re-introduction of traditional management by mowing or grazing have been demonstrated in several experiments reported from the region (Galvánek and Lepš, 2008; Valkó et al., 2011, 2012). For pastures, low intensity grazing (i.e., < 0.5 animal units per hectare) is recommended with a strong preference for traditional herding of local cattle breeds or free grazing by wild horses and cattle (Török et al., 2016a,b; Tóth et al., 2017). As re-introduction of traditional management practices is often not feasible or economically sustainable, conservation authorities are seeking alternative management practices, like prescribed burning during the dormant season. Valkó et al. (2013) suggest that prescribed burning with long fire-return periods (i.e., at least three consecutive years without burning) might be a cost-effective and appropriate tool in eliminating accumulated litter and sustaining grassland biodiversity. It was found that for recovery and sustainability of high biodiversity of various taxonomic groups of organisms, a mosaic management (i.e., a spatially and temporally dynamic combination of mown and abandoned grassland patches) would be most appropriate and cost effective. It became evident that not only performance of a single management activity, such as mowing or grazing, but adoption of the whole scheme of traditional management regimes is necessary to maintain the extraordinary grassland diversity of a particular region (Babai et al., 2014). The importance of small-scale, low-intensity farming in conservation of European biodiversity and the maintenance of cultural landscapes has been recognized for decades and led to the development of the High Nature Value (HNV) concept in the 1990s (Keenleyside et al., 2014). Similarly, the role of traditional ecological knowledge (TEK; multi-generational, culturally transmitted knowledge and ways of doing things) is increasingly appreciated nowadays and various recent studies (Babai and Molnár, 2014; Babai et al., 2014) have shown that there are many traditional rural cultures in Eastern Europe that use TEK in their agricultural practices. Its application in grassland conservation has a huge, still not sufficiently used, potential.

When grasslands are completely eliminated due to their transformation into croplands, forests, plantations or urban areas, their recovery can be based on spontaneous succession or technical reclamation (Prach and Hobbs, 2008). Spontaneous succession is increasingly involved in restoration and it is the most promising approach in landscapes where the proportion of target grassland communities is high. There are promising examples reported from central Europe in various grassland habitats (Ruprecht, 2006; Albert et al., 2014; Prach et al., 2015). The most frequently applied technical reclamation methods include sowing of regional seed mixtures and plant material transfer (Török et al., 2011), successfully used in large-scale grassland restoration projects in some countries in the region (Hungary: Lengyel et al., 2012; Czech Republic: Prach et al., 2015), while in the northernmost countries of the region no experience exists so far, or only the first attempts have been made in this direction (Gazenbeek, 2008; Metsoja et al., 2012, 2014; Rūsiņa, 2017).

In Poland, agri-environmental schemes are part of the EU Common Agricultural Policy (CAP) and provide payments to farmers for protecting the environment on their farmland by adopting environment-friendly farming practices or for maintaining habitats and species with certain management practices. Total financial expenditure on agri-environment payments in the EU during 2007–2013 was over 33 billion EUR (38 billion US$; Żmihorski et al., 2016). The effect of CAP payments on biodiversity in Eastern Europe is, however, ambiguous. On the one hand, the CAP-related payments together with other direct payments that are at least partly used for nature conservation (LIFE, LIFE+, structural and rural development funds) increased the available budget for activities related to sustainable grassland management and restoration in Eastern European countries (Mihók et al., 2017). On the other hand, CAP payments enabled in many regions increased intensification of agriculture, leading to a decrease in farmland biodiversity even in the short run (Tryjanowsky et al., 2011; Pe'er et al., 2014; Sutcliffe et al., 2015). One solution would be the extension and refinement of agri-environmental schemes, fine-tuned by considering local perquisites and differences in land management (Wegener et al., 2011; Báldi et al., 2013; Sutcliffe et al., 2015).

Résumé and Future Prospects

The importance of Eastern European grassland biodiversity for the whole of Europe and even in broader context is very high, as grasslands in the region harbor many relict species of high conservation value and a high proportion of the European and Mediterranean steppes are situated there. Evaluation of the monetary value of semi-natural and natural grasslands ecosystems is rather a neglected research area in Eastern Europe. Although they are key contributors of several ecosystem services, their area is either too small and declining or still very common and too familiar for local people, so that they do not recognize the importance of semi-natural grasslands and do not value them. The EU policy is a driving force to elaborate this approach at the national level and to raise public awareness about it, so it is a growing field both in science and nature conservation policy. Restoration of grassland habitats has given a rich ground to scientific research with importance for restoration ecology of grassland habitats globally; however, in many countries the accessibility to the results of grassland restoration projects is relatively poor. Although a conservation system is well established in terms of nature protected areas, the real conservation effort gives only negligible results in several countries because of negative demographic, economic and socio-political drivers. Although there are very promising examples of good practice in conservation and sustainable management, the future trends in conservation of grassland biodiversity in semi-natural and natural grasslands are not very promising in the region.

Acknowledgements

P. Török was supported by NKFIH K 119 225 grant during manuscript preparation. S. Rūsiņa was supported by the University of Latvia grant No. AAP2016/B041. M. Janišová was supported by the grant VEGA 2/0027/15.

Abbreviations

BP = Years Before Present; CAP = Common Agricultural Policy (of the European Union); HNV = High Nature Value; IUCN = International Union for the Conservation of Nature; TEK = Traditional Ecological Knowledge

Appendix: Table showing spatial Extent of Grasslands in Eastern Europe.

Country	Total Area of Permanent Grasslands	Area of High Quality Natural and Semi-grasslands	Area of Grasslands under Protection	Main Types of High Quality Grasslands	Threats	Source
Albania	4,500 km^2	No data	No data; 25 Emerald sites	No data	Overgrazing Overcutting Soil erosion Abandonment	Shundi (2006), Rupa (2013)
Belarus	29,748 km^2	No data, 412 km^2 mapped	No data	Of the total mapped grassland area: Rocky: 0.5% Dry and semi dry: 4.4% Mesic: 24.3% Wet: 70.8%	Abandonment	Witkowski (2006), Maslovski (2007)
Bosnia and Herzegovina	14,100 km^2	No data	No data; 28 Emerald sites	Most meadows are in the lowland, lower hilly area, also on flat areas in mountains regions	Lack of educated management Abandonment	Alibegovic-Grbic (2009)
Bulgaria	13,726 km^2	5,513 km^2	6,080 km^2 of all grasslands in Natura 2000	Mostly extensively managed pastures	Overgrazing near to settlements Uncontrolled burnings Decrease in livestock Abandonment	Hamnett (2006), Stefanova and Kazakova (2013)
Croatia	3,433 km^2	No data	~ 3,000 km^2 in Natura 2000	Traditional hay making Mediterranean grasslands historically used for sheep grazing	Abandonment Overgrazing	Beneš (2013)
Czech Republic	9,800 km^2	2,715 km^2	No data	Alpine: 1.9% Dry and semi dry: 2.8% Mesic: 14.2% Wet: 80.6% Halophytic < 1% Other: 1%	Intensification of agriculture Abandonment Lack of educated management in protected areas Urbanization	Veselý et al. (2011), Hönigová et al. (2012)

Estonia	2,961 km²	~ 1,300 km²	~ 750 km²	Rocky: < 1% Dry and semi dry: 18.2% Mesic: 23.6% Wet: 41.3% Halophytic 16.8%	Abandonment Intensification of agriculture Change in water regime Urbanization	Heinsoo et al. (2010), Talvi and Talvi (2012), EUROSTAT (2016b)
Hungary	~ 10,000 km² (7,840 km² managed and 2,500–3,000 km²) abandoned cropland with grassland vegetation	~ 2,300 km²	~ 68% of HNV protected, 31% protected only by Natura 2000	Steppes, sand steppes and alkaline: 43.3% Rocky: 3% Mesic: 11.4% Wet: 42%	Abandonment Decrease in livestock Invasive species encroachment Change in water regime	Tasi et al. (2014), KSH (2016), Mihók et al. (2017), Z. Molnár (pers. comm.)
Latvia	6,403 km²	~ 500 km²	~ 230 km²	Rocky: 0.05% Dry and semi dry: 6.5% Sandy: 1.9% Mesic: 57.7% Wet: 34.0% Halophytic: 0.4%	Abandonment Intensification of agriculture Change in water regime Conversion to arable land	Auniņš (2013), Rūsiņa (2017)
Lithuania	6,059 km²	744 km²	~ 177 km²	Dry and semi dry: 3.8% Sandy: 0.2% Mesic: 76.1% Wet: 19.9%	Intensification of agriculture Conversion to arable land Abandonment Change in water regime Afforestation	EUROSTAT (2016a), V. Rašomavičius (pers. comm.)
Macedonia	5,900 km²	~ 650 km²	No data; 35 Emerald sites	Pastures in Macedonia are mainly natural and semi-natural, divided into summer and winter pastures of low productivity or low quality High mountain pastures in western Macedonia are traditionally used for sheep grazing in the summer	Conversion to arable land Weed infestation Lack of access roads to the sheepfolds and pens Poor water supply Pasture abandonment Soil degradation	Kratovalieva and Milcevska (2013)

Table contd. ...

...Table contd.

Country	Total Area of Permanent Grasslands	Area of High Quality Natural and Semi-grasslands	Area of Grasslands under Protection	Main Types of High Quality Grasslands	Threats	Source
Moldova	3,510 km^2	No data meadows 21 km^2, pastures 3,489 km^2	No data Steppe: 11.3% of the country	Steppes Semi-dry grasslands Mesic grasslands	Abandonment Overgrazing Conversion to arable land	Anon. (2009b), Shabanova et al. (2014)
Montenegro	4,600 km^2	No data	No data, 32 Emerald sites	Traditionally, domestic production of meat and milk are far below the consumption Extensive or semi-extensive farming prevails	Abandonment Overgrazing	Dubljevic (2009)
Poland	39,390 km^2	No data	~ 3,783 km^2 in Natura 2000	Alpine Rocky Sandy (boreal zone) Semi-dry Mesic Wet grasslands	Abandonment	CBD National Report of Poland (2014)
Romania	45,319 km^2	4,991 km^2 mapped	No data	Mapped grasslands are Steppe: 34.7% Rocky: 0.9% Dry and semi-dry: 4.7% Mesic: 44.2% Wet: 15.4%	Abandonment Replacement of cattle grazing by sheep	Sârbu et al. (2009), Veen and Metzger (2009), EUROSTAT (2016c)
Slovakia	8,450 km^2 (2003)	3,200 km^2 (2002)	Of the HNV grasslands between 1,500 and 2,000 km^2 are covered	Mesic: 62% Wet: 15% Dry: 8% Alpine: 4% (11% cannot be classified in lack of characteristic species)	Intensification of agriculture Abandonment Lack of educated management in protected areas Urbanization	Šeffer et al. (2002)

Slovenia	4,000 km^2	No data	~ 2,000 km^2 in Natura 2000	Humid, marshy grasslands grazed by cattle or mown Hay meadows and pastures typical of karst areas Extensive pastures in hilly areas grazed by cattle and sheep Shepherded summer grazing on alpine pastures	Intensification of agriculture Tourism Uncontrolled grazing Intensification of agriculture	Seliškar (1996)
Serbia (with Kosovo)	14,245 km^2	No data	No data; 61 Emerald sites	Livestock-raising region includes mountain areas of semi-natural and natural grasslands: (a) Crop-farming and livestock-raising region, including lowlands and flat areas in river valleys; (b) Mixed farming region: Hilly land with different climates and soils – livestock production and grazing	Overgrazing Conversion to arable land Abandonment	Stošić and Lazarević (2009), Djordjevic-Milošević (2013)
Ukraine	78,400 km^2	No data	No data	All types	Abandonment (in some regions) Overgrazing (in some regions) Conversion to arable land Afforestation	Bogovin (2006), Burkovsky et al. (2013)

References

Albert, Á.-J., A. Kelemen, O. Valkó, T. Miglécz, A. Csecserits, T. Rédei, B. Deák, B. Tóthmérész and P. Török. 2014. Secondary succession in sandy old fields: A promising example of spontaneous grassland recovery. Appl. Veg. Sci. 17: 214–224.

Alibegovic-Grbic, S. 2006. Country Pasture/Forage Resource Profiles Bosna and Herzogovina. FAO, Rome.

Anders, I., J. Stagl, I. Auer and D. Pavlik. 2014. Climate change in Central and Eastern Europe. Adv. Global Change Res. 58: 17–30.

Anon. 1974. Latvijas PSR Arheoloģija [Archaeology of Latvian SSR.] Zinātne, Rīga.

Anon. 2009b. The Fourth National Report on Biological Diversity. Republic of Moldova, Chisinau.

Auniņš, A. (ed.). 2013. European Union Protected Habitats in Latvia. Interpretation Manual. Latvian Fund for Nature & Ministry of Environmental Protection and Regional Development, Riga.

Babai, D. and Z. Molnár. 2014. Small-scale traditional management of highly species-rich grasslands in the Carpathians. Agric. Ecosyst. Environ. 182: 123–130.

Babai, D., Á. Molnár and Z. Molnár. 2014. Ahogy Gondozza, Úgy Geszi Gasznát Hagyományos Ökológiai Tudás És Gazdálkodás Gyimesben. [Traditional Ecological Knowledge and Land Use in Gyimes (Eastern Carpathians).] MTA Bölcsészettudományi Kutatóközpont Néprajztudományi Intézet, Budapest & MTA Ökológiai Kutatóközpont Ökológiai és Botanikai Intézet, Vácrátót.

Báldi, A., P. Batáry and D. Kleijn. 2013. Effects of grazing and biogeographic regions on grassland biodiversity in Hungary—Analysing assemblages of 1200 species. Agric. Ecosyst. Environ. 166: 28–34.

Balevičienė, J., A. Balevičius, O. Grigaitė, D. Patalauskaitė, V. Rašomavičius, Z. Sinkevičienė and J. Stankevičiūtė. 2000. Lietuvos Raudonoji Knyga. Augalų Bendrijos [Lithuanian Red Data Book. Plant Communities]. Botanikos instituto leidykla, Vilnius.

Barker, G. 1985. Prehistoric Farming in Europe. Cambridge University Press, London.

Benayas, J.M., A.C. Newton, A. Diaz and J.M. Bullock. 2009. Enhancement of biodiversity and ecosystem services by ecological restoration: A meta-analysis. Science 325: 1121–1124.

Beneš, I. 2013. Common Grazing in Croatia. Report from the Best Practices for Sustainable Use of Common Grasslands in the western Balkans and Europe, SE Europe Round Table of Southeast Europe HNV Farming Network, 15 April, 2013, Sofia, Bulgaria. URL: http://see.efncp.org/networking/events/2013/20130415.

Bezák, P. and M. Bezáková. 2014. Landscape capacity for ecosystem services provision based on expert knowledge and public perception (case study from the north-west Slovakia). Ekológia (Bratislava) 33: 344–353.

Bobbink, R. 1991. Effects of nutrient enrichment in Dutch chalk grassland. J. Appl. Ecol. 28: 28–41.

Bogovin, A.V. 2006. Country Pasture/Forage Resource Profiles: Ukraine. FAO, Rome.

Boreiko, V., I. Parnikoza and A. Burkovskiy. 2013. Absolute 'zapovednost'—A concept of wildlife protection for the 21st century. Bull. Eur. Dry Grassl. Group 19/20: 25–30.

Botta-Dukát, Z. 2008. Invasion of alien species to Hungarian (semi-) natural habitats. Acta Bot. Hung. 50(Suppl.): 219–227.

Bullock, J.M., R.G. Jefferson, T.H. Blackstock, R.J. Pakeman, B.A. Emmett, R.F. Pywell, J.P. Grime and J. Silvertown. 2011. Semi-natural grasslands. pp. 161–196. *In*: The UK National Ecosystem Assessment Technical Report. UNEP-WCMC, Cambridge.

Burkovsky, O.P., O.V. Vasyliuk, A.V. Yena, A.A. Kuzemko, Y.I. Movchan, I.I. Moysienko and I.P. Sirenko. 2013. Ostanni Stepy Ukrainy: Buty Chy Ne Buty [Last Steppes of Ukraine: To Be or Not to Be]. Geoprynt, Kyiv.

CBD National Report of Belarus. 2014. Convention on Biological Diversity. Republic of Belarus. Fifth National Report [in Russian]. Ministry of Natural Resources and Environmental Protection of the Republic of Belarus, Minsk. URL: https://www.cbd.int/doc/world/by/by-nr-05-ru.pdf.

CBD National Report of Poland. 2014. Fifth National Report on the Implementation of the Convention on Biological Diversity. Poland. Warsaw. URL: https://www.cbd.int/doc/world/pl/pl-nr-05-en.pdf.

Chytrý, M., T. Dražil, M. Hájek, V. Kalníková, Z. Preislerová, J. Šibík, K. Ujházy, I. Axmanová, D. Bernátová, (...) and M. Vymazalová. 2015. The most species-rich plant communities in the Czech Republic and Slovakia (with new world records). Preslia 87: 217–278.

Davoudi, S., M. Wishardt and I. Strange. 2010. The ageing of Europe: Demographic scenarios of Europe's futures. Futures 42: 794–803.

Dengler, J., M. Janišová, P. Török and C. Wellstein. 2014. Biodiversity of Palaearctic grasslands: A synthesis. Agric. Ecosyst. Environ. 182: 1–14.

Dengler, J., I. Biurrun, I. Apostolova, E. Baumann, T. Becker, A. Berastegi, S. Boch, L. Cancellieri, I. Dembicz, (...) and F. Weiser. 2016. Scale-dependent plant diversity in Palaearctic grasslands: a comparative overview. Bull. Eurasian Dry Grassl. Group 31: 12–26.

Didukh, Y.P. (ed.). 2009. Green Book of Ukraine. Alterpres, Kyiv.

Didukh, Y.P. (ed.). 2009. Red Data Book of Ukraine. Plant Kingdom. Globalkonsalting, Kyiv.

Djordjevic-Milošević, S. 2013. Use of Grasslands in the Republic of Serbia. Report from the Best Practices for Sustainable Use of Common Grasslands in the Western Balkans and Europe, SE Europe Round Table of Southeast Europe HNV Farming Network, 15 April, 2013, Sofia, Bulgaria. URL: http://see.efncp.org/networking/events/2013/20130415.

Dubljević, R. 2009. Country Pasture/Forage Resource Profiles. Montenegro. FAO, Rome.

Emanuelsson, U. 2009. The Rural Landscapes of Europe—How Man has Shaped European Nature. Forskningsrådet Formas, Stockholm.

EUROSTAT. 2016a. Agricultural Census in Lithuania. URL: http://ec.europa.eu/eurostat/statistics-explained/index.php/Agricultural_census_in_Lithuania#Land_use.

EUROSTAT. 2016b. Agricultural Census in Estonia. URL: http://ec.europa.eu/eurostat/statistics-explained/index.php/Agricultural_census_in_Estonia#Land_use.

EUROSTAT. 2016c. Agricultural Census in Romania. URL: http://ec.europa.eu/eurostat/statistics-explained/index.php/Agricultural_census_in_Romania.

Fischer, M. and S. Wipf. 2002. Effect of low-intensity grazing on the species-rich vegetation of traditionally mown subalpine meadows. Biol. Conserv. 104: 1–11.

Galvánek, M. and J. Lepš. 2008. Changes of species richness pattern in mountain grasslands: Abandonment vs. restoration. Biodivers. Conserv. 17: 3241–3253.

Gavrilova, G. 2003. Introduction. pp. 12–17. *In*: G. Andrušaitis (ed.). Red Data Book of Latvia. Rare and Threatened Plants and Animals. Vol. 3: Vascular Plants. Institute of Biology, Riga.

Gavrilova, N.S. and L.A. Gavrilov. 2009. Rapidly aging populations: Russia/Eastern Europe. pp. 113–131. *In*: P. Uhlenberg (ed.). International Handbook of Population Aging. New York, Springer.

Gazenbeek, A. 2008. Boreālo zālāju atjaunošana un regulārā apsaimniekošana: LIFE-Daba projektu pieredze [Restoration and recurring management of boreal grasslands, seen through the lens of LIFE-Nature projects]. pp. 9–28. *In*: A. Auniņš (ed.). Aktuālā Savvaļas Sugu un Biotpu Apsaimniekošanas Problemātika Latvijā. Latvijas Universitāte, Rīga.

[GISD] Global Invasive Species Database. 2017. Global Invasive Species Database. URL: http://www.issg.org/database.

Habel, J.C., J. Dengler, M. Janišová, P. Török, C. Wellstein and M. Wiezik. 2013. European grassland ecosystems: threatened hotspots of biodiversity. Biodivers. Conserv. 22: 2131–2138.

Hamnett, R. 2006. Country Pasture/Forage Resource Profiles Bulgaria. FAO, Rome.

Harrison, P.A., M. Vandewalle, M.T. Sykes, P.M. Berry, R. Bugter, F. de Bello, C.K. Feld, U. Grandin, R. Harrington, (...) and M. Zobel. 2010. Identifying and prioritising services in European terrestrial and freshwater ecosystems. Biodivers. Conserv. 19: 2791–2821.

Heinsoo, K., I. Melts, M. Sammul and B. Holm. 2010. The potential of Estonian semi-natural grasslands for bioenergy production. Agric. Ecosyst. Environ. 137: 86–92.

Helm, A., T. Aavik, N. Ingerpuu, M. Ivask, R. Karise, L. Kasari, T. Kupper, R. Marja, M. Meriste, (...) and A. Tiitsaar. 2016. Monitoring changes in biodiversity patterns and in landscape structure during the large-scale grassland restoration in Estonia. p. 241. *In*: J. Kollmann and M. Hermann (eds.). Best Practice in Restoration. The 10th European Conference on Ecological Restoration. Abstract Volume. Chair of Restoration Ecology, Technische Universität München, Freising.

Hensgen, F., L. Bühle, I. Donnison, M. Fraser, J. Vale, J. Corton, K. Heinsoo, I. Melts, H. Herzon and M. Mikk. 2007. Farmers' perceptions of biodiversity and their willingness to enhance it through agri-environment schemes: A comparative study from Estonia and Finland. J. Nat. Conserv. 15: 10–25.

Hopkins, A. and B. Holz. 2006. Grassland for agriculture and nature conservation: Production, quality and multi-functionality. Agron. Res. 4: 3–20.

Hönigová, I., D. Vačkář, E. Lorencová, J. Melichar, M. Götz, G. Sonderegger, V. Oušková, M. Hošek and K. Hobot. 2012. Survey on Grassland Ecosystem Services. Report to the EEA—European Topic Centre on Biological Diversity. Nature Conservation Agency of the Czech Republic, Prague.

IPCC. 2014. Climate Change 2014: Impacts, Adaptation, and Vulnerability. Part B: Regional Aspects. Contribution of Working Group II to the Fifth Assessment Report of the Intergovernmental Panel on Climate Change. Cambridge University Press, Cambridge.

Jepsen, M.R., T. Kuemmerle, D. Müller, K. Erb, P.H. Verburg, H. Haberl, J.P. Vesterager, M. Andrič, M. Antrop, (...) and A. Reenberga. 2015. Transitions in European land-management regimes between 1800 and 2010. Land Use Policy 49: 53–64.

Kaltenborn, B.P., O.I. Vistad and S. Stanaitis. 2002. National parks in Lithuania: Old environment in a new democracy. Nor. J. Geogr. 56: 32–40.

Keenleyside, C., G. Beaufoy, G. Tucker and G. Jones. 2014. High Nature Value Farming throughout EU-27 and its Financial Support under the CAP. Report Prepared for DG Environment, Contract ENV B.1/ETU/2012/0035. Institute for European Environmental Policy, London.

Kelemen, E., G. Nguyen, T. Gomiero, E. Kovács, J.P. Choisis, N. Choisis, M.G. Paoletti, L. Podmaniczky, J. Ryschawy, (...) and K. Balázs. 2013. Farmers' perceptions of biodiversity: Lessons from a discourse-based deliberative valuation study. Land Use Policy 35: 318–328.

Klein, L. (ed.). 2008. Diversity of Nature in Estonia. Estonian Nature Conservation in 2007. Estonian Environment Information Centre, Tallin.

Kratovalieva, S. and T. Milcevska. 2013. Common Grazing in Macedonia. Report from the Best Practices for Sustainable Use of Common Grasslands in the Western Balkans and Europe, SE Europe Round Table of Southeast Europe HNV Farming Network, 15 April, 2013, Sofia, Bulgaria. URL: http://see.efncp.org/networking/events/2013/20130415.

[KSH] Központi Statisztikai Hivatal. 2016. 4.1.4. Földhasználat Művelési Ágak és Gazdaságcsoportok Szerint. URL: http://www.ksh.hu/docs/hun/xstadat/xstadat_eves/i_omf001a.html.

Leah, T. 2016. Grasslands of Moldova: Quality status, vulnerability to anthropogenic factors and adaptation measures. Sci. Pap., Ser. A Agron. 59: 100–105.

Lengyel, S., K. Varga, B. Kosztyi, L. Lontay, E. Déri, P. Török and B. Tóthmérész. 2012. Grassland restoration to conserve landscape-level biodiversity: A synthesis of early results from a large-scale project. Appl. Veg. Sci. 15: 264–276.

Lepasaar, H. and Ü. Ehrlich. 2015. Non-market value of Estonian seminatural grasslands: a contingent valuation study. Estonian Discuss. Econ. Policy 23: 135–141.

Löbel, S., J. Dengler and C. Hobohm. 2006. Species richness of vascular plants, bryophytes and lichens in drygrasslands: the effects of environment, landscape structure and competition. Folia Geobot. 41: 377–393.

Ložek, V. 2008. Vývoj v době poledové. pp. 24–28. *In*: I. Jongepierová (ed.). Louky Bílých Karpat [Grasslands of the White Carpathian Mountains]. ZO ČSOP Bílé Karpaty, Veselí nad Moravou.

Maslovski, O. (ed.). 2007. Grassland Inventory of Belarus. Belarus Botanical Society and Royal Duch Society for Nature Conservation, Minsk, Belarus.

Melluma, A. 1994. Metamorphoses of latvian landscapes during fifty years of soviet rule. GeoJournal 33: 55–62.

Melts, I. 2014. Biomass from Semi-natural Grasslands for Bioenergy. Ph.D. Thesis in Environmental Conservation, Estonian University of Life Sciences, Tartu.

Metsoja, J.-A., L. Neuenkamp, S. Pihu, K. Vellak, J.M. Kalwij and M. Zobel. 2012. Restoration of flooded meadows in Estonia—Vegetation changes and management indicators. Appl. Veg. Sci. 15: 231–244.

Metsoja, J.-A., L. Neuenkamp and M. Zobel. 2014. Seed bank and its restoration potential in managed and abandoned flooded meadows. Appl. Veg. Sci. 17: 262–273.

Metzger, M.J., G.H. Bunce, R.H.G. Jongman, C.A. Mücher and J.W. Watkins. 2005. A climatic stratification of the environment of Europe. Global Ecol. Biogeogr. 14: 549–563.

Mihók, B., M. Biró, Z. Molnár, E. Kovács, J. Bölöni, T. Erős, T. Standovár, P. Török, G. Csorba, (...) and A. Báldi. 2017. Biodiversity on the waves of history: Conservation in a changing social and institutional environment in Hungary, a post-soviet EU member state. Biol. Conserv. 211: 67–75.

Mucina, L., H. Bültmann, K. Dierßen, J.-P. Theurillat, T. Raus, A. Čarni, K. Šumberová, W. Willner, J. Dengler, (...) and L. Tichý. 2016. Vegetation of Europe: Hierarchical floristic classification system of plant, bryophyte, lichen, and algal communities. Appl. Veg. Sci. 19, Suppl. 1: 1–264.

Oppermann, R., G. Beaufoy and G. Jones (eds.). 2012. High Nature Value Farming in Europe. 35 European Countries—Experiences and Perspectives. Verlag Regionalkultur, Ubstadt-Weiher.

Paal, J. 1998. Rare and threatened plant communities of Estonia. Biodivers. Conserv. 7: 1027–1049.

Peel, M.C., B.L. Finlayson and T.A. McMahon. 2007. Updated world map of the Köppen-Geiger climate classification. Hydrol. Earth Syst. Sci. 11: 1633–1644.

Pe'er, G., L.V. Dicks, P. Visconti, R. Arlettaz, A. Báldi, T.G. Benton, S. Collins, M. Dieterich, R.D. Gregory, (...) and K. Henle. 2014. EU agricultural reform fails on biodiversity. Science 344: 1090–1092.

Poschlod, P. 2015. Geschichte der Kulturlandschaft. Ulmer, Stuttgart.

Prach, K. and R.J. Hobbs. 2008. Spontaneous succession versus technical reclamation in the restoration of disturbed sites. Restor. Ecol. 16: 363–366.

Prach, K., K. Fajmon, I. Jongepierová and K. Řehounková. 2015. Landscape context in colonization of restored dry grasslands by target species. Appl. Veg. Sci. 18: 181–189.

Price, D. (ed.). 2000. Europe's First Farmers. Cambridge University Press, Cambridge.

Protopopova, V.V., M.V. Shevera and S.L. Mosyakin. 2006. Deliberate and unintentional introduction of invasive weeds: A case study of the alien flora of Ukraine. Euphytica 148: 17–33.

Pullin, A.S., A. Báldi, O.E. Can, M. Dieterich, V. Kati, B. Livoreil, G. Lövei, B. Mihók, O. Nevin, (...) and I. Sousa-Pinto. 2009. Conservation focus on Europe: Major conservation policy issues that need to be informed by Conservation Science. Conserv. Biol. 23: 818–824.

Pyšek, P., M. Chytrý and V. Jarošík. 2010. Habitats and land use as determinants of plant invasions in the temperate zone of Europe. pp. 66–79. *In*: C. Perrings, H. Mooney and M. Williamson (eds.). Bioinvasions and Globalization. Ecology, Economics, Management, and Policy. Oxford University Press, Oxford.

Rabinovič, M.G., A.O. Viires, I.A. Leinesare and V.I. Morkunas (eds.). 1985. Historical-Ethnographic Atlas of the Baltics. Vol. 1. Agriculture [in Russian]. Mokslas, Vilnius.

Rodwell, J.S., J.H.J. Schaminée, L. Mucina, S. Pignatti, J. Dring and D. Moss. 2002. The Diversity of European Vegetation—An Overview of Phytosociological Alliances and their Relationships to EUNIS Habitats. National Reference Centre for Agriculture, Nature and Fisheries [Report No. EC-LNV 2002(054)], Wageningen.

Roleček, J., I. Čornej and A.I. Tokarjuk. 2014. Understanding the extreme species richness of semi-dry grasslands in east-central Europe: A comparative approach. Preslia 86: 13–34.

Rupa, M. 2013. Example from Albania. Report from the Best Practices for Sustainable Use of Common Grasslands in the Western Balkans and Europe, SE Europe Round Table of Southeast Europe HNV Farming Network, 15 April, 2013, Sofia, Bulgaria. URL: http://see.efncp.org/networking/events/2013/20130415.

Ruprecht, E. 2006. Successfully recovered grassland: A promising example from Romanian old-fields. Restor. Ecol. 14: 473–480.

Rūsiņa, S. 2008. Dabisko zālāju atjaunošanas pasākumu ietekme uz veģetāciju aizsargājamo ainavu apvidū Ziemeļgauja [Influence of semi-natural grassland restoration on the vegetation in the Protected Landscape Area Northern Gauja]. pp. 57–72. *In*: A. Auniņš (ed.). Aktuālā Savvaļas Sugu un Biotopu Apsaimniekošanas Problemātika Latvijā. Latvijas Universitāte, Rīga.

Rūsiņa, S. 2016. Latvijas Lauku Attīstības Programmas 2007–2013. Gadam Ietekme uz Bioloģisko Daudzveidību: Atbalstīto ES Nozīmes Aizsargājamo Zālāju Biotopu Botāniskā Daudzveidība [The influence of Latvian Rural Development Programme 2007–2013 on Biological Diversity: Botanical Diversity of Supported EU Importance Grassland Habitat Areas]. Report for Ex-post evaluation of Latvian Rural Development Programme 2007–2013, Riga.

Rūsiņa, S. (ed.). 2017. Protected Habitat Management Guidelines for Latvia. Vol. 3: Semi-natural Grasslands. Nature Conservation Agency, Sigulda.

Rūsiņa, S. and A. Auniņš. 2017. Biodiversity—The guarantee of grassland ecosystem services. pp. 38–39. *In*: S. Rūsiņa (ed.). Protected Habitat Management Guidelines for Latvia. Vol. 3: Semi-natural Grasslands. Nature Conservation Agency, Sigulda.

Ružičková, H. and H. Kalivoda. 2007. Kvetnaté Lúky—Prírodné Bohatstvo Slovenska. VEDA, Bratislava.

Sârbu, A., G. Negrean and I. Sârbu. 2009. The grasslands of the Dobrogea, Romania. pp. 219–225. *In*: P. Veen, R. Jefferson, J. de Smidt and J. van der Straaten (eds.). Grasslands in Europe of High Nature Value. KNNV Publishing, Zeist.

Šeffer, J., R. Lasák, D. Galvánek and V. Stanová. 2002. Grasslands of Slovakia—Final Report on National Grassland Inventory 1998–2002. Daphne, Bratislava.

Seliškar, A. 1996. Travišćna in močvirna vegetacija. pp. 99–106. *In*: J. Gregori, A. Martinčič, K. Tarman, O. Urbanc-Berčič, D. Tome and M. Zupančič (eds.). Narava Slovenije, Stanje in Perspektive. Društvo ekologov Slovenije, Ljubljana.

Shabanova, G.A., T.D. Izverskaya and V.S. Gendov. 2014. Flora i Rastitel'nost' Budzhatskikh Stepey Respubliki Moldova [Flora and Vegetation of the Budzhak Steppe of the Republic of Moldova]. Eco-Tiras, Chisinau.

Shundi, A. 2006. Country Pasture/Forage Resource Profiles Albania. FAO, Rome.

Stefanova, V. and Y. Kazakova. 2013. Common Grazing in Bulgaria. Report from the Best Practices for Sustainable Use of Common Grasslands in the Western Balkans and Europe, SE Europe Round Table of Southeast Europe HNV Farming Network, 15 April, 2013, Sofia, Bulgaria. URL: http://see.efncp.org/networking/events/2013/20130415/.

Stošić, M. and D. Lazarević. 2009. Country Pasture/Forage Resource Profiles Serbia. FAO, Rome.

Strazdiņa, B., D. Jakovels and A. Auziņš. 2015. Zālāju Biomasas Resursi Siguldas un Ludzas Novadā. Ziņojums [Resources of Grassland Biomass in Sigulda and Ludza Municipalities. Report]. LIFE Grassservice [No. LIFE12BIO/LV/001130], Riga.

Sutcliffe, L.M.E., P. Batáry, U. Kormann, A. Báldi, L.V. Dicks, I. Herzon, D. Kleijn, P. Tryjanowski, I. Apostolova, (...) and T. Tscharntke. 2015. Harnessing the biodiversity value of central and Eastern European farmland. Divers. Distrib. 21: 722–730.

Talvi, T. and T. Talvi. 2012. Semi-Natural Communities. Preservation and Management. Ministry of Agriculture, Viidumäe - Tallinn.

Tasi, J., M. Bajnok, A. Halász, F. Szabó, Z. Harkányiné Székely and V. Láng. 2014. Magyarországi komplex gyepgazdálkodási adatbázis létrehozásának els lépései és eredményei. Gyepgazdálkodási Közlemények 2014: 57–64.

Thuiller, W., S. Lavorel, M.B. Araujo, M.T. Sykes and I.C. Prentice. 2005. Climate change threats to plant diversity in Europe. Proc. Natl. Acad. Sci. USA 102: 8245–8250.

Török, P., E. Vida, B. Deák, S. Lengyel and B. Tóthmérész. 2011. Grassland restoration on former croplands in Europe: An assessment of applicability of techniques and costs. Biodivers. Conserv. 20: 2311–2332.

Török, P., N. Hölzel, R. van Diggelen and S. Tischew. 2016a. Grazing in European open landscapes: How to reconcile sustainable land management and biodiversity conservation? Agric. Ecosyst. Environ. 234: 1–4.

Török, P., O. Valkó, B. Deák, A. Kelemen, E. Tóth and B. Tóthmérész. 2016b. Managing for species composition or diversity? Pastoral and free grazing systems of alkali grasslands. Agric. Ecosyst. Environ. 234: 23–30.

Tóth, E., B. Deák, O. Valkó, A. Kelemen, T. Miglécz, B. Tóthmérész and P. Török. 2017. Livestock type is more crucial than grazing intensity: Traditional cattle and sheep grazing in short-grass steppes. Land Degrad. Dev. DOI: 10.1002/ldr.2514 (in press).

Tryjanowski, P., T. Hartel, A. Báldi, P. Szymański, M. Tobolka, I. Herzon, A. Goławski, M. Konvička, M. Hromada, (...) and K. Kujawa. 2011. Conservation of farmland birds faces different challenges in western and central-Eastern Europe. Acta Ornithol. 46: 1–12.

Turtureanu, P.D., S. Palpurina, T. Becker, C. Dolnik, E. Ruprecht, L.M.E. Sutcliffe, A. Szabó and J. Dengler. 2014. Scale- and taxon-dependent biodiversity patterns of dry grassland vegetation in Transylvania (Romania). Agric. Ecosyst. Environ. 182: 15–24.

Valkó, O., P. Török, B. Tóthmérész and G. Matus. 2011. Restoration potential in seed banks of acidic fen and dry-mesophilous meadows: Can restoration be based on local seed banks? Restor. Ecol. 19: 9–15.

Valkó, O., P. Török, G. Matus and B. Tóthmérész. 2012. Is regular mowing the most appropriate and cost-effective management maintaining diversity and biomass of target forbs in mountain hay meadows? Flora 207: 303–309.

Valkó, O., P. Török, B. Deák and B. Tóthmérész. 2013. Prospects and limitations of prescribed burning as a management tool in European grasslands. Basic Appl. Ecol. 15: 26–33.

Vanwambeke, S.O., P. Meyfroidt and O. Nikodemus. 2012. From USSR to EU: 20 years of rural landscape changes in Vidzeme, Latvia. Landsc. Urban Plan 105: 241–249.

Veen, P. and M. Metzger. 2009. Lowland grasslands and climate in Central Europe. pp. 43–51. *In*: P. Veen, R. Jefferson, J. de Smidt and J. van der Straaten (eds.). Grasslands in Europe of High Nature Value. KNNV Publishing, Zeist.

Veselý, P., J. Skládanka and Z. Havlíc□ek. 2011. Metodika hodnocení kvality píce travních porostu□ v chráne□ných krajinných oblastech. Mendelova univerzita v Brne□, Brno.

Walter, H. and S.-W. Breckle. 1991. Ökologische Grundlagen in globaler Sicht. Fischer, Stuttgart.

Wegener, S., K. Labar, M. Petrick, D. Marquardt, I. Theesfeld and G. Buchenrieder. 2011. Administering the common agricultural policy in Bulgaria and Romania: obstacles to accountability and administrative capacity. Int. Rev. Adm. Sci. 77: 583–608.

Wesche, K., D. Ambarlı, J. Kamp, P. Török, J. Treiber and J. Dengler. 2016. The Palaearctic steppe biome: A new synthesis. Biodivers. Conserv. 25: 2197–2231.

Wilson, J.B., R.K. Peet, J. Dengler and M. Pärtel. 2012. Plant species richness: The world records. J. Veg. Sci. 23: 796–802.

Witkowski, H. 2006. Country Pasture/Forage Resource Profiles: Belarus. FAO, Rome. URL: http://www.fao.org/ag/agp/agpc/doc/counprof/belarus/belarus.htm.

World Resources Institute. 2005. Millennium Ecosystem Assessment—Ecosystems and Human Well-being: Biodiversity Synthesis. Island Press, Washington, DC.

Young, J., C. Richards, A. Fischer, L. Halada, T. Kull, A. Kuzniar, U. Tartes, Y. Uzunov and A. Watt. 2007. Conflicts between biodiversity conservation and human activities in the central and Eastern European countries. Ambio 36: 545–550.

Żmihorski, M., D. Kotowska, Å. Berg and T. Pärt. 2016. Evaluating conservation tools in Polish grasslands: The occurrence of birds in relation to agri-environment schemes and NATURA 2000 areas. Biol. Conserv. 194: 150–157.

5

Grasslands of the Mediterranean Basin and the Middle East and their Management

Didem Ambarlı,[1,2,*] *Michael Vrahnakis*,[3] *Sabina Burrascano*,[4] *Alireza Naqinezhad*[5] and *Manuel Pulido Fernández*[6]

Introduction

Grasslands of the Mediterranean Basin (M) and the Middle East (ME) are probably the earliest examples of species-rich semi-natural grasslands as these two regions together (M&ME) represent birthplaces of farming and settled life as well as a giant biodiversity hotspot. This chapter focuses on the geographical area that stretches from the Atlantic Ocean in the west to Afghanistan in the east, from southern France and the Italian Po Plain in the north to the coastal belt of North Africa (north of the Sahara) in the south (Fig. 5.1). This area spreads across 21 countries, fully or partly. According to a global study, M&ME grasslands represent the 'Mediterranean Basin Dry Grassland' division within the 'Mediterranean Scrub, Grassland and Forb Meadow' formation (Dixon et al., 2014). FAO (2014) gives a general idea about the share of grasslands by countries with numbers on the permanent meadows and pastures, land used for five years or more to grow herbaceous forage crops, either cultivated or growing wild (Table 5.1): Algeria, Afghanistan and Iran have the largest share, followed by Morocco and Turkey.

The variety of climate types, orography and position with respect to floristic regions (Mediterranean, Euro-Siberian, Irano-Turanian and Saharo-Sindian) play an important role in the diversification and distribution of M&ME grasslands. According to Köppen's

[1] Faculty of Agriculture and Natural Sciences, Düzce University, Konuralp, Düzce, 81620, Turkey.
[2] Department of Ecology and Ecosystem Management, Technische Universität München, 85350 Freising-Weihenstephan, Germany
[3] Department of Forestry and Management of Natural Environment, Thessaly University of Applied Sciences, Mavromihali str., Karditsa, 43110, Greece.
[4] Department of Environmental Biology, Sapienza University, Rome, 00185, Italy.
[5] Department of Biology, Faculty of Basic Sciences, University of Mazandaran, Babolsar, P.O. Box 47416-95447, Mazandaran, Iran.
[6] Facultad de Filosofía y Letras, Universidad de Extremadura, Cáceres, 10071, Spain.
Emails: mvrahnak@teilar.gr; sabina.burrascano@uniroma1.it; a.naqinezhad@umz.ac.ir; mapulidof@unex.es
* Corresponding author: didem.ambarlı@gmail.com

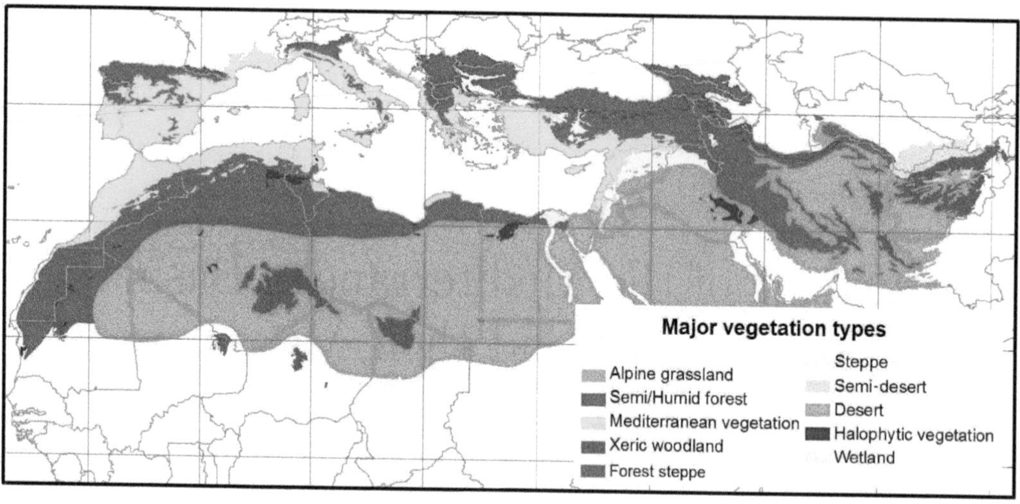

Fig. 5.1 Approximate delimitation of the region covered in this chapter ('Mediterranean Basin and the Middle East') and its climax vegetation types, adapted from the 'Terrestrial Ecoregions of the World' (Olson et al., 2001). Please note that some of the alpine grasslands of the region are missing in the original study, e.g., in Greece, Turkey and Iran.

system (Geiger, 1961), a high variety of climate types occurs in the region—Mediterranean along the sea coast and lowlands; arid steppe climate in rain shadows of Turkey, Spain, and transition to deserts in Levant and North Africa; temperate climate in the North of Spain, Greece, coastal Turkey and along Italian mountain chains; cold continental climate from east Anatolia to north of Iran and Afghanistan. The Mediterranean climate, dominant across the region, is a warm-temperate climate with annual average temperatures of 14–18°C, with monthly averages over 0°C and annual rainfall between circa 400–1,200 mm. Plants are exposed to severe drought stress for a period of two to five months in summer (Pignatti, 2003) when high temperatures are coupled with scarce precipitation mostly due to Bermuda-Azores High.

The complex geomorphology of the region, which supports remarkable grassland diversity, was formed by drastic tectonic and volcanic activities. Huge mountain ranges occur along the northern border (Alps) and within the region: Apennines (Italy), Sierra Nevada (Spain), Pindus and Rhodope (Greece) and North Anatolian Mountains, Aegean-cost mountains, Taurus (Turkey), Zagros and Alborz (Iran), Hindu Kush (Afghanistan) and Atlas of North Africa. The region is also characterized by nearly 5,000 islands, the largest being Cyprus, Crete, Sicily, Corsica and Sardinia. Several plateaus and plains are covered with steppe-like dry grasslands, such as East Anatolian high plateau and Mesopotamia. Also, substrates are extremely diverse. In the north-west, siliceous (slate and granite) and calcareous (limestone and dolomite) bedrocks are widespread. Tertiary sediments occur in the Iberian Peninsula and metamorphic and igneous substrates can be found in the southern Italian Peninsula (IGME, 1995). Hard limestone and flysch in the Pindus and schists and gneiss in the Rhodope massifs (Higgins and Higgins, 1996). Atlas ranges are based on carbonate sediments with basal units formed by continental conglomerates and sandstones and minor lacustrine limestone (Nedjraoui, 2006; Ellero et al., 2012). The Mediterranean part of Algeria and Egypt also displays sedimentary bedrocks (Elakkari, 2005; Siddall, 2013). Turkey has a great diversity of igneous, sedimentary and metamorphic rocks. The ME is covered with sedimentary rocks, mainly limestone and chalk.

Table 5.1 Area statistics for countries of the Mediterranean Basin and the Middle East (M&ME) and size of permanent meadows and pastures according to FAOstat (2014). (Note that some figures of the Table related to per cent area of the countries or pastures are out of date. The figures for France and Italy refer to the whole country despite only part of the country is included in the M&ME region. Palestine is not included as no statistics were available.)

Country	Area (km²)	Permanent Meadows and Pastures (PMP) (km²)	Fraction of Country Territory (%)	Fraction of Total PMP (%)
Afghanistan	652,860	300,000	46.0	15.9
Algeria	2,381,740	329,918	13.9	17.5
Cyprus	9,250	18	0.2	< 0.1
Egypt	1,001,450	NA	NA	NA
France	549,087	94,382	17.2	5.0
Greece	131,960	44,500	33.7	2.4
Iran	1,745,150	294,770	16.9	15.6
Iraq	435,050	40,000	9.2	2.1
Israel	22,070	1,400	6.3	0.1
Italy	301,340	40,410	13.4	2.1
Jordan	89,320	7,420	8.3	0.4
Lebanon	10,450	4,000	38.3	0.2
Libya	1,759,540	133,000	7.6	7.0
Malta	320	NA*	NA	NA
Morocco	446,550	210,000	47.0	11.1
Portugal	92,225	18,166	19.7	1.0
Spain	505,940	93,900	18.6	5.0
Syria	185,180	81,880	44.2	4.3
Tunisia	163,610	48,410	29.6	2.6
Turkey	785,350	146,170	18.6	7.7
Total	**11,268,442**	**1,888,344**	**16.8**	**100.0**

*NA: not available.

The total population living in the region is more than 643 million individuals and the labor force sums to 229 million (Table 5.2). Of these on average 35.7 per cent works in the agricultural sector, varying from 1.1 per cent for Israel to 78.6 per cent for Afghanistan. Rural population is generally low for the north-western Mediterranean countries and very high in the ME region. The annual gross domestic product (GDP) in power purchasing parity as a measure of size of an economy is highest for France (i.e., more than 2 trillion US$), and the lowest for Malta and Cyprus (i.e., less than 30 billion US$). The contribution of agriculture to the GDP ranges from 1.4 per cent (Malta) to 22.0 per cent (Afghanistan), and is 7.0 per cent on average.

Table 5.2 Population, labor force, gross domestic product (GDP), and their share in the agriculture sector for Mediterranean and Middle East Countries (Source: http://www.indexmundi.com/; accessed: 6 October 2017. Note that the figures for France and Italy refer to the whole country despite only part of the country is included in the M&ME region.)

Country	Population	Labor Force	Fraction of Working People (%)	Agriculture in Labor Force (%)	GDP* (billion US$)	Agriculture in GDP (%)
Afghanistan	31,822,848	7,512,000	23.6	78.6	45	22
Algeria	38,813,720	11,150,000	28.7	30.9	285	13.2
Cyprus	1,205,575	415,100	34.4	3.8	29	2.3
Egypt	86,895,096	27,690,000	31.9	29.2	551	11.3
France	66,259,012	29,940,000	45.2	2.4	2,276	1.7
Greece	10,775,557	4,918,000	45.6	12.6	267	4.1
Iran	80,840,712	27,720,000	34.3	16.3	987	9.1
Iraq	32,585,692	8,900,000	27.3	21.6	249	5.7
Israel	7,821,850	3,493,000	44.7	1.1	273	2.1
Italy	61,680,120	25,740,000	41.7	3.9	1,805	2.2
Jordan	7,930,491	1,898,000	23.9	2.0	86	4.2
Lebanon	5,882,562	1,481,000	25.2	1.6	64	5.7
Libya	6,244,174	1,644,000	26.3	17.0	74	1.9
Malta	412,655	190,400	46.1	1.6	16	1.4
Morocco	32,987,206	11,730,000	35.6	39.1	180	13.1
Palestine	4,550,368	NA**	NA**	NA**	NA**	NA**
Portugal	10,813,834	5,395,000	49.9	8.6	243	2.4
Spain	47,737,940	23,200,000	48.6	4.2	1,389	2.5
Syria	17,951,640	5,014,000	27.9	17.0	108	19.5
Tunisia	10,937,521	3,974,000	36.3	14.8	108	10.1
Turkey	79,841,871	27,910,000	35.0	18.4	1,167	6.1
Total	**643,990,444**	**229,914,500**	**35.7**	**16.2**	**10,204**	**7.0**

* Based on purchasing power parity (PPP); ** NA: not available.

Origin and Types of Grasslands

To summarize vegetation types of the M&ME and to maintain consistency, we adopted the zones, sub-levels and nomenclature provided by Mucina et al. (2016). Grasslands occur along three major vegetation zones found in the countries of the M&ME—Mediterranean, steppe and nemoral zone (Table 5.3). Intrazonal grasslands and herblands, semideserts, and oromediterranean grasslands and scrub are found in the Mediterranean zone. In the steppe zone, zonal steppe grasslands, mountain steppes with thorn-cushions, forest steppes (= steppe forests) and intrazonal saline vegetation occur, while the nemoral zone

Table 5.3 Main types of grassland vegetation of the M&ME region in terms of phytosociological classes. (Azonal classes, such as coastal salt marshes, screes and snowbed vegetation, as well as ruderal grasslands (see the glossary) are not listed. For European grasslands, the nomenclature and information synthesized by Mucina et al. (2016) were used.)

1. GRASSLANDS OF THE MEDITERRANEAN ZONE	
1.1 INTRAZONAL MEDITERRANEAN GRASSLANDS AND HERBLANDS	
Lygeo sparti-Stipetea tenacissimae Rivas-Mart, 1978 *nom. conserv. propos.*	Circum-mediterranean pseudosteppes on calcareous rocky substrates and relict edaphic steppes on deep clayey soils
Stipo giganteae-Agrostietea castellanae Rivas-Mart et al., 1999	Mediterranean thermo- to supramediterranean and humid submediterranean perennial acidophilous oligo-mesotrophic grasslands
Poetea bulbosae Rivas Goday et Rivas-Mart in Rivas-Mart, 1978	Mediterranean and Magrebinian seasonal perennial and ephemeroid pastures in the thermo- to oromediterranean belts
Helianthemetea guttati Rivas Goday et Rivas-Mart, 1963	Mediterranean and submediterranean-atlantic annual lowgrown ephemeral herb- and grass-rich vegetation on acidic substrates
Stipo-Trachynietea distachyae S. Brullo in S. Brullo et al., 2001a,b	Mediterranean calciphilous annual and ephemeroid swards and grasslands
1.2 INTRAZONAL MEDITERRANEAN SEMIDESERTS	
Pegano harmalae-Salsoletea vermiculatae Br.-Bl. et O. de Bolòs, 1958	Mediterranean and Macaronesian semi-desertic halo-nitrophilous scrub in hyperarid coastal habitats
1.3 VEGETATION OF OROMEDITERRANEAN GRASSLANDS AND SCRUB	
Festucetea indigestae Rivas Goday et Rivas-Mart, 1971	Iberian and North African xerophilous silicicolous fescue grasslands in the supra- to cryomediterranean belts
Saginetea piliferae Gamisans, 1975	Relict oromediterranean silicicolous swards of Corsica and Sardinia
Rumici-Astragaletea siculi Pignatti et Nimis in E. Pignatti et al., 1980	Siculo-Calabrian oromediterranean and upper mesomediterranean pulvinate scrub and related grasslands on siliceous substrates
Trifolio anatolici-Polygonetea arenastri Quézel, 1973	Oromediterranean, slightly chionophilous mat-grass swards of Eastern Anatolia, Sterea Hellas, Southern Macedonia and Bulgaria
Festuco hystricis-Ononidetea striatae Rívas-Mart et al., 2002	Submediterranean submontane-montane and oromediterranean dry grasslands and related dwarf scrub on calcareous substrates of the Iberian Peninsula, the Western Alps and the Apennines
Carici-Genistetea lobelii Klein, 1972	Cyrno-Sardean oromediterranean cushion-tragacanthic scrub and related grasslands
Daphno-Festucetea Quézel, 1964	Xeric oromediterranean grasslands and cushion-tragacanthic scrub on calcareous and ultramafic substrates of the Hellenic mainland and the Aegean region
Diantho troodi-Teucrietea cyprii S. Brullo et al., 2005	Oromediterranean scrub on ultramafic substrates of Cyprus
1.4 ZONAL MEDITERRANEAN FORESTS AND SCRUB	
Ononido-Rosmarinetea Br.-Bl. in A. Bolòs y Vayreda, 1950	Mediterranean scrub (*tomillar, espleguer, romeral, garrigue, phrygana, batha*) on base-rich substrates

Table 5.3 contd. ...

...Table 5.3 contd.

2. GRASSLANDS OF THE STEPPE ZONE	
2.1 ZONAL STEPPE GRASSLANDS	
Festuco-Brometea Br.-Bl. et Tx. ex Soó, 1947	Dry grassland and steppe vegetation of mostly base- and colloid-rich soils in the submediterranean, nemoral and hemiboreal zones of Eurasia
Prangetea ulopterae Klein, 1987*	Grassland communities physiognomically marked by the dominance of big *Apiaceae* and tall *Polygonum* species in the subalpine and lower alpine belts of the Irano-Anatolian massifs
Onobrychidetea cornutae Klein, 1987*	Grassland communities dominated by *Onobrychis cornuta* and thorny cushions of *Astragalus* and *Acantholimon* in the subalpine and lower alpine belts of the Irano-Anatolian massifs
2.2 INTRAZONAL SALINE VEGETATION OF THE STEPPE ZONE	
Festuco-Puccinellietea Soó ex Vicherek, 1973	Saline steppes and secondary saline steppic grasslands of the continental regions of Europe
Crypsietea aculeatae Vicherek, 1973	Pioneer ephemeral dwarf-grass vegetation in periodically flooded saline habitats of submediterranean and (sub) continental Eurasia
3. GRASSLANDS OF THE NEMORAL FOREST ZONE	
3.1 INTRAZONAL BOREO-TEMPERATE GRASSLANDS AND HEATH	
Nardetea strictae Rivas Goday et Borja Carbonell in Rivas Goday et Mayor López, 1966 *nom. conserv. propos.*	Secondary mat-grass swards on nutrient-poor soils at low and mid-altitudes of the temperate, boreal and subarctic regions of Europe and alpine parts of Turkey including secondary oligotrophic pastures of the *Nardetalia*
Koelerio-Corynephoretea canescentis Klika in Klika et Novák, 1941	Dry grasslands on sandy soils and on rocky outcrops of the temperate to boreal zones of Europe, the North Atlantic islands and Greenland
Molinio-Arrhenatheretea Tx., 1937	Anthropogenic managed pastures, meadows and tall-herb meadow fringes on fertile deep soils at low and mid-altitudes (rarely also high altitudes) of Europe, east and north Anatolia
3.2 GRASSLANDS OF THE NEMORAL OROSYSTEMS	
Mulgedio-Aconitetea Hadač et Klika in Klika et Hadač, 1944	Tall-herb vegetation in nutrient-rich habitats moistened and fertilized by percolating water at high altitudes of Europe, Siberia and Greenland
Juncetea trifidi Hadač in Klika et Hadač, 1944	Acidophilous grasslands in the alpine belt of the nemoral zone of Europe, the Caucasus and in the boreo-arctic and arctic zones of Northern Europe and Greenland
Elyno-Seslerietea Br.-Bl., 1948	Alpine and subalpine calcicolous swards of the nemoral mountain ranges of Europe
Alchemillo retinervis-Sibbaldietea parviflorae Vural, 1996*	Alpine grasslands extending from northeast Anatolia to sea-facing slopes of the Caucasus

* Non-European grasslands not covered by Mucina et al. (2016). Note here that a synthesis study for the Middle Eastern or North African units has not been conducted yet.

is characterized by intrazonal boreo-temperate grasslands and grasslands of the nemoral orosystems. Overall the grassland types of the M&ME region are very diverse, from lowlands to high altitudes and include, among others, low-grown ephemeral herb- and grass-rich Mediterranean vegetation, desert steppes, saline vegetation with succulent chenopod scrubs, anthropogenic grasslands, humid grass-rush meadows, tall-rush wetlands, chionophilous (snow-bed) vegetation, high-alpine and subnival herbaceous scree communities, patchy green wet meadows and fens in mountain slopes and riverbanks (see examples in Fig. 5.2). In this section, only those covering large areas in the M&ME region are explained briefly.

Dry rainfed (not-irrigated) grasslands dominate M&ME ecosystems from lowlands to subalpine zone. Guarino (2006) hypothesized that the floristic diversity of Mediterranean dry grasslands draws its origin from the Messinian Salinity Crisis, which occurred between 7.2 and 5.3 million years ago (Hsü et al., 1973). During this age, partly due to Antarctic cooling, a marine regression produced the connection of the known Iberian land and Africa, and led to the palaeo-Mediterranean sea-level depression (Drinia et al., 2007). Accordingly, large areas of Mediterranean Sea were partly or almost completely desiccated and revealed new ecological niches for pioneer plants on mountain ranges, canyons and other land formations, with the revealed landscape of deep floor being dominated by salty lands (saltmarshes, salty deserts) and alkaline lake spots (Hsü et al., 1977). Several driving forces governed speciation processes (Raven, 1973; Guarino, 2006), leading to the current species-rich flora with many neo-endemics especially for therophytes (annual plants) (Pignatti, 1979) as well as three major functional types of Mediterranean dry grasslands: (i) wintergreen perennial, dominated by large caespitose hemicryptophytes; (ii) wintergreen ephemeral, rich in thermo-xerophilous therophytes, and (iii) summergreen perennial, dominated by chamaephytes and hemicryptophytes. The first group is a form of open perennial grassland with many Mediterranean and Mesogean floristic elements (Guarino, 2006). Vegetation of wintergreen ephemerals is found on nutrient-poor soils where the climate is seasonally dry. Finally, the summergreen perennial dry grasslands include steppe communities of the Iberian mountains up to the Middle East. In addition, Mediterranean scrub (or in the languages of the region—*tomillar, espleguer, romeral, garrigue, phrygana, batha*) on base-rich substrates comprises many grassland elements though grasses usually do not dominate the communities. Major Mediterranean grassland types are further differentiated into subtypes, depending on bedrock, soil, climate and abundance of ephemeral species.

Steppes are defined as a vegetation type dominated by herbs, mainly grasses and other graminoids, sometimes with a significant admixture of chamaephytes, in arid and semi-arid climates (Wesche et al., 2016). They cover large areas in North Africa, the Middle East as well as parts of Mediterranean Europe. From the physiognomic point of view, steppes are classified as desert steppes, typical (tall-grass) steppes, forest steppes and alpine steppes following the humidity gradient (Fig. 5.1 for distribution of typical and forest steppes). Desert steppes occur in dry continental plains or areas neighboring deserts with generally less than around 250 mm annual precipitation. They are characterized by xeromorphic dwarf-scrublands and dominated by various species of *Artemisia* (*A. sieberi, A. fragrans, A. kopetdaghensis, A. quettensis* [= *Seiphidium quettense*], *A. lehmaniana, A. mesatlantica* and *A. santolina*), mixed with *Zygophyllum* spp., *Dorema ammoniacum, Salsola* spp., *Dendrostellaria lessertii* and various chamaephytes, such as *Thymus, Ziziphora* and *Astragalus* spp. Typical steppes are dominated by perennial grasses, such as *Stipa holosericea, Bromus tomentellus* and *Festuca valesiaca*. Such vegetation is now fragmented and degraded in various parts of M&ME due to conversion to other land cover types

Fig. 5.2 Examples of grasslands of the Mediterranean Basin and the Middle East. (A) A typical Mediterranean grassland with *Stipa tenacissima* and cushion forming shrubs, Cabo de Gata, Spain; (B) *dehesa* with managed oaks (see the glossary), province of Cáceres, south-west Spain; (C) *Artemisia* steppe mixed with *Pistacia* woodlands in the Taftan Mts., south-east Iran; (D) tragacanthic steppe with *Onobrychis cornuta* in the Alborz Mts., Iran; (E) Alpine grassland, Artvin, north-east Turkey; (F) wetland grazed by buffalos, Prespa, Greece. Photos by S. Burrascano, M. Pulido Fernández, A. Naqinezhad, A. Talebi, H. Ambarlı and M. Vrahnakis (in this sequence).

and grazing. Forest steppes characterize the montane lands in transition from steppes to either temperate forests or Mediterranean vegetation. These landscapes are marked with scattered trees or small stands of the genera *Pistacia, Amygdalus, Quercus, Juniperus, Crataegus, Rhamnus, Paliurus, Acer, Jasminium, Rubia, Colutea, Cornus* and *Lonicera*. The actual cover of the forest steppes is assumed to be lower than the climatic potential due to forest destruction. Tragacanthic, i.e., thorny cushion-forming shrubby vegetation, is one of the most important features of high mountain zone throughout the Middle East and south-west Asia, where summer rain is absent and where the water supply depends on melted snow (Breckle et al., 2013). These shrublands occur across a wide elevation range in Iran, Turkey and Afghanistan, from the timberline (1,500 m a.s.l.) up to 4,220 m a.s.l. and often also penetrate into the forest zone below (Zohary, 1973). They are used mainly as summer pastures (Breckle et al., 2013). The most important genera of these vegetation types are *Astragalus, Acantholimon, Acanthophyllum, Onobrychis, Gypsophylla, Cousinia* and *Arenaria*. Those habitats are particularyly endemic-rich. It is claimed that the present broad distribution of vegetation units of thorny shrubs, dwarf scrub, or thorn cushions must have resulted from overgrazing (Ariapour et al., 2017).

Agronomic Use of Grasslands

M&ME grasslands are among those with the longest history of human intervention, which affected ecosystem dynamics, productivity and structure (Grove and Rackham, 2003). The so-called Fertile Crescent, extending from the Zagros mountains in Iran to central Anatolia in Turkey, Nile Valley in Egypt, is the birthplace of settled life and farming. Initial steps of sheep and goat domestication as well as colonist farming date back more than 10,500 cal BP (Zeder, 2008). At that time the vegetation hosted wild cereal stands in extensive fire-maintained grasslands and enabled the first farming practices (Turner et al., 2010). Anthropogenic fires can be considered as the first man-made measure to massively control and exploit nature for ease of moving and traveling, hunting and food gathering. Then livestock husbandry started to replace grazing by wild herbivores and natural habitats started to be replaced by croplands. According to Pignatti (1978), during the Palaeolithic age a positive feedback was established between humans and Mediterranean grasslands—the surge in human population resulted in greater disturbances of grasslands, which in turn controlled competitive advantages and increased the human ecological fitness. During the Neolithic age, an extensive selection and domestication of the most productive livestock and plant breeds took place (see Squires and Feng, 2018). The mutual dependence of grassland resources and Mediterranean people led Waddington (1975) and Guarino (2006) to formulate the hypothesis of homeorhetic equilibrium (i.e., system returns to a trajectory rather than to a state). According to this hypothesis, there is a homeorhetic state of dynamic equilibrium between human and natural perturbations and climatic fluctuations that shaped Mediterranean ecosystems. However, this equilibrium is at the expense of soil stability and productivity, as humans—via agropastoral activities—uptake nutrients rapidly, while the return rates are very low. Over time, the new equilibrium is characterized by low energetic rates, while soil degradation leading to desertification appears as a severe threat for M&ME grasslands.

Livestock keeping is the major human activity on grasslands of the M&ME as animal husbandry often completely relies on grassland resources. Total number of livestock (excluding poultry) in the region was around 387 million by 2014 (Table 5.4). Sheep have the biggest share (58.4 per cent) followed by cattle (19.0 per cent) and goats (18.6 per cent).

Table 5.4 Livestock capital (number of heads, except for beehives) for Mediterranean and Middle East Countries in 2014 (Data from FAOstat, 2014). (Note that the figures for France and Italy refer to the whole country despite only part of the country is included in the M&ME region.)

Subregion/Country	Cattle	Goats	Sheep	Horses	Asses	Mules	Camels	Buffaloes	Beehives**
A. Southern Europe (SE)									
France	19,095,797	1,290,623	7,239,057	408,028	15,000*	30,742	–	–	789,056
Greece	659,000	4,255,000	9,072,000	33,000*	35,000*	20,000*	–	1,800*	1,457,000*
Italy	5,756,072	937,029	7,166,020	390,886	24,900*	9,000	–	369,352	500,000*
Malta	14,883	4,627	10,526	1,050*	480*	300*	–	–	NA
Portugal	1,549,000	382,000	2,032,000	20,000*	115,000*	55,000*	–	–	333,000*
Spain	6,078,700	2,704,250	15,431,800	250,000*	140,000*	110,000*	–	–	2,450,000*
Total (SE)	**33,153,452**	**9,573,529**	**40,951,403**	**1,102,964**	**330,380**	**2,25,042**	–	**371,152**	**5,529,056**
B. Northern Africa (NA)									
Algeria	2,049,652	5,129,839	27,807,734	42,010	134,920	30,190	354,465	–	400,000*
Egypt	4,762,491	4,185,761	5,502,637	74,616	1,277,363	1,165*	158,269	3,949,262	929,626
Libya	200,000*	2,580,000*	7,150,000*	46,500*	29,000*	NA	57,000*	–	37,000*
Morocco	3,238,688	6,147,225	19,230,835	140,000*	947,000	456,000	182,830	–	375,000*
Tunisia	671,200	1,248,200	6,805,700	57,100*	241,000*	82,100*	236,500*	–	589,000*
Total (NA)	**10,922,031**	**19,291,025**	**66,496,906**	**3,60,226**	**2,629,283**	**569,455**	**9,89,064**	**3,949,262**	**2,330,626**

C. Turkey and Levant (TL)									
Cyprus	60,874	3,22,400	322,400	660*	5,200*	1,500*	–	–	40,000*
Israel	461,000	108,000	574,000	4,000*	5,000*	1,600	5,500*	–	100,000
Jordan	69,800	929,000	2,818,000	2,430*	9,000*	1,200*	13,150*	100*	44,000*
Lebanon	87,000*	560,000*	452,000*	4,000*	15,000*	5,000*	200*	–	200,000*
Palestinian Territory	35,000*	210,000*	728,000*	–	–	–	–	–	47,000*
Syria	1,090,458	2,285,778	17,858,139	16,469	79,206	2,454	58,715	7,933	507,829
Turkey	14,223,109	10,344,936	31,140,244	131,480	170,503	41,397	1,442	121,826	7,082,732
Total (TL)	**16,027,241**	**14,760,114**	**53,892,783**	**1,59,039**	**2,83,909**	**53,151**	**79,007**	**129,859**	**8,021,561**
D. Middle East (ME)									
Afghanistan	5,349,000	7,059,000	13,485,000	171,000	1,441,000	24,000	171,000	–	NA
Iran	5,000,000*	20,120,000*	45,000,000*	140,000*	1,600,000*	175,000	110,000*	120,000*	3,200,000*
Iraq	2,800,000*	1,216,898	6,545,146	53,000*	380,000*	11,500	62,085	187,401	NA
Total (ME)	**13,149,000**	**28,395,898**	**65,030,146**	**364,000**	**3,421,000**	**210,500**	**343,085**	**307,401**	**3,200,000**
TOTAL (M&ME)	**73,251,724**	**72,020,566**	**226,371,238**	**1,986,229**	**6,664,572**	**1,058,148**	**1,411,156**	**4,757,674**	**19,081,243**

* FAO estimated; ** For *Apis mellifica, dorsata, florea* and *indica;* NA: Data not available.

Of the 226.4 million sheep, 29.4 per cent are found in North Africa, followed by a 28.7 per cent in the Middle East. Steppes of North Africa are widely used throughout the year for grazing and browsing, mainly by sheep and goats in the mountainous areas. Transhumance, seasonal migration of livestock on fixed historical routes to forage productive meadows, is a traditional practice that often occurred in the past in M&ME countries. Although less common now in southern Europe (Oteros-Rozas et al., 2013), it is still alive in Turkey and Levant, Middle East and Northern Africa, especially on the southern slopes of the Atlas mountains in Morocco (Akasbi et al., 2012). Similarly, pastoral nomadism, i.e., nomads and livestock, which do not have permanent settlements but use a large territory for forage, is on decline in various parts of the region due to pressures of centralized administrations, lack of education and health services.

Almost half (45.3 per cent) of the total number of cattle (73.3 million) is farmed in southern Europe, while a significant percentage is found in Turkey and Levant (21.9 per cent). Cattle grazing is confined to the more productive grasslands, i.e., grasslands representing intrazonal boreo-temperate grasslands or grasslands of the nemoral orosystems. Such mesic or wet grasslands are also managed for hay-cutting—some of them are mown one to three times during the growing season for feeding the domestic livestock in winter.

M&ME countries hold 72.0 million goats (39.4 per cent of them in the Middle East followed by 26.8 per cent in North Africa). Due to their ability to overcome difficulties generated by terrain steepness and climatic extremes, goats are basic elements to sustain rural economies in the mountainous remote areas. Traditional pastures of south-east Spain have mainly been grazed by goats (Correal et al., 2009). They also play an important ecological role in controlling shrub encroachment, thus avoiding the deterioration of priority habitat types.

There are various forms of animal farming practiced in the region. Agroforestry, a land-use system in which trees or shrubs are managed or cultivated in addition to crop production or livestock husbandry, is widely practiced in the Mediterranean with great heterogeneity in land management. Trees have been historically cleared for livestock husbandry purposes, either by removing all the trees (grasslands) or combining scattered trees with the management of a productive herbaceous layer (rangelands). These rangelands, called *dehesas* in Spanish and *montados* in Portuguese, occupy more than 20,000 km^2 in the southwestern part of the Iberian Peninsula. They represent one of the most valuable ecosystems for both countries (Spain and Portugal) for different reasons—biodiversity, meat production, landscape, preservation of traditional values and rural population fixation (Díaz et al., 1997; Escribano et al., 2015). In Portugal, the endangered system, called *campo-bouça*, is formed by mosaic-like from grasslands surrounded by fruit orchards and forests (e.g., cork oak). This system is grazed mainly by cattle and is typical of the northern region close to Rivers Minho and Douro. Other important grassland management types are the Portuguese *lameiros*, i.e., traditional grasslands (even meadows) of high mountains in north-east Portugal. Some autochthonous breeds of cows permanently use their evergreen forage resources. These grasslands and their traditional management are important cultural elements as they shape the idiosyncrasy of local people; yet they are at risk of disappearing (Oliveira et al., 2007). The majority of the M&ME grasslands are managed extensively. Some of the grasslands are rotationally cultivated—in Iberia, they are sown with fodder species, such as oat (*Avena sativa*) and vetch (*Vicia sativa*) every three to four years (traditional practice) if not abandoned (Joffre et al., 1988). In North Africa, crops of cereals, such as barley, are used for this purpose by sedentary people (Zucca et al., 2013).

Harvesting wild plants and animals is a very common practice in the M&ME grasslands. Removal of tragacanthic *Astragalus* species, such as *A. microcephalus*, is common in ME steppes. They have multiple uses as fuel, fodder, medicinal plant or even cleaning stuff. Tragacanth gum produced by *A. gummifer* (Milkvetch) is collected widely to be used as thickener in confections, salad dressings, sauces or for medicinal purposes. As a result, *Astragalus* steppes are degraded and open to erosion and desertification. Semi-arid steppes of south-east Spain have historically (until the 1970s) been used to harvest 'esparto fiber'—a fibre of great strength and flexibility produced from *Stipa tenacissima* and *Lygeum spartum* (Maestre et al., 2007). Peat extraction, harvesting orchids and *Allium* spp. are among the serious threats on wet grasslands of Iran (Jalili et al., 2014). They are considered as the most vulnerable ecosystems of Alborz and Zagros mountainous ranges (e.g., Naqinezhad et al., 2009). In addition, steppe grasslands are also an important source of game animals, such as, partridges (*Perdix perdix*) and wild hares (*Lepus* spp.).

Beekeeping is another important source of income in the region which is based on grasslands. Around 19.1 million beehives are recorded in M&ME countries; 42.0 per cent of them are found in Turkey and Levant followed by southern Europe (29.0 per cent) (Table 5.4). Beekeeping is the major source of income in most of the flower-rich meadows of northeast Anatolia, where the price of honey produced by traditional ways could be as high as 250 EUR/kg in the year 2017.

Value of Grasslands

Meat and dairy are the major products of the M&ME grasslands with a high spatial and temporal variability in productivity, depending on vegetation composition, climatic factors, particularly the annual rainfall, or soil nutrient contents. The mean productivity for Mediterranean grasslands in terms of dry matter (DM) is approximately 2,000 kg DM ha^{-1} (Pulido Fernández, 2014). Those numbers can be as low as 200 kg DM ha^{-1} for steppes in south-east Spain dominated by nanophanerophytes and chamaephytes, 210–520 kg DM ha^{-1} those dominated by *Stipa tenacissima* (Robles and González-Rebollar, 2006), and 5,000 kg DM ha^{-1} for oceanic grasslands of Cantabrian region (CIFA, 2007). In terms of milk production, Iberian *dehesas* have a mean productivity in terms of milk solids (MS) of 1,440 and 2,390 kg MS ha^{-1} (Olea et al., 1990; González et al., 2007), respectively. By 2015, a quantity of 18.653 million tons of milk and 1.149 million tons of meat were produced in Turkey (Turkish Statistical Institute, 2016).

M&ME grasslands are also particularly valuable for other ecosystem services without market price. According to Pardos (2010), temperate grasslands can store an average of 243 tons of carbon per hectare, both in soil and vegetation. These values are higher than those estimated in temperate forests, tropical savannas, tundra, alpine meadows, deserts, semideserts and croplands. Mediterranean grasslands play also a crucial role in prevention of soil loss and land degradation by water and wind erosion with soil retention capacity generally more than an order of magnitude compared to arable land (Cerdan et al., 2010). In Mediterranean landscapes, grasslands also play a key role as fire breaks, so reducing fire risk. *Dehesas* and *montados* produce several economic and social benefits, such as increase in farm income through multiple uses of natural resources, employment stability, increase security against price change and combat depopulation. They contribute to mitigate desertification and contribute resilience of the ecosystem against climate change, decreasing fire risk and serving as ecological corridors and refuges for large mammals and birds of prey (Sequeira, 2008).

Apart from the multiple ecological services (e.g., carbon sequestration, water cycling or erosion control) that grasslands can provide (commonly known), we must consider a wide range of ecotouristic and cultural activities. For instance, in Turkey there are specific places where grassland is perceived as of high cultural services. A good example is the significance of *yaylas* (highlands used as summer pasture) of northeast Anatolia that attract thousands of trekking tourists and mountaineers as well as 'Karadenizli' people (from the Black Sea) every year to join in *yayla* festivals even if they had abandoned their lands tens of years ago. Grasslands also play a role in recreation from daily picnic activities to summer vacations in alpine areas. It is also the case in many highlands of Iran (*yeylagh*) where summer pastures are suitable for feeding animals, a source for medicinal plants for local people and also resting places for temporary settlements.

The M&ME region is a biodiversity hotspot of hotspots. Its territory is covered by three of the world's biodiversity hotspots (Mittermeier et al., 2011)—Mediterranean Basin, Irano-Anatolian and Caucasus; the latter represented by northeastern highlands of Turkey in the region. The Mediterranean Basin is one of the species-rich regions in terms of plants and animals of global importance, as well as high endemism rate. It has more than 1,240 vertebrate species and circa 22,500 vascular plant species, half of which are endemics (CEPF, 2016). Steppes of the hotspot display a high ratio (circa 30 per cent) of plant endemism, extremely pronounced for forb species. Additionally, the region is the center of evolution and distribution for the species of the genera *Verbascum, Salvia, Acanthophyllum, Cicer, Dianthus, Onosma, Euphorbia, Gypsophila, Minuartia, Noaea, Onobrychis, Oxytropis, Scorzonera* and especially cushion-forming ones such as *Astragalus* and *Acantholimon* (Kürschner, 1986). Steppes of Anatolia form the center for secondary adaptive radiation, giving rise to numerous groups of closely related, often vicarious, neo-endemic species as well as monotypic genera, which is explained with the recent neogenesis of the land (Pils, 2013). Furthermore, the Fertile Crescent is important as a center of crop genetic diversity, known for hosting wild relatives of many legumes and cereals.

Mediterranean grasslands are among the most important habitats in Europe for various animal groups. Among them are the steppe birds, such as Great bustard (*Otis tarda*) and Lesser kestrel (*Falco naumanni*), Little bustard (*Tetrax tetrax*), Black-bellied sandgrouse (*Pterocles orientalis*) and raptors, such as the Hen Harrier (*Circus cyaneus*). A less popular bird of the region, Caucasian grouse (*Lyrurus mlokosiewiczi*) is endemic to alpine grasslands and related shrublands of the Greater and Lesser Caucasus. It is near-threatened globally due to an expected decline in the future owing to various factors, such as uncontrolled tourism development destroying and degrading of its habitats (BirdLife International, 2016). Various endangered species benefit from a mosaic of habitats, including both grasslands and shrublands or woodlands. For example, the threatened Spanish imperial eagle (*Aquila adalberti*) and Black vulture (*Aegypius monachus*) nest in silvopastoral *montados* and *dehesas* and use the open grassland patches for feeding (Bugalho and Abreu, 2008; Vrahnakis et al., 2009). Iberian lynx (*Lynx pardinus*) is one of the most popular and threatened large mammals of the region (Ferreras, 2001). Highland grasslands of protected areas of the Alborz Mts. in Iran are habitats for many threatened mammals, such as Brown bear (*Ursus arctos*), Leopard (*Panthera pardus*), Wild goat (*Capra aegagrus*) and Mouflon (*Ovis orientalis*).

The region is particularly diverse in insect fauna as the Mediterranean Basin holds 75 per cent of the European insect species, including more than half of the continent's butterflies (Blondel and Aronson, 1999). Among the countries, Turkey has the highest record of butterfly richness with more than 380 species. Most important habitats as prime butterfly areas are alpine and sub-alpine grasslands of northeast Anatolia managed in traditional ways of grazing and hay-cutting (Karaçetin et al., 2011).

Threats to Grasslands

The major threats widespread in the region as well as in priority grassland habitats are: (i) unregulated (unsustainable) grazing, (ii) changes in landuse intensity/type, (iii) abandonment of traditional human interventions, and (iv) climate change; finally, all results in habitat degradation, fragmentation and loss. Further threats appear from shrub encroachment, reforestation, afforestation with non-endemic tree species, invasion by weeds or non-native species, various types of construction works and airborne nitrogen deposition (Vrahnakis et al., 2010).

Livestock grazing is a key actor on Mediterranean biodiversity (Noor Alhamad, 2006; Zamora et al., 2007). Nevertheless, a clear relation between grazing, biodiversity and overall ecosystem functioning is still questionable (Troumbis and Memtsas, 2000). The Intermediate Disturbance Hypothesis (Sousa, 1984), which anticipates the highest levels of biodiversity under intermediate levels of disturbance, is supported by evidences from Mediterranean ecosystems with a long history of grazing (Alados et al., 2004; but see Fadda et al., 2008 for Coleoptera diversity). Overgrazing was the main source of decline in ecosystem functions in the past (Cortina et al., 2009). In northern Africa, high stocking rates and conversion into croplands are still major threats. In the Middle East, for all types of pastoralism (sedentary, semi-nomadic and nomadic) overgrazing has been the case along the past millennia. Together with fuel collection, overgrazing resulted in shifts in plant community composition, such as invasion by ruderals, i.e., ruderalization, by unpalatable species such as various *Artemisia* spp. or grazing-resistant forms such as tragacanthic species, i.e., tragacanthization (Zohary, 1973). Grasslands of the Middle East benefit from grazing exclosures for 10 to 30 years with increase in vegetation cover, increase in cover of perennial grasses (particularly *Agropyron* spp., *Stipa* spp., *Poa bulbosa*) and forbs and decrease in cover of dwarf shrubs (*Acantholimon* spp., *Noaea mucronata*, *Artemisia sieberi*) (Akbarzadeh, 2006).

Undergrazing in space and time (or both) or grazing abandonment is now a trend in the whole region (Fadda et al., 2008). The response of grasslands to abandonment varies from region to region: In Turkey, studies report that some of the overgrazed sites may benefit from abandonment to recover and succession to semi-natural woodlands is very slow compared to many European countries (Ambarlı and Bilgin, 2014). Abandonment in Mediterranean landscapes results in shrub encroachment and increase of catastrophic wildfires (Zomeni et al., 2008). Dramatic decline of traditional activities had detrimental effects on species-rich communities. Most of the Mediterranean rangelands that lack systematic grazing suffer from invasion of thorny shrubs, especially *Sarcopoterium spinosum*, which results in declines of biodiversity and forage quality (Reisman-Berman and Henkin, 2007). Feeding and survival of Griffon vulture (*Gyps fulvus*) populations in the uplands of the Cantabrian mountains, north-west Spain, is strongly linked to transhuman activity with sheep, which was reduced by 62 per cent (Olea and Mateo-Tomás, 2009). The species-rich flora of pseudo-alpine grasslands in Greece was strongly controlled by traditional livestock husbandry activities, like long-distance transhumance or short-distance vertical movements (Ispikoudis et al., 2004), which provides intermediate disturbance, seed dispersal (Poschlod et al., 1998) and landscape heterogeneity (Ispikoudis et al., 2004) and suitable habitats for wildlife, such as Egyptian vulture (*Neophron percnopterus*) and Golden eagle (*Aquila chrysaetos*).

Until the beginning of 21st century, grasslands of southern Europe suffered from conversion to croplands, forests or urban areas. This was mostly due to the mechanization of agricultural production, increasing food demands of humans and then the institutional framework imposed by the European Union's Common Agricultural Policy (CAP; Vrahnakis et al., 2010). Land use and cover are undergoing dramatic alterations in Mediterranean Europe (Serra et al., 2008) usually at the expense of grasslands (e.g., Chouvardas and Vrahnakis, 2009). One of the major driving forces for recent alterations may be found in European Commission's system of subsidies that favor agricultural intensification and expansion of arable lands. In France, grasslands have been considerably altered in the last 60 years by intensification, land abandonment or alteration or destruction by civil engineering works (Muller et al., 1998). Restoration efforts through mechanical removal of plant cover, improvement of soil chemistry, reseeding and grazing exclosure have promising results (Papanastasis, 2009).

Another emerging threat is climate change. Until the end of the current century, average annual temperature of the M&ME region will increase by 3.5–5°C and precipitation will decrease by 20 per cent at least (Christensen et al., 2007). In plant communities grown in regions of highly xeric character, like the majority of M&ME grasslands, the prospect of climate change with more severe drought episodes suggests a particular threat (Rodwell et al., 2007). Climate change is a serious threat to mesophilic vegetation of the region as well. According to recent models that use bioclimatic variables for predicting species distribution through the characterization of climatic envelopes (SPECIES artificial neural network, under the A2 SRES emission storyline and the HadCM3 and PCM climate change scenarios; Ruosteenoja et al., 2003), it was found that more mesophilous plant species, like the grasses *Arrhenatherum elatius* and *Alopecurus pratensis* as well as the orchid *Gymnadenia conopsea*, will face serious population decline from the Italian siliceous alpine grasslands in the next decades (Harrison et al., 2006). Under dry climatic scenarios, the species composition of calcareous grasslands will experience an increase in ruderal species (JNCC, 2007). The magnitude of change, at least for calcareous grasslands, seems to be related to the stage of natural succession, with late-successional grasslands, mostly composed of short-growing and long-lived plant species, experiencing less severe changes (Calaciura and Spinelli, 2008).

Grassland biodiversity is often threatened by the dominance of species considered as weeds, such as representatives of the genera *Carduus*, *Carlina*, *Cirsium*, *Cnicus*, *Echinops*, *Senecio* and *Cistus*. They show high plasticity and competitive advantages, supported by biological attributes, such as accumulation of toxic substances. The Middle East is a large center of many weeds of Mediterranean and tropics (Zohary, 1973) as there are about 1,500 segetal (see the glossary) and ruderal species, most of which are categorized as facultative segetals, i.e., growing both in cultivated lands and natural grasslands.

Conservation of Grasslands

The major tool for nature protection in the European part of the Mediterranean is the European Union's Natura 2000 network of protected areas (European Union, 2017). There are several grassland habitat types in Mediterranean with conservation priority (European Communities, 2009). The *6210 semi-natural dry grasslands and scrubland facies on calcareous substrates (with the priority element of *important orchid rich sites)* (*Festuco-Brometea* class of zonal steppes) is a priority habitat and the network covers 2,620 km^2 of it in the region. This habitat contributes almost to half (44 per cent) of the Natura 2000 sites of the Mediterranean Biogeographical region (Calaciura and Spinelli, 2008). A total of 6,937 km^2 included in the Natura 2000 network is covered by the *6220* Pseudo-steppe with grasses and annuals (Thero-Brachypodietea)*, corresponding to the *Lygeo sparti-Stipetea tenacissimae* (*Thero-Brachypodietea* used as the name of this vegetation) (San Miguel, 2008). Also, *6230* Species-rich Nardus grasslands* occur in six different bioregions of Europe, and in the Mediterranean they cover an area of 383 km^2, which corresponds to circa 20 per cent of their continental extent (Galvánek and Janák, 2008). The *6310* Dehesas with evergreen Quercus* spp. are also a priority habitat type with a stronghold in Spain and Italy. There is no exact figure for the total amount of this habitat in protected areas; it was estimated that the percent area of the habitat protected is around 33 per cent for Spain and 10 per cent for Italy (EEA, 2017a). In addition, the *5330 Thermo-Mediterranean and pre-desert scrubland* is a common habitat that is dominated by various scrub species as well as forbs and succulents. It is in favorable status in Portugal, whereas unfavorable in other parts of the European Mediterranean (EEA, 2017b).

Moving out of Europe, grasslands of Turkey are under-represented in the country's official protected area network. Only 1.5 per cent of dry grasslands and related xerophilous woodlands of inner Turkey occur within the network (Ambarlı et al., 2016). This figure is expected to be lower for all grassland types, as one of the most grassland-rich region, i.e., East Anatolia, is least protected and the protected areas are biased towards forests. Iran has a total of 203 protected sites, covering 10 per cent of Iranian territory (Department of Environment, 2017). Most of them are characterized by types of grasslands and forest steppes but their conservation effectiveness is questionable (Kolahi et al., 2012). In Syria, there are 14 rangeland protected areas (Barkoudah et al., 2000), but it is unclear how well they currently function. The protected areas cover 0.2 to 13.1 per cent of the terrestrial territory of the countries in the North African belt. We do not have detailed information, but most of them target conservation and management of game animals or wetlands of international importance (UNEP-WCMC, 2017).

Management and Restoration of High Nature Value Grasslands

European Mediterranean countries have the highest percentage of high natural value (HNV) farming systems in Europe (Hellegers, 1998), as Mediterranean pastures fulfil many characteristic conditions, like small-scale low-input farming, extensive grazing, different types and levels of disturbance creating a diverse mosaic of habitats, high quality products and multifunctional services and goods (Escribano et al., 2015). *Dehesas* and *montandos* are good examples of HNV farming for reasons, like biodiversity conservation, preservation of traditional values and rural population fixation (Díaz et al., 1997). They form small-scale landscape mosaics, composed of cereal cropping and livestock grazing under open canopy of *Quercus* spp. Another HNV example is the pseudo-steppe with grasses and annuals. An extensive grazing system takes place in these semi-natural grasslands in a mosaic habitat with cereal production on ancient croplands. In Turkey, three types of extensive farming on marginal grassland landscapes can be considered as HNV, i.e., (i) extensive rotational crop production with local cultivars of cereals, legumes and forage crops coupled with extensive livestock production (cross or local breeds) in rangelands for family subsistence (some of them involve agroforestry), (ii) small-scale hay production from meadows, forage crops and stubbles, and (iii) transhumance and nomadism, involving *yaylas* (Redman and

Hemmami, 2008). The above-mentioned land management types are mixed in farms at various places, depending on the climate and topography.

M&ME grasslands of HNV farming have been progressively degraded by many historical and recent processes, which involve a wide range of context-dependent driving forces, such as climate change, unsustainable land-use systems due to poverty, global or continental agricultural policies. Many grasslands have been restored through natural regeneration due to low intensity of land uses induced by economic changes (Ramírez and Díaz, 2008). On the other hand, a large surface of grasslands was lost due to shrub encroachment in the last decades, as a by-result of land abandonment where natural regeneration is not possible. Apparently, ecological restoration of grasslands involves shrub clearing and reintroduction of livestock management. Parameters important in determination of the restoration method involve abiotic factors, target plant and animal communities, breed of livestock suitable for the area, costs, national/regional laws, subsidies or land tenure. Its consequences will depend on the mechanisms utilized, such as stocking rates, type of labor or machinery. Livestock management aimed at promoting total biodiversity is a challenging task. While it stands as a proper ecological tool, it is broadly considered as a marginal activity of extensive grazing with low economic profitability. Increasing biodiversity and reducing fire risk are some of foreseen goals in this kind of restoration projects.

The calculation of proper stocking rate is crucial to maintain moderate grazing that will promote productivity, biodiversity and ecosystem services of grasslands (Vrahnakis, 2015). Among the proposed methods are keeping stocking rates always lower than the carrying capacity (Calaciura and Spinelli, 2008), mixing different types or livestock species/breeds (especially mixing with goats) and rotational grazing to control shrubs and create habitat heterogeneity (Lavado Contador et al., 2015). Management for conservation of grassland animals, such as insects or birds, requires highly structured vegetation for feeding and refuge. Suggested management methods include leaving uncut circa 10 per cent of the area before summer each year for hay meadows, arranging timing of the harvest in crop rotations and hay meadows to prevent destruction of animal populations and maintenance of patches of shrubs to promote feeding and nesting wild animals.

The recovery/restoration of over-utilized grasslands is usually attempted mainly by strategies like reseeding, establishment of pre-germinated plants (if necessary) and reduction of intensity or exclusion/prohibition of agricultural activities, especially uncontrolled grazing. In northern Africa, for instance, fodder shrub plantations (e.g., *Atriplex nummularia*) have been extensively introduced in dryland areas to rehabilitate degraded grasslands and to mitigate desertification processes since the 1980s (Zucca et al., 2013). The use of tree plantations for land restoration is often questionable. In the Mediterranean Basin, pine trees have been extensively used for land restoration since the late 19th century (Pausas et al., 2004). However, the restoration success of such plantations is questionable due to low resistance of planted pines to fires and reduced plant species richness as a result of afforestation methods (Pausas et al., 2004; Chirino et al., 2006). It was found that dry grassland communities are as successful as pine plantations in semi-arid areas to mitigate erosion and restore wildlife richness (Chirino et al., 2006). Temporary or permanent grazing exclusion has been one of the most extreme measures taken by several institutions, particularly in drylands, and is strongly criticized because of increased wildfire risk (Perevolotsky and Seligman, 1998).

For southern Europe, EU policies are promoting the adoption of agri-environmental measures by farmers as well as the creation of Nature 2000 network, among many other actions, since the 1980s. Their schemes are focused on conservation practices that can guarantee a low-intensity land use, such as grazing, plowing, tree conservation or shrub clearing (Peco et al., 2001). New institutional programmes, such as *Flowering Meadows* are promising to promote the adoption of agri-environmental practices by farmers in France (de Sainte Marie, 2014a), being implemented in circa 140 km^2 so far (de Sainte Marie, 2014b). Outside the EU, the first steps through HNV and grassland conservation are promising pilot projects, such as 'Developing a National Agri-Environmental Programme for Turkey' and 'Conservation and Sustainable Management of Turkey's Steppe Ecosystems'.

Résumé and Future Prospects

Mediterranean and Middle East (M&ME) grasslands are important resources that provide services and goods for man and nature. Their significance follows M&ME human life's evolution from the early start of history till today, while the general depletion of natural resources is signaling their contribution to human well-being in the future as considerably high. Their services and goods sustain local economies and societies, and, in many M&ME countries, they contribute significantly to the GDP. Also, M&ME grasslands are particularly valuable for ecosystem services, such as carbon sequestration, water cycling, erosion control, wildlife biodiversity and they serve as valuable places for ecotouristic and cultural activities. However, they are under pressure of unregulated (unsustainable) grazing, changes in landuse type/intensity, abandonment of traditional human interventions, shrub encroachment, reforestation and afforestation with non-native tree species and climate change. The authors of this chapter believe that it is essential for scientific community to formulate modern tools and practices to deal with these threats. Much research is needed to find the balance between local conditions and grazing capacity and stocking rate, especially for the extensively used grasslands of different environments. Also, land administration and management authorities must shift their thoughts towards understanding the values provided by grasslands and move away from considering grasslands as badlands or land without economic or social reference. At larger scales, mainstreaming biodiversity into natural resource use is needed through policy formulation and implementation. Networking and common initiatives between key stakeholders and policy makers from M&ME countries can help in fine-tuning decisions related to grassland resources. Finally, successful EU initiatives, like Natura 2000 and HNV certification, can offer their rationale or perhaps even can be expanded to cover all M&ME countries.

Glossary

Caespitose = Growing in dense tufts.

Campo-bouça = A Portuguese word for mosaic-like landscapes composed of grasslands surrounded by fruit orchards and forests.

Chamaephyte = Plant whose dormant buds or shoot-apices are above the surface of the ground but not exceeding 25 cm, a term used in Raunkiaer's plant life-form classification.

Chionophile = An organism that can thrive in cold winter conditions.

Dehesa = Spanish word for a multifunctional, agrosylvopastoral system in which animal husbandry takes place in landscapes with trees.

Ephemeral = Short-lived.

Favorable status	=	European Commission has a system for assessing conservation status of habitats as 'favorable' or 'unfavorable', based on evaluation of certain parameters, such as total surface area.
Flysch	=	A thick sequence of marine sediments made up of a marked alternation of fine sediments, such as shales, marls and siltstones with coarse sediments, such as sandstone.
Hemicryptophyte	=	A plant whose dormant buds at the soil surface and whose aerial parts are herbaceous, a term used in Raunkiaer's plant life-form classification.
Lameiros	=	Traditional grasslands of high mountains in north-east Portugal.
Mesogean	=	Recent use of the term "Mesogea" covers the east Mediterranean and the now-closed Bitlis-Zagros Ocean (the southern branch of Neo-Tethys). The differentiation of Mediterranean elements took place on the continental tectonic microplates (like Ibero-Mauritanic, Tyrrhenian, Cyrenaic, Balkanic, and Anatolian), while the origin of Mesogean elements are the Irano-Turanian and African regions.
Montados	=	Portuguese word for *dehesa* (see definition above).
Nemoral	=	Biogeographical zone whose zonal vegetation is deciduous broad-leaved forests.
Ruderal	=	Adjective to describe species or communities that are first to colonize disturbed lands.
Segetal	=	Adjective to denote weed species or communities growing among crops.
Therophyte	=	Annual plant, a term used in Raunkiaer's plant life-form classification.
Yayla/Yeylagh	=	Highlands used as summer pasture in the Middle East.

Abbreviations

cal BP = Calibrated Years Before Present; DM = Dry Matter; EU = European Union; FAO = Food and Agriculture Organization; GDP = Gross Domestic Product; HNV = High Nature Value; M&ME = Mediterranean Basin and the Middle East; MS = Milk Solids

References

Akasbi, Z., J. Oldeland, J. Dengler and M. Finckh. 2012. Social and ecological constraints on decision making by transhumant pastoralists: a case study from the Moroccan Atlas Mountains. J. Mt. Sci. 9: 307–321.

Akbarzadeh, M. 2006. The role of exclosure fencing in protection of grassland diversity. pp. 155–179. *In*: M.H. Asareh. (ed.). Plant Diversity of Iran. Research Institute of Forests and Rangelands Publication, Tehran.

Alados, C.L., A. El Aich, V.P. Papanastasis, H. Ozbek, T. Navarro, H. Freitas, M. Vrahnakis, D. Larrosi and B. Cabezudo. 2004. Change in plant spatial patterns and diversity along the successional gradient of Mediterranean grazing ecosystems. Ecol. Model. 180: 523–535.

Ambarlı, D. and C. Bilgin. 2014. Effects of landscape, land use and vegetation on bird community composition and diversity in Inner Anatolian steppes. Agric. Ecosyst. Environ. 182: 37–46.

Ambarlı, D., U.S. Zeydanlı, Ö. Balkız, S. Aslan, E. Karaçetin, M. Sözen, Ç. Ilgaz, A.G. Ergen, Y. Lise, (...) and M. Vural. 2016. An overview of biodiversity and conservation status of steppes of the Anatolian Biogeographical Region. Biodivers. Conserv. 25: 2491–2519.

Ariapour, A., H. Badripour and M.H. Jouri. 2017. Rangeland and pastureland of Iran: Problems and potentials. pp. 121–148. *In*: V.R. Squires, Z.H. Shang and A. Ariapour (eds.). Rangeland Along the Silk Road: Transformative Adaptation under Climate and Global Change. Nova Science Publishers, New York.

Barkoudah, Y., A. Darwish and M. Antoun. 2000. Biological Diversity: National Report. Biodiversity Unit, Syrian Arab Republic Ministry of Environment. URL: https://www.cbd.int/doc/world/sy/sy-nr-01-en.pdf.

BirdLife International. 2016. Lyrurus Mlokosiewiczi. The IUCN Red List of Threatened Species 2016. URL: http://dx.doi.org/10.2305/IUCN.UK.2016-3.RLTS.T22679483A92815595.en.

Blondel, J. and J. Aronson. 1999. Biology and Wildlife of the Mediterranean Region. Oxford University Press, Oxford.

Breckle, S.-W., I.C. Hedge and M.D. Rafiqpoor. 2013. Vascular Plants of Afghanistan: An Augmented Checklist. Scientia Bonnensis, Bonn.

Bugalho, M.N. and J.M.F. Abreu. 2008. The multifunctional role of grasslands. Options Méditerr. A 79: 25–30.

Calaciura, B. and O. Spinelli. 2008. Management of Natura 2000 Habitats. 6210 Semi-natural dry grasslands and scrubland facies on calcareous substrates (Festuco-Brometalia) (*important orchid sites). European Commission, Brussels.

[CEPF] Critical Ecosystem Partnership Fund. 2016. Mediterranean Basin. URL: http://www.cepf.net/resources/hotspots/Europe-and-Central-Asia/Pages/Mediterranean-Basin.aspx.

Cerdan, O., G. Govers, Y. Le Bissonnais, K. Van Oost, J. Poesen, N. Saby, A. Gobin, A. Vacca, J. Quinton, (...) and T. Dostal. 2010. Rates and spatial variations of soil erosion in Europe: A study based on erosion plot data. Geomorphology 122: 167–177.

Chirino, E., A. Bonet, J. Bellot and J.R. Sánchez. 2006. Effects of 30-year-old pine plantations on runoff, soil erosion, and plant diversity in a semi-arid landscape in south eastern Spain. Catena 65: 19–29.

Chouvardas, D. and M.S. Vrahnakis. 2009. A semi-empirical model for the near-future evolution of the lake Koronia landscape. J. Environ. Prot. Ecol. 3: 867–876.

Christensen, J.H., B. Hewitson, A. Busuioc, A. Chen, X. Gao, I. Held, R. Jones, R.K. Kolli, W.-T. Kwon, (...) and P. Whetton. 2007. Regional climate projections. pp. 847–940. *In*: S. Solomon, D. Qin, M. Manning, Z. Chen, M. Marquis, K.B. Averyt, M. Tignor and H.L. Miller (eds.). Climate Change 2007: The Physical Science Basis. Contribution of Working Group I to the Fourth Assessment Report of the Intergovernmental Panel on Climate Change. Cambridge University Press, Cambridge.

CIFA. 2007. Los Pastos en Cantabria y su Aprovechamiento. Centro de Investigación y Formación Agraria de Cantabria, Consejería de Desarrollo Rural, Ganadería, Pesca y Biodiversidad, Dirección General de Desarrollo Rural, Santander.

Correal, E., M. Erena, S. Ríos, A. Robledo and M. Vicente. 2009. Agroforestry systems in Southeastern Spain. pp. 183–210. *In*: A. Rigueiro-Rodríguez, J. McAdam and M.R. Mosquera-Losada (eds.). Agroforestry in Europe: Current Status and Future Prospects. Springer, Dordrecht.

Cortina, J., F.T. Maestre and D. Ramírez. 2009. Innovations in semiarid restoration. The case of *Stipa tenacissima* L. steppes. pp. 121–144. *In*: S. Bautista, S. Aronson and R.J. Vallejo (eds.). Land Restoration to Combat Desertification. Innovative Approaches, Quality Control and Project Evaluation. Fundación CEAM, Valencia.

de Sainte Marie, C. 2014a. Rethinking agri-environmental schemes. A result-oriented approach to the management of species-rich grasslands in France. J. Environ. Plan. Manag. 57: 704–719.

de Sainte Marie, C. 2014b. Result-Based Payments for Flowering Meadows in France: Support Tools and Collective Learning Process. Conference on Results Based Agri-environment Schemes: Payments for Biodiversity Achievements in Agriculture. Brussels.

Department of Environment. 2017. Department of Environment—Islamic Republic of Iran. ULR: https://en.doe.ir/.

Díaz, M., P. Campos and F.J. Pulido. 1997. The Spanish dehesas: a diversity in land-use and wildlife. pp. 178–209. *In*: D. Pain and M. Pienkowski (eds.). Farming and Birds in Europe: The Common Agricultural Policy and its Implications for Bird Conservation. Academic Press, London.

Dixon, A.P., D. Faber-Langendoen, C. Josse, J. Morrison and C.J. Loucks. 2014. Distribution mapping of world grassland types. J. Biogeogr. 41: 2003–2019.

Drinia, H., A. Antonarakou, N. Tsaparas and G. Kontakiotis. 2007. Palaeoenvironmental conditions preceding the Messinian Salinity Crisis: A case study from Gavdos Island. Geobios 40: 251–265.

Elakkari, T.S. 2005. Structural Configuration of the Syrt Basin. M.Sc. Thesis, University of Twente, Enschede.

Ellero, A., G. Ottria, M.G. Malusà and H. Ouanaimi. 2012. Structural geological analysis of the High Atlas (Morocco): Evidences of a transpressional fold-thrust belt. pp. 229–258. *In*: E. Sharkov (ed.). Tectonics—Recent Advances. InTech. URL: http://www.intechopen.com/books/tectonics-recent-advances.

Escribano, A.J., P. Gaspar, F.J. Mesías, M. Escribano and F. Pulido. 2015. Comparative sustainability assessment of extensive beef cattle farms in a high nature value agroforestry system. pp. 65–86. *In*: V.R. Squires (ed.). Rangeland Ecology, Management and Conservation Benefits. Nova Publishers, New York.

European Communities. 2009. Natura 2000 in the Mediterranean region. URL: http://ec.europa.eu/environment/nature/info/pubs/docs/biogeos/Mediterranean.pdf.

[EEA] European Environment Agency. 2017a. 6310 Report under the Article 17 of the Habitats Directive, Period 2007–2012: Dehesas with Evergreen *Quercus* spp. URL: http://art17.eionet.europa.eu/article17/reports2012/static/factsheets/grasslands/6310-dehesas-with-evergreen-quercus-spp.pdf.

[EEA] European Environment Agency. 2017b. Report under the Article 17 of the Habitats Directive Period 2007–2012: 5330 Thermo-Mediterranean and Pre-steppe Scrub. URL: https://forum.eionet.europa.eu/habitat-art17report/library/2001-2006-reporting/datasheets/habitats/sclerophyllous_scrub/sclerophyllous_scrub/5330-thermo-mediterranea/.

European Union. 2017. Natura 2000. URL: http://ec.europa.eu/environment/nature/natura2000/index_en.htm.

Fadda, S., F. Henry, J. Orgeas, P. Ponel, E. Buisson and T. Dutoit. 2008. Consequences of the cessation of 3,000 years of grazing on dry Mediterranean grassland ground-active beetle assemblages. C. R. Biol. 331: 532–546.

FAO. 2014. FAOstat. URL: http://www.fao.org/faostat/en/#data/.

Ferreras, P. 2001. Landscape structure and asymmetrical inter-patch connectivity in a metapopulation of the endangered Iberian lynx. Biol. Conserv. 100: 125–136.

Galvánek, D. and M. Janák. 2008. Management of Natura 2000 Habitats. 6230 *Species-rich *Nardus* grasslands. European Commission, Brussels.

Geiger, R. 1961. Überarbeitete Neuausgabe von Geiger, R.: Köppen-Geiger/Klima der Erde. (Wandkarte 1: 16 Mill.). Klett-Perthes, Gotha.

González, F., M. Murillo, J. Paredes and P.M. Prieto. 2007. Recursos pascícolas de la dehesa extremeña. Primeros datos para la modelización de su gestión. Pastos 37: 231–239.

Grove, A.T. and O. Rackham. 2003. The Nature of Mediterranean Europe: An Ecological History. 2nd ed. Yale University Press, New Haven.

Guarino, R. 2006. On the origin and evolution of the Mediterranean dry grasslands. Ber. Reinhold-Tüxen-Ges. 18: 195–206.

Harrison, P.A., P.M. Berry, N. Butt and M. New. 2006. Modelling climate change impacts on species' distributions at the European scale: Implications for conservation policy. Environ. Sci. Policy 9: 116–128.

Hellegers, P. 1998. The role of agricultural policy in maintaining High Nature Value farming systems in Europe. pp. 85–89. *In*: J.P. Laker and J.A. Milne (eds.). 2nd LSIRD Conference on Livestock Production in the European LFAs. Macaulay Land Use Research Institute, Bray, Ireland.

Higgins, M.D. and R. Higgins. 1996. A Geological Companion to Greece and the Aegean. Duckworth Publishers, London.

Hsü, K.J., M.B. Cita and W.B.F. Ryan. 1973. The origin of the Mediterranean evaporites. pp. 1203–1231. *In*: W.B.F. Ryan and K.J. Hsü (eds.). Initial Reports of the Deep Sea Drilling Project. Vol. 13. U.S. Government Printing Office, Washington.

Hsü, K.J., L. Montadert, D. Bernoulli, M.B. Cita, A. Erickson, R.E. Garrison, R.B. Kidd, F. Mèlierés, C. Müller and R. Wright. 1977. History of the Messinian salinity crisis. Nature 267: 399–403.

IGME. 1995. Mapa Geológico de la Península Ibérica, Baleares y Canarias a Escala 1:1.000.000. Instituto Geológico y Minero de España, Madrid.

Ispikoudis, I., M.K. Sioliou and V.P. Papanastasis. 2004. Transhumance in Greece: Past, present and future prospects. pp. 211–229. *In*: R.G.H. Bunce, M. Pérez Soba, R.H.G. Jongman, A. Gómez Sal, F. Herzog and I. Austad (eds.). Transhumance and Biodiversity in European Mountains. IALE Publications, Wageningen.

Jalili, A., A. Naqinezhad and A. Kamrani. 2014. Wetland Ecology, with a Special Approach on Wetland Habitats of Southern Alborz. University of Mazandaran Publication, Mazandaran.

[JNCC] Joint Nature Conservation Committee. 2007. Second Report by the United Kingdom under Article 17 on the Implementation of the Directive from January 2001 to December 2006. Supporting Documentation for Making Conservation Status Assessment: Technical Note IV Conservation Status Reporting and Climate Change in the UK. JNCC, Peterborough.

Joffre, R., J. Vacher, C. de los Llanos and G. Long. 1988. The dehesa: An agrosilvopastoral system of the Mediterranean region with special reference to the Sierra Morena area of Spain. Agrofor. Syst. 6: 71–96.

Karaçetin, E., H.J. Welch, A. Turak, Ö. Balkiz and G. Welch. 2011. Conservation Strategy for Butterflies in Turkey. Doğa Koruma Merkezi, Ankara. URL: http://www.dkm.org.tr/Dosyalar/YayinDosya_uJCt50XD.pdf.

Kolahi, M., T. Sakai, K. Moriya and M.F. Makhdoum. 2012. Challenges to the future development of Iran's protected areas system. Environ. Manag. 50: 750–765.

Kürschner, H. 1986. The subalpine thorn-cushion formations of western South Asia: Ecology, structure and zonation. Proc. R. Soc. Edinb. 89B: 169–179.

Lavado Contador, J.F., M. Pulido Fernández, S. Schnabel and E. Herguido Sevillano. 2015. Fragmentation of SW Iberian rangeland farms as assessed from fencing and changes in livestock management. Effects on soil degradation. pp. 183–192. *In*: H. Alphan, M. Atik, E. Baylan and N. Karadeniz (eds.). Proceedings of the International Congress on Landscape Ecology. PAD Publications No. 2, 23–25 October 2014, Antalya.

Maestre, F., D. Ramírez and J. Cortina. 2007. Ecología del esparto (*Stipa tenacissima* L.) y los espartales de la Península Ibérica. Rev. Ecosist. 16: 111–130.

Mittermeier, R.A., W.R. Turner, F.W. Larsen, T.M. Brooks and C. Gascon. 2011. Global biodiversity conservation: the critical role of hotspots. pp. 3–22. *In*: F.E. Zachos and J.C. Habel (eds.). Biodiversity Hotspots—Distribution and Protection of Conservation Priority Areas. Springer, Berlin.

Mucina, L., H. Bültmann, K. Dierßen, J.-P. Theurillat, T. Raus, A. Čarni, K. Šumberová, W. Willner, J. Dengler, (...) and L. Tichý. 2016. Vegetation of Europe: Hierarchical floristic classification system of vascular plant, bryophyte, lichen, and algal communities. Appl. Veg. Sci. 19, Suppl. 1: 3–264.

Muller, S., T. Dutoit, D. Alard and F. Grévilliot. 1998. Restoration and rehabilitation of species-rich grassland ecosystems in France: A review. Restor. Ecol. 6: 94–101.

Naqinezhad, A., A. Jalili, F. Attar, A. Ghahreman, B.D. Wheeler, J.G. Hodgson, S.C. Shaw and A. Maassoumi. 2009. Floristic characteristics of the wetland sites on dry southern slopes of the Alborz Mts., N. Iran: The role of altitude in floristic composition. Flora 204: 254–269.

Nedjraoui, D. 2006. Country Pasture/Forage Resource Profiles. Algeria. FAO, Rome.

Noor Alhamad, M. 2006. Ecological and species diversity of arid Mediterranean grazing land vegetation. J. Arid Environ. 66: 698–715.

Olea, P.P. and P. Mateo-Tomás. 2009. The role of traditional farming practices in ecosystem conservation: The case of transhumance and vultures. Biol. Conserv. 142: 1844–1853.

Olea, L., J. Paredes and M.P. Verdasco. 1990. Características y producción de los pastos de las dehesas del S. O. de la Península Ibérica. Pastos 20-21: 131–156.

Oliveira, F., G. Moreno, L. López and M. Cunha. 2007. Origem, distribuição e funções dos sistemas agro-florestais. Pastagens Forrag. 28: 93–115.

Olson, D.M., E. Dinerstein, E.D. Wikranamayake, N.D. Burgess, G.V.N. Powell, E.C. Underwood, J.A. D'Amico, I. Itoua, H.E. Strand, (...) and K.R. Kassem. 2001. Terrestrial ecoregions of the world: A new map of life on earth. BioScience 51: 933–938.

Oteros-Rozas, E., R. Ontillera-Sánchez, P. Sanosa, E. Gómez-Baggethun, V. Reyes-García and J.A. González. 2013. Traditional ecological knowledge among transhumant pastoralists in Mediterranean Spain. Ecol. Soc. 18(3): Article 33.

Papanastasis, V.P. 2009. Restoration of degraded grazing lands through grazing management: Can it work? Restor. Ecol. 17: 441–445.

Pardos, J.A. 2010. Los Ecosistemas Forestales y el Secuestro de Carbono ante el Calentamiento Global. INIA, Madrid.

Pausas, J.G., C. Bladé, A. Valdecantos, J.P. Seva, D. Fuentes, J.A. Alloza, A. Vilagrosa, S. Bautista, J. Cortina and R. Vallejo. 2004. Pines and oaks in the restoration of Mediterranean landscapes of Spain: New perspectives for an old practice—A review. Plant Ecol. 171: 209–220.

Peco, B., J.J. Oñate and S. Requena. 2001. Dehesa grasslands: natural values, threats and agri-environmental measures in Spain. pp. 37–44. *In*: EFNCP (ed.). Recognising European Pastoral Farming Systems and Understanding their Ecology. Proceedings of the Seventh European Forum on Nature Conservation and Pastoralism. EFNCP [EFNCP Occasional Publication No. 23], Argyll.

Perevolotsky, A. and N.G. Seligman. 1998. Role of grazing in Mediterranean rangeland ecosystems. BioScience 48: 1007–1017.

Pignatti, S. 1978. Evolutionary trends in Mediterranean flora and vegetation. Vegetation 37: 175–185.

Pignatti, S. 1979. Plant geographical and morphological evidences in the evolution of the Mediterranean flora (with particular reference to the Italian representatives). Webbia 34: 243–255.

Pignatti, S. 2003. The Mediterranean ecosystem. Bocconea 16: 29–40.

Pils, G. 2013. Endemism in mainland regions—Case studies: Turkey. pp. 311–321. *In*: C. Hobohm (ed.). Endemism in Vascular Plants. Springer, Dordrecht.

Poschlod, P., S. Kiefer, U. Tränkle, S. Fischer and S. Bonn. 1998. Plant species richness in calcareous as affected by dispersibility in space and time. Appl. Veg. Sci. 1: 75–90.

Pulido Fernández, M. 2014. Indicadores de calidad del suelo en áreas de pastoreo. Ph.D. Thesis, Universidad de Extremadura, Cáceres.

Ramírez, J.A. and M. Díaz. 2008. The role of temporal shrub encroachment for the maintenance of Spanish holm oak *Quercus ilex* dehesas. For. Ecol. Manag. 255: 1976–1983.

Raven, P.H. 1973. The evolution of Mediterranean floras. pp. 213–224. *In*: F. Di Castri and H.A. Mooney (eds.). Mediterranean-type Ecosystems. Origin and Structure. Springer-Verlag, Berlin.

Redman, M. and M. Hemmami. 2008. Developing a National Agri-Environment Programme for Turkey. Buğday Ekolojik Yaşamı Destekleme Derneği, İstanbul.

Reisman-Berman, O. and Z. Henkin. 2007. *Sarcopoterium spinosum*: Revisiting shrub development and its relationship to space occupation with time. Isr. J. Plant Sci. 55: 53–61.

Robles, A.B. and J.L. González-Rebollar. 2006. Pastos áridos y ganado del sudeste de España. Sci. Changements Planet–Sécheresse 17: 309–313.

Rodwell, J.S., V. Morgan, R.G. Jefferson and D. Moss. 2007. The European Context of British Lowland Grasslands. JNCC Report, No. 394. URL: http://jncc.defra.gov.uk/pdf/jncc394_webpt1.pdf.

Ruosteenoja, K., T.R. Carter, K. Jylhä and H. Tuomenvirta. 2003. Future Climate in World Regions: An Intercomparison of Model-based Projections for the New IPCC Emissions Scenarios. Finish Environment Institute, Helsinki.

San Miguel, A. 2008. Management of Natura 2000 Habitats. 6220 *Pseudo-steppe with Grasses and Annuals of the Thero-Brachypodietea. European Commission, Brussels.

Sequeira, E.M. 2008. Pasture and fodder crop as part of high natural value farm systems at Mediterranean dryland agro-ecosystems. Options Méditerr. A 79: 17–21.

Serra, P., X. Pons and D. Sauri. 2008. Land-cover and land-use change in a Mediterranean landscape: A spatial analysis of driving forces integrating biophysical and human factors. Appl. Geogr. 28: 189–209.

Siddall, R. 2013. Geology in the British Museum: The Monumental Stones of the Eastern Desert. UCL Earth Sciences. URL: http://www.ucl.ac.uk/~ucfbrxs/Homepage/walks/Egypt-BM.pdf.

Sousa, W.P. 1984. The role of disturbance in natural communities. Annu. Rev. Ecol. Syst. 15: 353–392.

Squires, V.R. and H. Feng. 2018. Brief History of Grassland Utilization and its Significance to Humans. pp. 2–12 (this volume).

Troumbis, A.Y. and D. Memtsas. 2000. Observational evidence that diversity may increase productivity in Mediterranean shrublands. Oecologia 125: 101–108.

Turkish Statistical Institute. 2016. Yıllık Hayvan Sayısı ve Hayvansal Üretim. URL: https://biruni.tuik.gov.tr/hayvancilikapp/hayvancilik.zul.

Turner, R., N. Roberts, W.J. Eastwood, E. Jenkins and A. Rosen. 2010. Fire, climate and the origins of agriculture: micro-charcoal records of biomass burning during the last glacial-interglacial transition in Southwest Asia. J. Quat. Sci. 25: 371–386.

UNEP-WCMC. 2017. World Database of Protected Areas. June 2017. URL: www.protectedplanet.net.

Vrahnakis, M. 2015. Rangeland Science. Hellenic Academic Electronic Publications – Kallipos, Athens.

Vrahnakis, M.S., M. Kadroudi, E. Kyriazi, D. Vasilakis, Y. Kazoglou and P. Birtsas. 2009. Variation of structural and functional characteristics of grasslands in the foraging areas of the Eurasian black vulture (*Aegypius monachus* L.). Grassl. Sci. Eur. 14: 269–272.

Vrahnakis, M.S., G. Fotiadis, T. Merou and Y.E. Kazoglou. 2010. Improvement of plant diversity and methods for its evaluation in Mediterranean basin grasslands. Options Mediterr. A 92: 225–236.

Waddington, C.H. 1975. A catastrophe theory of evolution. *In*: C.H. Waddington. (ed.). The Evolution of an Evolutionist. Cornell University Press, Ithaca, New York.

Wesche, K., D. Ambarlı, J. Kamp, P. Török, J. Treiber and J. Dengler. 2016. The Palaearctic steppe biome: A new synthesis. Biodivers. Conserv. 25: 2197–2231.

Zamora, J., J.R. Verdu and E. Galante. 2007. Species richness in Mediterranean agroecosystems: Spatial and temporal analysis for biodiversity conservation. Biol. Conserv. 134: 113–121.

Zeder, M.A. 2008. Domestication and early agriculture in the Mediterranean Basin: Origins, diffusion, and impact. Proc. Natl. Acad. Sci. USA 105: 11597–11604.

Zohary, M. 1973. Geobotanical Foundations of the Middle East (Volume I and II). Gustav Fischer Verlag, Stuttgart.

Zomeni, M., J. Tzanopoulos and J.D. Pantis. 2008. Historical analysis of landscape change using remote sensing techniques: An explanatory tool for agricultural transformation in Greek rural areas. Landscape and Urban Planning 86: 38–46.

Zucca, C., M. Pulido-Fernández, F. Fava, L. Dessena and M. Mulas. 2013. Effects of restoration actions on soil and landscape functions: *Atriplex nummularia* L. plantations in Ouled Dlim (Central Morocco). Soil Tillage Res. 133: 101–110.

6

Land Use of Natural and Secondary Grasslands in Russia

Jennifer S.F. Reinecke,[1,*] *Ilya E. Smelansky*,[2] *Elena I. Troeva*,[3] *Ilya A. Trofimov*[4] and *Lyudmila S. Trofimova*[4]

Introduction

The Russian Federation covers a territory of 17.125 million km^2, which represents 31.5 per cent of the Eurasian continent. It stretches from European Russia in the west to the Pacific Ocean in the east, with the Ural Mountains serving as a conventional border between European Russia and Siberia. Over 70 per cent of the territory is lowland, but mountain ranges line the southern border (Caucasus, Altai and Sayan mountains) and prevail in NE Siberia (Verkhoyansk, Chersky and Stanovoy ranges). The climate of this vast country ranges from the polar to the subtropical climate zone and is further influenced by the distance from the oceans (continentality) and particular landforms. The mean monthly temperatures range from –50°C (Yakutia) to +5°C (northern Caucasus) in January and from +1°C (northern coastline of Siberia) to +25°C (Caspian Depression) in July. Annual precipitation ranges from 200–250 mm at the lower Volga to 800 mm in the forest zones of European Russia and the Far East and exceeds 1,600 mm along the coast of the Black Sea in the Caucasus. Consequently, a broad range of biomes is found in Russia—polar deserts and tundra in the tundra geographical zone; forest-tundra, taiga, and bogs in the boreal zone; different deciduous forests in the nemoral zone; forest-steppe, steppe and desert-steppe in the steppe zone. This latitudinal zonation is most obvious in European Russia,

[1] Senckenberg Museum of Natural History, Am Museum 1, 02826 Görlitz, Germany.
[2] Sibecocenter LLC, P.O. Box 547, 630090, Novosibirsk, Russia.
[3] Laboratory of Genesis and Ecology of Soil-Vegetation Cover, Institute for Biological Problems of Cryolithozone, Siberian Division of the Russian Academy of Sciences, 41 Lenin Ave., 677980, Yakutsk, Republic of Sakha (Yakutia), Russia.
[4] Laboratory of Geobotany, All-Russian Williams Fodder Research Institute, 1 Nauchny Gorodok Str., 141055, Lobnya, Moscow Oblast, Russia.
Emails: ilya@savesteppe.org; troeva.e@gmail.com; lstrofi@mail.ru; lstrofi@mail.ru
* Corresponding author: jennifer.reinecke@senckenberg.de

while it becomes less clear towards the more continental climate of eastern Russia. Nearly 51 per cent of Russia's territory includes forests (i.e., boreal taiga and broad-leaved forests: 8.67 million km²), about 21 per cent are mountain regions (3.63 million km²), tundra and forest-tundra zones comprise over 17 per cent (3.00 million km²). The steppe zone covers only about 10 per cent (1.65 million km²) (Ogureyeva, 1999; Trofimov et al., 2010, 2016).

An accurate inventory of the area of temperate grasslands in Russia does not exist currently. Estimates, based on rather coarse land inventory data, refer to a total area of 679,923 km² of pasture lands (Rosreestr, 2013), but this does not take cultivated grasslands, alpine grasslands and arctic tundra pasture into account (Smelansky and Tishkov, 2012). Twenty provinces, mostly along the southern borders of the country, comprise both the largest amount and the largest fractions of pastures (Table 6.1, Fig. 6.1). The zonal steppes of southern Siberia and the northern Caspian Lowland represent the highest share of natural grasslands. Alpine and subalpine meadows at high altitudes and azonal grasslands in floodplains, lakesides and depressions are other natural grasslands. A significant part of Russian grasslands, and the majority of grasslands in the forest zone, are semi-natural—long-term fallows of abandoned cropland, mostly in the steppe, but also in the forest zone; hayfields and pastures resulting from deforestation and bog/lake drainage in the forest and tundra zone; and irrigated and cultivated dry steppes in the steppe zone for use as hayfields.

Grasslands are among the most important natural resources of Russia. The socio-economic settings of Russia are largely determined by physiographic factors. The vast territory, with its diversity of climatic conditions, landforms and soils offers abundant natural resources, but also creates a need for sustainable management strategies and technologies that take these intrinsic natural features of the land into account. Almost

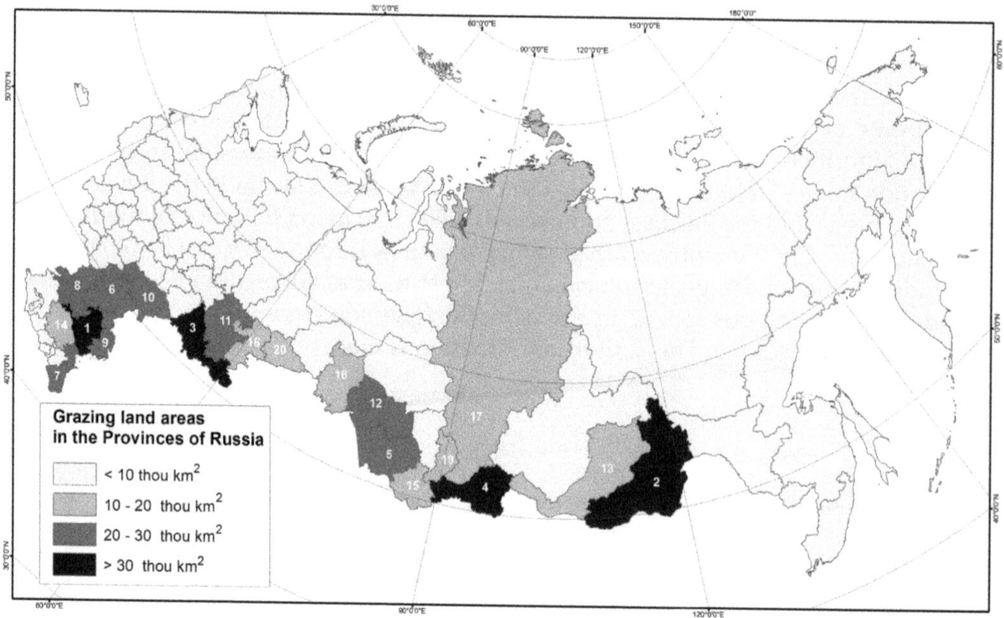

Fig. 6.1 Area of grazing lands per administrative province of Russia. Based on the National Land Inventory on January 1, 2013 (Rosreestr, 2013). (For the names of the provinces, see Table 6.1.)

Table 6.1 Top 20 provinces with the highest estimated area of grazing land (pastures), i.e., more than 10,000 km^2 each. Based on the National Land Inventory on January 1, 2013 (Rosreestr, 2013).

Rank	Name of the Province	Area of Grazing Land [km^2]	Fraction of Total Land Area [%]	Prevalent Types of Grasslands	Type of Steppe
1	Kalmykia Republic	53,281	71.3	Sagebrush northern desert and desertified steppe	Pontic-Kazakh
2	Trans-Baikal Territory (Zabaikalskii Krai)	44,813	10.4	True steppe and meadow steppe	Inner Asian
3	Orenburg	39,810	32.2	True steppe and dry steppe	Pontic-Kazakh
4	Tuva Republic	34,008	20.2	Dry and true montane steppes and cryophyte montane steppe	Inner Asian
5	Altai Territory (Altaiskii Krai)	27,911	16.6	Meadow steppe, true steppe, dry steppe, salt grasslands	Pontic-Kazakh
6	Volgograd	26,550	23.5	True and dry steppes, desertified steppe	Pontic-Kazakh
7	Daghestan Republic	25,887	51.5	Dry and true montane steppes, montane steppe with Mediterranean elements	Pontic-Kazakh
8	Rostov	24,948	24.7	True and dry steppes	Pontic-Kazakh
9	Astrakhan	24,186	49.3	Sagebrush northern desert and floodplain meadows	Small proportion of Pontic-Kazakh, but mainly no steppe
10	Saratov	24,030	23.7	True and dry steppes	Pontic-Kazakh
11	Bashkortostan Republic	23,579	16.5	Meadow steppe, true steppe, steppe meadows	Pontic-Kazakh
12	Novosibirsk	23,156	13.0	Salt grasslands, meadow steppe, true steppes	Pontic-Kazakh
13	Buryatia Republic	18,577	5.3	Meadow montane steppe, true montane steppe, dry montane steppe	Inner Asian
14	Stavropol Territory	16,263	24.6	Dry and true montane steppes, dry steppe, desertified steppe	Pontic-Kazakh
15	Altai Republic	15,231	16.4	Meadow montane steppe, dry montane steppe, true montane steppe, cryophyte montane steppe	Mainly Inner Asian, Pontic-Kazakh
16	Chelyabinsk	13,572	15.3	meadow steppe, true steppe, salt grasslands	Pontic-Kazakh
17	Krasnoyarsk Territory	13,365	5.6	Meadow steppe, true steppe, dry steppe	Inner Asian
18	Omsk	12,658	8.9	Salt grasslands, meadow steppe, true steppe	Pontic-Kazakh
19	Kurgan	10,250	14.3	Salt grasslands, meadow steppe, meadows	Pontic-Kazakh
20	Khakassia Republic	10,250	16.7	Meadow steppe, true steppe, dry steppe, salt grasslands	Mainly Inner Asian, Pontic-Kazakh

two-thirds of lowland areas are faced with poor climate and soil conditions, which are unfavorable for intensive crop farming and cattle breeding. Average human population density is low (8.6 per km^2, ranging from 33–86 in European Russia to 1–8 in Ural, Siberia and the Far East). Efforts are made to increase productivity and life quality in these regions through sustainable development. The All-Russian Williams Fodder Research Institute is one of the leading coordinators of agricultural research and educational activities related to forage production across the country. Scientists base their work on a wealth of data from 100 years of research, including nation-wide maps and databases on land-use cover, vegetation relevés, plant species characteristics, best-practice management techniques and long-term fertilization experiments (50–80 years). The fundamental publications stemming from this research, including monographs, reference books, and recommendations, are the basis of this chapter (e.g., Larin et al., 1950, 1951, 1956; Ramensky et al., 1956; Rabotnov, 1984; Savchenko et al., 1989, 1990; Kosolapov and Trofimov, 2014). Here the plant nomenclature follows Cherepanov (1995).

Origin and Types of Grasslands

Natural Grasslands

Steppes

Steppes represent the largest area of natural grasslands in Russia and form the zonal vegetation of the steppes. These dry grasslands first appeared when the climate became arid and continental about 230–300 million years ago. They developed in co-evolution with herbivorous animals, which found a unique ecological niche and food source in the steppe (Zherikhin, 1993). Ungulates (especially perissodactyles and antelopes) and burrowing rodents and lagomorphs act as 'edificators' (ecological engineers) of these steppe ecosystems (Mordkovich, 2014). During the Pleistocene, steppe landscapes underwent profound changes due to alternating glacials and interglacials (Protopopov and Protopopova, 2010; Kajotch et al., 2016). Cryophilic variants of the steppe appeared throughout the continent as an adaptation to the harsh environmental conditions. With the beginning of the Holocene (about 10,000 BP) the current sequence of latitudinal vegetation zones developed, and steppe vegetation retreated from its original extent (Markov et al., 1965a,b). The boundary between broadleaved forest and steppe was positioned similar to today during the period 7,000–4,800 BP, though temperature decline and increase in precipitation that followed, shifted the boundary southwards between 4,800–2,500 BP. A subsequent decline in woodland cover was caused by both climate change and human impact (Novenko et al., 2016). Another factor that determined steppe formation is the presence of carbonate-bearing loess soils, which are usually covered with steppe vegetation, if no leaching occurred (Tanfilyev, 1896; Sakalo, 1963). Today, the steppe belt stretches as a narrow strip (150–700 km) south of 55° northern latitude along the southern border of Russia (Fig. 6.2). Steppe is the dominating vegetation here, covering plain watersheds, hills and dry areas of river valleys. Additionally, specific types of steppe are found in the alpine region of the main mountain systems of southern Russia, i.e., Caucasus, Altai, Sayans, etc. About 2,300,000 km^2 of 35 administrative provinces are part of the steppe biome with an actual steppe estimated as 500,000 km^2. The four most important provinces with more than 30,000 km^2 of grasslands each are the Kalmyk Republic, Trans-Baikal territory (Zabaikalskii Krai), Orenburg Province and Tyva Republic (Table 6.1; Smelansky and Tishkov, 2012).

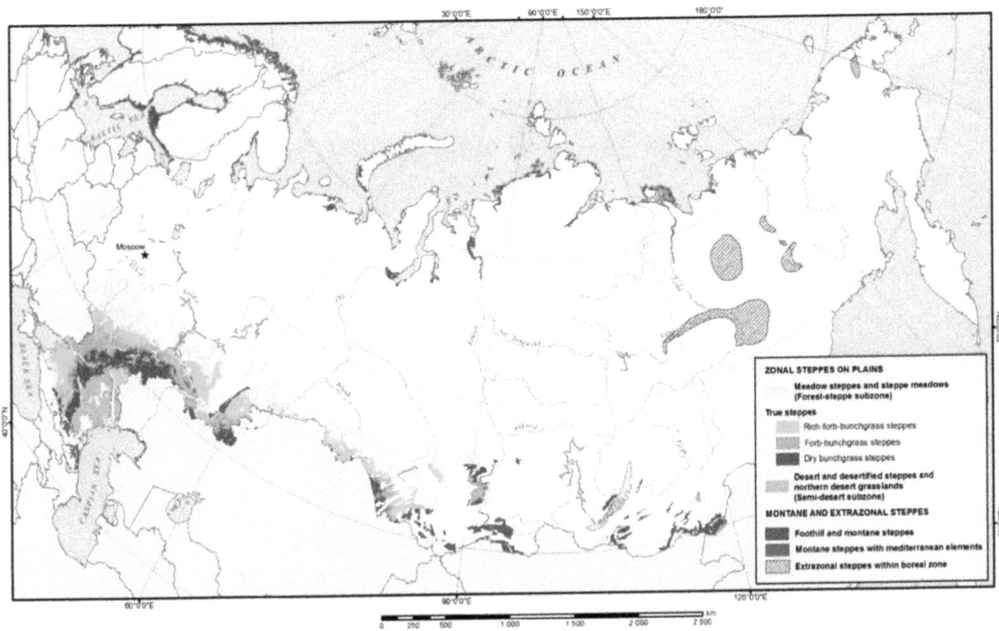

Fig. 6.2 Distribution of steppes in Russia. Note that the area of extrazonal steppes is not continuous, but reflects the broad distribution of very small-scale occurrences. Map adapted from Isachenko (1990), with additional information on extrazonal steppes from Yurtsev (1982) and Troeva and Cherosov (2012).

Russia encompasses a significant part of the diversity of the Eurasian steppe. In general, species richness, total biomass, fraction of belowground biomass, productivity and vegetation cover decrease towards the south (Smelyansky and Tishkov, 2012). At the margins of their range, steppes form landscape mosaics with other vegetation types. In the forest-steppe and semi-desert, steppe communities occur in close proximity to forest, desert, or shrubland communities. Steppes also fringe on saline and freshwater wetlands, grasslands and halophytic succulent communities, or are intermingeled with groves of birch and aspen throughout the vast plains of Western Siberia (Smelyansky and Tishkov, 2012). According to the scheme by Lavrenko (1970a,b) and Lavrenko et al. (1991), the steppe zone in the Eurasian plains is divided into the following latitudinal bands:

(A) Meadow Steppes are the most mesophytic steppe grasslands (Fig. 6.3a), typically forming the grassland component of the forest-steppe landscape, i.e., intermingling with trees and shrubs in floodplains. They grow on deep chernozem soils, which also make these sites favorable for agriculture. Herb cover is dense and dominated by loose bunch grasses (*Poa angustifolia, P. stepposa, Koeleria cristata, Helictotrichon schellianum, Festuca valesiaca, F. lenensis, Carex pediformis*, etc.) and colorful forbs (*Adonis vernalis, Medicago falcata, Salvia nutans, Phlomoides tuberosa, Veronica spicata, V. incana, Pulsatilla patens* s.l., *Aster alpinus*, etc.). The vegetation is rather species-rich (on average 64–67 species on 100 m^2) and productivity is high (18–25 t ha^{-1} yr^{-1}). A relatively large proportion of the biomass stands above the ground.

(B) True Steppes are dominated by xero- and mesophytic graminoids (*Stipa zalesskii, S. tirsa, S. pulcherrima, Helictotrichon desertorum, Carex supina*, etc.) (Fig. 6.3b). The characteristic soils are typical and southern chernozem, dark castanozem and castanozem. Plant growth

Fig. 6.3 Grassland vegetation in Russia: (a) meadow steppe with *Stipa dasyphylla*, Kursk; (b) true steppe with feather grasses, Kishkentay, Orenburg; (c) true petrophytic steppe, Altai Territory; (d) dry mountain steppe with grazing sheep and goats, Bar-Burgazy valley, Altai; (e) extrazonal steppe on SW-exposed slopes, Verkhoyansk, Sakha Republic; (f) floodplain meadows, Oka river, Oryol drop it. Photos by I. Smelansky (a, b, c, d), J. Reinecke (e) and I. Trofimov and L. Trofimova (f).

is interrupted by a short summer break in dry years. Vegetation cover is less dense and most of the biomass is below the ground. Depending on moisture conditions, these steppes may be intermixed with a significant proportion of mesophytic forbs (forb-bunchgrass steppes) or (meso-) xerophytic graminoids (e.g., *Stipa lessingiana, S. krylovii, Festuca valesiaca, Koeleria cristata, Agropyron pectinatum*) and xerophytic semi-shrubs (*Artemisia austriaca, A. glauca* and other *Artemisia* spp., some other *Asteraceae, Chenopodiaceae*) in more drought-adapted communities (dry bunchgrass steppes).

(C) Desertified and Desert Steppes are the most arid steppe types. They form the grassland component of semi-desert landscapes, where they intermingle with desert dwarfshrub communities, saline and xerophytic (semi-)shrub communities. Characteristic soils are light castanozem and castanozem. Aboveground vegetation cover is rather sparse (not more than 25–40 per cent), but belowground root systems are densely packed. Xerophytic grasses and semi-shrubs form the usual communities. Typical dominant species are *Stipa sareptana, S. glareosa, S. caucasica, S. gobica, Cleistogenes squarrosa, Artemisia* subgen. *Seriphidium.*

The steppe region (Steppe Geographical Zone) in Russia can be divided longitudinally into three major subregions: the Pontic-Kazakh Steppe Subregion, from which the western Siberian Steppe Subregion can be separated and the Daurian-Mongolian (Inner Asian) Steppe Subregion (Lavrenko et al., 1991; Mordkovich, 1994; Korolyuk, 2002). The principal boundary line passes across the Altai and Sayan Mountains separating the steppe biome into the western and eastern parts. The Pontic-Kazakh Steppe Subregion is characterized by appreciable winter precipitation and mild springs, which make meltwater available for plants in spring. As a result, western steppe vegetation has a spring peak in its seasonal development and is rich in spring ephemeroids and ephemers (Lavrenko et al., 1991; Wesche et al., 2016). On the contrary, late-summer annuals and cold dry weather in spring are characteristic for the Daurian-Mongolian steppes, where climate is more continental, but the growing season is limited by low and variable rains in summer (Lavrenko et al., 1991; Wesche et al., 2016). The ranges of many plant taxa are restricted to only one steppe subregion (Lavrenko et al., 1991). Thus, *Cymbaria, Saposhnikovia, Filifolium, Panzeria, Schizonepeta, Stellera, Lespedeza,* and two sections of *Stipa* (*Capillatae* and *Smirnovia*), etc. are recorded only in eastern steppes, while *Trinia, Seseli, Crambe, Salvia, Verbascum, Tulipa, Ornithogalum,* etc. and two other sections of *Stipa* (sect. *Stipa* and *Barbata*) are characteristic for western Steppes. Vegetation classification units are also related to this division in plant species ranges—the western steppes belong to the class *Festuco-Brometea,* while the eastern steppes are part of the class *Cleistogenetea squarrosae.*

Steppes form many types of edaphic variants, especially in the mountains along the southern border of Russia. Sand (psammophytic) steppes, stony (petrophytic) steppes (Fig. 6.3c), different variants of salt steppes, and calcareous steppes are recognized across all latitudinal and longitudinal zones (Lavrenko et al., 1991). In the subtropical climate of the Caucasus mountains, a type of Central Asian subtropical steppe with *Andropogon ischaemum* and *Elytrigia* species can be found (Lavrenko et al., 1991; Smelansky and Tishkov, 2012). Particularly noteworthy are the cold cryophytic steppes in the Altai (Fig. 6.3d) and Sayan Mountains. These extremely continental arid mountain steppe are in direct contact with and gradually transform into alpine tundra grasslands, but are also contiguous with quasi-zonal dry steppes of lower altitudes (Lavrenko et al., 1991; Korolyuk, 2002; Namzalov, 2015).

Isolated small areas of steppe vegetation occur far beyond the steppe zone as far as northeastern Siberia (central and northeast Yakutia, Chukotka) (Fig. 6.3e). These extrazonal 'islands' of steppe vegetation within the boreal taiga are considered remnants of the Pleistocene mammoth steppe (Lavrenko, 1981). They are confined to the prominent parts of terraces above floodplains and to south-facing slopes of river valleys (Zakharova et al., 2010; Troeva and Cherosov, 2012; Reinecke et al., 2017). They resemble species-poor local vicariants of true steppes and meadow steppes similar to the Transbaikalian-Mongolian steppes (class *Cleistogenetea*), with a unique hemicryophytic variant (e.g., Reinecke et al., 2017). Steppe vegetation in the Indigirka and Kolyma river basins have floristically common features with the xeric vegetation of north-western North America (Zakharova et al., 2010; Troeva and Cherosov, 2012). Characteristic species are *Stipa krylovii*, *Festuca lenensis*, *Koeleria cristata*, *Poa botryoides*, *Carex duriuscula*, *Pulsatilla flavescens*, *Artemisia commutata*, *A. frigida*, *Helictotrichon schellianum*, *H. krylovii* and *Veronica incana*. Species diversity ranges from 10 species on 100 m² in driest bunchgrass variants to up to 50 species on 100 m² in meadow steppes (Zakharova et al., 2010). Tundra steppes (class *Carici-Kobresieta*), occurring further north in the tundra zone, are similar in physiognomy but not considered actual steppe vegetation. Extrazonal steppes and tundra-steppes only cover small areas and are of minor economic interest, but represent special habitats with a unique biodiversity.

Alpine Grasslands

Mountain massifs are situated in all climate zones except for the semi-desert zone. More than half of the mountain area is occupied by forests and shrubs, but also supports 223,000 km² of grasslands on mountain soils. Within the territory of Russia, alpine and subalpine grasslands are most common in the Central Caucasus mountains and in the humid sector of the Central Altai, forming altitudinal belts above the tree line, which is located at circa 2,000 m a.s.l. (Belonovskaya, 2010; Zibzeev, 2012). Alpine grasslands are formed by short-grass communities with participation of cushions and rosettes and are confined to acidophilous mountain meadow soils on prominent and rather steep slopes. They are attributed to the class *Juncetea trifidi*. In the Caucasus, these short-grass graminoid communities belong to the order *Caricetalia curvulae*, including *Carex bigelowii* s.l., *Festuca ovina*, *Helictotrichon versicolor*, *Luzula spicata*, *Lupinaster polyphyllus*, *Salix glauca*, *Minuartia recurva*, *Anemone speciosa*, *Eritrichium caucasicum* in combination with lichens (Korotkov and Belonovskaya, 2001; Ermakov, 2012). The alpine grasslands of the Altai belong to the phytosociological order *Violo altaicae-Festucetalia kryloviana*e of the abovementioned class. The diagnostic species are *Carex ledebouriana*, *Dracocephalum grandiflorum*, *Festuca kryloviana*, *Gentiana grandiflora*, *Minuartia arctica*, *Patrinia sibirica*, *Viola altaica* (Ermakov, 2012). Subalpine medium- and tall-grass communities, which prefer rather warm and humid conditions, are classified into class *Mulgedio-Aconitetea*. Caucasian grass-dominated communities are grouped into the order *Calamagrostietalia villosae* (*Anemonastrum narcissifolium* s.l., *Bistorta major*, *Dianthus superbus*, *Solidago virgaurea*, etc.), while nitrophilous grasslands are represented by the order *Senecioni rupestris-Rumicetalia alpini* (*Carduus personata*, *Cerastium davuricum*, *Geranium sylvaticum*, *Heracleum asperum*, etc.) (Ermakov, 2012). Subalpine tall-grass coenoses of lower and medium altitudes in the Altai are attributed to the order *Trollio-Crepidetalia sibiricae* (*Aconitum septentrionale*, *Bupleurum longifolium*, *Chamaenerion angustifolium*, *Cirsium heterophyllum*, *Hedysarum theinum*, *Lilium pilosiusculum*, *Milium effusum*, *Phlomoides alpina*, *Pleurospermum uralense*, *Saussurea latifolia*, *Thalictrum minus*, *Veratrum lobelianum*). Medium-grass communities of the upper part of subalpine belt belong to the order *Schulzio crinitae-Aquilegietalia glandulosae* (*Anthoxanthum alpinum*, *Aquilegia glandulosa*, *Deschampsia altaica*, *Dracocephalum grandiflorum*, *Schulzia crinita*, *Trisetum altaicum*, etc.) (Ermakov, 2012; Zibzeev, 2012). These grasslands represent valuable pastures for alpine wildlife and livestock.

Azonal Grasslands

Floodplain meadows account for 250,000 km² of grasslands and can be found all across Russia. Meadows occur naturally where perennial, mesic herbs may compete with xerophylous and riparian species—in the floodplains (Fig. 6.3f), sea- and lake-sides, depressions and lagoons. Composition and species diversity of floodplain meadows is determined by duration of flooding and soil type. The central part of floodplain deposits is formed by sand and silt, producing nutrient-rich and well-drained soils, thus favoring the development of very productive and valuable grass stands. In the upper part of floodplains, deposits are basically composed of clay and silt. These soils are heavier, but also rich in nutrients and humus. Poor drainage and run-off from slopes, as well as high ground water tables may, however, lead to swamping in depressions (Parakhin et al., 2006).

The vegetation of these grasslands belongs to the classes *Molinio-Arrhenatheretea* and *Calamagrostetea langsdorffii* and is dominated by *Festuca pratensis, Dactylis glomerata, Phleum pratense, Elytrigia repens, Bromopsis inermis, Alopecurus pratensis, Agrostis stolonifera, Deschampsia cespitosa, Trifolium pratense, Lathyrus pratensis, Vicia cracca, Geranium pratense, Sanguisorba officinalis, Bistorta major, Rumex confertus, Ranunculus repens* and *Veratrum lobelianum* (Trofimov et al., 2010, 2016; Ermakov, 2012). The greatest floristic differences are observed in meadows growing under long-term (more than 15 days) flooding versus short-term (less than 15 days) flooding. Short-term flooded meadows occur on prominent parts of central and upper parts of floodplains. Their plant communities are close to zonal vegetation due to similar moisture and soil conditions, which is podsolic and least alluvial. Steppe meadows can develop here, in wet grass-forb meadows with *Carex* growing in depressions. Long-term flooded meadows are confined to lower floodplains and flat areas of central and upper floodplains. They contain mesic grass and forb-grass communities with an admixture of legumes, as well as wet grass- and forb-sedge meadows (Parakhin et al., 2006).

Secondary Grasslands

The majority of grasslands in the forest zone of Russia and in some part of the steppe zone are of anthropogenic origin. In the forest (boreal and nemoral) zone deforestation, bog and lake drainage have been carried out in order to create pastures and hayfields. Grazing and mowing can turn meadows into stable and long-standing ecosystems, which can exist for centuries without returning to their original state as long as management continues (Rabotnov, 1984; Parakhin et al., 2006). These include dry grass or grass-forb meadows on podsolic and sod-podsolic soils (50–60 per cent) in the plains and grass or grass-sedge meadows in lowlands and depressions with meadow, bog-podsolic and sod-gley soils (20–30 per cent; Trofimov et al., 2010, 2016). Similar to the natural azonal grasslands, secondary grasslands are part of the classes *Molinio-Arrhenatheretea* and *Calamagrostetea langsdorffii* (Ermakov, 2012). In the steppe zone irrigation has led to the transformation of natural grasslands to more productive meadows, but such transformation is rather rare.

Agronomic Use of Grasslands

History of Land Use

In the forest zone of European Russia people started burning forests to create and enlarge fodder lands as early as 4,000–11,000 BP. Nomadic pastoralism (Fig. 6.4) emerged in the Russian steppe zone at the beginning of the Bronze Age and continued to be a dominating life strategy up to Russian colonization in the 16th–19th centuries (Khodarkovsky, 2002; Sunderland, 2006; Frachetti et al., 2012). Steppes were extensively grazed by sheep, horses, cattle and camels of numerous tribes of mobile herders. Typically, the life strategy of

Fig. 6.4 Winter camp of mobile herders in the Altai Republic, at an elevation of about 2,300 m a.s.l. (Photo by I. Smelansky, 2016.)

Eurasian mobile pastoralists was based on year-round grazing with regular movements between two or more (up to dozens) seasonal pastures. Dominating strategies were transhumance in mountain areas and latitudinal moving on vast plains. Herders and their families moved with their livestock. Haymaking, crop production or irrigation played no or only a minor role; dung collection and use of fire for pasture management were common as well as massive 'surround' hunting for wild ungulates, but the environmental impact on grasslands was generally low and imitated effects of migrating wild ungulates (Khazanov, 2002; Kradin, 2007).

In the 10th–12th century, the golden age of Kievan Rus', cattle breeding and grassland farming rapidly developed along with arable farming. Cattle grazed on natural pastures all-year round, and grazing intensity depended on the forage quality of local pastures. In the 18th century, people started to prepare hay for winter, though northern Slavs had practiced mowing already in the 11th–12th century due to snow cover impeding winter grazing (Larin et al., 1950, 1951, 1956; Trofimov and Kravtsova, 1998; Trofimov, 2001; Kosolapov and Trofimov, 2014). At the beginning of the 20th century, grasslands were used both for grazing and mowing with systematic development of grassland farming in the mid-20th century. Complex arrangements for improvement of fodder land were made, including land clearing, drainage, irrigation and other intensification practices. Thus, largely natural grasslands were transformed into highly productive hayfields and pastures, reaching a productivity of 4,000–6,000 fodder units/ha.[1]

[1] A fodder unit (FU) is a measure for the nutritional value of fodder, conventionally used in Russian agronomy. 1 FU is equitable to the nutritional value of 1 kg of average dry oats (*Avena sativa*), resulting in 150 g of accumulated fat in livestock (Medvedev and Smetannikova, 1981).

European-style agriculture began to penetrate into steppes in the 16th century but remained local and mainly unproductive until the 19th century. Grasslands ware massively turned into arable lands in the 18–19th century for market-driven wheat production in different parts of the Russian steppe region, which resulted in almost complete destruction of the steppe landscape to the early 1900s (Khodarkovsky, 2002; Sunderland, 2006; Moon, 2013). However, up to the 1930s shifting long-fallow agriculture was prevalent across the steppe region, leaving approximately two-thirds of all arable lands unplowed for 10–20 years. These transformed grasslands were extensively used by peasants for grazing and hay making (Moon, 2013). In the 1920s–1930s this relatively extensive land use rapidly intensified with the development of modern machinery and the administrative will to implement mechanized agriculture. Finally, the last extensive areas of virgin steppes and secondary steppe grasslands were destroyed during the Soviet 'Virgin Land Campaign' of 1954–1963 when 197,000 km^2 (Russia only; 452,000 km^2 across the USSR) were turned into permanent arable lands (Semenov, 2012; Elie, 2015). The massive reduction of grassland area available for grazing without any decrease in livestock numbers resulted in a catastrophic increase of grazing pressure, exceeding the carrying capacity of steppe grasslands nationwide (Smelansky, 2007).

After 1991, the state farm system and entire Soviet agriculture collapsed. As a result, vast areas of both arable lands and grasslands were abandoned (Prishchepov et al., 2012; Smelansky, 2012b; Alcantara et al., 2013; Schierhorn et al., 2013). According to Lyuri et al. (2010), the decrease in cropland area across Russia was 440,000 km^2 (22 per cent of the whole arable area); across European Russia alone about 310,000 km^2 cropland were abandoned (Schierhorn et al., 2013). Some 260,000 km^2 of this abandoned cropland are scattered across 37 Russian provinces where steppes occur naturally (southern Russia: Smelansky, 2012b; Smelyansky and Tishkov, 2012). The majority of this abandoned land has spontaneously transformed into (more or less steppe-like) secondary grasslands (Fig. 6.5), whereas abandoned land in the forest zone is mainly subject to spontaneous afforestation (Lyuri et al., 2010; Ioffe et al., 2012; Kämpf et al., 2016).

The majority of Russian livestock is concentrated in the steppe region. During the agricultural crisis following the collapse of the USSR (1991–1999), economic reforms, which were not targeted at the maintenance of previous agricultural production potential, led to a sharp decline in livestock numbers. By 1999 cattle numbers had dropped to half of the numbers of 1990, while sheep and goats stabilized at a third of that level (circa 1,900,000 in 2006–2009 compared to 5,150,000 in 1990) (Kara-Murza et al., 2008; Smelyansky and Tishkov, 2012). Subsequently, overgrazing in natural pastures was reduced and became largely restricted to the local level, mainly around settlements. On the contrary, undergrazing became common in a majority of Russian grasslands (with the exception of a few mountainous or arid provinces, like Chechnya, Daghestan, Kalmykia, Tyva and Altai) (Smelyansky and Tishkov, 2012). Massive land abandonment and rising rural poverty caused a dramatic increase in the frequency and area of grassland wild fires, which have become one of the most powerful factors affecting grasslands in Russia since the mid-1990s (Dubinin et al., 2010; Smelansky, 2015; Tkachuk, 2015; Pavleichik, 2016).

The trends in land use abandonment were stopped in 2005–2008 and partially reversed. During the last decade both crop area and livestock numbers did not decrease further and even increased in many Russian provinces (Smelyansky and Tishkov, 2012; Meyfroidt et al., 2016). Systematic studies and inventories of grasslands during the 20th century have resulted in guidelines on classification and evaluation of fodder lands throughout the former USSR (Ramensky et al., 1956; Gaidamaka et al., 1984; Savchenko et al., 1987, 1989, 1990, etc.).

Fig. 6.5 Re-plowed fallow land with secondary steppe in Samara province. (Photo by I. Smelansky, 2014.)

Current Practice of Grassland Management

Real mobile pastoralism does not any longer exist in the Russian steppes. The mode most commonly used in steppe grasslands is extensive grazing in summer combined with wintering in barns. The grazing period for cattle decreases from the semi-desert zone towards the forest zone in the north to 130–160 days. Here, winter food, such as hay and silage, become increasingly important. Unlike cattle and sheep, horses can roam the steppe rangeland freely without everyday herders attending to them and also are more often left grazing in winter. Transhumance is still vital in the south-eastern Altai and Tuva mountains and to a lesser extent in the Caucasus, where steppes are used in rotation with alpine grasslands and open woodlands. Common practice here is the division into summer and winter rangelands. Transhumance grazing and extensive grazing in distant pastures are the grazing modes most compatible with the steppe ecosystem and rather compatible with meat and wool production. The production of milk and dairy products, however, is less effective under these regimes (Smelansky and Tishkov, 2012).

A typical grazing system in the majority of Russian stock-breeding farms includes three elements: natural pastures, sown pastures and perennial or annual crop cultivation for grazing or feeding bunks (Kutuzova, 2007; Kosolapov et al., 2012, 2014). Hay making is the keystone element in this system. Nationwide about 200,000–250,000 km^2 of natural and sown fodder lands are mown annually. Fodder biomass and quality is highest in spring and deteriorates in quality during summer and in biomass in early autumn. Unfavorable years can create a two- to three-fold loss in productivity and fodder reserves in natural grasslands, and up to five- to seven-fold under extreme conditions in the south of Russia. On the contrary, in favorable years productivity and fodder reserves may increase by 1.5–2 times (Kutuzova, 2007; Kosolapov et al., 2012, 2014). In pastures and hayfields nitrogen, phosphorus and potassium fertilization as well as chalking are applied (Trofimova et al., 2008) but this treatment is quite rare in the last decades. Common treatment of cultivated pastures also includes mowing of leftover vegetation, undersowing and irrigation.

Sustainable use is essential for the longevity of productive grasslands. Approaches to increase the productivity of pastures include reclamation and re-grassing. Reclamation aims at returning impoverished grass stands back to agricultural use, thus increasing fodder productivity (three to five times) and quality, especially regarding protein content. Hayfields and pastures are sown in places where re-grassing is not effective (grasslands with strongly reduced grass stands and/or proportion of shrubs and tussocks). Accelerated re-grassing is preferable on slope land forms and floodplains which are prone to soil erosion (Kosolapov and Trofimov, 2009, 2016). Long-term experiments have shown that the high productivity (50–74 GJ of metabolizable energy) of sown grasslands can be maintained over 60–70 years (Trofimova et al., 2008). Proper selection of herb-and-grass mixtures is critical for effective reclamation. This depends on regional climate conditions, as well as on slope aspect, expected productivity, intended use (hayfields, pastures or both), and artificial fertilization. Pure grass mixtures work best on landforms with extreme environmental conditions (lower floodplains, steep slopes, strongly salinized depressions, etc.). Monodominant crops are cultivated under severe conditions of deserts and semi-deserts (Trofimova et al., 2008). Sod tillage is used for grass stand rejuvenation in long-fallowed and natural grasslands to increase the portion of nutritious species, as well as to improve nutrition and moisture conditions of the soil. However, long-term studies show that grassland plant communities rely on a well-developed sod in order to be protected from droughts, erosion and nutrient leaching. Sustainable agriculture should thus consider the process of sod formation, which relies on the sod being unplowed during a long period of time (60 years). This way, soil fertility is increased mainly due to accumulation of humus and nitrogen, making further cost-intensive treatment unnecessary (Trofimova et al., 2008).

Ecological and Economic Value of Grasslands

Ecosystem services of Russian grasslands play an important, and in several regions even vital, provisioning role for both agricultural industry and the livelihood of rural people (Smelansky, 2003; Smelansky and Tishkov, 2012). The entire set of ecosystem services includes resource provisioning, soil fertility, climate and water regulation, carbon sequestration, erosion control, pollution control, biodiversity maintenance, securing human well-being, recreation and hedonistic values (Tishkov, 2005; Smelansky and Tishkov, 2012).

The most important ecosystem service of grasslands, which is actually marketable in the national economy, is the production of fodder as a resource for the stock-breeding sector of agriculture. Fodder production is the most expensive item (50–60 per cent of total input) in animal breeding. The aboveground phytomass of all fodder-producing ecosystems, including reindeer pastures, amounts to 130–180 Mt dry matter (DM). The overall fodder reserve that can be sustainably used is 60–80 Mt DM or 25–40 Mt fodder units per year (Trofimov et al., 2010, 2016; Kosolapov et al., 2014). This is enough to produce 15–20 Mt milk, 7–10 Mt meat and 0.1–0.15 Mt wool per year with a total annual income of 250–500 billion RUB (circa 4.35–8.70 billion US$). Thus, fodder production in natural and secondary grasslands is an important economic factor, but it differs widely across climatic zones (Trofimov et al., 2010, 2016; Kosolapov et al., 2014). Fodder productivity in natural grasslands varies from 1,400 to 8,100 kg ha^{-1} yr^{-1} in different provinces, with highest productivity in the steppe region (Bukvareva and Zamolodchikov, 2016). The overall fodder reserve in the steppe region is estimated at 20–25 Mt DM (Trofimov et al., 2010, 2016; Kosolapov et al., 2014). Steppe grasslands are the main base for national beef, sheep and horse meat production as well as some other products of cattle and sheep. In 2008, nine of the 10 top provinces of cattle meat production were situated in the steppe region; and of the 20 provinces with the highest number of cattle only two were outside the steppe region. The 35 'steppe provinces' in total produced more than 70 per cent of the total number of cattle in Russia and more than 90 per cent of the nation-wide wool yield (Smelansky and Tishkov, 2012).

Azonal floodplain meadows are the most valuable fodder lands, as pastures and especially as hayfields. Hayfields are usually highly productive, inter-annually stable and yield high-quality fodder. The average productivity of floodplain meadows is 1.5–2.5 t ha^{-1} yr^{-1} DM (Trofimov et al., 2010, 2016). Grasslands in the uplands of the forest zone are of average productivity, yielding 1.0–1.1 t ha^{-1} yr^{-1} DM. They are good pastures and hayfields for cattle, sheep and horses. The general fodder reserve in the forest zone is 25–35 Mt DM (Trofimov et al., 2010, 2016). The tundra zone is the fodder base for domestic reindeer, with 1,670,000 km^2 (over 51 per cent) of reindeer pastures and 4,000 km^2 (nearly one per cent) of other fodder grasslands, such as floodplain meadows and depressions, which are suitable for fodder production for dairy cattle. The productivity of tundra pastures is 0.8–1.2 t ha^{-1} yr^{-1} DM and the general fodder reserve in the tundra zone is 0.12–0.25 Mt DM (Trofimov et al., 2010, 2016). Alpine grasslands cover about 223,000 km^2 (over 25 per cent) and, with a fodder reserve of 10.0–25.0 Mt DM (Trofimov et al., 2010, 2016), reach an average productivity of 0.7–0.9 t ha^{-1} yr^{-1} DM. They are used as pastures for sheep, cattle and horses. Extrazonal steppes serve as natural pastures for cattle and horses though their productivity is low, ranging from 0.1 to 1.2 t ha^{-1} yr^{-1}, depending on moisture conditions (Andreyev et al., 1987).

On the other hand, only a small portion of the fodder produced is currently consumed by livestock. The consumption rates of total forage production across federal provinces usually reach 4–6 per cent (up to a maximum of 19 per cent in Kalmykia; Bukvareva and Zamolodchikov, 2016). Both the forage consumption rate and the livestock production index (based on the annual production of livestock grazing on unimproved natural grasslands; FAO) are highest in the steppe region (Naidoo et al., 2008; Bukvareva and Zamolodchikov, 2016). The carrying capacity of steppe pastures for all herbivores is usually estimated as 75–80 per cent of primary production. The majority of this is acquired by wild herbivores and only a small portion is consumed by livestock (Abaturov, 1984). Therefore, Russian grasslands generally have a significant potential to support increased livestock numbers, but no more than twofold, if these grasslands are used sustainably. At the same time, grazing pressure on the grasslands of Kalmykia and several other regions seems to be very close to the upper limit of their carrying capacity.

Fertile chernozem and chestnut soils are characteristic of steppe ecosystems and are the most important resource for Russian agrarian economy (Tishkov, 2005). These soils form the basis of Russian crop production, especially of cereals and oil-bearing seeds, which still hold the highest economic value. Russia has become one of the global top producers and exporters of grains (especially wheat) during recent years (Swinnen et al., 2017). An assessment in 2008 showed that the top 20 provinces on the list of national grain production were 'steppe provinces', and that the 35 Russian provinces with at least some steppe area (and steppe soils), contributed 86 per cent to the national total of grain production (Smelansky and Tishkov, 2012). Accordingly, the highest agricultural value lies in the steppe region, with an estimated national flow value of 29.3–46.0 billion US$ per year (Smelansky and Tishkov, 2012).

Grasslands also play an important role in carbon sequestration and the reduction of greenhouse gas emissions, thus benefitting society beyond sectorial and national boundaries. In grassland ecosystems, carbon accumulates as organic matter and 'black carbon' (inorganic carbon, such as charcoal, etc.) in the soil and may mitigate rising levels of atmospheric carbon dioxide (CO_2) (Jones, 2010). The total carbon pool in Russian steppe soils was assessed as 35,000 Mt (Smelansky, 2012a). It is estimated that Russian grasslands overall are a sink of circa 700 Mt C yr^{-1}, approximately as much as the entire Russian forests (Tishkov, 2005). Hayfields and pastures (compared to arable land) may prevent the emission of greenhouse gases of estimated 176 Mt CO_2 equivalents per year (Romanovskaya et al., 2014). Newly formed secondary grasslands in place of abandoned croplands act as a significant carbon sink of atmospheric CO_2: net ecosystem production is estimated as 2.96 ± 0.90 t C ha^{-1} yr^{-1} in the forest-steppe zone and 1.46–1.74 t C ha^{-1} yr^{-1} in the steppe zone (Kurganova et al., 2015). The amount of carbon sequestered in the abandoned lands over the period 1990–2009 amounts to 42.6 ± 3.8 Mt C yr^{-1} (Kurganova et al., 2014) and 470 Mt C for the period 1991–2009 in total (Schierhorn et al., 2013).

Grassland ecosystems of Russia also contribute significantly to biodiversity and harbor a great genetic diversity of plants (Smelansky, 2003). Russian steppes serve as habitats for 5,000 or more vascular plant species (40 per cent of the national flora), about 100 species of mammals and 150–180 species of birds (Tishkov, 2005). The total value of ecosystem services of the Russian steppe biome (other grasslands not included) is estimated at 234–445 US$ per hectare and year, or 11.7–22.2 billion US$ per year for the entire country (Smelansky and Tishkov, 2012).

Threats to Grasslands

Steppe is the most important natural grassland type in Russia in terms of conservation. The main threats to Russian steppes were recently analyzed under the framework of the UNDP/GEF project, 'Improving the Coverage and Management Efficiency of Protected Areas in the Steppe Biome of Russia' (2010–2016). A mixture of direct and indirect threats to the extent and ecosystem functioning of the steppe biome has been identified nationwide. Most of the threatening factors were associated with some kind of human activity and linked mostly with the intensification of land use (Deák et al., 2016).

Direct destruction by plowing for permanent crop cultivation was historically the main threat for natural and extensively used steppes and is still important (Smelansky, 2007; Smelansky and Tishkov, 2012). For centuries, the agricultural strategy had been oriented towards maximal expansion of arable land for production of marketable crops, and economic interests took a dangerous priority over ecological issues (Kiryushin, 1996; Dobrovolsky, 2008; Kashtanov, 2008; Trofimova et al., 2010; Trofimov et al., 2011, 2012; Kosolapov et al., 2012; Dengler et al., 2014; Ivanov, 2014). Plowing rates exceeded all reasonable limits, impacting even low-productive land on sands, salt-influenced and humus-poor soil. Meadow steppes of European Russia have been most affected; only small and highly fragmented remnants survived as pastures and hayfields for the last decades but are now mainly abandoned (Smelansky, 2007; Smelansky and Tishkov, 2012; Wesche et al., 2016). Cryophytic steppes in alpine regions and edaphic variants on stony slopes, sand steppes and salt habitats often remain the only untouched pieces of steppe grassland (Smelansky, 2007; Smelansky and Tishkov, 2012; Wesche et al., 2016). In the last decade, long-lived secondary grasslands on fallow lands in the steppe zone and some steppe remnants are threatened by crop encroachment due to pressure by both the Russian government policy directives and economic forces (Prishchepov et al., 2012; Smelansky, 2012b; Alcantara et al., 2013; Schierhorn et al., 2013).

Both over- and undergrazing are important threats for steppe ecosystems today. Since the mid-20th century, many steppes have been over-grazed and are characterized by strongly degraded grass stands and low productivity (Figs. 6.6a and 6.6b). Rotational grazing was not continued on many sites, which caused grassland deterioration, specifically threatening extrazonal steppes and grasslands in Yakutia affected by year-round grazing (Ivanova, 1967, 1981; Ivanova and Perfilyeva, 1972; Mironova, 1992; Gavrilyeva, 1998; Ivanov et al., 2004). After 1991, livestock numbers decreased dramatically, and overgrazing became restricted to the local level, mainly around settlements (Fig. 6.6c). At the same time sheep and goat numbers only slightly decreased or even increased in the dry steppes of the Caucasus region and later in the Tyva and Altai Republics (Smelansky and Tishkov, 2012). Absence of grazing during a decade or more was a key factor for spontaneous restoration of formerly disturbed grasslands (Smelansky, 2007; Kandalova, 2009; Kandalova and Lysanova, 2010; Morozova, 2012; Rusanov, 2013) and generally the recovered steppes become more mesic, with higher biodiversity. However, lack of grazing is not appropriate for long-term maintenance of steppe grasslands in Russia (while this may be the case under more arid environments, like in some parts of Mongolia and Kazakhstan) (Gavrilyeva, 1998; Ivanov et al., 2004; Smelansky, 2007; Kandalova, 2009; Kandalova and Lysanova, 2010). For a significant part of Russian grasslands, vegetation composition is unstable and shrub invasion and litter accumulation negatively affect the dominance of grasses (Troeva and Cherosov, 2012). In many areas, under-grazing led to the deterioration of pastures and hayfields. Specifically, one-third to half of grasslands are threatened by undergrazing and grass cutting, especially the meadow steppes in European Russia and West Siberia (Mathar et al., 2016). Over 20–30 per cent of the remaining area was covered with shrubs and forests, and 10 to 15 per cent of pastures contain hillocks (Kosolapov et al., 2015).

Anthropogenic desertification historically was a threat for the desert steppes of Kalmykia and dry steppes of central and eastern Siberia (Altai, Tyva, Khakassia Republics, and Dauria region) (Smelansky, 2007; Smelansky and Tishkov, 2012). Year-round overgrazing and irrational agricultural activity resulted in intensive disturbance of vegetation and deterioration of the soil. As a consequence, wind-blown sands cover pastures and settlements, threatening the economy and people's well-being. Irrigation and diversion of water also contributed to desertification, as is best known from the case of massive irrigation in the Black Lands area, Kalmyk Republic, in the 1980s. After

Fig. 6.6 Effects of overgrazing on steppe vegetation in Russia. (a) winter grazing on overgrazed dry steppe, Kazanovka Museum-Zapovednik, Khakassia; (b) overgrazed dry steppe, Kishkentay, Orenburg; (c) overgrazed true steppe near summer corral, Altai. (Photos by I. Smelansky, 2014 (a), 2012 (b), 2015 (c).)

the 1990s, the area of anthropogenic 'deserts' in Russia (and Kazakhstan) collapsed and became a scene of dramatic spontaneous restoration of steppe ecosystems (Neronov and Tchabovsky, 2003; Smelansky, 2007; Dubinin et al., 2010, 2011; Hölzel et al., 2012; Smelansky and Tishkov, 2012). These areas now face the risk of desertification again because sheep and goat numbers have been restored, while grazing regimes have not been improved.

Steppes are also directly threatened by the mining, oil and gas industries, which form an important basis of Russian economy and cover substantial territory, including steppe areas. Mining is specifically problematic in the southern Urals, Altai, Khakassia and Trans-Baikal (Dauria) areas, where extensive areas of cryophytic and meadow steppes have survived until now (Smelansky, 2007; Smelansky and Tishkov, 2012).

The threat posed to steppes by afforestation is rather high, but in recent decades it is limited to a few regions. The Belgorod province is an exceptional case of massive afforestation for political reasons (Titova et al., 2014).

Wild fires more or less strongly affect almost all grassland regions in Russia, but their effect on grassland ecosystems is inconsistent (Fig. 6.7). Steppe fires became numerous and frequent after 2000, when livestock numbers decreased (Smelansky, 2015). Surplus of dry, ungrazed biomass and unregulated agricultural burning are among the main causes for wild fires in steppes. Steppe fires are specifically extensive and frequent in Kalmykia, the Volga-Ural area, West Siberia, Khakassia, and Dauria (Trans-Baikal), where they are threatening the people and the economy. However, the ecological role of fires is not so clear. They can be important for maintaining grasslands in the absence of grazing and facilitate the development of steppe in former fallows. Wild fires are especially favorable for steppes in arid environments (e.g., in Kalmykia), where they facilitate a shift from *Artemisia* semi-

Fig. 6.7 Patches of petrophytic and typical meadow steppes after fire: green patches have burnt while yellow patches have not; Dauria. (Photo by I. Smelansky, 2015.)

shrubs to steppe with bunchgrasses (*Stipa* and others). Nevertheless, repeated burning may be detrimental to many regions and lead to the degradation of steppe ecosystems.

Several characteristic steppe species are endangered because of overharvesting for illegal international and/or domestic trade (e.g., early spring wild flowers, such as *Crocus* spp., *Bulbocodium versicolor*, *Pulsatilla* spp., *Tulipa suaveolens*, etc.) (Smelansky and Tishkov, 2012).

Other threats, such as invasion of alien species, are of lower priority. Effects of global climate change and permafrost degradation interact with local grazing patterns (see Wesche et al., 2016).

Conservation of Grasslands

Legal Aspects

The IUCN's World Commission on Protected Areas (WCPA) recently recognized temperate grasslands as the most disturbed and worst protected terrestrial biome (Henwood, 2010), which is also true for the Russian situation (Smelansky, 2007; Smelansky and Tishkov, 2012). In Russia, neither grasslands in general nor steppes in particular are officially recognized as a conservation priority. There is neither legislation nor any governmental policy regulating grassland conservation or protecting grasslands for their natural value. Russian conservationists (non-governmental organizations and experts) recently recognized grasslands as a first conservation priority (Smelansky, 2007) but so far were unable to translate their awareness into governmental policies and management rules.

Grasslands in Russia are legally considered agricultural land, either as pasture (grazing land) or hay-making land. In contrast to forests and tundra, the majority of grasslands lie on private land, municipal land and land of collective property (a temporary legal form of land property under the privatization process existing since 1993), but not on federal land (Smelansky, 2003, 2007; Allina-Pisano, 2007; Intigrinova, 2011). Grassland conservation on the one hand and grassland use and management on the other, are strongly separated from each other in Russia—both are regulated by different sectoral legislation and managed by different governmental institutions. Environmental legislation regulates all conservation-related issues, including protected areas, special protection of species and protection from damage under agricultural (and other industrial) use. The Federal Ministry of Natural Resources and Environment is the main governmental actor on conservation issues. Grassland use and management are regulated within agricultural and land legislation, and in part, within legislation on territorial planning. The main actor here is the Federal Ministry of Agriculture. Conservation laws and laws regulating the use of agricultural land and development of rural areas are not integrated at the field level. Both ministries and their agencies have practically no communication on any conservation issue. Thus, management decisions and actual use of grasslands do not take biodiversity or nature values into account (Smelansky, 2007). Vice versa, grassland conservation practice faces problems when conservation measures like regulative grazing and mowing or prescribed burning are to be implemented. Conservation instruments, like agro-environmental schemes, would be extremely useful for grassland conservation in Russia but have no chance under such a framework. The current legislation can only provide minimal protection, preventing transformation to other land uses and limiting selected risky management practices (Smelansky, 2007).

Furthermore, grasslands have no legal definition in Russian legislation, making them 'invisible' to Russian law. Formally being agricultural lands, grasslands could be protected under a more general framework of protection of agricultural biodiversity. Nevertheless the term 'agricultural biodiversity' or similar expressions are not even considered as legal terms and are certainly not being defined or applied in current legislation. Another instrument to introduce grassland as a specific conservation object could be an official *Green Book* or a *Red List* of ecosystems; unfortunately in Russia they do not exist.

On the federal level, the *Land Code of the Russian Federation* sets the following principles for managing farmland: agricultural lands are distinguished by their productivity and include pastures as well as hayfields; they should primarily be used for farming and are generally protected against improper use; weak constraints are imposed on non-farming use and on transformation of highly productive agricultural land. Punishment is imposed when violating these principles or when damaging the environment during agricultural practices. Theoretically this can be applied to the conversion of grassland into cropland or other forms of intensification. However, significant gaps in these federal regulations make them completely dependent on the goodwill of officials or judges. As a rule, this legislation does not work. Federal legislation also regulates potentially dangerous management practices, such as burning, fertilizing and the use of pesticides. For example, it is prohibited by law to burn dry plant residues and to store or apply pesticides, herbicides, fertilizers and other materials that are hazardous to wildlife and their habitats without taking necessary measures for their protection. However, none of these legal acts specifically discusses or addresses the impact of fires, pesticides, or agricultural chemicals on the grasslands themselves.

Several Russian provinces have their own legislation that bridge these gaps and thus make the law work. The *Provincial Law On Ensuring Soil Fertility on Agricultural Land in the Stavropol Territory* (of 15 May 2006), for example, includes a legal definition of land transformation, which is lacking in federal legislation and which was approved by the Stavropol Ministry of Agriculture. Provincial laws of several Russian provinces also designate limitations on grazing to prevent overgrazing.

Prioritization

As stated above, grasslands are not recognized as a high priority for national conservation in Russia. Nevertheless, some grassland areas are included in formally designated conservation sites of international importance within a range of conservation frameworks. Thus, of the 11 eco-regions in Russia listed in the WWF Terrestrial Global 200 (Olsen and Dinerstein, 2002), one is recognized specifically for its steppe ecosystem—the Daurian/Mongolian Steppe (transboundary with Mongolia and China). Two other eco-regions also contain important steppe tracts: the Altai-Sayan Montane Forests and Ural Mountains Taiga and Tundra (Smelansky and Tishkov, 2012). Furthermore, Russia holds 11 UNESCO World Nature Heritage sites (the last one was adopted on 7 July 2017), among which four, all located in southern Siberia, include significant tracts of steppe grasslands: Lake Baikal (1996), Golden Mountains of Altai (1998), Uvs Nuur Basin (2003) and Landscapes of Dauria (2017) (Smelansky and Tishkov, 2012; Butorin, 2013). Russia also designated 35 Ramsar sites, of which six contain some tracts of steppes and other grasslands (Smelansky and Tishkov, 2012). The BirdLife and Russian Bird Conservation Union (RBCU) recognizes 746 Important Bird Areas (IBAs) in Russia from which 462 IBAs are situated in the steppe region and about 170 IBAs include steppe ecosystems (Smelansky and Tishkov, 2012). From these, 88 IBAs contain significant steppe areas and/or are important for characteristic steppe birds, and 25 IBAs could be marked as very important to steppe conservation (Smelansky and Tishkov, 2012). For the European part of the country, Russia recently designated 1,257 Emerald Network sites, of which 345 are situated in the steppe region and many of them include steppe tracts and are important for conserving steppe species and communities (Sobolev and Belonovskaya, 2011–2013; Titova, 2016).

Sustainable use of steppe ecosystems strongly depends on the respective agricultural practices. High Nature Value Farmlands (HNVF) could be a very appropriate format for conserving the characteristic biodiversity of the steppe. Unfortunately, HNVFs are not formally recognized in Russia (Smelansky and Tishkov, 2012). A pilot study ('Identification of High Nature Value Farmlands for EECCA: Results of Assessment and Recommendations', 2005–2006) did not result in a respective federal policy.

Conservation of Grasslands: Protected Areas (PAs)

Regarding the above-mentioned scarcity of legal instruments for grassland conservation in Russia, protected areas are the most important instrument, having both a solid legal base and extensive practice. The total area of steppes and other grasslands and related shrublands in Russian PAs (all types, federal and provincial) is at least 11,000 km^2. We estimate that Russia holds about a quarter to one-third of the total area of Eurasian steppes that is under protection. Wesche et al. (2016), by contrast, estimate the area of steppes under protection even as circa 600,000 km^2 for entire Eurasia (excluding the grasslands in Tibet).

The area of protected steppe is not proportional to the extant steppe area. Protection of steppes has a long history—one of the first PAs in imperial Russia was 'Askania Nova' steppe (in Ukraine today) and one of the oldest still-existent Russian PAs is 'Central Chernozem Zapovednik', founded in 1935 to conserve few remnants of meadow steppe of European Russia. A significant portion of natural steppes has been saved on military training grounds, which represent the largest and (paradoxically) least disturbed areas of steppe. After 1990, many military grounds were abandoned and many grasslands were given the status of PA (Chibilev, 2016). Nonetheless, the steppe biome always was and still is drastically underprotected in Russia's system of PAs. Towards the end of the 20th century, only 0.11 per cent of the steppe biome area was estimated as formally protected in *Zapovedniks* and this represented the lowest coverage among all biomes of Russia (Nikolsky and Rumyantsev, 2002). Since the 1990s, the steppe area under strict protection increased significantly but still does not exceed 2 per cent of the biome area. Furthermore, within those few protected areas, different non-steppe ecosystems occupied most of the area. Later, especially in 2010–2016, Russian provincial authorities and the Federal Ministry of Nature Resources and Environment made a significant contribution to "improve the coverage and management efficiency of PAs in the steppe biome of Russia", but protection remains insufficient.

There are three levels of PAs in Russia: national (federal), provincial, and local (municipal). Accurate data on the size of PAs is only available for federal protected areas (Smelansky and Tishkov, 2012). All PAs are state-owned and state-managed. NGOs have no right to own or manage PAs; they may only take part in establishing new ones. Many Russian PAs have legally designated buffer zones, which are subject to significant restrictions, despite not being formally recognized as PAs. A majority of federal PAs are well established and protected, have a financing and management infrastructure and have legal rights on their land tracts and natural resources. At the same time, many provincial PAs only have formal status as PA, but no staff, budget, or even correctly designated borders. Federal PAs are organized as Federal State Strict Nature Reserves (*Zapovedniks*; IUCN category Ia), Federal Wildlife Refuges (Federal *Zakazniks*, IUCN categories IV and VI) and Federal Natural Monuments (IUCN category III). Nature Parks (IUCN category II) are analogous to National Parks but are established and controlled at the provincial level (see also Smelansky and Tishkov, 2012). Landscape Museum Reserves (*Landshaftny Muzey-Zapovedniks*, IUCN category V) are designed to conserve cultural and historical heritage in its natural landscape and managed by federal or provincial Ministries of Culture (see Smelansky and Tishkov, 2012).

About 10 (of 103) *Zapovedniks*, only two (of 50) National Parks, six (of 64) Federal *Zakazniks*, and nine (of 96) museum-reserves protect relatively large and/or important grassland tracts, mainly steppes (Smelansky and Tishkov, 2012, updated in 2017). However, the proportion of steppe and grassland habitats within these PAs is rather low. In total, grasslands comprise not more than 2 per cent of the total area of the federal PA system (Smelansky and Tishkov, 2012, with modifications). The greatest contribution to grassland conservation is made in six or seven *Zapovedniks*, altogether covering meadow steppes, true steppes, dry steppes, mountain steppes and saline meadows (Smelansky and Tishkov, 2012).

Management strategies for conserving steppes in PAs are subject to fierce debates lasting several decades. As a rule, the regime of Russian *Zapovedniks* prohibits many agricultural practices including grazing, mowing, prescribed burning, etc. Only very few PAs routinely use these practices as regulating measures. The best known case is the 'Centralno-Chernozemnii' *Zapovednik*, where grazing and mowing have been applied in different combinations during its entire history. The requirement of active management for the conservation of meadow steppes facing tree invasion and mesophytization has recently been recognized in many PAs.

Résumé and Future Prospects

Russia's natural and secondary grasslands, steppes and meadows, represent a significant proportion of the agricultural landscape and economy. Grasslands occur under diverse climatic and edaphic conditions, covering significant longitudinal and altitudinal ranges. Extrazonal relic steppes are evidence for a long history of vegetation dynamics of Palaearctic steppes in northern Siberia. All this is reflected in the current diversity of grassland plant species and vegetation types. They have great potential to provide sustainable, low-cost and productive sources of income (fodder production, livestock production, recreation, tourism) for rural development. On a global scale, Russia's grasslands could also be linked to the increasing demand for energy production and renewable resources. Gradually shrinking areas of intact dry grasslands, with their unique species and gene pool, are of primary concern in terms of conservation. However, conservation policy fails to integrate agricultural needs with conservation aims and the remaining areas of near-natural steppes are not well covered by protected areas and their legislation. Much stronger efforts are needed to actually integrate production and grassland biodiversity conservation.

Acknowledgements

We thank Jürgen Dengler for the invitation to write this chapter and for the thorough revision and helpful comments on our manuscript. We also are grateful to Svetlana Titova for her invaluable assistance in preparing the maps.

Abbreviations

BP = Years Before Present; DM = Dry Matter; IBA = Important Bird Area; IUCN = International Union for Conservation of Nature; Mt = Megaton = 1 Million Tons; NGO = Non-governmental Organization; PA = Protected Area

References

Abaturov, B.D. 1984. Mammals as a component of ecosystems (Case of herbivorous mammals in semidesert). [in Russian]. Nauka, Moscow, 286 pp.

Alcantara, C., T. Kuemmerle, M. Baumann, E.V. Bragina, P. Griffiths, P. Hostert and A. Sieber. 2013. Mapping the extent of abandoned farmland in Central and Eastern Europe using MODIS time series satellite data. Environ. Res. Lett. 8: Article 035035.

Allina-Pisano, J. 2007. The Post-Soviet Potemkin Village: Politics and Property Rights in the Black Earth. Cambridge University Press, New York, 24 pp.

Andreyev, V.N., T.F. Galaktionova, V.I. Perfilyeva and I.P. Scherbakov. 1987. Basic Features of Vegetation Cover of the Yakutian ASSR [in Russian]. Izd-vo YaF SO AN SSSR, Yakutsk, 156 pp.

Belonovskaya, E.A. 2010. Modern condition of pastoral ecosystems of Central Caucasus [in Russian]. Izv. RAN, Ser. Geogr. 1: 90–102.

Bukvareva, E.N. and D.G. Zamolodchikov (eds.). 2016. Ecosystem Services of Russia: Prototype of the National Report. Vol. 1. Services of terrestrial ecosystems. [in Russian]. Biodiversity Conservation Center Publishers, Moscow, 148 pp.

Butorin, A. (ed.). 2013. Landscapes of Dauria: Potential Serial Transnational World Heritage Property (the Russian Federation and Mongolia). ANNIE, Moscow, 32 pp.

Cherepanov, S.K. 1995. Vascular Plants of Russia and Neighboring Countries [in Russian]. Mir i Semya, Saint-Petersburg, 992 pp.

Chibilev, A.A. 2016. Steppe's Eurasia: A Regional Review of Natural Diversity [in Russian]. Institute of Steppe RAS, RGS, Moscow and Orenburg, 324 pp.

Deák, B., B. Tóthmérész, O. Valkó, B. Sudnik-Wójcikowska, I.I. Moysiyenko, T.M. Bragina, I. Apostolova, I. Dembicz, N.I. Bykov and P. Török. 2016. Cultural monuments and nature conservation: A review of the role of kurgans in the conservation and restoration of steppe vegetation. Biodivers. Conserv. 25: 2473–2490.

Dengler, J., M. Janišová, P. Török and C. Wellstein. 2014. Biodiversity of Palaearctic grasslands: A synthesis. Agric. Ecosyst. Environ. 182: 1–14.

Dobrovolsky, G.V. 2008. Soil degradation—A threat of global ecological crisis. Vek Glob. 2: 54–65.

Dubinin, M., P. Potapov, A. Lushchekina and V.C. Radeloff. 2010. Reconstructing long time series of burned areas in arid grasslands of southern Russia by satellite remote sensing. Remote Sens. Environ. 114: 1638–1648.

Dubinin, M., A. Luschekina and V. Radeloff. 2011. Climate, livestock, and vegetation: What drives fire increase in the arid ecosystems of Southern Russia? Ecosystems 14: 547–562.

Elie, M. 2015. The Soviet dust bowl and the Canadian erosion experience in the new lands of Kazakhstan, 1950s–1960s. Global Environ. 8: 259–292.

Ermakov, N.B. 2012. Prodromus of higher vegetation units of Russia. pp. 377–483. *In*: B.M. Mirkin and L.G. Naumova (eds.). Modern State of Basic Concepts of Vegetation Science. Gilem, Ufa.

Frachetti, M.D., D.W. Anthony, A.V. Epimakhov, B.K. Hanks, R.C.P. Doonan, N.N. Kradin and N. Shishlina. 2012. Multiregional emergence of mobile pastoralism and nonuniform institutional complexity across Eurasia. Current Anthr. 53: 2–38.

Gaidamaka, E.I., N.Y. Derkaeva, A.M. Cherkasov and T.A. Frieva. 1984. All-Union Instruction on Geobotanical Survey of Natural Fodder Lands and Preparing Large-scaled Geobotanical Maps [in Russian]. Kolos, Moscow, 104 pp.

Gavrilyeva, L.D. 1998. Pasture Degradation and Rational Use of Alas Vegetation in the Lena-Amga Interfluve [in Russian]. Abstract of Candidate Thesis, Institute of Applied Ecology of the North AS RS (Ya), Yakutsk, 20 pp.

Henwood, W.D. 2010. Toward a strategy for the conservation and protection of the world's temperate grasslands. Great Plains Res. 20: 121–134.

Hölzel, N., C. Haub, M.P. Ingelfinger, A. Otte and V.N. Pilipenko. 2012. The return of the steppe large-scale restoration of degraded land in southern Russia during the post-Soviet era. J. Nat. Conserv. 10: 75–85.

Intigrinova, T. 2011. Property Regimes for Pastoral Resources: Discussions, Practices and Problems [in Russian]. Gaidar Institute Publ., Moscow, 156 pp.

Ioffe, G., T. Nefedova and D.B. Kirsten. 2012. Land abandonment in Russia: A case study of two regions. Eurasian Geogr. Econ. 53: 527–549.

Isachenko, T.I. (ed.). 1990. The Map of Vegetation of the USSR [in Russian]. GUGK, Moscow; 4 sheets.

Ivanov, A.A., S.I. Mironova and D.D. Savvinov. 2004. Alas Meadows of the Lena-Viluy Interfluve of Central Yakutia under Various Regimes of Use [in Russian]. Nauka, Novosibirsk, 110 pp.

Ivanov, A.L. 2014. Scientific arable farming in Russia: outcomes and perspectives [in Russian]. Zemledelie 3: 25–29.

Ivanova, V.P. 1967. Grazing effect on steppe vegetation in the Lena River valley [in Russian]. pp. 86–93. *In*: M.V. Popov (ed.). Lyubite i Okhranyaite Prirodu Yakutii [Proceedings of 4th Republican Meeting on Nature Conservation in Yakutia]. Yakutskoye knizhnoye izdatelstvo, Yakutsk.

Ivanova, V.P. and V.I. Perfilyeva. 1972. Conservation of *Stipa* steppes of Yakutia [in Russian]. pp. 116–121. *In*: L.G. Elovskaya, I.P. Scherbakov, F.N. Kirillin and M.V. Popov (eds.). Priroda Yakutii i Eyo Okhrana [Proceedings of 6th Republican Meeting on Nature Conservation in Yakutia]. Yakutskoye knizhnoye izdatelstvo, Yakutsk.

Ivanova, V.P. 1981. *Festuca lenensis* steppes as one of stages of pasture digression in the Middle Lena Valley [in Russian]. pp. 37–56. *In*: V.N. Andreyev (ed.). Rastitelnost Yakutii i Eyo Okhrana. YaF SO AN SSSR, Yakutsk.

Jones, M.B. 2010. Potential for carbon sequestration in temperate grassland soils. Integr. Crop Manag. 11: 1–18.

Kajtoch, Ł., E. Cieślak, Z. Varga, W. Paul, M.A. Mazur, G. Sramkó and D. Kubisz. 2016. Phylogeographic patterns of steppe species in eastern Central Europe: a review and the implications for conservation. Biodivers. Conserv. 25: 2309–2339.

Kandalova, G.T. 2009. Steppe Pastures of Khakassia: Transformation, Restoration and Prospects of Use [in Russian]. Rosselkhozacademia, Novosibirsk, 163 pp.

Kandalova, G.T. and G.I. Lysanova. 2010. Restoration of steppe pastures in Khakassia [in Russian]. Geogr. Prir. Resur. 4: 79–85.

Kämpf, I., W. Mathar, I. Kuzmin, N. Hölzel and K. Kiehl. 2016. Post-Soviet recovery of grassland vegetation on abandoned fields in the forest steppe zone of western Siberia. Biodivers. Conserv. 25: 2563–2580.

Kara-Murza, S.G., S.A. Batchikov and S.Y. Glazyev. 2008. Where Russia Goes. The White Book of Reforms [in Russian]. Algoritm, Moscow, 448 pp.

Kashtanov, A.N. 2008. Arable Farming. Selected Texts. Rosselkhozakademia, Moscow, 685 pp.

Khazanov, A.M. 2002. Nomads and the Outside World [in Russian]. 3rd ed. Dayk-Press, Almaty, 603 pp.

Khodarkovsky, M. 2002. Russia's Steppe Frontier: The Making of a Colonial Empire, 1500–1800. Indiana University Press, Bloomington, 290 pp.

Kiryushin, V.I. 1996. Ecological Principles of Arable Farming. Kolos, Moscow, 367 pp.

Korolyuk, A.Y. 2002. Vegetation [in Russian]. pp. 45–94. *In*: I.M. Gadjiev, A.Y. Korolyuk and A.A. Titlyanova (eds.). Steppes of Inner Asia. SB RAS Publisher, Novosibirsk.

Korotkov, K.O. and E.A. Belonovskaya. 2001. The Great Caucasus alpine belt syntaxonomy. I. Alpine meadows with restricted distribution. Rastit. Ross. 1: 17–35.

Kosolapov, V.M. and I.A. Trofimov (eds.). 2009. Agrolandscape-Ecological Regionalization and Adaptive Intensification of Fodder Production in the Volga Region. Theory and Practice [in Russian]. Dom pechati – VYATKA, Moscow, 751 pp.

Kosolapov, V.M., I.A. Trofimov, L.S. Trofimova and E.P. Yakovleva. 2012. Fodder production as an important factor of productivity growth and arable farming sustainability. Zemledelie 4: 20–22.

Kosolapov, V.M. and I.A. Trofimov (eds.). 2014. Reference Book on Fodder Production [in Russian]. Rosselkhozakademia, Moscow, 717 pp.

Kosolapov, V.M., I.A. Trofimov and L.S. Trofimova. 2014. Fodder Production in Agriculture, Ecology and Rational Use of Natural Resources (Theory and Practice) [in Russian]. Rosselkhozakademia, Moscow, 135 pp.

Kosolapov, V.M., I.A. Trofimov, L.S. Trofimova and E.P. Yakovleva. 2015. Agrolandscapes of Central Black Earth Region. Regionalization and Management [in Russian]. Nauka, Moscow, 198 pp.

Kosolapov, V.M. and I.A. Trofimov (eds.). 2016. Forage Ecosystems of Central Black Earth Region of Russia: Agrolandscape and Technological Approaches [in Russian]. Rosselkhozakalemia, Moscow, 649 pp.

Kradin, N.N. 2007. Nomads of Eurasia [Russian]. Dayk-Press, Almaty, 416 pp.

Kurganova, I., V.L. de Gerenyu, J. Six and Y. Kuzyakov. 2014. Carbon cost of collective farming collapse in Russia. Global Change Biol. 20: 938–947.

Kurganova, I., V.L. de Gerenyu and Y. Kuzyakov. 2015. Large-scale carbon sequestration in post-agrogenic ecosystems in Russia and Kazakhstan. Catena 133: 461–466.

Kutuzova, A.A. 2007. Perspectives of Grassland Management Development [in Russian]. Kormoproizvodstvo 5: 12–15.

Larin, T.M., S.M. Agababyan, T.A. Rabotnov, A.F. Lyubskaya, V.K. Larina, M.A. Kasimenko, V.S. Govorukhin and S.Y. Zafren. 1950. Fodder Plants of Hayfields and Pastures. Part 1 [in Russian]. Selkhozgiz, Leningrad, 688 pp.

Larin, T.M., S.M. Agababyan, T.A. Rabotnov, A.F. Lyubskaya, V.K. Larina and M.A. Kasimenko. 1951. Fodder Plants of Hayfields and Pastures. Part 2 [in Russian]. Selkhozgiz, Leningrad, 948 pp.

Larin, T.M., S.M. Agababyan, T.A. Rabotnov, A.F. Lyubskaya, V.K. Larina and M.A. Kasimenko. 1956. Fodder Plants of Hayfields and Pastures. Part 3 [in Russian]. Selkhozgiz, Leningrad, 880 pp.

Lavrenko, E.M. 1970a. Provincial division of the Pontic-Kazakhstanian subregion of the Steppe region of Eurasia [in Russian]. Bot. Zhurnal 55: 609–625.

Lavrenko, E.M. 1970b. Provincial division of the Central-Asian subregion of the Steppe region of Eurasia [in Russian]. Bot. Zhurnal 55: 1734–1747.

Lavrenko, E.M. 1981. On vegetation of the Late Pleistocene periglacial steppes of the USSR [in Russian]. Bot. Zhurnal 66: 313–327.

Lavrenko, E.M., Z.V. Karamysheva and R.I. Nikulina. 1991. Steppes of Eurasia [in Russian]. Nauka, Leningrad, 146 pp.

Lyuri, D.I., S.V. Goryachkin, N.A. Karavaeva, E.A. Denisenko and T.G. Nefedova. 2010. Dynamics of Agricultural Lands of Russia in the XXth Century and Postagricultural Restoration of Vegetation and Soils [in Russian]. GEOS, Moscow, 415 pp.

Markov, K.K., G.I. Lazukov and V.A. Nikolaev. 1965a. The Quaternary. Part 1 [in Russian]. Izdatelstvo MGU, Moscow, 372 pp.

Markov, K.K., G.I. Lazukov and V.A. Nikolaev. 1965b. The Quaternary. Part 2 [in Russian]. Izdatelstvo MGU, Moscow, 435 pp.

Mathar, W.P., I. Kämpf, T. Kleinebecker, I. Kuzmin, A. Tolstikov, S. Tupitsin and N. Hölzel. 2016. Floristic diversity of meadow steppes in the Western Siberian Plain: Effects of abiotic site conditions, management and landscape structure. Biodivers. Conserv. 25: 2361–2379.

Medvedev, P.F. and A.I. Smetannikova. 1981. Fodder Plants of the European Part of the USSR: Reference-book [in Russian]. Kolos, Moscow, 336 pp.

Meyfroidt, P., F. Schierhorn, A.V. Prishchepov, D. Müller and T. Kuemmerle. 2016. Drivers, constraints and trade-offs associated with recultivating abandoned cropland in Russia, Ukraine and Kazakhstan. Global Environ. Change 37: 1–15.

Mironova, S.I. 1992. Pasture digression of alas grasslands [in Russian]. pp. 73–80. *In*: I.D. Zakharov (ed.). Intensification of Grassland Fodder Production in Yakutia [Collection of Articles]. Izd-vo SO RASKHN, Novosibirsk.

Moon, D. 2013. The Plough that Broke the Steppes: Agriculture and Environment on Russia's Grasslands, 1700–1914. OUP, Oxford, 319 pp.

Mordkovich, V.G. 1994. Originality of Siberian steppes, level of their disturbance and safety [in Russian]. Sib. Ecol. J. 1: 475–481.

Mordkovich, V.G. 2014. Steppe Ecosystems [in Russian]. 2nd ed. GEO Academic Publishing House, Novosibirsk, 170 pp.

Morozova, L.M. 2012. Spatio-temporal analysis of steppe vegetation dynamic in Southern Urals [in Russian]. Izv. Samar. Nauchnogo Tsentra RAN 14: 1328–1331.

Naidoo, R., A. Balmford, R. Costanza, B. Fisher, R.E. Green, B. Lehner, T.A. Malcolm and T.H. Ricketts. 2008. Global mapping of ecosystem services and conservation priorities. Proc. Natl. Acad. Sci. 105: 9495–9500.

Namzalov, B.B. 2015. The Steppes of Tuva and South-east Altai [in Russian]. Academic Publish. House 'GEO', Novoisibirsk, 294 pp.

Neronov, V.V. and A.V. Tchabovsky. 2003. Semidesert turned to steppe again [in Russian]. Priroda 4: 72–79.

Nikolsky, A.A. and V.Y. Rumyantsev. 2002. Zonal representativeness of the Zapovedniks system of the Russian Federation [in Russian]. pp. 160–165. *In*: Y.A. Izrael (ed.). Scientific Issues of Ecological Problems of Russia. Proceedings of the National Conference ad Memoriam A.L. Yashin. Part 1. Nauka, Moscow.

Novenko, E.Y., A.N. Tsyganov, O.V. Rudenko, E.V. Volkova, I.S. Zuyganova, K.V. Babeshko, A.V. Olchev, N.I. Losbenev, R.J. Payne and Y.A. Mazei. 2016. Mid- and late-Holocene vegetation history, climate and human impact in the forest-steppe ecotone of European Russia: new data and a regional synthesis. Biodivers. Conserv. 25: 2453–2472.

Ogureyeva, G.N. (ed.). 1999. Map Zones and Altitudinal Zonality Types of the Vegetation of Russia and Adjacent Territories. Scale 1: 80,000,000. Ekor, Moscow, 64 pp.

Olson, D.M. and E. Dinerstein. 2002. The Global 200: Priority ecoregions for global conservation. Ann. Mo. Bot. Gard. 89: 199–224.

Parakhin, N.V., I.V. Kobozev, I.V. Gorbachev, N.I. Lazarev and S.S. Mikhalev. 2006. Fodder Production [in Russian]. KolosS, Moscow, 218 pp.

Pavleichik, V.M. 2016. Long-term dynamics of wild fires in steppe regions (case of the Orenburg Province) [in Russian]. Orenbg. Univ. Her. 6(194): 74–80.

Prishchepov, A.V., V.C. Radeloff, M. Baumann, T. Kuemmerle and D. Müller. 2012. Effects of institutional changes on land use: agricultural land abandonment during the transition from state-command to market-driven economies in post-Soviet Eastern Europe. Environ. Res. Lett. 7: Article 024021.

Protopopov, A.V. and V.V. Protopopova. 2010. History of vegetation development. pp. 151–156. *In*: E.I. Troeva, A.P. Isaev, M.M. Cherosov and N.S. Karpov (eds.). The Far North: Plant Biodiversity and Ecology of Yakutia. Springer, Dordrecht.

Rabotnov, T.A. 1984. Grassland Science [in Russian]. 2nd ed. Izdatelstvo MGU, Moscow, 320 pp.

Ramensky, L.G., I.A. Tsatsenkin and O.N. Chizhikova. 1956. Ecological Evaluation of Fodder Lands Based on Vegetation Cover. VNII kormov, Moscow, 471 pp.

Reinecke, J., E. Troeva and K. Wesche. 2017. Extrazonal steppes and other temperate grasslands of northern Siberia—Phytosociological classification and ecological characterization. Phytocoenologia 47: 167–196.

Romanovskaya, A.A., V.N. Korotkov, N.S. Smirnov and A.A. Trunov. 2014. Land use contribution to the anthropogenic emission of greenhouse gases in Russia in 2000–2011. Russ. Meteorol. Hydrol. 39: 137–145.

Rosreestr. 2013. Lands of the Russian Federation on 1 January 2013 [in Russian]. The Federal Service for State Registration, Cadastre and Cartography, Moscow, 694 pp.

Rusanov, A.M. 2013. Modern transformation of natural vegetation in steppe biogeocoenoses [in Russian]. Orenbg. Univ. Her. 6(155): 122–126.

Sakalo, D.I. 1963. Ecological nature of steppe vegetation of Eurasia and its origin [in Russian]. pp. 407–425. *In*: V.L. Komarov (ed.). Material on the History of Flora and Vegetation of the USSR. Issue 4. Izdatelstvo AN SSSR, Moscow.

Savchenko, I.V., S.A. Dmitrieva and N.A. Semyonov. 1987. Methodology Guidelines on Classification of Hayfields and Pastures of the Plain Territory of the European Part of the USSR [in Russian]. VASKHNIL, VNII kormov, Moscow, 148 pp.

Savchenko, I.V., S.A. Dmitrieva and N.A. Semyonov. 1989. Methodology Guidelines on Classification of Natural Fodder Lands of the Plain Territory of Siberia and the Far East [in Russian]. VASKHNIL, Moscow, 122 pp.

Savchenko, I.V., S.A. Dmitrieva and N.A. Semyonov. 1990. Methodology Guidelines on Classification of Natural Fodder Lands of Melkosopochniks and Mountainous Regions of Caucasus, Siberia and the Far East [in Russian]. VASKHNIL, VNII kormov, Moscow, 136 pp.

Schierhorn, F., D. Müller, T. Beringer, A.V. Prishchepov, T. Kuemmerle and A. Balmann. 2013. Post-Soviet cropland abandonment and carbon sequestration in European Russia, Ukraine, and Belarus. Global Biogeochem. Cycles 27: 1175–1185.

Semenov, E.A. 2012. Virgin lands development in Russia and Kazakhstan: Background and economic results. Orenbg. Univ. Her. 13(149): 318–322.

Smelansky, I.E. 2003. Biodiversity of Agricultural Lands in Russia: Current State and Trends. IUCN, Moscow, 52 pp.

Smelansky, I.E. (ed.). 2007. Russian Steppe Conservation Strategy: NGO's Position. Biodiversity Conservation Center's Print, Moscow, 32 pp.

Smelansky, I. 2012a. The role of steppe ecosystems of Russia in carbon sequestration [in Russian]. Stepnoi Bull. 35: 4–8.

Smelansky, I. 2012b. How many abandoned farmlands are in the steppe region of Russia? [in Russian]. Steppe Bull. 36: 4–7.

Smelansky, I. (ed.). 2015. Steppe Fires and Fire Management in Steppe Protected Areas: Environmental and Conservation Aspects. Analytical Survey [in Russian]. BCC Press, Moscow, 144 pp.

Smelansky, I.E. and A.A. Tishkov. 2012. The steppe biome in Russia: ecosystem services, conservation status, and actual challenges. pp. 45–101. *In*: M.J.A. Werger and M.A. van Staalduinen (eds.). Eurasian Steppes: Ecological Problems and Livelihoods in a Changing World. Springer, Dordrecht.

Sobolev, N.A. and E.A. Belonovskaya (eds.). 2011–2013. Emerald Book of Russian Federation: Areas of Special Conservation Importance in European Russia. Proposals for Nominating. Institute of Geography, RAS, Moscow, 308 pp.

Sunderland, W. 2006. Taming the Wild Field: Colonization and Empire on the Russian Steppe. Cornell University Press, Ithaca, New York, 239 pp.

Swinnen, J., S. Burkitbayeva, F. Schierhorn, A.V. Prishchepov and D. Müller. 2017. Production potential in the 'bread baskets' of Eastern Europe and Central Asia. Global Food Secur. 14: 38–53.

Tanfilyev, I.G. 1896. The prehistoric steppes of European Russia [in Russian]. Geography 3: 72–92.

Tishkov, A.A. 2005. Biosphere Functions of Natural Ecosystems in Russia [in Russian]. Nauka, Moscow, 309 pp.

Titova, S. 2016. Workshop on Emerald network in the Steppe region of Russia, Ukraine, and Moldova [in Russian]. Stepnoi Bull. 47-48: 55–59.

Titova, S.V., K.N. Kobyakov, N.I. Zolotukhin and A.V. Poluyanov. 2014. White Hills without White Hills? Threats to Steppe Ecosystems in Belgorod Province [in Russian]. Institut Geographii RAN, Centralno-Chernozemnii Zapovednik, Kursky Gosudarstvenny Universitet, Moscow, 40 pp.

Tkachuk, T.E. 2015. Multi-year dynamics of steppe fires in Dauria [in Russian]. Uchenye Zap. Zabaikalskogo Gos. Univ., Ser. Est.-Nauch. 60: 72–79.

Troeva, E.I. and M.M. Cherosov. 2012. Transformation of steppe communities of Yakutia due to climatic change and anthropogenic impact. pp. 371–396. *In*: M.J.A. Werger and M.A. van Staalduinen (eds.). Eurasian Steppes: Ecological Problems and Livelihoods in a Changing World. Springer, Dordrecht.

Trofimov, I.A. and V.I. Kravtsova. 1998. Monitoring of natural fodder lands dynamics. Kalmykia [in Russian]. p. 56. *In*: V.I. Kravtsova (ed.). Space Methods of Geoecology. Atlas. Geographical Faculty of MSU, Moscow.

Trofimov, I.A. 2001. Methodological Principles of Aerospace Mapping and Monitoring of Natural Fodder Lands [in Russian]. Rosselkhozakademia, Moscow, 74 pp.

Trofimov, I.A., L.S. Trofimova and E.P. Yakovleva. 2010. Geography of fodder lands productivity in the natural zones of the Russian federation [in Russian]. pp. 154–156. *In*: G.V. Dobrovolsky, V.N. Kudeyarov and A.A. Tishkov (eds.). Geography of Productivity and Biogeochemical Cycles of Terrestrial Landscapes: to 100th Anniversary of Prof. N.I. Bazilevich. Conference Proceedings. Institute of Geography RAS, Moscow.

Trofimov, I.A., L.S. Trofimova and E.P. Yakovleva. 2011. Agrolandscape management for productivity increase and agricultural lands sustainability in Russia [in Russian]. Adapt. Kormoproizvod. 3: 14–15.

Trofimov, I.A., V.M. Kosolapov, L.S. Trofimova and E.P. Yakovleva. 2012. Global ecological processes, strategy of nature use and agrolandscape management [in Russian]. pp. 107–114. *In*: V.V. Snakin (ed.). Globalnye Ecologicheskie Processy [Proceedings of Scientific Conference]. Academia, Moscow.

Trofimov, I.A., L.S. Trofimova and E.P. Yakovleva. 2016. Productivity of natural fodder lands of Russia [in Russian]. Ispol'z. Okhrana Prir. Resur. Ross. 1: 42–50.

Trofimova, L.S., V.F. Kulakov and S.A. Novikov. 2008. Productive and environmental-forming potential of grassland agrophytocoenoses and the ways of its increase [in Russian]. Kormoproizvodstvo 9: 17–19.

Trofimova, L.S., I.A. Trofimov and E.P. Yakovleva. 2010. Significance, functions and potential of fodder ecosystems in biosphere, agrolandscapes and agriculture [in Russian]. Adapt. Kormoproizvod. 3: 23–28.

Wesche, K., D. Ambarlı, J. Kamp, P. Török, J. Treiber and J. Dengler. 2016. The Palaearctic steppe biome: a new synthesis. Biodivers. Conserv. 25: 2197–2231.

Yurtsev, B.A. 1982. Relics of the xerophyte vegetation of Beringia in northeastern Asia. pp. 157–177. *In*: D.M. Hopkins, J.V. Matthews, C.E. Schweger and S.B. Young (eds.). Palaeoecology of Beringia, Academic Press, New York.

Zakharova, V.I., M.M. Cherosov, E.I. Troeva and P.A. Gogoleva. 2010. Steppes. pp. 193–198. *In*: E.I. Troeva, A.P. Isaev, M.M. Cherosov and N.S. Karpov (eds.). The Far North: Plant Biodiversity and Ecology of Yakutia. Springer, Dordrecht.

Zherikhin, V.V. 1993. The nature and history of grass biomes [in Russian]. pp. 29–49. *In*: Z.V. Karamysheva (ed.). The Steppes of Eurasia: Problems of Conservation and Recovery. Nauka, Saint-Petersburg.

Zibzeev, E.G. 2012. Landscape forming high mountain communities of southern macroslope of the Terektinskiy Ridge (Central Altai): Classification, ecological and phytocoenotic characteristics. Turczaninowia 15: 83–108.

7

Grasslands of Kazakhstan and Middle Asia: Ecology, Conservation and Use of a Vast and Globally Important Area

Tatyana M. Bragina,[1,2] *Arkadiusz Nowak*,[3] *Kim André Vanselow*[4] and *Viktoria Wagner*[5,6,*]

Introduction

Delimitation and Size of Study Region

Located in the center of Eurasia, the region of Kazakhstan and Middle Asia (Kyrgyzstan, Tajikistan, Turkmenistan, and Uzbekistan; regional delimitation *sensu* Rachkovskaya et al., 2003) straddles two continents—Asia (85 per cent) and Europe (15 per cent; Fig. 7.1). The respective countries shared a similar natural, cultural and political history, including a membership in the Soviet Union, in the 20th century. Together, the region has a size of 4,008,139 km^2 and spans approximately 2,000 km from north to south and 2,800 km from west to east. Its northern border is aligned with European Russia, Siberia and the Altai mountains. The mountain systems of the Altai, Tarbagatai, Dzhungarian Alatau and the eastern ranges of the Tian Shan and Pamir form its eastern border, with the latter two mountain systems separating the region from the neighboring Tarim Basin. Its western edge borders on the Caspian Sea, while its southern edge aligns with the borders of Afghanistan and Iran.

[1] Department of Natural Science, Kostanay State Pedagogical Institute, Tauelsizdik str., 118, 110000 Kostanay, Kazakhstan.

[2] Federal State-Financed Scientific Institution "AzNIIRKH", Beregovaya str., 21B, 344002, Rostov-on-Don, Russia.

[3] Botanical Garden – Center for Biological Diversity Conservation in Powsin, Polish Academy of Sciences, Prawdziwka 2, 02-973 Warsaw, Poland.

[4] Institute of Geography, University of Erlangen-Nuremberg, Wetterkreuz 15, 91058 Erlangen, Germany.

[5] Department of Botany and Zoology, Masaryk University, Kotlářská 2, 611 37 Brno, Czech Republic.

[6] Department of Biological Sciences, University of Alberta, Edmonton, AB, T6G 2R3, Canada.
Emails: tm_bragina@mail.ru; a.nowak@obpan.pl; kim.vanselow@fau.de

* Corresponding author: viktoria.wagner@ualberta.ca

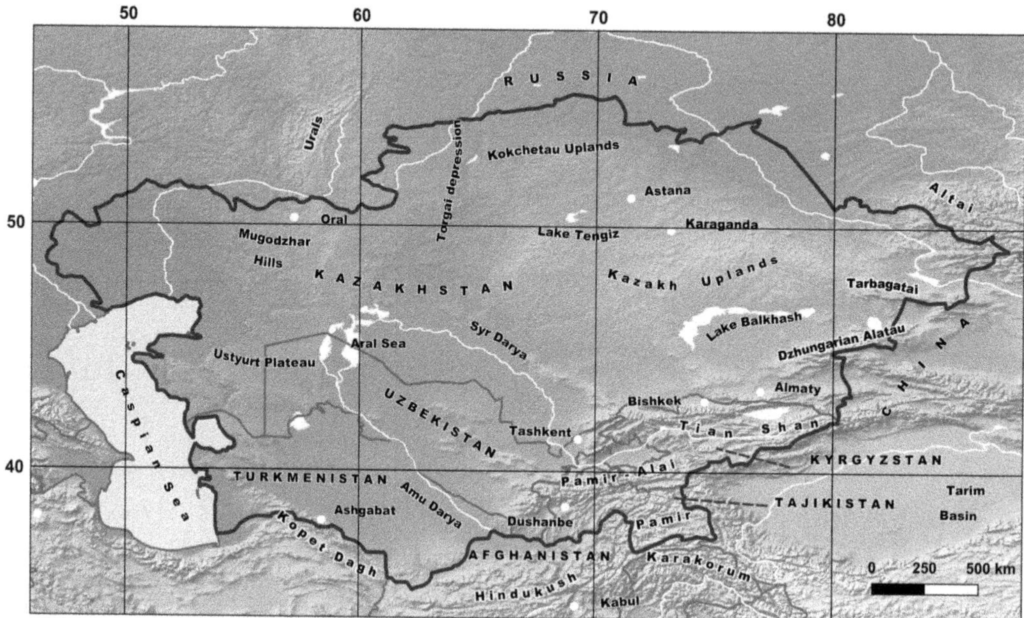

Fig. 7.1 Topographic map of Kazakhstan and Middle Asia. (The borders of the region are highlighted by the bold black line; country names are in capital letters; country borders are shown as gray lines. Authors' own work.)

The region is divided into three biogeographic zones, aligned as latitudinal belts from north to south—the forest-steppe, steppe and desert zones. In addition, a mountain zone frames the region in the east and south (Fig. 7.2). According to the botanical-geographical zoning of Eurasia (Lavrenko, 1970), the Kazakh steppe is part of the Black Sea-Kazakhstan subregion and the larger Eurasian phytogeographic steppe zone. In this framework, the Kazakh steppe is further divided into three provinces—the west Siberian steppe, the Zavolzhskaya Kazakhstan steppe (5 subprovinces) and the Altai mountain steppe (Rachkovskaya and Bragina, 2012).

Compared to the Mediterranean region (Ambarlı et al., 2018), the climate is generally colder (mean annual temperature 3.1°C vs. 13.3°C, Wesche et al., 2016). It differs from eastern-neighboring Central Asia (Pfeiffer et al., 2018) by its physiographical layout and an Irano-Turanian climate, which is in general characterized by winter- rather than summer-precipitation (Djamali et al., 2012; Wesche et al., 2016). The plains of this region are dominated by deserts, while the tall mountains of the Tian Shan, Pamir-Alai and Pamir, as well as geologically older highlands in their foreland, are covered by steppe, meadows, coniferous and deciduous forests as well as tundra vegetation.

The Tian Shan encompasses mountain ranges that stretch in an east-west direction across China, Kyrgyzstan, Kazakhstan and Uzbekistan. The highest mountains are the Peak Pobedi (7,439 m a.s.l.) and Khan Tengri (7,010 m a.s.l.). The Pamir mountains are predominantly located in Tajikistan and are surrounded by the Tian Shan and Pamir-Alai in the north, as well as extensions of the Kunlun Shan in the east. Due to this situation, they are often considered to be a mountain knot (Agakhanjanz, 1979; Walter and Breckle, 1986; Breckle and Wucherer, 2006). This vast mountain area can be differentiated into the Pamir-Alai in the north and the west and the east Pamir further southeast. The Pamir-Alai mountain system encompasses a number of distinct ranges that stretch in an east-

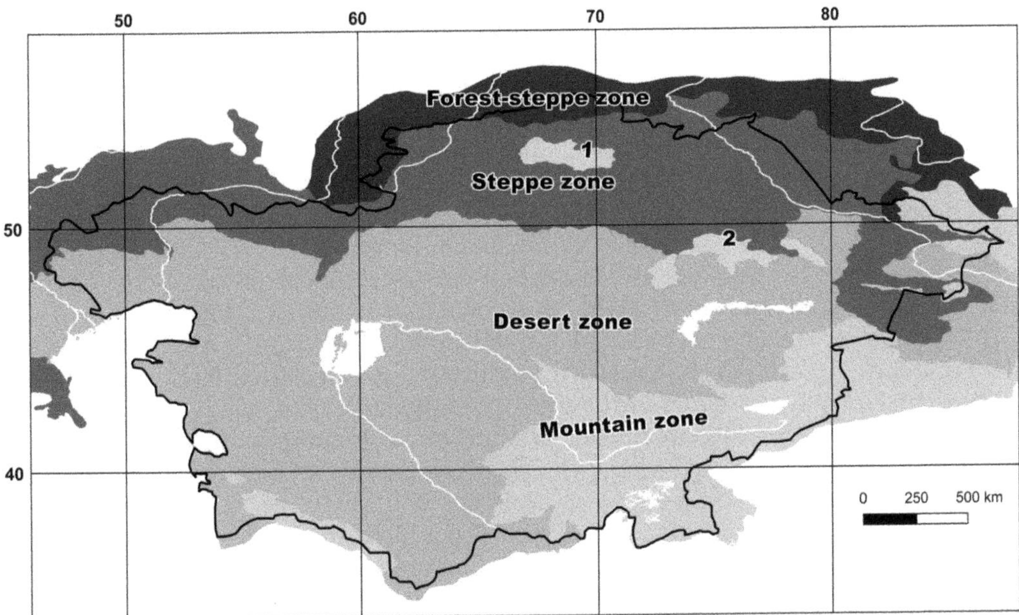

Fig. 7.2 Ecozones of Kazakhstan and Middle Asia (modified after Olson et al., 2001). (The Kokshetau (1) and Kazakh (2) Uplands were treated as part of the steppe and desert zones.)

west direction and reach heights of 5,069 m a.s.l. (Peak Skobelew). The Ferghana valley separates the western ranges of the Tian Shan from the Pamir-Alai. The west and east Pamir are usually divided at a longitude of circa 72° 45′ E. The former is characterized by narrow and deeply incised valleys and high peaks. The latter represents a high plateau with wide valleys between 3,500 and 4,000 m a.s.l., filled with alluvial sediments that are separated by relatively low ranges with peaks mostly around 5,000 to 5,500 m a.s.l. Both are predominantly situated in the Gorno-Badakhshan Autonomous Oblast (GBAO) of Tajikistan, while smaller parts belong to Afghanistan, China, and Kyrgyzstan.

Physiogeographic Setting

The continental location shapes the region's terrain, climate and hydrogeography, and consequently, its plant and animal life. The collision of the Indian and Eurasian plates in the Cenozoic marked the beginning of a new and ongoing orogenesis in Middle Asia, which deprived the region of southern precipitation and converted subtropical to temperate arid vegetation (Kamelin, 2017). Today, the region is dominated by steppe and desert, framed by high mountain ranges in the south and east (Tian Shan, Pamir-Alai, Pamir).

The northern part of this region encompasses the plains of western and northern Kazakhstan (60 per cent of the country's territory) and the hills and uplands of central Kazakhstan (30 per cent of the territory). A strong continental climate prevails across all subregions, with hot summers and cold winters, pronounced air dryness, westerly and northerly winds, and small amounts of rainfall.

The ecozones of Kazakhstan follow latitudinal biogeographic belts, in response to contrasting climatic conditions (Nikolayev, 1999; Fig. 7.2): the forest-steppe zone (2.4 per cent of Kazakhstan's territory in the northernmost part), the steppe zone (41.4 per cent) and the desert zone (more than 44 per cent). Mountain regions make up 12.6 per cent of Kazakhstan. The forest-steppe and steppe zones are extensively covered by grassland vegetation and differ in their climate (mean temperate in January: −16 to −19°C in the forest-steppe zone vs. −14 to −19°C in the steppe zone; mean temperate in July: +19 to +20°C vs. +19.5 to +25°C; mean annual precipitation: 310–350 mm vs. 150–300 mm) (Rachkovskaya and Bragina, 2012).

Deserts occupy large areas of the lowlands of Middle Asia and southern Kazakhstan. They also extend into low elevation mountain ranges. They are characterized by < 200 mm rainfall per year and are subdivided into three climate-related ecoregion types (the northern, middle and south desert ecoregions) (Lioubimtseva, 2002).

Like much of the adjacent desert plains, the mountains of Middle Asia are shaped by their land-locked position in the center of the continent. A geological uplift since the Cenozoic and low precipitation is responsible for the rugged topography of many mountain ranges. The geological profile of the mountains of Middle Asia is complex. In general, much of the surface of the northern and central Tian Shan is composed of granite and granodiorite, whereas sediments, including limestone, dolomite and sandstone, dominate the surface of the western Tian Shan (Anon., 1979, 1980). The middle and higher parts of the Hissar mountains are largely composed of extrusive rocks, mainly granite, granitoid and syenite. Some igneous outcrops also occur in the Darvaz mountains, Kuraminian mountains and in the west Pamir. In the Zeravshan and Turkestan mountains, the Cambrian and Sylurian sediments predominates. The rocks here are generally limestone (micrite limestone, bitumic limestone, marly limestone and dolomitic coral limestone), marble, dolomite, dolomitic shale, clay shale, phyllitic schist and argillaceous slate. Also several metamorphic rocks are present within the study area. The most common are migmatic gneiss, conglomerates and metamorphic mudstones (Nedzvedskiy, 1968).

Similar to the adjacent desert plains, the tall mountains of the Tian Shan, Pamir-Alai and Kopet Dagh receive high solar insolation (2,090–3,160 sunshine hours) and have a high-amplitude of annual temperatures. The Tian Shan is mostly shaped by a temperate climate, while the Kopet Dagh and Pamir-Alai are influenced by a subtropical climate (Alisov, 1956; Borisov, 1965). Precipitation sums vary greatly across the individual mountain systems. The outer northern and western chains of the Tian Shan and Pamir-Alai are fairly humid and benefit from Atlantic (annual precipitation in the western Tian Shan, at middle elevations: 900 mm; Wagner, 2009) and southern remnant air masses (600 mm in the Hissar Range at 800–1,200 m a.s.l.; Latipova, 1968; Narzikulov and Stanyukovich, 1968; Safarov, 2003), respectively. Meanwhile, the inner ranges of the Tian Shan (annual precipitation sum: 200–300 mm) and Pamir-Alai (400–500 mm) are much more continental (Sidorenko, 1971; Agakhanjanz, 1979; Vanselow, 2012). The west Pamir is influenced by winter rains and has a mean annual precipitation of circa 230 mm in the valleys. As it intercepts the humid airmasses from the westerlies, the neighboring east Pamir is much more arid, with a mean annual precipitation of < 100 mm in the valleys.

Several major rivers dissect Kazakhstan and Middle Asia. In the northern part, the Irtysch und Ischym rivers drain water from the Dzhungarian Alatau and Kazakh Uplands. They flow into the Ob river, which in turn empties into the Arctic Sea. By contrast, runoff from the Tian Shan, Pamir-Alai and Pamir discharges into the Amu Darya, Ili and Syr Darya rivers, which end in the endorheic basins of the Aral and Balkhash Seas. Some rivers vanish in the region's desert zone (Chu, Talas). A characteristic feature of the region is the high number of natural fresh- and saltwater lakes. Kazakhstan's water resources alone, for example, are accumulated in more than 48,000 lakes with a total surface of 45,000 km^2.

The biggest lakes include the Caspian Sea (with the Garabogazköl), the Aral Sea, Lake Balkhash, Issyk-Kul, Alakol, Zaisan and Tengiz.

The soil cover of Kazakhstan and middle Asia is marked by a clearly expressed geographical and latitudinal zonality. Soil scientists subdivide the soils of Kazakhstan into black soils (chernozems) (257,000 km^2), chestnut soils (kastanozems) (904,000 km^2), brown and grey-brown soils (calcisols) (1,192,000 km^2), and mountain soils (370,000 km^2; Faizov et al., 2006). The chernozems and dark chestnut soils have been mostly cultivated (Saparov, 2014).

Towards the south, the light chestnut soils of the plain and foothill sub-zones are gradually replaced by the grey-brown soil of the lower altitudes of the western and northern Tian-Shan and Pamir-Alai. In the mountains of the northern Tian Shan, Saur, Tarbagatai and western Altai, dark chestnut soil and mountain chernozems dominate the montane belt. In addition, the mountains of the northern Tian Shan show a belt of leached mountain chernozems, while the western Altai features a belt of mountain-meadow chernozem-like soils. Tajikistan can be divided into four main soil zones—soils of uplands and submountain areas (mainly chernozems and brown calcareous), soils of moderately high mountains (generally brown acidophilous), soils of alpine mountain belts with steppe and glaciers and soils of high mountain deserts (Leontieva, 1968).

The phytogeographical division of Central and Middle Asia are based on the distribution of plants and vegetation (Takhtajan, 1986; Grubov, 2010) and a bioclimatic division (Djamali et al., 2012). The western part belongs to the western Asiatic subregion and is generally included in the Irano-Turanian bioclimatic zone. In the east, the elevated plateaus of the east Pamir, Tibet and central Tian Shan are part of the Central Asiatic subregion and the Central Asiatic bioclimatic zone. This apparent division becomes more diffuse along the 73° eastern meridian between Bishkek and Khorogh due to the complex relief of the south-western ranges of the Tian Shan, the eastern ranges of the Pamir-Alai and the Ferghana Valley.

Socioeconomic Setting

Kazakhstan and Middle Asia are home to a population of approximately 68.3 million people (http://data.uis.unesco.org/). Population density is highest along the western foothills of the Tian Shan and their adjacent plains, especially in the Ferghana Valley. Lowest population densities are found in the deserts. The percentage of the population that is rural is high in all the countries and ranges between 42.3 per cent in Kazakhstan and 73.6 per cent in Tajikistan (Asian Development Bank, 2010). Officially, governments in all respective countries are secular and presidential democracies, although criticism of authoritarian rule is common. Governments develop and implement environmental legislation, which includes nature protection (Squires and Lu, 2018).

Origin and Types of Grasslands

Approximately 2 million km^2 of grasslands cover the region of Kazakhstan and Middle Asia (Wesche et al., 2016). A comprehensive description of this vegetation, including a classification, has yet to be compiled. This is due to the fact that biology and geography as sciences have been practiced in this region only since the 19th century. Early Russian and European botanists at the end of the 19th century and beginning of the 20th provided the first accounts of grasslands in the region (see Lipsky, 1903 for an overview of early botanical explorations). Meanwhile, in the Soviet Union, research on plant and animal systematics surged and led to the publication of regional species catalogues. Given the vast extent of the region, the dominant or physiognomic approach of Soviet scientists provided the fastest mean to classify vegetation. Krasheninnikov (1925), Baranov (1925), Sobolev (1948), Pavlov (1948) and Rubtsov (1952) collected first data on steppe vegetation of Kazakhstan. Lavrenko (1940, 1956) and Lavrenko et al. (1991) provided a biogeographic overview and classification of Eurasian steppe, including the steppe of Kazakhstan. Korovin (1961, 1962) published the first overview of vegetation in Middle Asia. More recently, Rachkovskaya et al. (2003) compiled an overview and map of the biogeographical zones of southern Kazakhstan and Middle Asia.

An assemblage of the scattered literature suggests that the structure and plant species composition of grasslands in the region, as well as the associated soils and fauna, differ foremost across latitudinal ecoregions (forest-steppe, steppe desert) and across mountain ranges (e.g., Tian Shan, Pamir-Alai, west and east Pamir). Hence, in the following paragraphs, the origin and types of grasslands are described separately for these zones:

(1) Forest-steppe zone

The forest-steppe zone stretches across the northern tip of Kazakhstan and is otherwise found in European Russia and the plains of the western Siberian lowland (Nosova, 1973; Fig. 7.2). In this zone, islands of birch and aspen stands (*Betula pendula, B. pubescens, Populus tremula*) are interspersed in a mosaic-manner with steppe, meadow steppe, wetland meadows and marshes (Rachkovskaya and Bragina, 2012; Fig. 7.3A). At a landscape scale, the average cover of forest patches is 20–30 per cent of watershed plains but can be as high as 50 per cent in some places. Steppe communities in this area are associated with leached chernozem and meadow-chernozem soils.

(2) Steppe zone

The Steppe zone of Kazakhstan covers the northern part of the Caspian Lowlands, the Torgay Plateau, the lower Ural piedmont plateau, the Trans-Ural, the west Siberian Lowland and the Kazakh Uplands (Russian: Melkosopochnik, Kazakh: Sary-Arka; Fig. 7.2). It occupies 41.4 per cent of the Kazakhstan territory (Rachkovskaya and Bragina, 2012). Along decreasing latitude, this zone is further divided into five subzones that differ in climate and vegetation:

(2a) Temperate and drought-tolerant, rich forb-feather grass steppe on chernozem

This vegetation grows on chernozem soil, along the southern periphery of the west Siberian Plain. It is dominated by *Stipa zalesskii* and *Festuca valesiaca*, and occasionally, *Poa angustifolia* and *Bromus inermis*. These species grow alongside more mesophytic forbs, like *Artemisia latifolia, Hieracium virosum, Lathyrus tuberosus, Pulsatilla multifida*, and *Veronica spuria*. Species richness can be comparatively high in these communities (45–60 species on 100 m^2). Palatable species make up to 100 per cent of the biomass (Rachkovskaya and Bragina, 2012).

(2b) Drought-tolerant forb-feather grass steppe on southern chernozem

This steppe is confined to the flat alluvial plains of the west Siberian Plain, the denudated plains of the Trans-Ural Plateau and the Kokchetav Uplands. Zonal soil types encompass southern chernozem soils and their carbonate variants, as well as saline soils and salt licks, which give rise to a complex mosaic of vegetation. In addition, rocky southern chernozem

Fig. 7.3 Grasslands of Kazakhstan: (A) meadow in forest-steppe zone, Kostanay Oblast; (B) steppe with *Stipa zalesskii* and *Phlomis tuberosa*, two characteristic steppe species, Karaganda region, Kazakh Uplands; (C) *Pulsatilla flavescens*, Naurzum Reserve; (D) *Tulipa schrenkii*, a red-listed steppe plant in Kazakhstan, Naurzum Reserve; (E) steppe with feather grasses, Naurzum Reserve, Dokuchaev Plataeu; (F) steppe with *Stipa borysthenica*, a characteristic species of sandy soils, Torgay Depression (all from Kazakhstan); photos by T. Bragina (A, C, D, E) and V. Wagner (B, F).

can be found in the Trans-Urals and the periphery of the central Kazakhstan Uplands. The most common plant communitiy is species-rich forb-feather grass steppe (mean total cover: 70–80 per cent), with *Stipa zalesskii* and *Festuca valesiaca* as the most dominant species and *Koeleria cristata* as a co-dominant. Forbs are frequent in the north but become less common towards the south, where they are represented mainly by meso-xerophytic and eury-xerophytic species (e.g., *Eryngium planum*, *Galium ruthenicum*, *Medicago romanica*, *Salvia stepposa*, *Seseli ledebourii*) and, more rarely, xero-mesophytic species (e.g., *Thalictrum minus*, *Xanthoselinum alsaticum*). In addition, forb-feather-grass steppe with *Stipa lessingiana* (a co-dominant; Bragina, 2016) and *Stipa korshinskyi* (an indicator of carbonate) can be found on calcareous soils. Currently, about 60 per cent of the zone's forb-grass steppe is plowed because it is one of the most important agricultural regions of the country.

(2c) Temperate dry bunchgrass steppe on dark chestnut soils

These plant communities are found on dark chestnut (castanozem) soils, which are often calcareous and loamy. They occupy the plains of the Turgay Plateau and the plains separating the central Kazakhstan Uplands (Rachkovskaya and Bragina, 2012). The main vegetation includes dry steppe (mean total cover: 50–60 per cent) with xerophytic bunchgrasses (*Festuca valesiaca, Koeleria cristata, Stipa capillata, S. lessingiana*) and creeping grasses (*Agropyron pectinatum, Cleistogenes squarrosa, Leymus ramosus*) (Fig. 7.3B). Forbs are rare and include xerophytic hemicryptophytes, like *Dianthus leptopetalus, Galatella divaricata, Jurinea multiflora, Phlomoides agraria* and *Pulsatilla flavescens* (Fig. 7.3C) and geophytes, like *Tulipa schrenkii* (Fig. 7.3D).

(2d) Dry xerophytic forb-bunchgrass steppe on chestnut soils

This vegetation type dominates large areas of the Sub-Ural and Turgay Plateau and represents the southern-most outlier of dry steppe in central Kazakhstan. The main zonal plant communities are made up of xerophytic forbs (*Galatella divaricata, G. tatarica, Phlomoides agraria, Tanacetum achilleifolium*), fescue (*Festuca valesiaca*), and feather grasses (Fig. 7.3E). *Stipa lessingiana* is characteristic of loamy soils and *S. borysthenica* of sandy soils (Fig. 7.3F). In total, this vegetation type occupies 530,000 km^2, representing about 20 per cent of the country, of which 33 per cent is cultivated.

(2e) Desert sagebrush-bunchgrass steppe on light chestnut soils

Desert-sagebrush-bunchgrass steppe is the most southern type of steppe in Kazakhstan. Its occurrence coincides roughly with the boundary between the steppe and desert, at 48° N latitude. They occur on light chestnut soils in the Caspian lowlands, Subural, Turgai Plateau, Mugodzhary, the central Kazakhstan Uplands and the Kulunda Plain. *Stipa lessingiana* is a characteristic species in the northern part, whereas *S. sareptana* is typical for southern communities. Compared to the dry steppe described above, desert-steppe has a more complex vegetation cover. The characteristic species of these communities are *Stipa lessingiana, S. sareptana* and *Festuca valesiaca*, the euxerophytic dwarf-shrubs *Artemisia lerchiana* (western Kazakhstan) and *A. semiarida* (central Kazakhstan) and *A. gracilescens, A. sublessingiana* (central and eastern Kazakhstan). Forbs are generally rare and include xerophytes, like *Kochia prostrata*, perennials with a short vegetative period (e.g., semi-ephemeroids like *Ferula* spp., ephemeroids like *Tulipa* spp.), and ephemerals (*Alyssum desertorum, Ceratocarpus arenarius*).

(3) Desert zone

Deserts cover the lower plains and mountain foothills of southern Kazakhstan and Middle Asia. They can be found on contrasting substrates (clay, sand, rocks) and are dominated by shrubs, ephemeroids and, more rarely, small trees. Although they are used as rangelands, they are not grasslands per se and thus not treated in this chapter. Azonal grasslands occupy the seasonally or permanently flooded shores of watercourses and waterbodies of the desert zone. They are comprised of reed beds with perennial rhizomatous grasses, sedges and rushes as dominants.

(4) Mountain zone

Grasslands cover extensive areas in the tall mountains of Kazakhstan and Middle Asia. Their structure, phenology and species composition varies along altitudinal gradients and bioclimatic conditions. They encompass steppe, meadows, grassy fens and mires, xerothermophilous swards, and the so-called semi-savanna (Korovin, 1961, 1962; Ovchinnikov, 1971; Stanyukovich, 1982; Agakhanjanz and Breckle, 2003). The following summary focuses on the grasslands of the region's highest mountain systems, the Tian Shan, Pamir-Alai, and west and east Pamir (for descriptions of grasslands in the region's other mountain systems see Rubtsov, 1948; Stepanova, 1962; Kamakhina, 2005; Baitulin and Kotukhov, 2011; Rachkovskaya and Bragina, 2012).

(4a) Tian shan

Grasslands can be found across all altitudes of the Tian Shan, except for the nival belt (Korovin, 1961, 1962). Their total extent has never been estimated but they are likely the most dominant vegetation type of this mountain system, covering a larger area than forests. Their vast spatial extent is favored by a humid-continental climate, thick loess deposits, a gentle topography, and thousands of years of human land use, most importantly grazing by livestock, forest suppression and tree removal. Grassland communities in the Tian Shan vary in their structure, phenology and species composition across altitude and phytogeographic provinces. Some studies have highlighted the importance of bedrock, soil pH and land use for local species composition (Karmysheva, 1961; Wagner, 2009).

In the western Tian Shan, lower elevations are dominated by semi-desert communities with annual grasses and sedges, like *Bromus danthoniae*, *B. tectorum*, *Carex pachystylis* and *Poa bulbosa* (Karmysheva, 1982). The upper foothills (and sometimes the montane belt) are covered by tall-forb communities, like *Elytrigia hispidus*, *Ferula tenuisecta*, *Hordeum bulbosum*, and *Prangos pabularia* (Demurina 1972). This vegetation type is a characteristic feature of the western Tian Shan, Pamir-Alai, and Pamir; similar communities occur to some extent only in southern Iran, Afghanistan and Kashmir (Korovin, 1962). Botanists have struggled to apply a consistent name to this vegetation type, referring to it as 'savannoides' or 'semi-savanna'. Some have used the term 'dry forb-rich steppe', although its species composition and phenology bears little similarity to the vegetation of the Kazakh steppe in the plains and the mountain steppe described below (Korovin, 1962).

Mountain steppe is one of the most widespread grassland types of the Tian Shan. It can be found across all phytogeographic provinces and along a wide altitudinal range, from the foothills to the alpine belt. This vegetation type occurs on slopes and soils with relatively low water availability, strong micro-climatic fluctuations and a high exposure to winds (Korovin, 1962). Unlike the foothill grasslands described in the previous paragraph, it bears a strong resemblance to the Kazakh steppe further north, as reflected by the dominance of the xerophytic bunchgrasses *Festuca valesiaca*, *Helictotrichon hookeri*, *Koeleria cristata* and *Stipa* spp. Mountain steppe is used extensively for livestock grazing (Korovin, 1962; Karmysheva, 1982).

The relatively precipitation-rich and gentle ridges of the northern and western Tian Shan are home to mesic montane and subalpine meadows, at approximately 1,400–3,000 m a.s.l., on deep loamy soils (Korovin, 1962; Karmysheva, 1982; Wagner, 2009). In some areas, they form parkland with *Juniperus semiglobosa*, *J. polycarpos* var. *seravschanica* (western Tian Shan, Fig. 7.4A) or *Picea schrenkiana* (northern Tian Shan, Fig. 7.4B) (Korovin, 1962). They are a regional variant of the more widely distributed Euro-Siberian grasslands (Wagner, 2009), as evidenced by the prevalence of *Bromus inermis*, *Dactylis glomerata*, *Galium verum*,

Fig. 7.4 Grasslands of Middle Asia: (A) montane meadows interspersed with *Juniperus semiglobosa*, *J. seravschanica* and *J. turkestanica*, Aksu-Jabagly Nature Reserve, north-western Tian Shan, Kazakhstan; (B) *Picea schrenkiana* parkland with montane meadows, Almaty region, north Tian Shan, Kazakhstan; (C) pastures in the Ferghana valley, Pamir-Alai, Tajikistan; (D) steppe with *Stipa kirghisorum*, Alai valley, Pamir-Alai, Tajikistan; (E) meadow with *Koeleria macrantha*, Panj, West Pamir, Tajikistan; (F) haymaking in floodplain meadows, Murghab valley, east Pamir, Tajikistan. (Photos by V. Wagner (A, B), A. Nowak (C, D, E) and K. Vanselow (F).).

Poa angustifolia, and *Veronica spuria*. Montane meadows are a secondary vegetation of forests that were removed in favor of pastures (Wagner, 2009). Since the last century, these meadows have been used mostly for hay-making (Korovin, 1962; Karmysheva, 1982).

Alpine grasslands occur on gentle slopes with good snow coverage in winter, across much of the Tian Shan. Typical plant species include *Gentiana algida*, *Phleum alpinum*, and *Poa alpina*. In the central and eastern Tian Shan, *Kobresia* mats cover sites with a fluctuating water regime (Korovin, 1962).

Finally, the Tian Shan is also home to mires and wet grasslands (*sazy*) occurring along rivers and lakes, and in depressions with constant water availability. In the northern and central Tian Shan, they can cover large areas of plateaus and gentle ridges in the subalpine and alpine belts, where groundwater is high and the mineral soil layer is rich in clay. The most dominant plant species of this vegetation type are sedges (e.g., *Carex melanostachya*, *C. pseudocyperus*) as well as brown-mosses. In the high mountains, they can form peat layers that are 50–80 cm deep (Korovin, 1962).

(4b) Pamir-alai

The grasslands of the Pamir-Alai are similarly favored by a continental climate and a long nomadic tradition (Field and Price, 1950; Figs. 7.4C, D). In the foothills and at the colline and montane belts of some regions, steppe and sward vegetation prevails. The rough terrain of Tajikistan has given rise to vast 'stony' mountainous steppe. Despite its patchy character and restricted surface, it forms a distinct vegetation type on the gentle slopes across elevations of 1,500–3,000 m a.s.l. In the most humid Hissaro-Darvasian range, steppe has a more limited vertical distribution (generally 3,000–3,500 m a.s.l.; Safarov, 2003). Formerly, steppe in Tajikistan was classified by the physiognomic, floristic or dominant approach. The most accepted division differentiates between: (a) Grass-steppe (with several graminoid species and forb or pasture species such as *Artemisia dracunculus* or *Carex stenophylloides*); (b) *Tipczakovy* steppe, named after its dominant, *Festuca valesiaca* (common Russian name: *tipczak*), with *Artemisia dracunculus*, *Poa litvinovii* and *Stipa kirghisorum;* and (c) Semi-desert formation with *Christolea pamirica*, *Stipa glareosa* and *S. subsessiliflora* that is typical of the highly elevated, dry Pamirian Plateau (Ovchinnikov, 1971; Stanyukovich, 1982; Agakhanjanz and Breckle, 2003).

Recent studies in the western Pamir-Alai revealed a considerable variety of plant community types and high plant diversity in the mountain feathergrass-steppe vegetation (Nowak et al., 2016a). The most thermophilous steppe type can be found in the western Pamir-Alai, especially in the montane belt (1,100–1,650 m a.s.l.) of the Zeravshan and Turkestan ranges. It includes *Stipa lipskyi* as a diagnostic species and several plants that also occur in xerothermophilous swards (e.g., *Boissierra squarrosa*, *Eremopyrum bonaepartis* and *Taeniatherum crinitum*). By contrast, stony steppe with *Stipa drobovii* features a considerable proportion of scree species (e.g., *Elaeosticta hirtula*, *Ferula foetidissima*, *Veronica intercedens*). Soils have a low organic topsoil cover and a considerable amount of fine-grained debris. This community type occurs at 1,300–2,300 m a.s.l. and is used as an inferior pasture for goats. The endemic steppe communities of western Pamir-Alai, mainly of the Fan and Zeravshan ranges, are characterized by the presence of *Stipa richteriana* subsp. *jagnobica*, *Astragalus sarytagius* and *Rochelia cardiosepala* as diagnostic taxa. They occur on gentle, stable slopes with a shallow topsoil layer in the upper montane and subalpine belt at altitudes from 1,700 to 2,500 m a.s.l., in an area frequently used as a pasture for goats and sheep. *Stipa margellanica*-steppe occupies stable ground with a rich organic soil layer and is found at subalpine slopes of U-shaped valleys, in the northernmost ranges of the Pamir-Alai. It includes Euro-Siberian and Irano-Turanian species (e.g., *Allium barszczewskii*, *Carex stenophylloides*, *C. turkestanica*, *Elaeosticta allioides*, *Poa bulbosa*). Furthermore, the eastern ranges of the Pamir-Alai harbor additional steppe communities that differ considerably in terms of altitudinal amplitude, floristic richness, distribution, soil preferences and management intensity (A. Nowak et al., unpubl. data).

Alpine pastures and meadows are the dominant vegetation type of the subalpine and alpine belts in Tajikistan. They form a vast green zone across all mountain ranges where the precipitation is relatively high. High-elevation areas are suitable habitats for these vegetation types due to sufficient water availability and deep loamy soils. They are dominated by dense, small, cryophilous swards with a great share of perennials like *Aster serpentimontanus, Eritrichium villosum, Gentiana algida, G. tianschanica, Potentilla flabellata* and *Trollius dzhungaricus* and a considerable number of *Oxytropis* and *Astragalus* spp. At the subalpine and montane interface (1,000–2,500 m a.s.l.), these communities become taller and more species rich. Typical plant species include *Aulacospermum simplex, Biebersteinia multifida, Ranunculus laetus, Stubendorfia orientalis, Tulipa ingens* and *T. praestans*. Meadows with *Rhinopetalum bucharicum, Koelpinia macrantha* and *Primula fedtschenkoi* cover gentle slopes at lower elevations (Fig. 7.4E).

With the increasing intensity of grazing, the species composition changes and local species diversity decreases. Alpine to subnival elevations (up to 4,500 m a.s.l.) are occupied by *Kobresia* mats with *Kobresia pamiroalaica, K. persica, K. capillifolia* and graminoids *Carex pseudofoetida, P. supina* and *Stipa regeliana*. The montane and subalpine belts give rise to more diverse pastures. Typical plants of the subalpine elevations are *Alfredia acantholepis, Lagotis korolkovii, Ligularia alpigena, Schmalhausenia nidulans* and *Tulipa hissarica*. At lower elevations, pastures typically include *Achillea bucharica, Anemone bucharica* and *Muscari bucharica*.

Xerothermophilous swards occur in the southernmost hills at the colline belt and foothills (circa 500–1,500 m a.s.l.). Dominant species encompass graminoids (e.g., *Avena trichophylla, Hordeum bulbosum, Vulpia persica*), interspersed with colorful herbs (e.g., *Anemone verae, Delphinium biternatum, Hypogomphia bucharica, Salvia bucharica*). This vegetation type has been poorly studied in Middle Asia and is almost ignored in summaries of vegetation literature (Stanyukovich, 1982). This is probably due to its diverse species composition and patchy distribution among the mosaic of dry meadows and pastures, xerothermophilous shrubs and ruderal vegetation.

Pamir-Alai is also home to grassy fens and mires (Koroleva, 1940; Afanasjev, 1956; Stanyukovich, 1982; Nowak et al., 2016b). They are composed of graminoid species (e.g., *Carex orbicularis* subsp. *hissaro-darvazica, Eleocharis quinqueflora, Poa supina*). A plant community with *Carex pseudofoetida* is widely distributed and forms flat fens on neutral, peaty soil, in highly elevated river valleys and lake basins. A fen community dominated by *Carex orbicularis* subsp. *hissaro-darvazica* (*Eleocharido quinqueflorae-Primuletum iljinskii*) is frequently found at lower elevations, mainly in the western Pamir-Alai. The mire communities of the eastern Pamir-Alai have much lower plant species diversity and consist mainly of dicotyledons. However, even at their highest elevations, graminoid species (*Alopecurus mucronatus, Calamagrostis neglecta, Catabrosa capusii, Deschampsia pamirica* and *Poa calliopsis*) can be found on marshy, sometimes inundated and saline substrates.

(4c) West and East Pamir

The west and east Pamir are more continental than the adjacent Pamir-Alai and Tian Shan. Therefore, deserts and semi-deserts predominate and grasslands are scarce. Nevertheless, about two-thirds of the income in this region is generated from agriculture and animal husbandry (Mountain Societies Development Support Programme, 2004).

Grasslands are economically important because they are used as pastures and hay meadows. This aspect has become particularly important for the local population after the collapse of the Soviet Union. Common exploitations include the collection of fuel, logging and overgrazing (Breckle and Wucherer, 2006). Consequently, the pressure on natural resources and human-induced land cover change have significantly increased (Breu and

Hurni, 2003). Pastures near villages and winter pastures are heavily overgrazed (Vanselow et al., 2012a).

Given the economic significance of pasturelands for the local human population, most previous studies focused on pasture productivity and degradation (Crewett, 2012; Dörre and Borchardt, 2012; Vanselow et al., 2012a; Shigaeva et al., 2013). By comparison, studies on the diversity, composition and ecological functions of grasslands are scarce (Lebedeva and Ionov, 1994; Borchardt et al., 2011) and those on grassland ecology and classification almost neglected. According to Russian botanists (Stanyukovich, 1982) the meadows and pastures of Tajikistan can be divided according to their altitudinal distribution, physiognomy and management. Almost all types of graminoid vegetation are grazed here with mowing restricted only to some forb and forest vegetation. Unfortunately, vegetation research in the region is hampered by an unclear botanical taxonomy. Many species names are obsolete and the existing determinations are disputed among taxonomists.

The west Pamir is a region of high biodiversity with circa 1,500 vascular plant species, including 160 endemics (Agakhanjanz and Breckle, 1995; Akhmadov et al., 2006; Kassam, 2009). Grasslands in this region can be divided into mountain steppe, mountain meadows and floodplain meadows.

Mountain steppe covers approximately 4 per cent of the west Pamir and occurs predominantly at 3,000–4,000 m a.s.l. Three main vegetation types can be distinguished: grass steppe, herbaceous steppe with *Cousinia*, and *Artemisia* steppes. Grass steppes are dominated by *Poa*, *Stipa*, or *Festuca* spp. and are usually associated with xerophytic mountain-desert plants (e.g., *Krascheninnikovia ceratoides*) and cryophytic species. The dominant species of the feather grass steppe are *Stipa badachschanica*, *S. caucasica*, *S. kirghisorum*, *S. orientalis*, *S. ovczinnikovii*, and *S. rubiginosa*. *Festuca* steppes are dominated by *Festuca alaica* or *F. valesiaca* associated with *Stipa* spp., *Poa relaxa*, and *Piptatherum alpestre*, whereas meadow-grass steppe is dominated by *Poa relaxa*. The region is also home to *Bromus paulsenii* steppe, tall grass steppes with *Elymus* and *Piptatherum* species and steppe communities with *Piptatherum laterale* and *Elymus* spp.

Mountain meadows are rare in the west Pamir and confined to small islands (Fig. 7.4E). The most frequent meadows are those dominated by *Aconogonon* species that sometimes occur with grasses (e.g., *Hordeum bulbosum*, *Dactylis glomerata*, *Calamagrostis* spp., *Stipa* spp.) and herbs (Agakhanjanz, 1985; Vanselow et al., 2016).

Floodplain meadows are widespread but constitute only 2 per cent of the west Pamir cover. They are limited to riparian sites and are strongly affected by groundwater, melting snow and overgrazing. Sedge meadows are dominated by *Carex griffithii* and *C. orbicularis*, sometimes in association with *Blysmus compressus*, *Kobresia* spp., *Primula* spp., as well as other herbs and grasses. Bog sedge meadows are predominantly composed of *Kobresia humilis*, *K. pamiroalaica* and *K. schoenoides* accompanied by many species of sedges, grasses, *Trifolium repens* and other herbs. *Triglochin palustris* occurs frequently in swamps (Vanselow et al., 2016).

The east Pamir shows a less diverse flora, with only 738 different species of higher plants (Agakhanjanz and Breckle, 2004). Most vegetation types are dominated by dwarf shrubs and cushion plants, in particular *Krascheninnikovia ceratoides*, and resemble deserts or semi-deserts. These formations are vital grazing land and provide fuel material. Riparian floodplain meadows are the most important type of meadows in the east Pamir (circa 5 per cent of the regional cover). They are dominated by sedges and bog sedges (e.g., *Carex pseudofoetida*, *C. melanantha* and *Kobresia royleana*) and are subject to groundwater influence. Halophytic plants like *Glaux maritima* and *Triglochin maritima* can be found in brackish soils (Vanselow, 2012). At high elevations (usually > 4,200 m a.s.l.), small patches of mountain meadows can be found that are also characterized by sedges and grasses but associated with more forbs, e.g., *Smelowskia calycina*. There is little mountain steppe on the Pamir plateau. Exceptions are desert steppe and tall grass steppe. The latter are dominated by *Leymus secalinus* or *Hordeum turkestanicum* on silty terraces and by tussocks of *Stipa splendens* on gravely river terraces. The desert steppe can be separated into herbaceous and grass steppe. The latter occurs predominantly on sand deposits in valleys and is characterized by *Festuca, Poa* and *Stipa* spp. (Walter and Breckle, 1986; Vanselow, 2012).

Steppe Fauna

The intersection of several biogeographic regions has given rise to a unique fauna in Kazakhstan and Middle Asia. The forest-steppe, steppe and desert animals of the Kazakh plains and uplands form a faunistic subregion of the greater Holarctic zoogeographical region. In general, the steppe fauna includes taxa confined to the adjacent forest and steppe zone and elements of the more southerly semi-desert and desert.

The steppe of Kazakhstan is at the crossroads of important migratory flyways that stretch from the Siberian tundra and taiga to South Asia and Africa. Water reservoirs are rich in fish. In spite of intensive economic activity, the number of animals has not declined in the steppe, except for the Great bustard (*Otis tarda*), Willow grouse (*Lagopus lagopus*) and Saiga antelope (*Saiga tatarica*) (Bragina, 2015a; Kamp et al., 2016). Large-scale plowing of steppe has limited the natural habitat and the population of large mammals. Uncontrolled poaching of birds of prey, reptiles and saiga has become a serious problem for conservation.

Middle Asia is home to several rare animal species. River lowlands provide habitat for Persian gazelle (*Gazella subgutturosa*), Eared hedgehog (*Paraechinus hynomelus*), African wild cat (*Felis libyca*), Steppe agama (*Agama sanguinolenta*), Desert monitor (*Varanus griseus*), Snake-arrow (*Taphrometopon lineolatum*) and the snake *Echis carinatus* (Safarov et al., 2014). Its mountains are inhabited by critically endangered animals, such as the Snow leopard (*Uncia uncia*) and the Argali (Marco-Polo sheep, *Ovis ammon polii*) (Beg, 2009).

Agronomic Use of Grasslands

The Kazakh steppe is subject to conflicts between agricultural interest and nature conservation because many types of steppe occur on soils that are excellent for crop farming (Baitulin et al., 1999; Hölzel et al., 2002; Rachkovskaya and Bragina, 2012). Since the last century, the steppe zone has become dominated by crop farming (mostly wheat). Due to its geographical location in the Eurasian wheat belt, Kazakhstan is the sixth largest wheat producer in the world. The Oblasts of Kostanay, north Kazakhstan, Akmola and parts of Pavlodar, Karaganda Oblast, west Kazakhstan and Aqtobe Oblasts are the primary wheat-producing regions in the country. In the mountains of Middle Asia, crop agriculture is confined to valleys. Mountain grasslands have been used for millennia as summer pastures (*zhailau*).

Historical Development

The first human activity in Kazakhstan and Middle Asia can be traced back to the Palaeolithic. Humans hunted, fished and gathered plants. Cattle breeding and primitive farming appeared in the late Stone Age around 8,000–3,000 BC. In the Bronze Age

(2,000–1,000 BC), steppe cultures appeared. During the period of 1,000 BC–AD 1,000, the Kazakh steppe was invaded by nomadic tribes, who used their pastures for cattle-breeding. Since then, a (semi-) nomadic lifestyle was practiced in most parts of Kazakhstan and Middle Asia (Tolstov et al., 1963).

Until the last century, zonal steppe was generally not used for crop agriculture. Nomads exerted a low to moderate influence on the steppe through livestock grazing. By comparison, the resettlement of nomads in combination with migrations of European peasants at the end of the 19th century, affected the landscape and wildlife to a much larger extent. In the 20th century, the intensity of plowing, logging, irrigation, hunting, fishing and road construction increased further. The Soviet government banned nomadism and private property. The Soviet Virgin Lands Programme (1954–1960) opened up the vast steppe to farming in order to alleviate food shortage among the Soviet population. During this period, 254,000 km^2 of natural steppe in Kazakhstan were plowed for crop farming. In contrast to the high expectations, high crop yields dropped initially. The use of old agricultural technology in a dry-climate terrain proved to be unsuitable for intensive agro-economic activities. Most virgin lands of the region were prone to wind and water erosion and lost 30–50 per cent of humus on average (Peterson, 1993). The problem of reduced soil fertility was aggravated by rising water-tables, which resulted in large-scale soil salinization. In some areas, overgrazing resulted in habitat deterioration and desertification. As a consequence, a number of small lakes disappeared in the region and many large man-made reservoirs were built for irrigation. The groundwater level dropped. The most spectacular loss has been the extinction of larger animals (*Panthera tigris virgata, Acinonyx jubatus venaticus, Cervus elaphus bactrianus*), the desiccation of the Aral Sea and the almost total clearance of river gallery forests. In addition, mining operations increased and fires became more frequent. A higher number of vehicles impacted the natural vegetation in a negative manner. As a result, the environment of the region has become increasingly unfavorable for the livelihood of the population. In fact, almost the entire region faces lack of freshwater (Bekchanov et al., 2018) and desertification (Baitulin et al., 1999; Bragina, 1999, 2007a). Permanent pastures in the vicinity of settlements are seriously degraded since steppe is only well adapted to periodical grazing by migrating herds (Akiyanova et al., 2017).

After the breakup of the Soviet Union in the 1990s, intensive crop production was no longer an economically and ecologically viable option. Most of the land on former state farms in the dry steppe zone was left fallow. About 15–20 years after abandonment, arable fields in central Kazakhstan were transformed back to steppe by spontaneous succession. However, steppe vegetation communities on abandoned fields are far from being similar to near-natural steppe (Brinkert et al., 2016). Moreover, the number of livestock (the second most important source of income in rural areas besides wheat) decreased dramatically (Esenova and Dobson, 2000; Khazanov, 2002; Kerven, 2003). Livestock farming played only a minor economic role after privatization. This situation has endangered the meat supply to the local human population and exacerbated the conservation of wild ungulates, especially the Saiga antelopes. So far, only a small percentage of primary producers have opted for, or have been provided with, the opportunity to start small farms.

Humans have inhabited Middle Asia at least since the Paleolithic (Tolstov et al., 1963). In the Bronze Age, when wild ungulates were domesticated, locals practiced crop cultivation in river valleys and transhumance in the mountains. The most recent archaeological evidence suggests that the indigenous population practiced a mixed agropastoral system in foothills (Spengler, 2015). Till the 1930s, semi-nomads migrated with their livestock—sheep, horses and more rarely yaks, cows and camels—vertically throughout the year (Dakhshleyger, 1980). In spring and summer, pastoralists kept their herds on montane and subalpine pastures (*zhaylau*); in winter, they descended to the lower foothills and plains (*kstau*) where livestock browsed for forage under the thin snow cover. To expand their pastures, locals burned and logged the native forests and cleared the shrub cover (Dakhshleyger, 1980).

The colonization by European and Russian settlers, in the 19th century, who competed for land in the lower foothills and plains and the mass-collectivization and forced sedentarization during the Soviet era, triggered a shift in life-style (Karmysheva, 1981). As a consequence, the semi-nomadic life-style almost vanished from Middle Asia. In the Soviet Union, only state-owned *kolchoz* and *sovchoz* groups were allowed to operate livestock farms. Since the beginning of the 1990s, following political independence, livestock farms in Middle Asia have become privately owned again.

While livestock grazing was present in Middle Asia for millennia, mowing was introduced as late as in the 19th century, as a technique borrowed from European settlers (Dakhshleyger, 1980). Today, both livestock grazing and mowing are practiced in the region, although the former is still more common than the latter. Despite the drastic changes in life-style, meadows continue to be used foremost in a traditional manner; ameliorations like fertilization, seeding and multiple hay cuttings per year, are still rare (Wagner, 2009).

Current Practices of Grassland Management in Kazakhstan and Middle Asia

The grasslands of Kazakhstan and Middle Asia are important for the region's economy and food security. During the first years of their independence, Kazakhstan and the countries of Middle Asia focused their policies on maximizing the production of single crops, such as wheat and cotton. In Kazakhstan, this strategy strengthened the crop economy; in the other countries it was a step towards self-sufficiency in crop production. This single-crop production strategy became widespread but had an obvious negative impact on crop diversification. The wheat and cotton production increased while livestock farming became a neglected sector. In recent years, livestock farming has gained more governmental support but livestock productivity could only improve if measures to enhance rangeland management and increase forage production are undertaken (Squires, 2012; Squires et al., 2017a,b). Land degradation and desertification processes prevail in many localities across the region, with deleterious consequences for soils and vegetation (Lal, 2007; Saparov, 2014). A change in policy towards ecologically sound rangeland improvement and integrated crop and livestock production might also benefit soil conservation, in the long-term.

Grazing intensity is one of the most important factors for grassland conservation and management in Middle Asia. Traditional livestock husbandry is still a main income for the local population. Sheep, goats and horses, and more sporadically cattle, donkeys and camels, are common grazing animals throughout Middle Asia. It is hardly possible to find any credible statistical data for the grasslands of Middle Asia related to management, biodiversity threats, large-scale productivity or environmental policy. Some projects reveal considerable flaw even in the basic data on livestock population densities (Anon., 2013). However, all authors of this chapter have observed overgrazing in Middle Asia. Overgrazing has deteriorated grasslands in many localities of this region, including steppe and meadows. Grazing intensity has fluctuated over the last two decades due to an economic depression that followed the collapse of the Soviet Union. The last years have been characterized by a considerable increase in livestock numbers (Safarov et al., 2014).

In Tajikistan, pastures and hayfields cover approx. 32,000 km² (70 per cent of the total agriculturally used land) and are rather stable in extent. Only the amount of arable fields has decreased in the last decade. However, the number of livestock numbers has significantly changed, with the highest increase in sheep, goats (32.7 per cent) and cattle (11.4 per cent; Safarov et al., 2014). In Kyrgyzstan, the number of cattle increase from 1.05 to 1.36 million, yaks from 21,900 to 31,500, sheep and goats from 3.06 to 5.42 million and horses from 3,45,000 to 3,98,000 during the period 2006–2012 (Anon., 2013).

The increased livestock density will most likely have detrimental consequences for soil and vegetation. In Tajikistan, circa 25 per cent of pastures are already designated as degraded. Under heavy grazing, weeds can increase to cover 70–90 per cent in meadows and pastures. Currently, about 37,000 km² of pasture (i.e., one-third of the territory) are classified as polluted, deserted or covered by thorny shrubs or invasive species and therefore, are regarded as unsuitable for grazing (Safarov et al., 2014). The density of livestock population per unit area exceeds its optimal level by more than 10 times (Safarov et al., 2014).

A similar situation prevails in Kyrgyzstan. The pasture load reached 2.14 sheep per hectare in the last few years, which exceeds the environmentally sound standards by 3.3 times. In Uzbekistan, pastures as most widespread agricultural land cover circa 46 per cent of the country territory; crops are less widespread (circa 11 per cent) but economically important (Sadikov et al., 2015). Most pastures are situated in the country's desert zone, in the west (78.1 per cent), and are similarly threated by degradation. Pasture productivity declines by 1.5 per cent per year, on an average (Sadikov et al., 2015).

Economic Value of Grasslands

The amount of crop-cultivated area in Kazakhstan has gradually increased from 184,000 km² in 2013 to 2,15,000 km² in 2015. In 2020, it is estimated to reach 352,000 km². In 2005, the cultivated area of forage crops was 1,845 km², 1,652 km² in 2008, 1,639 km² in 2010, 2,062 km² in 2013 and according to projections, will reach 10,290 km² in 2020.

The Republic of Kazakhstan has vast pasture resources (1,820,000 km²), including hayfields (48,000 km²) and cropland for fodder production (25,000 km²). Currently, the productivity of pastures does not exceed 250 feed units per hectare. It is especially low in Mangistau, Kyzylorda, Atyrau Oblasts, with only 110–120 feed units per hectare due to overgrazing (264,000 km² of degraded pasture).

The productivity of natural grasslands varies across biogeographic zones and vegetation types. Highest values are achieved by meadow steppe in the forest-steppe zone (500–1,000 kg/ha dry mass; in wet years: 1,500 kg/ha). In the northern part of the steppe zone, productivity of natural steppe on chernozem in the subzone of forb-feather grass steppe reaches 500–800 kg/ha. In the southern dry steppes, productivity in the forb-feather grass natural steppe on carbonate soils decreases to 300–500 kg/ha. Further south, in the dry steppe, it reaches 150–350 kg/ha (but remains high on sandy soils: 300–500 kg/ha). In the most arid steppe regions of Kazakhstan, productivity is variable (250–550 kg/ha, deserted natural steppes) but generally low (100–300 kg/ha, saline soils on solonetz) (Rachkovskaya and Bragina, 2012). The productivity of natural riparian meadows reaches 700–900 kg/ha. A slightly higher productivity is found in floodplain meadows (800–1,100 kg/ha), while dry meadows are less productive (400–600 kg/ha).

Livestock numbers have undergone substantial changes in Kazakhstan due to a reform of agricultural production methods and operation in market relations in the 1990s. The population of farm animals was drastically reduced from 9.5 to 4.2 million cattle in 1999. In 1991, 723,600 t of beef were produced but in 2009, beef production reached only 396,00 t. Currently, 82 per cent of cattle are concentrated in private farms, 6–7 per cent remains in private (peasant) farms and the remaining are the property of agricultural enterprises. This makes it difficult to improve mechanical technology in livestock husbandry (USDA, 2013). From 2007 to 2012, the number of livestock species significantly increased in agricultural enterprises of Kazakhstan, namely sheep and goats from 16,080,000 to 17,633,300, horses from 1,291,100 to 1,686,200 and camels from 143,200 to 164,800, while cattle decreased from 5,840,900 to 5,690,400 (Omarkhanova, 2015).

Threats to Grasslands

Humans have deteriorated the natural landscapes in Kazakhstan and Middle Asia (Orlovosky and Orlovosky, 2018). The grasslands of this region are particularly vulnerable to degradation and desertification. The major industrial and mining centres in central and eastern Kazakhstan and in the Ferghana valley are facing high levels of technogenic disturbances and industrial pollution by toxic waste, such as lead, mercury, chromium, and uranium. Large sections of west Kazakhstan and Turgay are contaminated with oil and radioactive material, high levels of salinity with industrial wastewater and technological transformation of the soil landscape, leading to accumulation of toxic heavy metals (lead, cobalt, nickel, vanadium, etc.) and radionuclides (thorium, barium, radium). These pollutants are spread over 59 per cent of the area in the Atyrau region, 19 per cent in Aktobe and 13 per cent in west Kazakhstan (Saparov et al., 2006; Saparov, 2014).

Many vertebrate (rodents) and beetle species, which are of prime importance for maintaining natural dynamics of steppe ecosystems, have decreased alarmingly; some are on the verge of extinction. The decrease of livestock numbers and the loss of large Saiga herds have resulted in large areas being undergrazed. As a consequence, steppe fires have become more severe (Abaturov et al., 1982; Fadeev and Sludskii, 1982; Bekenov et al., 1998). Hovewer, large areas of dry steppe remain in a natural or semi-natural condition. Some natural grasslands have been protected in government-owned areas since the first half of the last century (see also section on conservation, below).

In Tajikistan, pasture quality is crucial for ensuring livelihood of the rural population. During the years 2010–2013, grazing intensity drastically increased (Strong and Squires, 2012). On the one hand, the growth of livestock numbers to almost 8 million animals (one million cattle, six million sheep and goats, 0.5 million horses) ensures the survival of the local human population. However, on the other hand, this trend led to severe grassland degradation (Lal, 2007; Safarov et al., 2014). Moreover, the lack of the sustainable usage policy of grazing areas promotes a considerable change in plant composition, loss of fodder mass productivity and soil erosion (Halimova, 2012; Robinson et al., 2012). Species typical for semi-deserts have increasingly colonized degraded pasturelands (e.g., *Peganum harmala*). It is estimated that overgrazing, water shortage and soil erosion, have degraded 57 per cent of the rural area (around settlements, winter pastures) and 40 per cent of natural grasslands (distant, alpine summer pastures). Approximately 30–35 per cent of this land is extremely degraded (Safarov et al., 2014).

Changes in Management Type and Intensity

During the Soviet period, crop farming in the steppe zone largely concentrated on wheat production without taking into account the ecological vulnerability of dry steppe. This practice has harmed the region's biodiversity and ecosystem functioning. In future, the fate of locally available natural resources will depend on the newly evolving balance between crop and animal farming. As Kazakhstan continues on its path as a market economy, new forms of organization are beginning to emerge within, among, and outside the former collective farms. This trend might lead to a dynamic reconstruction of the livestock sector. Socio-economic and ecological research will have to pay particular attention to the ongoing changes in land use, especially the relative magnitude of crop and livestock farming.

In Tajikistan, pastures and hay meadows harbor a number of red-listed and threatened species. Any significant transformation in their management could have crucial effects on species diversity. The country's increased population growth rate in the recent decade will likely influence traditional farming and pastoralism. Distant-pasture cattle breeding with consequent migration will probably decrease and many distant pastures are likely be abandoned. Meanwhile, pastures near settlements, temporary residential areas and rivers, may expand and become more prone to overgrazing. It is estimated that by 2020, the area of mountain pastures will be reduced from 35,000 to < 30,000 km^2 (Safarov et al., 2014). In the Pamir, pastures, including grasslands, face currently the threats of growing livestock numbers, increasing socio-economic disparities among livestock holders and unclear land-use regulations. Unlike in the past, the spatial distribution of livestock changed considerably during the post-Soviet transformation. Whereas pastures were used mono-seasonally by collective and state farms, meadows were often excluded from grazing and used for hay making (Fig. 7.4F). Today, the majority of pasture users are smallholders who cannot afford the relocation of their herds. Therefore, pastures, including hay meadows, suffer under multi-seasonal use. Pastures near settlements (where user claim is weak) and winter pastures are heavily overgrazed. By contrast, extensive areas claimed by powerful livestock owners are used less frequently. The inequality of user claims is the result of legal pluralism. Most collective farms were transformed into private farmer associations in 1999. They inherited the land titles for the pastures and allocated pasture rights to their members according to the forage needs of their livestock. In practice, however, many pasture users evade this system by justifying their claim based on customary and inherited law dating back to the tenancy period shortly before the breakdown of the Soviet Union. The coexistence of these two competing forms of rights—the official land titles and the informal user rights—lead to inappropriate claims and overgrazing (Kraudzun, 2012; Vanselow et al., 2012a, 2012b).

Climate Change

For the past few years, the temperature regime of the area has considerably changed. The mean annual temperature increased by 0.6°C. The months of March, July and November became colder, while the mean monthly temperatures of December, January, February, April and August increased. Winters became warmer by 2–4°C, with frequent thaw, haze and mist. As a result, the snow depth became lower. In summer and autumn, droughts became more frequent. However, despite this manifestation of climate change, the state of natural populations, communities and ecosystems remain comparatively stable (Lioubimtseva and Henebry, 2009; Squires and Feng, 2017). In the light of a changing climate and the oil-based economy, Kazakhstan aims to focus on low-carbon development (Decree of the President of the Republic of Kazakhstan dated May 30, 2013, No. 577), as demonstrated during the EXPO fair in Astana, in 2017.

Conservation of Grasslands

Rare Flora and Fauna

The flora of the forest-steppe and steppe zones of Kazakhstan comprise approximately 2,000 vascular plant species, most of them perennial grasses and forbs, including geophytes, semi-shrubs and annuals (Rachkovskaya and Bragina, 2012). About 30 plant species (1.5 per cent of the total flora) are endemic to the steppe zone (Rachkovskaya et al., 1999). Rare plants are particularly confined to rocky sediments and chalky soils. Some isolated low mountains (Chingiztau, Karatau, Mugodzhary, Ulytau) are known for rare plants (e.g., *Clausia kasakorum, Lepidium eremophyllum, Megacarpaea mugodsharica, Potentilla kasachstanica, Scorzonera diacanthoides, Tanacetum ulutavicum* and *Vincetoxicum mugodsharicum*). The central and eastern Kazakh Uplands are home to a group of local endemics (*Astragalus kasachstanicus, Erysimum kasakhstanicum, Euphorbia andrachnoides,* and *Seseli eriocarpum, Thymus cerebrifolius*) and regional (*Oxytropis gebleriana, Thymus kasakhstanicus, T. kirghisorum*). In addition, *Astragalus kustanaicus* (on the Turgay Plateau) and *Linaria dolichocarpa* (in western and central Kazakhstan) are associated with sandy substrates and *Leymus akmolinensis* with *liman* vegetation. Western Kazakhstan is also home to several endemic species, such as *Artemisia lessingiana, Astragalus zingeri, Jurinea mugodsharica, J. fedtschenkoana*. Endemic trees and shrubs occur in the Kazakh Upland, such as *Berberis karkaralensis, Betula kirghisorum* and *Caragana bongardiana*. Typical Trans-Volga-Kazakhstan endemics are *Astragalus macropus, Dianthus acicularis, Scabiosa isetensis, Silene suffrutescens* and *Stipa korshinskyi. Tulipa schrenkii* is a red-listed species that grows in the northern steppe of Kazakhstan.

The large territory of virgin steppe provides habitat for > 700 species of vascular plants, including feather-grasses, tulips, and wild onions and a rare fauna that includes Bustard (*Ardeotis*) species, Sociable lapwing (*Vanellus gregarious*), Black lark (*Melanocorypha yeltoniensis*), White-winged lark (*M. leucoptera*), Steppe eagle (*Aquila nipalensis*), Pallid harrier (*Circus macrorus*), Bobak marmot (*Marmota bobak*), Steppe pika (*Ochotona pusilla*), Corsac fox (*Vulpes corsac*). The steppe and deserts of Kazakhstan are vital for seasonal migrations and calving grounds for the critically endagered migratory Saiga antelope (*Saiga tatarica*; Kamp et al., 2016). In addition, some steppe areas are interspersed with pine forests that provide habitat for rare predatory birds, including the Imperial eagle (*Aquila heliaca*), Saker falcon (*Falco cherrug*), Red-footed falcon (*Falco vespertinus*), White-tailed eagle (*Haliaeetus albicilla*) and Golden eagle (*Aquila chrysaetos*). Given their position at the cross road of Central Asian and Siberian-south European flyways, the steppe wetlands are internationally significant for migratory birds.

About 1,500 invertebrate species, representing 160 families and > 15 orders, have been reported for the steppe zone (Anon., 1950; Bragina, 2004, 2009b and others). The Emperor moth (*Saturnia pavonia*) is listed in the *IUCN Red Book*. Four invertabrate species are recorded in the *Kazakh Red Book* (the Emperor dragonfly *Anax imperator,* the Mantis species *Bolivaria brachyptera,* the Scollid wasp *Scolia hirta* and the predatory Bush cricket *Saga pedo*).

There is no credible information as to how many vascular plant species, especially endemics, inhabit the grasslands of Middle Asia. In Tajikistan, recent research estimates the number of endemics and subendemics in grasslands as circa 370 (Nowak et al., 2011). Some examples of narrowly distributed, so-called national endemics, in Tajikistan are: *Allium oreodictyum, Astragalus lancifolius, Ferula violacea, Lagochilus kschtutensis, Onobrychis baldshuanica, Polygonum ovczinnikovii, Stipa richteriana* subsp. *jagnobica, S. ovczinnikovii, S. zeravshanica* and *Zeravshania regeliana*. Most of them are red-listed and protected by the country's Species Protection Act. This list could be effectively amended after completion of studies in the desert steppe of east Pamir and *Artemisia*-steppe across the country.

Protected Areas

Protection of Zonal Forest-steppe and Steppe in Kazakhstan

In Kazakhstan, all protected areas are public and managed by the government (Bragina, 2007b). They differ in their degree of protection, with the most important protected categories being Zapovedniks (hence: Nature Reserve, Category 1a according to the International Union of Conservation of Nature), National Nature Parks (hence: National Park, Category 2) and State Nature Reservations (Category 1b or 2; Fig. 7.5). In general, protected areas have offices with full-time administrative and scientific staff, rangers and employees for education, public outreach and ecotourism. Large-scale Category 1 or 2 protected areas are scattered throughout all regions of Kazakhstan, except for the country's western part. The flora and fauna of zonal steppe is also protected in smaller sanctuaries of national importance (forest-steppe: 6 sanctuaries, steppe zone: 21).

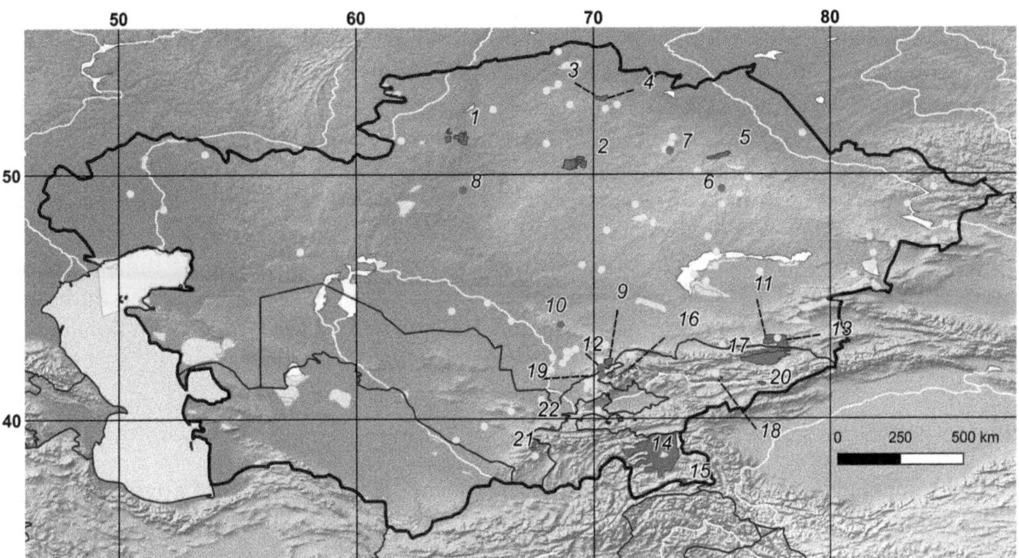

Fig. 7.5 Protected areas (colored) in Kazakhstan and Middle Asia. Red colours and numbers indicate protected areas that are mentioned in the text: 1–Naurzum Nature Reserve; 2–Korgalzhin Nature Reserve; 3–Kokshetau National Park; 4–Burabay National Park; 5–Bayanaul National Park; 6–Karkaraly National Park; 7–Buyratau National Park; 8–Altyn Dala Nature Reserve; 9–Aksu Jabagly Nature Reserve; 10–Karatau Nature Reserve; 11–Almaty Nature Reserve; 12–Sairam-Ugam National Park; 13–Ile-Alatau National Park; 14–Pamir (Tajik) National Park; 15–Zorkul Nature Reserve; 16–Besh—Aral Nature Reserve; 17–Issy-kul Nature Reserve; 18–Karatel-Zhapyrk Nature Reserve; 19–Karabuurinskiy Nature Reserve; 20–Naryn Nature Reserve; 21–Hissar Nature Reserve; 22–Zaaminsky Nature Reserve. *Data source*: UNEP-WCMC and IUCN (2017) and author's own supplements.

Category 1a guarantees the highest protection under the Kazakh Law on Protected Areas. Zonal steppe is protected under this regime in the Naurzum (1,914 km^2) and Korgalzhin (5,432 km^2, incl. 2,000 km^2 of wetland surface) Nature Reserves. The former reserve protects feather grass-steppe on sand and loam deposits along the vast slopes of the Turgai Plateau, saline plains and southern isolated populations of *Pinus sylvestris* and *Betula kirghisorum*, as well as shrub communities and wetlands. The latter reserve protects fescue-feather grass steppe, sagebrush-fescue-feather grass steppe, stony steppe of the Kazakh Uplands, complex steppe on saline plains and wetlands. Recently, these two nature reserves were the first natural objects in Kazakhstan and Middle Asia to be included in the UNESCO World Heritage List (Committee Decision No. 1102 from 7 July 2008). Together, they form the Heritage site 'Saryarka-Steppe and Lakes of Northern Kazakhstan' (Bragina, 2009a,b).

Zonal steppe is also protected in seven national parks, in Kazakhstan. The Burabay (1,299 km^2) and Kokshetau (1,820 km^2) National Parks are both located in the forest-steppe zone and were established already under the Soviet government in order to protect meadow steppe, species-rich forb-feather grass steppe and *Pinus sylvestris*-forests on granite. Furthermore, Bayanaul (685 km^2) and Karkaraly (1,121 km^2) National Parks protect open *Pinus sylvestris*-forests and stony steppe in the Kazakh Uplands. A comparatively recent addition, Buyratau National Park (890 km^2), was established in the Ereymentau mountains, north-central Kazakhstan, in order to protect relict southern *Betula kirghisorum* and *Alnus glutinosa*-forest patches, as well as grassland communities on the transition of arid and dry steppe. Altyn Dala (4,900 km^2) is the most important State Nature Reserve for zonal steppe. Located in the Turgay plain, it protects the large and mostly undisturbed desert steppe of central Kazakhstan.

In recent years, scientists, non-governmental organizations and the government have teamed up to expand the network of protected areas in the steppe and desert-steppe zones of Kazakhstan (Chibilev, 1998; Bragina, 2007a,b, 2014, 2015b; Bragina et al., 2013; Demina and Bragina, 2014; Deák et al., 2016). Furthermore, recent environmental legislation provides that, in the future, strictly protected natural areas shall be connected by ecological corridors; the latter surrounded by buffer zones with limited human disturbance. The concept of an ecological network (Chibilyev, 1998; Demina and Bragina, 2014) in the wider region of Kazakhstan and Middle Asia was developed in the course of the project for "creating an ecological network for long-term conservation of biological diversity in eco-regions of Central Asia", supported by UNEP/GEF/WWF in 2003–2006 (Balbakova et al., 2006).

Protection of Grasslands in Middle Asia

Many protected areas in Middle Asia were established under the Soviet government and their protection categories follow the ones outlined above for the Kazakh plains. About 220,000 km^2 of land is protected in Middle Asia (UNEP-WCMC and IUCN, 2017) and the density of conservation sites is particularly high in the mountain region (Fig. 7.5). Extensive grasslands (meadows, steppe or alpine mats) are protected in the Aksu-Jabagly (approximate overall protected area: 751 km^2, Kazakhstan), Karatau (400 km^2), and Almaty (733 km^2) Nature Reserves, the National Parks of Sayram-Ugam (1,491 km^2) and Ile-Altau (1,644 km^2, all in Kazakhstan), in Pamir National Park (26,000 km^2) and Zorkul Nature

Reserve (877 km², all in Tajikistan), Besh-Aral (1,120 km²), Issy-kul (197 km²), Karatel-Zhapyrk (213 km²), Karabuurinskiy (590 km²) and Naryn Nature Reserves (1,055 km², in Kyrgyzstan), Hissar (1,629 km²) and Zaaminskiy Nature Reserve (537 km², in Uzbekistan; Sokolov and Syroechkovskiy, 1990; UNEP-WCMC and IUCN, 2017).

The Pamir contains large protected areas. In practice, however, they are insufficiently managed. With circa 26,000 km², the Pamir National Park is the largest among the protected areas in Tajikistan. It was founded in 1992 and added to UNESCO World Heritage in 2013. The area comprises predominantly unsettled, mountainous landscape, including the Fedchenko Glacier, with a length of 77 km. It is the biggest glacier in Pamir and one of the longest valley glaciers outside the polar regions (Haslinger et al., 2007). Another important protected area encompasses the Zorkul depression with its lakes, wetlands and surrounding mountains. This area was designated as early as 1972 as a nature sanctuary (*zakaznik*). In the year 2000 it was extended to an area of 877 km² and upgraded to a strict nature reserve (*zapovednik*, IUCN Category I). Because of its outstanding importance for birds, in 2001, this protected area was added to the Ramsar list of wetlands. Meanwhile, it is also on the proposal list for future world heritage nominations (Diment et al., 2012).

Resumé and Future Prospects

For millenia, the grasslands of Kazakhstan and Middle Asia have been shaped by natural forces and by a long history of human land-use, particularly livestock grazing. Their vast extent, their role as habitat for a distinct flora and fauna and the fact that most land is public and comparatively little disturbed, make them a globally significant territory. Given stable economic and political conditions, they have the potential to be used in a multi-functional way in the future, particularly through sustainable agriculture and (eco-) tourism (Akiynova et al., 2017; Squires et al., 2017a). Science and technology may provide an overall knowledge shift when it comes to recognizing processes and initiating sustainable development (Akramkhanov et al., 2018). A proper conservation policy will be key to maintaining the distinct diversity of the Kazakh and Middle Asian grasslands and promote a sustainable economic development among the rural population.

Specific priorities for grassland conservation encompass (Rachkovskaya and Bragina, 2012):

- Identification and conservation of the remaining northern areas of the species-rich forb—feather grass steppe (as sanctuaries, natural monuments at the national or local level)
- Identification and conservation of natural steppe areas
- Creation of a trans-boundary ecological network that should link the region's grasslands with those of neighbouring countries.

In future, the grasslands of Kazakhstan and Middle Asia will continue to play an important role as pastures and hay fields and ensure the livelihood of the region's population. Their fate will depend on the balance between traditional management practice and economic growth. Monitoring, conservation and restoration will be crucial to sustain their ecological integrity.

Glossary

Feed unit	=	Amount of feed corresponding to its nutritive value to 1 kg of oat grains of average quality, i.e., 5.95 MJ energy content.
Kazakh Uplands	=	Series of small and intermediate mountains interspersed across the centre of Kazakhstan (also called *Melkosopochnik*).
Kolchoz	=	Widespread type of a collective farm in the Soviet Union, formally owned by local members.
Kstau	=	Pastures of lower foothills and plains.
Liman	=	Mesic meadows in seasonally wet depressions, in the steppe and desert zone.
Melkosopochnik	=	See definition of Kazakh Uplands above.
Middle Asia	=	Region that is defined here *sensu* Rachkovskaya et al. (2003) as the countries of Kyrgyzstan, Tajikistan, Turkmenistan and Uzbekistan.
Sazy	=	Mires and wet grasslands.
Sovchoz	=	Widespread type of a (large) collective farm in the Soviet Union, formally owned by the state.
Steppe	=	vast natural dry grasslands of Eurasia.
Tipczakovy steppe	=	Dry grassland vegetation named after its dominant, *Festuca valesiaca* (common Russian name: *tipczak*).
Virgin Lands Campaign	=	A major political and agricultural movement in the Soviet Union, in the middle of the last century, in which steppe that was never plowed before was converted to cropland.
Zakaznik	=	Type of protected area in Russia and former Soviet republics that is equivalent to *sanctuary* in the UNESCO World Heritage terminology.
Zapovednik	=	Strictest protected areas in the region, often established when the respective countries were part of the Soviet Union.
Zhaylau	=	Traditional summer pasture used by semi-nomads in the mountains.

References

Abaturov, B.D., M.V. Kholodova and A.E. Subbotin. 1982. Intensity of feeding and digestibility of forages in saiga [in Russian]. Zool. Zh. 61: 1870–1881.

Afanasjev, K.S. 1956. Vegetation of the Turkestan Ridge [in Russian]. Izd. Akademii Nauk, Moscow.

Agakhanjanz, O.E. 1979. Besonderheiten in der Natur der ariden Gebirge der UdSSR. Petermanns Geogr. Mitt. 123: 73–77.

Agakhanjanz, O. 1985. Ein ökologischer Ansatz zur Höhenstufengliederung des Pamir-Alai. Petermanns Geogr. Mitt. 129: 17–23.

Agakhanjanz, O.E. and S.-W. Breckle. 1995. Origin and evolution of the mountain flora in Middle Asia and neighbouring mountain regions. pp. 63–80. *In*: F.S. Chapin III and C. Körner (eds.). Arctic and Alpine Biodiversity. Springer, Berlin.

Agakhanjanz, O.E. and S.-W. Breckle. 2003. Vegetation of the Pamirs—Classification, cartography, altitudinal belts. Bielef. Ökol. Beitr. 18: 17–19.

Agakhanjanz, O.E. and S.-W. Breckle. 2004. Pamir. pp. 151–157. *In*: C.A. Burga, F. Klötzli and G. Grabherr (eds.). Gebirge der Erde. Ulmer, Stuttgart.

Akhmadov, K.M., S.-W. Breckle and U. Breckle. 2006. Effects of grazing on biodiversity, productivity, and soil erosion of alpine pastures in Tajik Mountains. pp. 239–247. *In*: E.M. Spehn, M. Liberman and C. Körner (eds.). Land Use Change and Mountain Biodiversity. CRC, Boca Raton.

Akiyanova, F.Z., R.K. Temirbayeva, A.K. Karynbayev and K.B. Yegemberdiyeva. 2017. Assessment of the scale, geographical distribution and diversity of pastures along the Kazakhstan's part of the route of the great Silk Road. pp. 257–298. *In*: V.R. Squires, Z.H. Shang and A. Ariapour (eds.). Rangelands along the Silk Road: Transformative Adaptation under Climate and Global Change. Nova Science Publishers, New York.

Akramkhanov, A., U. Djanibekov, N. Nishanov, N. Djanibekov and S. Kassam. 2018. Barriers to sustainable land management in Greater Central Asia with special reference to the five former Soviet republics. pp. 113–130. *In*: V.R. Squires and Q. Lu (eds.). Sustainable Land Management in Greater Central Asia: An Integrated and Regional Perspective. Routledge, London.

Alisov, B.P. 1956. Climate of the USSR [in Russian]. Izd. Moskovskogo Univ., Moscow.

Ambarlı, D., M. Vrahnakis, S. Burrascano, A. Naqinezhad and M. Pulido Fernández. 2018. Grasslands of the Mediterranean Basin and the Middle East and their Management (pp. 89–111 this volume).

Anon. 1950. Animal world of the USSR. Vol. 3. Steppe zone [in Russian]. Izd. Akad. Nauk., Moscow.

Anon. 1979. Geological Map of the Kazakh SSR. South-Kazakhstan Series. Scale 1:500,000. K-42-B Chimkent, K-43-B Alma-Ata [in Russian] Ministerstvo Geologii SSSR, Moscow.

Anon. 1980. Geological Map of the Kyrgyz SSR. Scale 1:500,000 [in Russian]. Ministerstvo Geologii SSSR, Leningrad.

Anon. 2013. Fifth National Report on Conservation of Biodiversity of the Kyrgyz Republic. State Agency on Environment Protection and Forestry under the Government of the Kyrgyz Republic, Bishkek.

Asian Development Bank. 2010. Central Asia Atlas of Natural Resources. Asian Development Bank, Manila.

Baitulin, I.O., N.P. Ogar and T.M. Bragina (eds.). 1999. Transformation of Natural Ecosystems and their Components under Desertification [in Russian]. Gridan, Almaty.

Baitulin, I.O. and Y.A. Kotukhov. 2011. Flora of Vascular Plants of the Kazakh Altai [in Russian]. Ministry of Education and Sciences of the Republic of Kazakhstan, Institute of Botany and Phytointroduction., Almaty.

Balbakova, F., T. Berkeliev, T. Bragina, Y. Chikin, M. Mirutenko, I. Onufrenya, E. Rachkovskaya, R. Sadvokasov, N. Safarov and O. Tsaruk. 2006. Econet – Central Asia. E-Poligraph, Moscow.

Baranov, V.I. 1925. The Southern Boundary of the Chernozem Steppes in the Kustanai Province (According to the Data of the 1923 Studies) [in Russian]. Kazglavli, Orenburg.

Beg, G.A. 2009. Cross-border cooperation for biodiversity conservation and sustainable development. Case studies on Karakoram, Hindukush and Pamir. pp. 184–211. *In*: H. Kreutzmann, G.A. Beg and J. Richter (eds.). Experiences with and Prospects for Regional Exchange and Cooperation in Mountain Areas. International Conference, Kathmandu/Nepal, Nov 30 - Dec 2, 2007. InWEnt—Internationale Weiterbildung und Entwicklung gGmbH, Kathmandu.

Bekchanov, M., N. Djanibekov and J.P.A. Lamers. 2018. Water in Central Asia: A cross-cutting management issue. pp. 211–236. *In*: V.R. Squires and L. Qi (eds.). Sustainable Land Management in Greater Central Asia: An Integrated and Regional Perspective. Routledge, London.

Bekenov, A., J.A. Grachev and E.J. Milner-Gulland. 1998. The ecology and management of the Saiga antelope in Kazakhstan. Mamm. Rev. 28: 1–52.

Borchardt, P., U. Schickhoff, S. Scheitweiler and M. Kulikov. 2011. Mountain pastures and grasslands in the SW Tian Shan, Kyrgyzstan—Floristic patterns, environmental gradients, phytogeography, and grazing impact. J. Mt. Sci. 8: 363–373.

Borisov, A.A. 1965: Climates of the U.S.S.R. Oliver & Boyd, Edinburgh.

Bragina, T.M. 1999. Regularities of transformation of animal communities in soils under desertification. pp. 24–25. *In*: B.R. Striganova (ed.). Problems of Soil Zoology. Mat-II (XII) All-Russian Meeting on Soil Zoology. 15–19 November 1999 [in Russian]. KMK Publishing House, Moscow.

Bragina, T.M. 2004. The Regularities of Changes of Animal Communities in Soils under Desertification. Doctoral Thesis [03.00.16] [in Russian]. A.N. Severtsov Institute of Ecology and Evolution, Moscow. URL: http://earthpapers.net/zakonomernosti-izmeneniy-zhivotnogo-naseleniya-pochv-pri-opustynivanii.

Bragina, T.M. 2007a. Influence of pasture degression on communities of soil invertebrates in dry steppes [in Russian]. Isv. Orenbg. State Agrar. Univ. 3: 32–33.

Bragina, T.M. 2007b. Protected Areas of Kazakhstan and Prospects of Establishing an Ecological Network (with a Legislative Framework for Protected Areas) [in Russian]. Kostanai Printing House, Kostanai.

Bragina, T.M. 2009a. State and perspectives of conserving the steppes of Kazakhstan. pp. 39–43. *In*: A.A. Chibilev (ed.). Steppes of Northern Eurasia: Materials of the V International Symposium [in Russian]. IPK Gazprompechat, LLC Orenburggazpromtrans, Orenburg.

Bragina, T.M. 2009b. Naurzum Ecological Network (History of Study, Current State and Long-term Biodiversity Conservation in a Region Representing a UNESCO World Heritage Site) [in Russian]. Kostanaypoligrafiya, Kostanai.

Bragina, T.M., A.D. Asylbekov, A.K. Agazhaeva and Z. Kuragulova. 2013. On the concept of developingf special protected natural areas in steppes of Kazakhstan [in Russian]. Stepnoy Byull. 39: 30–35.

Bragina, T.M. 2014. The concept of developing special protected natural areas (PAs) for steppes in the Republic of Kazakhstan until 2030 [in Russian]. Bull. Altai Sci. 4: 181–185.

Bragina, T.M. 2015a. Saiga migratory routes in Kazakhstan are taken under protection. Saiga News 19: 3–4.

Bragina, T.M. 2015b. The present status of the steppes of Kazakhstan and the prospects for the formation of a steppe ecological network. pp. 43–45. *In*: A.A. Chibilev (eds.). Steppes of Northern Eurasia: Materials of the VII International Symposium, IP UB RAS [in Russian]. Dimur, Orenburg.

Bragina, T.M. 2016. Soil macrofauna (invertebrates) of Kazakhstanian *Stipa lessingiana* dry steppe. Hacquetia 15: 105–112.

Breckle, S.-W. and W. Wucherer. 2006. Vegetation of the Pamir (Tajikistan): Land use and desertification problems. pp. 225–237. *In*: E. Spehn, M. Liberman and C. Körner (eds.). Land-use Change and Mountain Biodiversity. CRC, Boca Raton.

Breu, T. and H. Hurni. 2003. The Tajik Pamirs. Challenges of Sustainable Development in an Isolated Mountain Region. Centre for Development and Environment (CDE), Bern.

Brinkert, A., N. Hölzel, T.V. Sidorova and J. Kamp. 2016. Spontaneous steppe restoration on abandoned cropland in Kazakhstan: Grazing affects successional pathways. Biodivers. Conserv. 25: 2543–2561.

Chibilyev, A.A. 1998. Steppes of Northern Eurasia (Ecological and Geographical Outline and Bibliography) [in Russian]. Urals Branch of RAS, Ekaterinburg.

Crewett, W. 2012. Improving sustainability of pasture use in Kyrgyzstan: The impact of pasture governance reforms on livestock migration. Mt. Res. Dev. 32: 267–274.

Dakhshleyger, G.F. (ed.). 1980. Kazakh husbandry on the turn of the XVIIII-XX century [in Russian]. Akademiya Nauk Kaz. SSR, Alma-Ata.

Deák, B., B. Tóthmérész, O. Valkó, B. Sudnik-Wójcikowska, I.I. Moysiyenko, T.M. Bragina, I. Apostolova, I. Dembicz, N.I. Bykov and P. Török. 2016. Cultural monuments and nature conservation: A review of the role of kurgans in the conservation and restoration of steppe vegetation. Biodivers. Conserv. 25: 2473–2490.

Demina, O. and T. Bragina. 2014. Fundamental basis for the conservation of biodiversity of the Black Sea-Kazakh steppes. Hacquetia 13: 215–228.

Demurina, E.M. 1972. The vegetation of dry mixed-forb steppes of Middle Asia. FAN, Tashkent.

Diment, A., P. Hotham and D. Mallon. 2012. First biodiversity survey of Zorkul reserve, Pamir Mountains, Tajikistan. Oryx 46: 13–17.

Djamali, M., S. Brewer, S.-W. Breckle and S.T. Jackson. 2012. Climatic determinism in phytogeographic regionalization: A test from the Irano-Turanian region, SW and Central Asia. Flora 207: 237–249.

Dörre, A. and P. Borchardt. 2012. Changing systems, changing effects—Pasture utilization in the post-Soviet transition: Case studies from Southwestern Kyrgyzstan. Mt. Res. Dev. 32: 313–323.

Esenova, S. and W.D. Dobson. 2000. Changing Patterns of Livestock, Meat, and Dairy Marketing in Post-communist Kazakhstan. Babcock Institute for International Dairy Research and Development, University of Wisconsin, Madison.

Fadeev, V.A. and A.A. Sludskii. 1982. The Saiga in Kazakhstan [in Russian]. Akad. Nauk, Alma-Ata.

Faizov, K.S., S.B. Kenenbaev, Zh.U. Mamutov and M.B. Esimbekov. 2006. Soil Geography and Ecology of Kazakhstan [in Russian]. Institute of Soil Science named after Uspanov, Scientific-production. Center of Agriculture and Plant Growing, Almaty, 1–348 pp.

Field, H. and K. Price. 1950. Early history of agriculture in Middle Asia. Southwest J. Anthropol. 6: 21–31.

Grubov, I.V. 2010. Schlussbetrachtung zum Florenwerk 'Rastenija Central'noj Azii' (Die Pflanzen Zentralasiens) und die Begründung der Eigenständigkeit der mongolischer Flora. Feddes Repert. 121: 7–13.

Halimova, N. 2012. Land tenure reform in Tajikistan: Implications for land stewardship and soil sustainability : A case study. pp. 305–329. *In*: V.R. Squires (ed.). Rangeland Stewardship in Central Asia: Balancing Improved Livelihoods, Biodiversity Conservation and Land Protection. Springer, Dordrecht.

Haslinger, A., T. Breu, H. Hurni and D. Maselli. 2007. Opportunities and risks in reconciling conservation and development in a post-Soviet setting: The example of the Tajik National Park. Int. J. Biodivers. Sci. Manag. 3: 157–169.

Hölzel, N., C. Haub, M.P. Ingelfinger, A. Otte and V.N. Pilipenko. 2002. The return of the steppe—large-scale restoration of degraded land in southern Russia during the post-Soviet era. J. Nat. Conserv. 10: 75–85.

Kamakhina, G.L. 2005. Flora and vegetation of the central Kopet Dagh [in Russian]. WWF, Ashkhabad.

Kamelin, R.V. 2017. The history of the flora of Middle Eurasia [in Russian]. Turczaninowia 20: 5–29.

Kamp, J., M.A. Koshkin, T.M. Bragina, T.E. Katzner, E.J. Milner-Gulland, D. Schreiber, R. Sheldon, A. Shmalenko, I. Smelanskiy, (...) and R. Urazaliev. 2016. Persistent and novel threats to the biodiversity of Kazakhstan's steppes and semi-deserts. Biodivers. Conserv. 25: 2521–2541.

Karmysheva, B.K.C. 1981. Versuch einer Typologisierung der traditionellen Formen der Viehwirtschaft Mittelasiens und Kasachstans am Ende des 19. Anfang des 20. Jh. pp. 91–96. *In*: R. Krusche (ed.). Die Nomaden in Geschichte und Gegenwart. Beiträge zu einem internationalen Nomadismus Symposium am 11. und 12. Dez. 1975 im Museum für Völkerkunde, Leipzig. Akademie Verlag, Berlin.

Karmysheva, N.K. 1961. The effects of haying on dry meadows of the Aksu Jabagly Nature Reserve [in Russian]. Tr. Inst. Bot. AN Kaz. SSR 11: 49–57.

Karmysheva, N.K. 1982. Flora and vegetation of the western ranges of the Talass Alatau [in Russian]. Nauka, Alma-Ata.

Kassam, K.A. 2009. Viewing change through the prism of indigenous human ecology: Findings from the Afghan and Tajik Pamirs. Hum. Ecol. 37: 677–690.

Kerven, C. 2003. Prospects for Pastoralism in Kazakhstan and Turkmenistan: From State Farms to Private Flocks. Routledge, London.

Khazanov, A.M. 2002. Nomads and the Outside World [in Russian]. 3rd ed. Dayk-Press, Almaty.

Koroleva, A.S. 1940. Outline of the vegetation of the central part of the southern slope of the Hissar range and its natural feeding resources. pp. 1–140. *In*: Vegetation of Tajikistan and its Exploitation [in Russian]. Akad. Nauk USSR, Trud. Tadzhik. Bazy, Vol. 8. Moscow.

Korovin, E.P. 1961. Vegetation of Middle Asia and Southern Kazakhstan. Book 1 [in Russian]. 1. 2nd ed. Izd. Akad. Nauk Uzbek. SSR, Taschkent.

Korovin, E.P. 1962. Vegetation of Middle Asia and Southern Kazakhstan. Book 2 [in Russian]. 2nd ed. Izd. Akad. Nauk Uzbek. SSR, Taschkent.

Krasheninnikov, I.M. 1924. Vegetation Cover of the Kyrgyz Republic [in Russian]. Trudy Ob-va Izuch. Kirgiz. Kraya 5: 1–104.

Kraudzun, T. 2012. Livelihoods of the 'new livestock breeders' in the eastern Pamirs of Tajikistan. pp. 89–107. *In:* H. Kreutzmann (ed.). Pastoral Practices in High Asia. Agency of 'Development' Affected by Modernisation, Resettlement and Transformation. Springer, Dordrecht.

Lal, R. 2007. Soil and environmental degradation in Central Asia. pp. 127–136. *In*: R. Lal, M. Suleimenov, B.A. Stewart, D.O. Hansen and P. Doraiswamy (eds.). Climate Change and Terrestrial Carbon Sequestration in Central Asia. Taylor & Francis, New York.

Latipova, W.A. 1968. Amount of precipitation. pp. 68–69. *In*: I.K. Narzikulov and K.W. Stanyukovich (eds.). Atlas of the Tajik SSR [in Russian]. Akad. Nauk Tadzh. SSR, Dushanbe.

Lavrenko, E.M. 1940. Steppes of the USSR. pp. 1–265. *In*: Keller, B.A., Komarov, N.F., Lavrenko, E.M. and A.V. Prosorovskiy (eds.). Vegetation of the USSR. Vol. II [in Russian]. Akad. Nauk SSSR, Moscow-Leningrad.

Lavrenko, E.M. 1956. Steppes and agricultural lands on former steppe land. pp. 595–730. *In*: Lavrenko, E.M. and V.B. Soczava: Explanatory Text to 'Geobotanical Map of the USSR'. Scale 1:4000000. Vol. 2 [in Russian]. Akad. Nauk SSSR, Moscow.

Lavrenko, E.M. 1970. The provincial division of the Black Sea-Kazakhstan subregion of steppe region of Eurasia [in Russian]. Bot. Zh. 55: 609–625.

Lavrenko, E.M., Z.V. Karamysheva and R.I. Nikulina. 1991. Steppes of Eurasia [in Russian]. Nauka, Leningrad.

Lebedeva, L.P. and R.N. Ionov. 1994. Problems of plant cover protection in the Tien Shan of the Kyrgyz Republic [in Russian]. Izv. NAN KR 1: 63–72.

Leontieva, R.S. 1968. Soils. pp. 93–95. *In*: I.K. Narzikulov and K.W. Stanyukovich (eds.). Atlas of the Tadzhik SSR [in Russian] Akad. Nauk Tadzh. SSR, Dushanbe.

Lioubimtseva, E. 2002. Arid environments. pp. 267–283. *In*: M. Shahgedanova (ed.). The Physical Geography of Northern Eurasia. Oxford University Press, Oxford.

Lioubimtseva, E. and G.M. Henebry. 2009. Climate and environmental change in arid Central Asia: Impacts, vulnerabilities and adaptations. J. Arid Environ. 73: 963–977.

Lipsky, V.I. 1903. Flora of Middle Asia, i.e. Russian Turkestan and Khanates of Bukhara and Khiva. Vol. II. History of the Botanical Exploration of Middle Asia [in Russian]. Gerol'd, St. Petersburg.

Mountain Societies Development Support Programme. 2004. The 2003 Baseline Survey of Gorno-Badakhshan Autonomous Oblast, Tajikistan. Aga Khan Foundation. Khorog.

Narzikulov, I.K. and K.W. Stanyukovich. 1968. Atlas of the Tajik SSR [in Russian]. Akad. Nauk Tadzh. SSR, Dushanbe.

Nedzvedskiy, A.P. 1968. Geological composition. pp. 14–15. *In*: I.K. Narzikulov and K.W. Stanyukovich (eds.). Atlas of the Tajik SSR [in Russian]. Akad. Nauk Tadzh. SSR, Dushanbe.

Nikolayev, V.A. 1999. Landscapes of the Asian Steppes [in Russian]. MGU, Moscow.

Nosova, L.M. 1973. Floristic-geographical analysis of the northern steppes of European USSR [in Russian]. Nauka, Moscow.

Nowak, A., S. Nowak and M. Nobis. 2011. Distribution patterns, ecological characteristic and conservation status of endemic plants of Tadzhikistan—A global hotspot of diversity. J. Nat. Conserv. 19: 296–305.

Nowak, A., S. Nowak, A. Nobis and M. Nobis. 2016a. Vegetation of feather grass steppes in the western Pamir Alai Mountains (Tajikistan, Middle Asia). Phytocoenologia 46: 295–315.

Nowak, A., M. Nobis, S. Nowak and V. Plášek. 2016b. Fen and spring vegetation in western Pamir-Alai Mts. in Tajikistan (Middle Asia). Phytocoenologia 46: 201–220.

Olson, D.M., E. Dinerstein, E.D. Wikramanayake, N.D. Burgess, G.V.N. Powell, E.C. Underwood, J.A. D'Amico, I. Itoua, H.E. Strand, (...) and K.R. Kassem. 2001. Terrestrial ecoregions of the world: A new map of life on Earth. BioScience 51: 933–938.

Omarkhanova, Z.M. 2015. Problems of livestock sustainability in Republic of Kazakhstan [in Russian]. Russ. Electron. Sci. J. 1: 34–40.

Orlovosky, L. and N. Orlovosky. 2018. Biogeography and natural resources of Greater Central Asia: An overview. pp. 23–47. *In*: V.R. Squires and L. Qi (eds.). Sustainable Land Management in Greater Central Asia. An Integrated and Regional Perspective. Routledge, London.

Ovchinnikov, P.N. 1971. Species composition in the vegetation cover of the Varzob ravine. 1. Higher plants. pp. 151–213. *In*: P.N. Ovchinnikov (ed.). Flora and Vegetation of the Varzob Ravine [in Russian]. Nauka, Leningrad.

Pavlov, N.V. 1948. Botanical Geography of the USSR [in Russian]. Izd. Akad. Nauk Kaz. SSR, Alma-Ata.

Peterson, D.J. 1993. Troubled Lands. The Legacy of Soviet Environmental Destruction, Westview Press, Boulder, US.

Pfeiffer, M., C. Dulamsuren, Y. Jaschke and K. Wesche. 2018. Grasslands of China and Mongolia: Spatial extent, land use and conservation (pp. 170–198 this volume).

Rachkovskaya, E.I., N.P. Ogar and O.V. Marynich. 1999. Rare plant communities of the steppes of Kazakhstan and their protection [in Russian]. Stepnoy Byull. 3-4: 41–46.

Rachkovskaya, E.I., E.A. Volkova and V.N. Khramtsov (eds.). 2003. Botanical Geography of Kazakhstan and Middle Asia (Desert Region) [in Russian]. Komarov Botanical Institute, St. Petersburg.

Rachkovskaya, E.I. and T.M. Bragina. 2012. Steppes of Kazakhstan: Diversity and present state. pp. 103–148. *In*: M.J.A. Werger and M.A. van Staalduinen (eds.). Eurasian Steppes: Ecological Problems and Livelihoods in a Changing World. Springer, Dordrecht.

Robinson, S., C. Weidemann, S. Michel, Y. Zhumabayeva and N. Singh. 2012. Pastoral tenure in Central Asia: Theme and variation in the five former Soviet Republics. pp. 239–274. *In*: V.R. Squires (ed.). Rangeland Stewardship in Central Asia: Balancing Improved Livelihoods, Biodiversity Conservation and Land Protection. Springer, Dordrecht.

Rubtsov, N.I. 1948. Plant cover of the Dzhungar Alatau. Izd. Akad. Nauk Kaz. SSR, Alma-Ata.

Rubtsov, N.I. 1952. Vegetation cover of Kazakhstan. pp. 385–451. *In*: I.P. Gerasimov (ed.). Essays on the Physical Geography of Kazakhstan [in Russian]. Izd. Akad. Nauk SSSR, Alma-Ata.

Sadikov, K., A. Abdurahmanov and I. Bekmirzaeva. 2015. Fifth National Report of the Republic of Uzbekistan on Conservation of Biodiversity. State Committee for Nature Protection of the Republic of Uzbekistan. Tashkent.

Safarov, N. 2003. National Strategy and Action Plan on Conservation and Sustainable Use of Biodiversity. Governmental Working Group of the Republic of Tajikistan. Dushanbe.

Safarov, N., T. Novikova and K. Shermatov. 2014. Fifth National Report on Preservation of Biodiversity of the Republic of Tajikistan. National Center on Biodiversity and Biosafety of the Republic of Tajikistan, Dushanbe.

Saparov, A. 2014. Soil resources of the Republic of Kazakhstan: Current status, problems and solutions. pp. 61–73. *In*: L. Mueller, A. Saparov and G. Lischeid (eds.). Novel Measurement and Assessment Tools for Monitoring and Management of Land and Water Resources in Agricultural Landscapes of Central Asia. Springer International Publishing Switzerland, Basel.

Saparov, A.S., F.E. Kozybayeva and G.A. Saparov. 2006. Status of soil ecology in Kazakhstan/current status of potato and vegetable growing and their scientific support. pp. 694–698. *In*: Proceeding of the International Scientific and Practical Conference 'Scientific and Innovative Basis for the Development of Potato, Vegetable and Melon-growing in the Republic of Kazakhstan' Dedicated to the 70th Anniversary of the Founding of the Institute, Kazakh Research Institute of Potato and Vegetable Growing [in Russian], Almaty.

Shigaeva, J., B. Wolfgramm and C. Dear. 2013. Sustainable Land Management in Kyrgyzstan and Tajikistan: A Research Review. University of Central Asia, Mountain Societies Research Institute, Bishkek.

Sidorenko, A.V. 1971. Hydrology of the USSR. Vol. XL. Kirghiz SSR [in Russian]. Nedra, Moscow.

Sobolev, L.N. 1948. Natural Fodder Lands in Kazakhstan. Proceedings of Expedition on the Study of Land Funds of the Kazakh SSR. Vol. 9 [in Russian]. Akad. Nauk Kaz. SSR, Alma-Ata.

Sokolov, V.E. and E.E. Syroechkovskiy. 1990. Zapovedniks of the USSR. Zapovedniks of Middle Asia and Kazakhstan [in Russian]. Mysl, Moscow.

Spengler, R.N. III. 2015. Agriculture in the Central Asian Bronze Age. J. World Prehist. 28: 215–253.

Squires, V. (ed.). 2012. Rangeland Stewardship in Central Asia: Balancing Improved Livelihoods, Biodiversity Conservation and Land Protection. Springer, Dordrecht.

Squires, V.R. and H.Y. Feng. 2017. Rangeland and grassland in the region of the former Soviet Republics: Future implications for Silk Road countries. pp. 17–31. *In*: V.R. Squires, Z.H. Shang and A. Ariapour (eds.). Rangelands along the Silk Road: Transformative Adaptation under Climate and Global Change. Nova Science Publishers, New York.

Squires, V.R., Z.H. Shang and A. Ariapour. 2017a. Rangelands along the Silk Road: Transformative Adaptation under Climate and Global Change. Nova Science Publishers, New York.

Squires, V.R., Z.H. Shang and A. Ariapour. 2017b. Unifying perspectives: future problems and prospects for people and natural resource base on which they depend. pp. 343–357. *In*: V.R. Squires, Z.H. Shang and A. Ariapour (eds.). Rangelands along the Silk Road: Transformative Adaptation under Climate and Global Change. Nova Science Publishers, New York.

Squires, V.R. and L. Qi. (eds.). 2018. Sustainable Land Management in Greater Central Asia. An Integrated and Regional Perspective. Routledge, London.

Stanyukovich, K.W. 1982. Vegetation. pp. 358–435. *In*: C.M. Saidmuradov and K.W. Stanyukovich (eds.). Tajikistan. Flora and Natural Resources [in Russian]. Izd. Donish, Dushanbe.

Stepanova, E.F. 1962. Vegetation and Flora of the Tarbagatai [in Russian]. Izd. Akad. Nauk Kaz. SSR, Alma-Ata.

Strong, P.J.H. and V. Squires. 2012. Rangeland-based livestock: A vital subsector under threat in Tajikistan. pp. 213–235. *In*: V. Squires (ed.). Rangeland Stewardship in Central Asia: Balancing Improved Livelihoods, Biodiversity Conservation and Land Protection. Springer, Dordrecht.

Takhtajan, A. 1986. Floristic Regions of the World. University of California Press, Berkeley.

Tolstov, S.P., T.A. Zhdanko, S.M. Abranzona and N.A. Kislyakova (eds.). 1963. Peoples of Middle Asia and Kazakhstan. Vol. II [in Russian]. Izd. Akad. Nauk, Moscow.

UNEP-WCMC and IUCN. 2017. Protected Planet: Kazakhstan, Kyrgyztan, Tajikistan, Turkmenistan, Uzbekistan; The World Database on Protected Areas (WDPA)/The Global Database on Protected Areas Management Effectiveness (GD-PAME). URL: www.protectedplanet.net.

USDA. 2013. Republic of Kazakhstan. Agricultural Development Programme 2013–2020. Gain Report. URL: goo.gl/f2wKEI.

Vanselow, K.A. 2012. The High-mountain Pastures of the Eastern Pamirs (Tajikistan)—An Evaluation of the Ecological Basis and the Pasture Potential. Südwestdeutscher Verlag für Hochschulschriften, Saarbrücken.

Vanselow, K.A., T. Kraudzun and C. Samimi. 2012a. Grazing practices and pasture tenure in the Eastern Pamirs. Mt. Res. Dev. 32: 324–336.

Vanselow, K.A., T. Kraudzun and C. Samimi. 2012b. Land stewardship in practice—An example from the Eastern Pamirs of Tajikistan. pp. 71–90. *In*: V. Squires (ed.). Rangeland Stewardship in Central Asia: Balancing Improved Livelihoods, Biodiversity Conservation and Land Protection. Springer, Dordrecht.

Vanselow, K.A., C. Samimi and S.-W. Breckle. 2016. Preserving a comprehensive vegetation knowledge base—An evaluation of four historical Soviet vegetation maps of the Western Pamirs (Tajikistan). PLoS One 11: Article e0148930.

Wagner, V. 2009. Eurosiberian meadows at their southern edge: Community patterns and phytogeography in the NW Tian Shan. J. Veg. Sci. 20: 199–208.

Walter, H. and S.-W. Breckle. 1986. Ökologie der Erde. Band 3: Spezielle Ökologie der gemäßigten und arktischen Zonen Euro-Nordasiens. Fischer, Stuttgart.

Wesche, K., D. Ambarlı, J. Kamp, P. Török, J. Treiber and J. Dengler. 2016. The Palaearctic steppe biome: A new synthesis. Biodivers. Conserv. 25: 2197–2231.

8

Grasslands of China and Mongolia: Spatial Extent, Land Use and Conservation

Martin Pfeiffer,[1,2,3,*] *Choimaa Dulamsuren*,[4] *Yun Jäschke*[5,6] and *Karsten Wesche*[6,7,8]

Introduction—Physiographic and Socio-economic Settings of the Region

Mongolia and China host the world's largest continuous grasslands. These grasslands are natural, while secondary, human-made grasslands are practically absent in Mongolia and of limited extent in China. The present chapter focuses on the natural grasslands and we can thus broadly follow biogeographic concepts to delimit our study area (Fig. 8.1 A). The grasslands of China and Mongolia represent the eastern part of the Palaearctic/Eurasian steppe biome ('Central Asia'; Wesche et al., 2016), they border to Middle Asia westwards in Kazakhstan and surroundings (see Bragina et al., 2018). The major biogeographic and climatic divide along the Tian Shan–Altay mountain system separates Mongolia and China from the western steppes and also demarcates the western political borders of our focus countries. In total, Mongolia and the relevant part of China (including the Chinese Provinces of Qinghai, Gansu and the Autonomous Regions Inner Mongolia, Xinjiang and Tibet) cover circa 7.9 million km^2 (plus 0.9 million km^2 in adjacent Russia, all estimated after Olson et al., 2001). Estimates differ, but grasslands in Mongolia may account for 1.0 million km^2 and in China for circa 2.8 million km^2 (Russia 0.5 million km^2).

[1] Department of Biogeography, University of Bayreuth, Universitätsstr. 30, 95447 Bayreuth, Germany.

[2] Helmholtz-Zentrum für Umweltforschung (UFZ), Brückstr. 3a, 39114 Magdeburg, Germany.

[3] Mongolian National University, Ikh Surguuliin gudamj-1, Ulaanbaatar, Mongolia.

[4] Department of Plant Ecology and Ecosystems Research, Albrecht von Haller Institute for Plant Sciences, University of Göttingen, Untere Karpüle 2, 37073 Göttingen, Germany.

[5] Née: Yun Wang.

[6] Senckenberg Museum of Natural History, Görlitz, PO Box 300 154, 02806 Görlitz, Germany.

[7] German Centre for Integrative Biodiversity Research (iDiv) Halle-Jena-Leipzig, Deutscher Platz 5e, 04103 Leipzig, Germany.

[8] International Institute Zittau, Technische Universität Dresden, Markt 23, 02763 Zittau, Germany.
Emails: dulamsuren.choimaa@biologie.uni-goettingen.de; yun.wang@senckenberg.de; karsten.wesche@senckenberg.de

* Corresponding author: martin.pfeiffer@uni-bayreuth.de

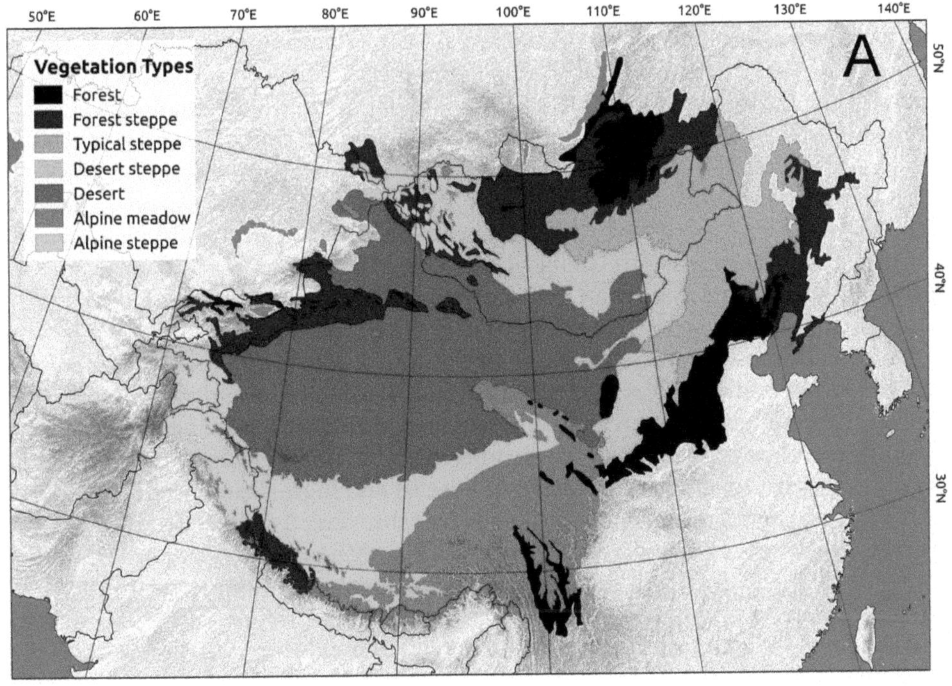

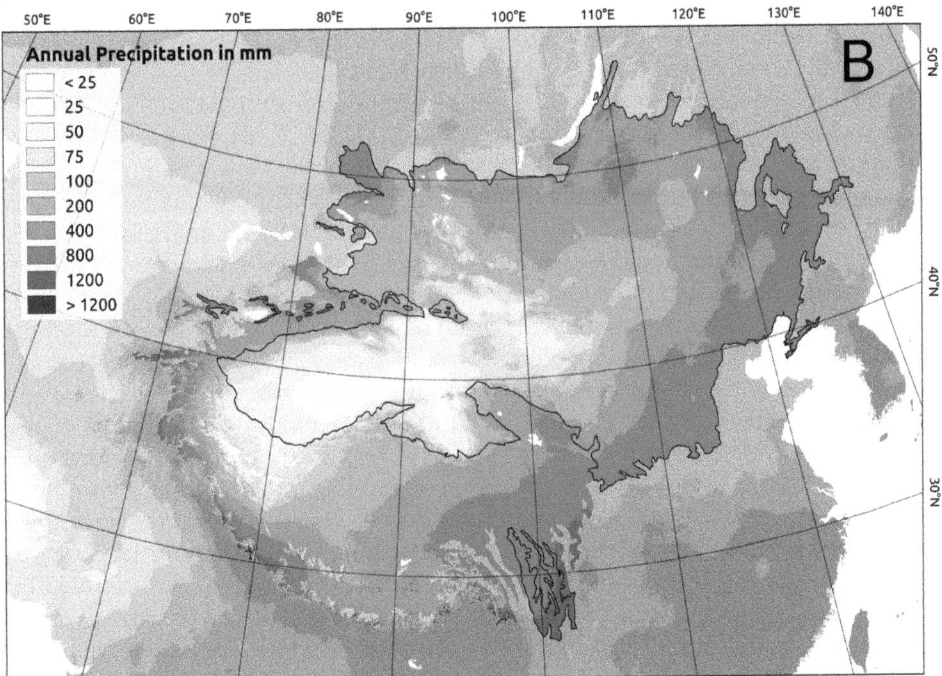

Fig. 8.1 Region covered in this chapter. A: Geography of the Mongolian and Tibetan grassland region with the major vegetation formations (lines indicate country boundaries). B: Mean annual precipitation in the Mongolian and Tibetan grasslands (modeled data from Hijmans et al., 2005) (the line indicates the border between the Mongolian and Tibetan subregions, draft J. Treiber).

The grasslands are environmentally diverse (Table 8.1). The topography is dominated by the world's highest mountains (Himalaya-Altay-Sayan system), but here are also some of the deepest depressions (northern Tarim basin). Elevation allows to distinguish two main grassland subregions: (1) the Mongolian subregion in Mongolia (and adjacent Russia) and northern China at elevations typically below 2,000 m a.s.l. and (2) the Tibetan subregion, i.e., mainly the Qinghai-Tibetan Plateau (QTP), which typically lies above 4,000 m a.s.l. The grasslands owe their existence to the large mountains screening monsoonal precipitation from the south and Mediterranean disturbances from the west, while the Great Khinggan in north-eastern China blocks rain even in the east. Xinjiang and Gansu lie in the rain shadow of Tian Shan and Kunlun mountains and host some of the driest deserts (Taklamakan, Tengger, Badain Jaran and the Chinese part of Gobi). Mean annual precipitation is below 400 mm across much of the region (Fig. 8.1 B), restricting tree growth to the northern boundaries (towards Siberian Taiga), mountains and water surplus sites. The Mongolian subregion receives rain from the north-east, resulting in progressively drier conditions towards the southwestern basins. The Tibetan subregion shows similar gradients, with relatively moist conditions in the east and high-altitude cold semi-desert in the north-west. Patterns in mean temperatures are more directly controlled by topography, with mean temperatures around zero across the Mongolian subregion and locally lower values on the Qinghai-Tibetan high plateau in spite of the more southern latitude.

Mean climate conditions are, however, of limited relevance in these extremely seasonal regions. Winters are generally cold with temperatures dropping well below –20°C, while summers are warm, and even high-altitude grasslands in Tibet often have day-time temperatures close to 10°C or higher (Babel et al., 2014). Changes are rapid and springs and autumns are relatively short. The growing season is controlled by the strong seasonality in precipitation. Unlike, Middle Asia (Bragina et al., 2018), winters and springs are typically dry with limited snow and moisture availability remains low until monsoonal rains fall in late May and June. More than 50 per cent of the annual totals are confined to the three summer months of June, July and August (Table 8.1); as a consequence, each of these months receive 30 mm or more. This is not so different from much more beneficial climates in, e.g., Central Europe and explains why Mongolia and China host appreciable grasslands at mean annual precipitation values well below 150 mm. In that sense, extreme continentality is an advantage compared to the much more 'Mediterranean' climates of, e.g., many North America prairies (Starrs et al., 2018).

The second important aspect in terms of management is the tremendous interannual variability. The same site in the Mongolian Gobi may experience a relatively decent annual total of 180 mm in one year and desert-like conditions with < 50 mm in another (Fig. 8.2). The interannual variability, as commonly summarized in the coefficient of variation (CV), makes land use especially difficult. At CV > 30 per cent commonly found in, e.g., the Mongolian Gobi (Fig. 8.2), environmental variability may seriously hamper land use resulting in relatively low grazing impact due to frequent droughts and livestock mortality ("non-equilibrium conditions", *sensu* Fernández-Giménez and Allen-Diaz, 1999). Interannual changes are less pronounced in the moister north-east as demonstrated by the station Zuunmod in the northern Mongolian forest steppes (Fig. 8.2), where relatively stable rains allow herds to gradually build up. Interannual variability in winter precipitation is also characteristic—closed snow cover may occur in extreme years resulting in mass mortality of herds that are typically freely herded without or with very limited additional winter forage supply. The phenomenon is locally known as *dzuud* (Mongolian) or *kengschi* (Tibetan) and can result in the loss of up to 80 per cent of livestock in an affected region (Middleton et al., 2015).

Table 8.1 Major natural grassland regions in China and Mongolia and key environmental indicators as well as land use characteristics (for key climate variables, median and interquartile range based on Hijmans et al. [2005] are given). Steppe types are distinguished according to physiognomy/formation, i.e., forest-steppe mosaics, typical grass steppes and dry steppes at lower altitudes and alpine steppes and alpine meadows in the extensive mountain systems. Not listed are relatively small areas of alpine steppes in Mongolia (< 0.1 million km²). Exact calculation of spatial extent is difficult due to lack of detailed maps, we used original area estimates for TEOW (Terrestrial Ecosystems of the World; Olson et al., 2001) weighted by an estimated share of grasslands in the TEOW. The share of formally protected reserves was estimated by overlaying the main protected areas as registered in the IUCN, UNEP-WCMC world database of protected areas (http://www.protectedplanet.net, categories I–IV).

Steppe Subregion	**Mongolian Subregion**						**Tibetan Subregion**	
Country	**Mongolia (Russia)**	**Mongolia**	**Mongolia**	**China**	**China**	**China**	**China (Russia[1])**	**China**
Steppe Formations	**Forest Steppes**	**Grass Steppes**	**Desert Steppes**	**Forest Steppes**	**Grass Steppes**	**Desert Steppes**	**Alpine Steppes**	**Alpine Meadows[2]**
Physical setting								
Estimated grassland extent (million km²)	0.3(0.4)	0.3	0.4	0.4	0.6	0.1	0.9(0.1)	0.8
Elevation (m a.s.l.)								
Median	1,850(900)	1,100	1,400	3,350	900	1,300	4,750(2,350)	4450
25–75 per cent quartile	1,400–2,300	850–1,300	1,150–1,700	2,350–4,050	550–1,200	1,150–1,450	4,300–5,100	3,900–4,800
Mean annual temperature (°C)								
Median	–2.4(–2.1)	0.4	1.4	2.0	1.8	6.8	–3.3(–3.5)	–1.3
25–75 per cent quartile	–5.1–0.7	–0.3–1.0	–1.8–3.3	–0.6–4.9	0.1–4.1	4.6–7.8	–5.3–0.9	–3.3–1.3
Mean annual precipitation (mm)								
Median	260(340)	220	130	310	380	280	230(270)	440
25–75 per cent quartile	200–330	190–250	110–170	220–440	310–430	200–410	160–300	330–570
Mean precipitation warmest quarter (mm)								

Table 8.1 contd. ...

...*Table 8.1 contd.*

Steppe Subregion	**Mongolian Subregion**						**Tibetan Subregion**	
Country	**Mongolia (Russia)**	**Mongolia**	**Mongolia**	**China**	**China**	**China**	**China (Russia[1])**	**China**
Steppe Formations	**Forest Steppes**	**Grass Steppes**	**Desert Steppes**	**Forest Steppes**	**Grass Steppes**	**Desert Steppes**	**Alpine Steppes**	**Alpine Meadows[2]**
Median	170(240)	150	90	260	270	170	120(140)	270
25–75 per cent quartile	120–220	140–180	80–110	220–310	210–300	130–240	70–170	220–330
Characteristic forage graminoids	*Koeleria macrantha, Agropyron cristatum*	*Stipa krylovii, Cleistogenes squarrosa*	*Stipa caucasica*	*Festuca valesiaca, Poa* sect. *Stepposa*	*Leymus chinensis, Stipa grandis*	*Stipa caucasica*	*Stipa purpurea*	*Kobresia pygmaea*
Socioeconomy								
Typical land tenure	State	State	State	State	Private	Private	State	Private/state
Grazing system	Semi-mobile	Mobile	Mobile	Sedentary	Sedentary	Semi-mobile	Mobile	Semi-mobile
Characteristic livestock species	Cattle, goats, sheep, horses	Goats, sheep, cattle, horses	Goats, sheep, camels, horses	*Sheep, goat, cattle*	*Goats, sheep, cattle*	*Goats, sheep, camels*	Goats, yaks sheep	Yaks sheep
Percentage of reserves	11(3)	8	10	5	5	1	36(27)	40

[1] 'Russia' refers to the alpine steppes in the Russian Altai mountains that are actually a part of the Mongolian subregion.

[2] Montane forest-steppe mosaics are known as 'forest meadow' in China, alpine pastures of Tibet are widely known under the term 'alpine meadows'.

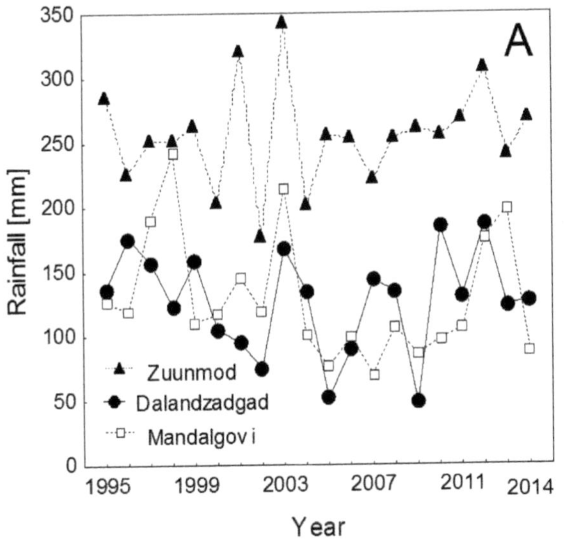

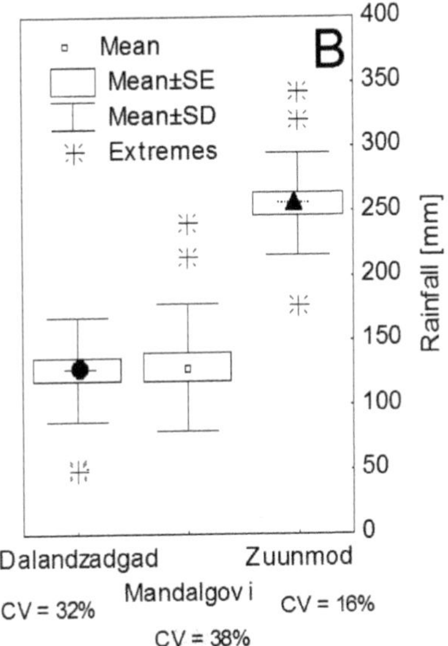

Fig. 8.2 Twenty years of rain records in the southern Mongolian Gobi (Dalandzadgad, 43°34′ N 104°24′ E), in central Mongolia (Mandalgovi, 45°35′ N 106°25′ E) and in the vicinity of Ulaanbaatar (Zuunmod, 4°32′ N 106°57′ E). (A. Annual totals showing high variability. B. Means, extremes and standard deviation calculated over years, SD. The coefficient of variation CV gives standard deviation as fraction of the mean in per)

Low and variable precipitation constrains sedentary farming to water surplus sites, such as oases and to the northern (Siberia), eastern (Inner Mongolia) and southern (Loess plateau, Tibetan valleys) margins of Central Asia. Instead, animal husbandry is the prevailing form of land use. Mongolia still hosts extensive grazing systems based on mobile animal pastoralism (Addison et al., 2012), although the transition to market economy led to privatization of herds and reduced mobility (Lkhagvadorj et al., 2013a,b). Recently, herders became partly organized in pasture user groups that jointly develop (informal) agreements on sustainable use (Leisher et al., 2012; Dorligsuren and Dorjgotov, 2017). The majority of the range is, however, still open, and ownership rests with the state.

In Inner Mongolia, in contrast, traditional semi-nomadic pastoralism was replaced by sedentary animal husbandry during the 1950s to 1970s, when human population and livestock numbers strongly increased and many grasslands were transformed to croplands (Wu et al., 2015). Livestock in Inner Mongolia is typically kept in fenced paddocks, where herders have long-term user rights. Taking account of spatial and temporal variability, herders try to maintain some limited flexibility and spread their paddocks across large regions, necessitating negotiations with neighbors or costly investments into transport.

The situation in the QTP and Xinjiang is intermediate between Inner Mongolia and (Outer) Mongolia. Grassland is owned by the state, but a large share of the range is contracted to individual households. Herders are still partly mobile, but sedentarization schemes have gained momentum (Ptackova, 2011). Herders typically have winter settlements, but move with their herds in summer. Fences are established due to the householder pastoral land contract system or ecological construction projects across pastoral land of China, and the conversion of cropland to forest and grassland program (*'Grain for Green' or 'Sloping Land Conversion'*) aimed at grassland restoration enforces this (Hua and Squires, 2015).

The literature on the grasslands is vast but scattered. Much has been published in Chinese, Russian or Mongolian (Yunatov, 1950; Tuvshintogtokh, 2014) and is thus not easily accessible to an international readership. Chinese authors have rapidly embraced English as a publication language, but relevant literature is still being published in Russian. English summaries on vegetation and major habitats include works by Lavrenko and Karamysheva (1993), Zhu (1993), Hilbig (1995) and Gunin et al. (1999), while Wesche and Treiber (2012) gave a more recent account. Schaller (1998) described the ecology of the large mammals of the region. The socio-economical setting as well as environmental and political histories of the rangelands have been summarized by Fernández-Giménez (1999), Squires et al. (2010), Squires (2012) and by a number of contributions to edited volumes such as those by Kreutzmann (2012) and Squires et al. (2012, 2017). In the last decade, a multitude of single studies has been published, and we are thus increasingly able to draw specific inferences and make locally adapted recommendations; some of which will be reflected in the following subchapters.

Origin and Types of Grasslands

Environmental History of Mongolian and Chinese Grasslands

The grasslands formed as a consequence of the region getting drier with the uplift of the Qinghai-Tibetan plateau since the Tertiary. Standard palynological studies and the analysis of fossil soils in the Mongolian steppes suggest the onset of a more recent drying trend with the start of the Subboreal, circa 3,800 years BP (Dinesman et al., 1989). From the end of the last glacial period until the mid-Atlantic permafrost contributed much to the water supply of the vegetation. In the mid-Atlantic and Subboreal, permafrost retracted to high mountain ranges leading to the aridization of formerly herb-rich meadow steppes. Climate has become increasingly arid and continental during the last 4,000 years (Dinesman et al., 1989). Pastoral livestock husbandry got a wider distribution in Mongolia during the Bronze Age (Volkov, 1967), yet in the dry steppes of Mongolia livestock keeping left major imprints on the landscape not earlier than 900–600 years ago (Dinesman et al., 1989).

Hunter-gatherers have roamed this region since the early Holocene, yet the current pastoral livelihood system probably is not that old. Today's pastoralists depend on farmers for exchange of food (like the barley-based staple *tsampa* in Tibetean) and their societies should be younger than the Neolithic revolution (Chen et al., 2015a). This view has, however, been recently challenged with evidence for permanent occupation being dated at 7,000 BP or older (Meyer et al., 2017). There is, however, no conclusive evidence of strong human impact on vegetation during the early Holocene. Assessing when humans have actually altered grasslands (and not merely used them) is difficult, because of (a) limited availability of suitable archives (pollen profiles), (b) long-term trends towards drier conditions (since circa 3,000–4,000 BP, i.e., the Subboreal) parallel increases in human populations and (c) the difficulty to devise grazing indicator species that are not merely responsive to increasing aridity. Traces like charcoal are, however, widespread since the Subboreal period, and at least partly climatically driven recession of forest cover may have been accelerated by humans' burning and clearing forests for pastures (Dinesman et al., 1989; Gunin et al., 1999; Miehe et al., 2014).

The controversy about human impact on the region's grasslands is not settled. It is perhaps fair to state that most of today's grasslands occur in regions where the climate conditions are not suitable for forest development. The widespread occurrence of forests on north-facing slopes and grasslands on south-facing slopes and in dry valleys of forest-steppe ecotones in Mongolia and in Xinjiang suggest that water constraints are the main driver for the present vegetation distribution even at the northern fringe of the Central Asian steppe region (Dulamsuren, 2004). Nevertheless, the spatial distribution of forests and grasslands has been modified as the result of land-use at many places.

The extensive natural steppes and high-altitude grasslands have been grazed over evolutionary time-scales as proven by the rich native grassland fauna. It is, however, unclear if current grazing pressure exceeds the (theoretical) natural level. Neither data on natural ungulate densities nor large ungrazed reserves as reference are available, and the strong climatic controls render detection of relatively subtle degradation effects very difficult.

Major Grassland Ecoregions

For the purpose of this short overview, we—somewhat simplistically—distinguish five main types of grassland landscapes (Table 8.1, see also Fig. 8.3). Aridity controls the zonation in the Mongolian subregion (Fig. 8.1): Forest steppe landscapes ('forest meadows' in Chinese), i.e., mosaics of forest and dense steppes, are mainly found in the large mountain ranges of Tian Shan, Altay, Khentey and similar mountains. In Mongolia, these mosaics are extensive in the north towards the Siberian Taiga, while the country's central parts are occupied by typical grass steppes. They become progressively taller towards the moister east of Mongolia, i.e., in transition to the tall grass steppes, which once had dominated over northern and central Inner Mongolia. Open desert steppes constitute the margin towards the large dryland region internationally known as Gobi and its surroundings. Tibet hosts alpine steppes in the dry (north-) west and sedge-dominated (mostly *Kobresia pygmaea*) pastures in the moister east. The latter are—although grazed rather than mown—known as alpine meadows in the relevant literature.

The vastly different socioeconomic context described above led us to consider Mongolian and Chinese grasslands separately. Moving from the landscape physiognomy to the realized vegetation, available classification schemes become more complex. In China, generally four main grassland ecosystem types are distinguished—meadow steppe, typical steppe, desert steppe and alpine steppe (Kang et al., 2007). For Mongolia, Karamysheva and Khramtsov (1995) distinguish four main types of grassland vegetation—meadow steppe in the most precipitation-rich regions of the north (mostly in our forest steppe class), followed by true (or typical) grass steppes, desertified steppe and finally desert steppe with increasing aridity. At high elevations, mountain steppes that are drier than meadow steppes as well as alpine grasslands with *Kobresia* diversify the latitudinal vegetation pattern.

Giving exact data on the spatial distribution of these units is difficult, because unified specific international maps are not readily available. We build on coarse-scale maps from the Terrestrial Ecoregions of the World (TEOW; Olson et al., 2001) and reclassified available polygons. Potential extent of grasslands in the given region was estimated using crude classes (10, 25, 50, 75, 100 per cent). Our rough estimates indicate that the Mongolian subregion hosts circa 2.5 million km^2 and the Tibetan subregion circa 1.8 million km^2 of grasslands or other habitats with a large share of graminoids (Table 8.1). In Mongolia, there are three main types—forest steppe, typical steppe and desert steppe account for some 0.3, 0.3 and 0.4 million km^2 of grasslands each. Russia contributes another 0.4 million km^2 of grasslands in forest steppes. Grasslands in plains and mountains of northern China broadly

Fig. 8.3 Grassland types in Mongolia and China. A. Mountain steppes in the Baytag Bogd mountains at the Chinese-Mongolian border: grasslands form mosaics with Larch forests in northern exposures. B. Pastoralists shifting camp in the southern forelands of the Mongolian Altay: Mobility always has been and still is a key element of sustainable land use in dry grasslands. C. Forest steppe in Khuvsgul area: trees can only grow on the northern slopes where soil moisture is higher. D. Tall grass steppes are the natural grasslands of eastern Mongolia (here: W of Choybalsan) and neighboring China. E. Winter camp in the middle Gobi Aymak of Mongolia: Camps are used every year, leading to the formation of piospheres with a sacrifice zone in the immediate surroundings of the camp. F. Desert steppe east of Bayankhongor, Mongolia: Sparse grasslands with the feather grass *Stipa glareosa* cover extensive regions in the Gobi, the soil between the bunch grasses is sealed by deflation pavements. G. Typical summer camp with yak hair tent on alpine meadow from the eastern part of the QTP. H. Sheep and goats grazing on alpine steppe dominated by *Caragana versicolor* in western TAR.

resemble those of Mongolia in both structure and spatial extent. The largest part of Chinese grasslands is, however, situated in the Tibetan subregion. The Tibet Autonomous Region (hereafter TAR), the eastern and northern Qinghai-Tibetan Plateau (QTP), the Qilian Shan and the Tian Shan together host 0.8 million km^2 of *Kobresia* alpine pastures, shrublands and meadows, while alpine steppes contribute another 0.9 million km^2.

Trends in Agronomic Use of Grasslands

Nomadic pastoralism is the traditional form of land use. Animal husbandry plays an important role in nomads' livelihood and livestock production is the only agricultural activity in many areas. The proportion of different livestock species strongly depends on conditions of a given grazing land (Table 8.1), and the trends over time differ across the region (Fig. 8.4).

In Mongolia, the privatization of the livestock sector in 1992 has reduced herders' mobility. Since governmental support for the marketing of livestock products and for the organization of movements effectively ceased, pastoralists have been forced to seek proximity to local markets and often lack the economic potential to cover the costs for seasonal movements with livestock and households (Lkhagvadorj et al., 2013a). Reduced mobility has resulted in more uneven use of the available range and in local overgrazing around towns, water sources and roads. Strongly increased livestock numbers have aggravated this problem (Fig. 8.4A–B). Mongolia's total livestock number has more than doubled since the early 1990s to more than 50 million animals in 2015. Goats and sheep account for most of this increase. In the early 1990s, this increase in livestock numbers was associated with an increase in the number of herder households as a consequence of soaring unemployment in the cities. After 2000, numbers have decreased again, partly because of unfavorable climate conditions hitting inexperienced herders hard and due to the development of more attractive economic alternatives and massive immigration of people into the cities (Lkhagvadorj et al., 2013b). The increase in livestock numbers was at first favored by a series of moist years (Retzer and Reudenbach, 2005), but the overall trend remained in spite of several events of drought- and cold-related mass mortality (Fig. 8.4A–B). The unprecedented increase in goat and (with a delay) also sheep numbers after the 1990s has triggered a scientific debate about overgrazing. Mongolia's livestock sector is exposed to little competition for pasture area by other land uses; the significance of crop farming has been subjected to fluctuations in the past decades and arable fields have occupied less than one per cent of Mongolia's total land area. In contrast to the recent trend for reduced livestock mobility, the directives for Mongolia's planned livestock economy during socialism instructed very frequent pasture rotations with up to thirty moves during summer, which might have had positive effects for the grasslands, but was detrimental for livestock productivity (Shagdarsuren, 2005).

Land use change in Chinese Inner Mongolia has been reviewed in detail by Wu et al. (2015). Since the early 1950s, large areas of grassland were converted to croplands, and the cultivated area increased from 40,000 to 70,000 km^2 until 2001. More than 10 per cent of steppes have been plowed (John et al., 2009). Number of livestock, including horses, cattle, sheep, goats and camels rose to 40 million in 1995 and exceeded 60 million in 2013 (Fig. 8.4C). In parallel, human population rose from 6.1 million in 1953 to 24.7 million in 2010. New urban developments, human infrastructure and rapid expansion of open pit mining fragmented the Inner Mongolian grassland. By now, most of Inner Mongolian herders are sedentarized.

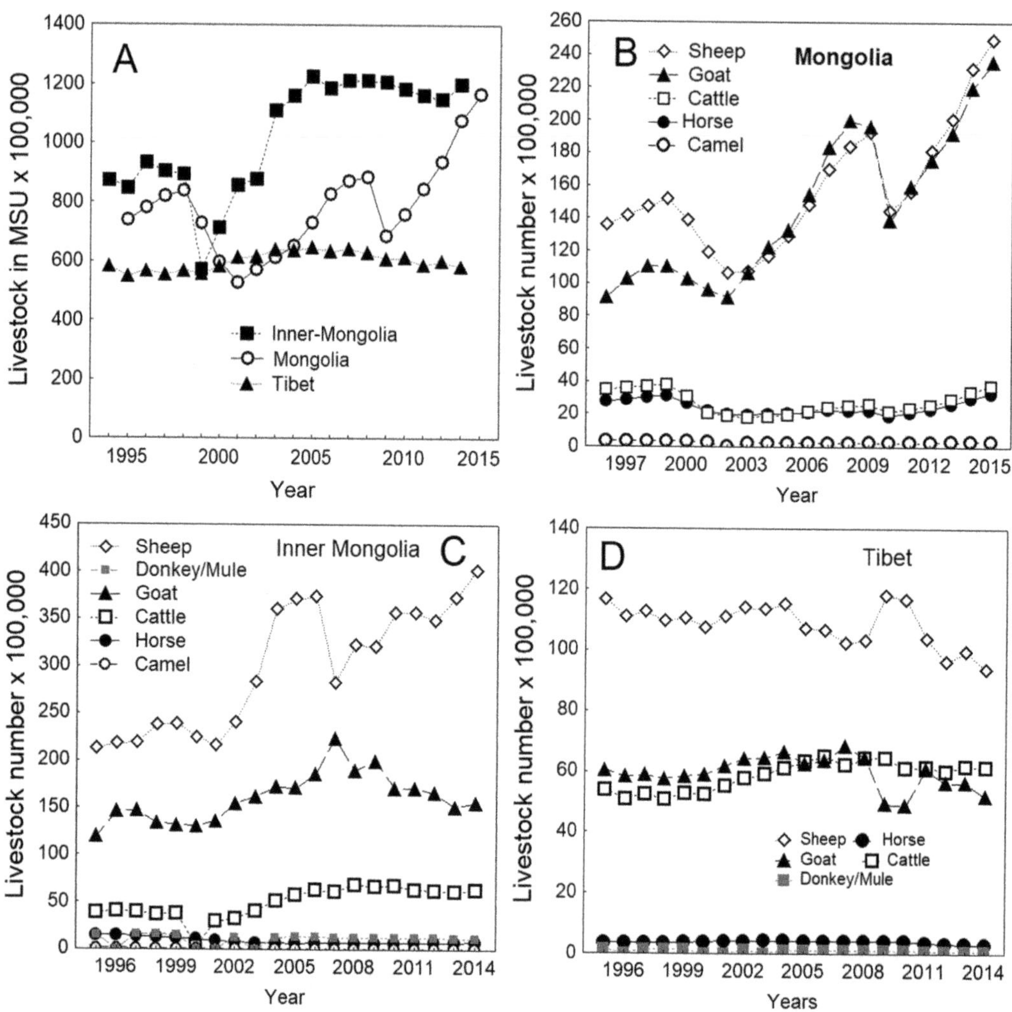

Fig. 8.4 Population growth of livestock in Mongolian and Chinese grasslands 1995–2015. A. Development of livestock in 'Mongolian Sheep Units' (MSU) for better comparison of the ecological impact. One MSU is equivalent to 1 kg dry forage uptake a day (1 horse = 7 MSU, 1 cattle or yak = 6 MSU, 1 donkey or mule = 3 MSU, 1 sheep = 1 MSU, 1 goat = 0.9 MSU). The total population of livestock has risen sharply in Mongolia and Inner Mongolia, while in Tibet livestock numbers have been stable over the years. While total livestock numbers in Inner Mongolia were 23 per cent higher in than in Mongolia in 2014 (64 million vs. 52 million), MSUs differed only by 11 per cent due to differing livestock composition. B–D. Real population numbers. B. In Mongolia, increase in goats has been disproportionally large and numbers reached those of sheep; both groups together have doubled their population. Horse number is almost equal to that of cattle. C. In Inner Mongolia, sheep and cattle populations showed pronounced increase, while horse and camel populations declined. Donkeys and mules still account for more than 50 per cent of the equids' population, in contrast to Mongolia, where these animals are rare. Pigs are not listed here as they are held in pigpens. D. In Tibet, population numbers of sheep, goats and horses were declining, while the number of cattle (including yaks) rose. (Mind the differing scaling of the Y-axes. Data are from the end of the year 2016 and were retrieved from the Mongolian Statistical Information Service (http://1212.mn/en/), the National Bureau of Statistics of China (http://data.stats.gov.cn) and the *China Statistical Year Book* 1996–2016.)

In Xinjiang, grassland area decreased from 1990 to 2010 by 14,557 km^2, the largest decrease within China's provinces (Liu et al., 2014). From 2000 to 2015 the human population grew from 18.5 million to 23.6 million mainly in the cities, while the rural population slightly decreased; in the same time period livestock numbers increased by 5 per cent to 45.8 million, irrigated area increased by 60 per cent and overall farmland area by 70 per cent (http://data.stats.gov.cn). As in the other Chinese provinces, the government has implemented a series of sedentarization, fencing and ecological restoration initiatives to address pasture degradation (Liao et al., 2015). The implementation of land use policies varies among different regions in China, resulting in different social and ecological impacts (Table 8.2). The most influential policy, named as *household responsibility system*, is aimed to change grassland user rights from collective to private (household based, for details see Hua and Squires, 2015). It has been started in the early 1980s in Inner Mongolia and then followed in other regions in China. More recently, herders are encouraged to manage their pastures and livestock in a sustainable way by adopting the *payment for ecosystem service programs*. Based on a 'tragedy of commons' scenario, sedentarization policies were established with the aim to reduce grassland degradation and to improve the socio-economic situation of pastoralists (Ptackova, 2011). However, private land tenure reduces flexibility in grazing management and settling herders resulted in further degradation of rangelands due to fencing, spatial concentration of livestock, continuous year-round grazing and livestock trampling near key-infrastructure (Taylor, 2006; Li et al., 2015; Squires and Hua, 2017). At the same time pressure on the few remaining unfenced areas was enhanced (Yeh, 2009; Hua et al., 2015). It has been estimated that the current stocking rates exceed the carrying capacity of Inner Mongolia by 3.2 times, making supplementary winter feeding necessary, while still failing to provide sufficient net household income to herders (Briske et al., 2015). Sedentarization policies are thus increasingly questioned, especially where forced re-settlement has occurred (Ptackova, 2011; Du, 2012; Li and Li, 2012; Conte and Tilt, 2014), and when livelihood security of (former) pastoralists became threatened (Liao et al., 2015).

Since China's annexation in 1959, significant changes in agricultural structure and land use were implemented in the TAR. Sedentarization efforts have been part of China's grassland policy since that time (Ptackova, 2011). After a period of collectivization, both privatization and fencing policies gained momentum from the 1990s and have effectively settled the herders (Table 8.3), while efforts have been made to reduce livestock densities and restore degraded pastures. These include implementation of National Parks and resettlement of nomads from degraded areas into other regions. The situation thus increasingly resembles the Inner Mongolian case and raises similar skepticism about the ecological, social and cultural consequences (Wang et al., 2014). On the QTP, conversion of grasslands to farmlands is of limited importance due to the high altitude. The main agricultural regions are located in central and southern of Tibet, along the Yarlung Tsangpo river valley and also along the main river valleys in western Sichuan and eastern Qinghai. High elevation and extreme climate conditions result in a short growing season, which restricts the crop production. Barley and wheat are main crops and give high yields because of the intensive solar radiation. Rapeseed (canola), potatoes, peas, radish and occasional fruit trees are also planted (Miller, 2005).

Table 8.2 Important grassland policies and their general impacts on ecosystem health, animal husbandary and pastoral livelihoods in grasslands of the QTP and Inner Mongolia. (+) and (–) mean positive and negative effects. The time when a certain policy started to be implemented differs among different administrative regions (adopted from Wang, 2016).

Period	Policies/Projects	Changes of Livelihood	General Effects
1960s–early 1980s (QTP and Inner Mongolia)	Collectivized ownership of agricultural capital after the People's Republic of China was established	Livestock and other resources became collective property; pastoralists lost decision-making rights on pasture management, although many communes still followed traditional management patterns	Debated: 'Tragedy of commons', or still sustainable
Since early 1980s (Inner Mongolia), since 1990s (QTP)	Decentralization/ Decollectivization; *Householder Pastoral Land Contract System*; by 2013, the contracted grassland covered 71 per cent of total grassland area in China	Grassland user rights began to be contracted to individual households, although grassland remained state-owned; fences were introduced to separate different grazing units; inequality caused social conflicts	(–) on ecosystem functioning, animal husbandry and pastoralist livelihoods
Since 2000 (Inner Mongolia), since 2006 (QTP)	Sedentarization of pastoralists (Nomadic settlement, erection of fences, planting of grass, construction of permanent houses and animal sheds at winter pastures)	Old people are likely to stay in the settlements for the entire year; migration radius of livestock was reduced depending on the distance to winter settlements	(+/–) pastoral livelihoods (+) animal husbandry (–) ecosystem functioning
Since 2003 (Inner Mongolia), since 2004 (QTP)	*Conversion of Cropland to Forest and Grassland Program* ('in Chinese')	Livestock number and the corresponding output of dairy products were reduced	(+) on ecosystem functioning (–) on pastoral society
Since 2011 (QTP and Inner Mongolia)	Projects for grassland ecological constructions, 'subsidy' system for grassland conservation	Cause social problems	No information

Table 8.3 Traded livestock products in Inner Mongolia, TAR and Mongolia. Given are total weight of products for the years 2005 and 2015 (Data from Inner Mongolia and TAR were extracted from the National Bureau of Statistics of China (http://data.stats.gov.cn)).

	Inner Mongolia		TAR		Mongolia	
	2005	2015	2005	2015	2005	2015
Meat (in 1,000 tons)	2,294.6	2,457.1	214.6	280.2	384.0	373.0
– Beef + yak (in 1,000 tons)	336.0	528.9	127.6	165.1	91.0	93.3
– Mutton + goat (in 1,000 tons)	724.3	925.9	74.7	82.3	204.0	221.0
– Horse (in 1,000 tons)	NA	NA	NA	NA	76.0	51.0
– Pork (in 1,000 tons)	875.9	708.1	12.2	15.3	N/A	00.6
Milk (in 1,000 tons)	6,910.5	8,032.0	212.1	300.4	425.9	874.4
Sheep wool (in tons)	94,902	127,187	8,718	7,687	14,164	25,821
Cashmere (in tons)	6,395	8,380	1,226	962	3,671	8,858

Value of Grasslands

Economy and Ecosystem Services

In Mongolia, there is no direct political pressure on herd size. High livestock numbers are seen as a buffer to overcome natural disasters, and market economy with rising cashmere prices has given incentives to enhance the number of goats (Fig. 8.4B; see Berger et al., 2013 and the discussion following it). The number of goats is highly correlated with household income as shown by an example from a forest steppe site in Mongolia, where 60–70 per cent of the herders' cash income was from cashmere sales (Lkhagvadorj et al., 2013a,b). Meat also contributes to herders' subsistence both through the sale in local markets and due to direct own consumption. In terms of cash income, dairy products are less significant, since organized milk collection stopped with the privatization of the livestock sector. Trends in Inner Mongolia were somewhat different. Total livestock populations have increased by 47 per cent from 1995 to 2014, especially in sheep and cattle (Fig. 8.4C). Goat numbers also tended to increase, but the trend was far less rapid than in neighboring Mongolia, implying a less strong dependence on the Cashmere market. In TAR, sedentarization of herders even led to a reduction in the number of goats, sheep and horses, with a growth of cattle population (Fig. 8.4D). Overall grazing pressure in TAR seems stable or even slightly declining (Fig. 8.4A; see also Harris, 2010).

Governmental statistics (Table 8.3) demonstrate that pastoral agriculture is an important part of the economy and essential for food security. Although the total number of livestock in Mongolia and Inner Mongolia now differs by only 20 per cent, the meat industry in Inner Mongolia has much higher productivity than in Mongolia. There, marketed production of meat from cattle/yaks and sheep/goats is about 10 per cent and 16 per cent of that in Inner Mongolia, milk trade is only 10 per cent. Pork is an essential part (one-third) of the huge meat production in Inner Mongolia, but not in TAR and Mongolia. To the contrary, in Mongolia, horse meat is widely consumed. One possible reason for the much lower productivity in terms of traded goods in Mongolian agriculture is the large size of the country, the low population density and the missing infrastructure (transport, abattoirs and cold storage). Foot and Mouth Disease still breaks out in parts of the country, preventing access to foreign markets (e.g., European Union). Most of the Mongolian food products are still consumed within the country and many of these do not even reach the domestic market and are thus directly consumed for household's own subsistence. In TAR, wool and cashmere production as well as meat production from sheep and goats is declining as a consequence of governmental policies (Table 8.3).

With respect to the importance of livestock, grassland primary production and thus forage provision is a key grassland ecosystem service (ES). Others include nutrient cycling, carbon and nitrogen sequestration, protection of soil and water quality, climate regulation and biodiversity conservation. Not least, grasslands maintain cultural traditions of herder people and there is the steppe's potential for tourism. Estimates of monetary values for ES are always problematic and grasslands are no exception. ES for non-degraded alpine grassland on the QTP have been calculated as 98,359 US$/ha with 309 US$/ha for provision of above ground biomass, 400 US$/ha for maintenance of biodiversity and 33,070 US$/ha and 64,580 US$/ha for sequestration of carbon and nitrogen (Wen et al., 2013). According to these estimates, severe degradation of grassland in QTP until 2008 caused losses in biomass and biodiversity valued 198 US$/ha and 200 US$/ha and of 8,003 US$/ha and 13,315 US$/ha by carbon emission and nitrogen loss (Wen et al., 2013). Absolute values clearly are highly questionable and cannot be seen as real 'market prices' rather than as 'minimum replacement value of natural assets and services' (Dong et al., 2012), but they show the general point that grasslands provide a broad range of ES, some of them with great significance even for global aspects (carbon storage). Light grazing increases sequestration of carbon and nitrogen in the QTP (Wen et al., 2013) and in Inner Mongolia (Liu et al., 2012). Investigation of ES by remote sensing data in Inner Mongolia demonstrated that from 1995 to 2011, 96 per cent of the area experienced declining ES as a consequence of land use change, while in 33 per cent of the area ES also declined because of climate change. At the same time, ES increased because of land use or climate change in 11 per cent of the (overlapping) area (Wang et al., 2016). In northern Xinjiang, the decrease of ecosystem service value (ESV) derived from natural pastures was calculated in a simulation model to 29 per cent in the years 1990 to 2010 (Dong et al., 2012). Given the vast extent of the grasslands, the huge environmental heterogeneity and the still relatively limited availability of spatially explicit data, figures on ES are always subject to huge potential errors. Care should be taken with calculations merely based on ecological modeling (Dong et al., 2012), while evaluation of ESV from remote sensing date could be a way to derive more realistic data. Reliable ESV can be a useful tool for decision makers. Uncertainty is exemplified by the case of Mongolia, where the topic of ES has so far been indirectly studied only with a focus on forage productivity and restoration (Jamsran, 2010). Sustainable tourism is, however, of growing economic importance in China and Mongolia and should be included in ESV calculations.

Ecological footprint (EF) estimations are an additional tool to assess the sustainability of land use policies. A recent investigation in the northern Chinese drylands showed that the EF increased 2.63 times from 0.35 to 1.26 billion global hectares between 1990 to 2010, thus demonstrating that the whole area has become unsustainable within this time (Li et al., 2016).

Relevance for Biodiversity Conservation

With a total of circa 5,000–6,000 vascular plant species, the flora of the drylands and steppes in Mongolia, China and neighboring regions is not particularly diverse (compared to 27,000 species in the western drylands of Kazakhstan and surroundings; Manafzadeh et al., 2016). It is, however, very distinct and the number of endemics is high as shown for Mongolia, where some 160 of a total of circa 3,130 species are considered country endemics (i.e., 5 per cent; Urgamal et al., 2014). A further 460 species (15 per cent) are subendemics that are shared with neighboring countries, mainly China. The northern Chinese drylands host further, often quite old endemics (Ge et al., 2003). The Himalayas and associated mountain regions are also distinct, with parts of eastern Tibet being among the top 20 regions for endemism (mainly neo-endemism) in China (López-Pujol et al., 2011). The species are, however, often woody perennials and relate to the subtropical diversity hotspot of Yunnan and surrounding areas.

Steppes are most well-known for the enigmatic fauna, which not only has value in terms of biodiversity conservation but also for tourism (Samiya and Mühlenberg, 2006). The steppes and arid lands in Mongolia, Tibet and to a lesser extent Xinjiang still host large herds of freely migrating ungulates (Schaller, 1998; Mallon and Jiang, 2009; Batsaikhan et al., 2014). Twelve species of large herbivores and five species of caprins are living in the area (Table 8.4), the majority of them highly threatened migratory species that require large space for foraging and reproduction (Mallon and Jiang, 2009). Some of them still occur in vast herds emblematic of different rangeland systems in the region. The Mongolian gazelle

Table 8.4 Main species of large herbivores in the Mongolian subregion and the QTP and their 2016 Red List status according to IUCN (2016).

Species	*Scientific name*	Species Red List Status 2016
Przewalski's horse	*Equus przewalskii*	Endangered (EN)
Kiang or Tibetan wild ass	*Equus kiang*	Least Concern (LC)
Asian wild ass	*Equus hemionus*	Near Threatened (NT)
Wild camel	*Camelus ferus*	Critically Endangered (CR)
Wild yak	*Bos mutus*	Vulnerable (VU)
Chiru or Tibetan antelope	*Pantholops hodgsonii*	Near Threatened (NT)
Saiga antelope	*Saiga tatarica mongolica*	Critically Endangered (CR)
Mongolian gazelle	*Procapra gutturosa*	Least Concern (LC)
Tibetan gazelle	*Procapra picticaudata*	Near Threatened (NT)
Przewalski's gazelle	*Procapra przewalskii*	Endangered (EN)
Goitered gazelle	*Gazella subgutturosa*	Vulnerable (VU)
White-lipped deer	*Cervus albirostris*	Vulnerable (VU)
Mongolian argali	*Ovis ammon*	Near Threatened (NT)
Asiatic ibex	*Capra sibirica*	Least Concern (LC)
Bharal, Blue sheep	*Pseudois nayaur*	Least Concern (LC)

(*Procapra gutturosa*) on the Daurian steppe in eastern Mongolia, Inner Mongolia and Russia is still estimated to number over a million animals (Batsaikhan et al., 2014; Wingard et al., 2014). The Chiru or Tibetean antelope (*Pantholops hodgsonii*) is endemic to the Tibetan Plateau. It has been decimated by intensive poaching to less than 150,000 individuals down from over a million; now enforced protection has resulted in increasing populations (Schaller, 1998; IUCN, 2016). Goitered gazelles (*Gazella subgutturosa*) are widespread throughout arid Asia in a wide range of semi-desert and desert habitats, but have their largest population in the eastern Mongolian steppes (circa 28,500 individuals according to Buuveibaatar et al., 2017). In recent years, their Mongolian population declined severely due to ongoing poaching (IUCN, 2016). The Tibetan gazelle (*Procapra picticaudata*) roams around the whole QTP with an estimated population of around 100,000 (Schaller, 1998; Wingard et al., 2014). The Saiga antelope (*Saiga tatarica*), a nomadic herding species that generally inhabits the open dry steppe grasslands, is found in five populations across Middle and Central Asia and has undergone an even more dramatic population decline from an estimated global population of around 1.5 million (mostly in Kazakhstan) just 20 years ago to about 50,000 individuals in 2016 (Wingard et al., 2014). The subpopulation of the Mongolian Saiga (*S. tatarica mongolica*) declined from 5,200 individuals in 2000 to only 750 in 2004 due to poaching, severe winters and summer drought (IUCN, 2016). Przewalski's gazelle (*Procapra przewalskii*) lives in a few remnant populations in the area around the Qinghai lake.

Viable populations of the Khulan or Asian wild ass (*Equus hemionus*) are found in southern Mongolia, Inner Mongolia and Xinjiang/western Gansu, while the larger Kiang or Tibetan wild ass is restricted to the QTP. Population estimates for the Khulan in South Gobi in 2013 were about 35,900 individuals (Buuveibaatar et al., 2016). The Przewalski's horse (*Equus ferus przewalskii*) has recently been reintroduced in Mongolia and Xinjiang, after being extinct in the wild for almost 50 years (Mallon and Jiang, 2009). Wild yak (*Bos mutus*) and Wild Bactrian camel (*Camelus ferus*) are other prominent, yet rare, species (Table 8.4). Infrastructure, including roads, rail lines, pipelines and fences, disrupts the habitats of these migrating herbivores, thus impeding access to forage and breeding grounds or causing direct mortality (Wingard et al., 2014). Illegal poaching for meat and horns and other body parts for the traditional Chinese medicine trade is a threat to Saiga and other gazelle species. Increased wildfire frequency due to climate change is another source of concern. Ancestors of domestic animals, i.e., wild camel, wild yak and Przewalski horse are threatened by hybridization with domestic varieties (Mallon and Jiang, 2009).

Small mammals dominate the steppes in species number and abundances; many of the burrowing herbivores species are eco-engineers with high impact on the ecosystem (Samiya and Mühlenberg, 2006). The burrows and mounds of their colonies increase landscape heterogeneity and diversity, influence water drainage and soil fertility, provide new habitat for animals and plants as well a highly nutritious forage for large herbivores (Wesche et al., 2007). Marmots (*Marmota* spp.), ground squirrels (*Spermophilus* spp.), hares (*Lepus* ssp.), pikas (*Ochotona* spp.) and voles (*Microtus* ssp.) are characteristic taxa in the steppes, while Zokors (*Myospalax baileyi*) are fossorial rodents of the QTP. Some of these species experience cyclic population fluctuations with rapid population growth. At peak densities, they can compete with livestock for forage; moreover, they can carry human diseases, both making them subject to intensive research and state run pest control. The benefit of these eradication programs has been questioned, as species provide important ecosystem services at medium population densities (Wesche et al., 2007; Lai and Smith, 2003). Marmots and hares are hunted by humans for their meat and fur, and all species are important prey species for avian and mammal predators.

Small carnivores of the region comprise foxes (*Vulpes* spp.), as well as martens (*Martes* spp.) and other Mustelidae. The top predators are wolves (*Canis lupus*), lynx (*Lynx lynx*), bears (*Ursus* spp.) and snow leopard (*Panthera uncia*), all subject to (illegal) poaching.

A large diversity of raptors benefits from the abundance of prey in the steppes. Steppe eagle (*Aquila nipalensis*) and Saker falcon (*Falco cherrug*) are widespread endangered avian predators, while Cinereous vultures (*Aegypius monachus*) live on carrion of wild and domestic ungulates. Larks (*Alaudidae*) are dominant birds of the steppe region which show high species richness and abundance (Wesche et al., 2016). Other, more emblematic species are cranes (*Grus* spp.) and pheasants. A high number of passage migrants pass the steppes on their way to and from Siberia.

The steppes are also rich in invertebrates: ants (*Formicinae*) play a major role in the soil fauna (Bayartogtokh et al., 2014), while grasshoppers (*Orthoptera*), moths and butterflies (*Lepidoptera*) are abundant in many places. Invertebrate herbivores are in direct competition with grazers and can be used as bioindicators for overgrazing (Enkthur et al., 2016). Ground-dwelling beetles are a diverse group of species with numerous ecological roles in steppe ecosystems (Khurelpurev and Pfeiffer, 2017). Dung beetles (*Scarabeidae*) and darkling beetles (*Tenebrioniadae*) are essential for the recycling of dung, carrion and decaying plant material. Tenebrionid beetles are especially adapted to arid climate and underwent a remarkable radiation in the Mongolian subregion; the dominant genera include *Anatolica*, *Blaps*, *Cyphogenia* and *Microdera* (Paknia et al., 2013).

Threats to Grasslands

Compared to prairies of North America or steppes of southern Russia, steppes in parts of the study region are still relatively intact, representing what has been called the world's finest grasslands (Batsaikhan et al., 2014). However, there have been major changes, but remoteness and sheer size of the area does not allow us to draw simple conclusions on change of spatial extent for its steppe regions. Remote sensing remains the tool-box of choice, but different satellite-based inventories arrived at different results concerning spatial trends in the steppes. This does not only refer to magnitude, but even to direction of change in terms of steppe decline or increase (see e.g., Lehnert et al., 2016).

Climate Change Effects

As a consequence of differences across the two subregions (Table 8.1), climate change effects are likely to differ between and even within major steppe regions of the Mongolian and Tibetan subregions. Trends in temperature are relatively uniform (Vandandorj et al., 2017). In line with the highly continental location, temperature increases exceed global averages (Yue et al., 2013; IPCC, 2014). It is, however, unclear if potentially increasing drought stress may be offset by rising CO_2 levels and thus increase water use efficiency (Piao et al., 2015). Even more important are trends in moisture availability, but the high temporal variability renders detection of long-term trends difficult (Yue et al., 2013). As a consequence, climate modeling faces severe difficulties with low consensus among precipitation models (IPCC, 2014). Compared to annual totals, trends in seasonal patterns are more relevant yet even more difficult to derive. For the Mongolian subregion, studies have described a tendency for a shift from summer to winter or spring precipitation (Dagvadorj et al., 2009; Peng et al., 2010). Snow in deep winter would evaporate without any favorable effects on plant growth and may instead pose serious problems for overwintering animals (Middleton et al., 2015). An increase in spring precipitation, however, would extend the growing season as shown by a correlation between snow cover and growth (Peng et al., 2010). The potentially increasing frequency of extreme rain and drought events in summer (Vandandorj et al., 2017) is often seen as detrimental, yet may lead to overall improved growth in the drier rangelands where small rains cannot yield plant-effective precipitation (Knapp et al., 2008). Warming experiments in northern Mongolia stressed the importance of landscape scale variation (Liancourt et al., 2012).

An increase in water availability, due to melting of glaciers in western Tian Shan and a rise of precipitation of 6 mm per decade, has been documented between 1962–2004 for the Aksu and Tarim basins (Rumbaur et al., 2015) in Xinjiang. In this province, climate change generally has a recent positive effect on primary productivity, with richer harvests and an increase in forest cover from 2.9 to 4.2 per cent between 2004 to 2015 (http://data.stats.gov.cn).

In the Tibetan steppe region, temperatures have also strongly and almost uniformly increased (Chen et al., 2013). Higher temperatures negatively affect growth of *Kobresia* pastures at their lower altitudinal (and thus upper thermal) limit (Klein et al., 2004), while effects in the more typical and more extensive alpine range are complex (Dorji et al., 2013). Across most parts of the QTP, precipitation patterns are again more relevant for ecosystem functioning than temperature, yet trends in precipitation are heterogeneous (Lehnert et al., 2016), resulting in declining vegetation cover in the southern part and increasing cover in the north-eastern part of the plateau. With respect to seasonal patterns, some authors found a reduced length of growing season (Yu et al., 2010), while others asserted that increased snow precipitation in spring may result in a longer growing season and increased growth (Wan et al., 2014; Chen et al., 2015b). Again, snowfall in deep winter would pose a problem in term of livestock mortality. Thus, climate interacts with grazing patterns, even if manipulated in small-scale experiments (Klein et al., 2007; Hua et al., 2018). Effects of climate change on soil carbon are even more complicated with losses due to permafrost degradation being potentially compensated by increased growth and sequestration under warmer conditions (Chen et al., 2013).

Grazing Degradation

Given that degradation is a function of livestock numbers, herder mobility and climate, there should be marked differences between individual regions (Fig. 8.2; Table 8.1). Moist steppes are generally more susceptible to degradation than drier sites, a pattern often explained in the framework of non-equilibrium dynamics, i.e., frequent droughts (or snow catastrophes) keeping livestock numbers relatively low (Fernández-Giménez and Allen-Diaz, 1999).

Grassland degradation is related to the increase in unpalatable species, the spread of shrubs and halfshrubs, the reduction of vegetation cover and biomass and finally the increase in bare soil resulting in increased soil erosion and losses of soil organic carbon, reduced water-holding capacity and increased soil compaction (Squires et al., 2010). These changes have high impact on the invertebrate fauna associated with the original plant communities (Enkhtur et al., 2017). Another indirect effect is that heavy grazing is associated with a dramatic increase in small mammal populations, such as the Plateau pika (*Ochotona curzoniae*) on the QTP, or the Brandt's vole (*Lasiopodomys brandti*) in Mongolia and Inner Mongolia, a rodent that aggravates grazing pressure and thus promotes degradation (Zhang et al., 2003).

The spread of pastoral nomadism in the Mongolian-Chinese steppe region was associated with livestock replacing the wild herbivores. Bazha et al. (2008) assumed that the steppe vegetation of Mongolia was probably not fundamentally changed by humans and their livestock until the 20th century, when livestock numbers started to increase strongly. Geomorphological results, however, contradict this view. Lehmkuhl et al. (2007) attributed the occurrence of sand dunes that were deposited in the Russian Altai 1,500 years ago to overgrazing, assuming that livestock opened the vegetation cover of grasslands and thus caused wind erosion.

For Mongolia, estimates of spatial extent of grazing degradation differ. Liu et al. (2013) found a general decrease in vegetation biomass across Mongolia between 1998 and 2008, yet land use effects ranked second to effects of climate to explain these losses. Grassland degradation is not pervasive in Mongolia (Addison et al., 2012) despite the doubling of livestock numbers since 1990 (Fig. 8.4), possible because livestock keeping in Mongolia has remained largely mobile. Severe changes are found in the direct vicinity of nomad camps, small livestock enclosures, roads (where pastoralists concentrate for reasons of accessibility) and near wells and watercourses (Sasaki et al., 2008). Locally strongly degraded areas contrast with extended grassland regions with limited signs of degradation. Overgrazing may still lead to increase in species from drier environments, e.g., increases in the halfshrub *Artemisia frigida* and the shrub *Caragana pygmaea* in meadow and mountain steppes, spread of *A. frigida* and *C. microphylla* in true steppes and the dominance of *C. stenophylla* and *C. korshinskii* in desert steppes (Bazha et al., 2012).

Grazing degradation is more widespread in grasslands of China (Li and Huntsinger, 2011; Squires et al., 2010), but its magnitude also depends on the local context (review by Wang and Wesche, 2016) in terms of vegetation composition, soil nutrient content, local grazing intensity, climate and productivity. While grazing effects on vegetation cover and aboveground standing biomass tend to be negative throughout (as a direct consequence of biomass removal), negative responses in species richness are reported in less than a third of the reviewed studies ($N > 60$). Almost equally common are unimodal responses, i.e., richness may be higher at a moderate grazing intensity than under non-grazing. For several vegetation indicators, grazing effects tend to become more negative with increasing precipitation and thus primary productivity. This hints at land-use climate interaction. Responses in soil indicators are even less homogenous. For pH, bulk density and plant available phosphorus, grazing effects tend to be close to zero on average. Impacts on soil organic carbon and total nitrogen contents tend to be negative, but with large variances between studies and between major grassland types.

In spite of claims of widespread degradation, evidence of grazing impact in the Tibetan high altitude pastures are often based on questionable assessment methods (Harris, 2010). A more recent remote sensing study again arrived at a complex picture, demonstrating that the consequence of increasing livestock numbers on vegetation cover were restricted to certain parts of the QTP (Lehnert et al., 2016). This renders governmental efforts to reduce livestock numbers in general questionable. Contraction of pasture user rights to individual households or communities, establishing fences and sedentarization schemes (Ptackova, 2011; Hua and Squires, 2015) reduced mobility and flexibility of the local herders (Squires and Hua, 2017). This is a risky strategy in view of the inherent environmental variability described above. Local herder communities seek alternative sources of income and the collection of medicinal fungus *Ophiocordyceps sinensis* has increased. The potential impact of this practice needs to be critically evaluated. Similarly, in Xinjiang pastoralists are also forced to diversify their income in response to socio-ecological changes and policy pressures, which often leads to a reduced welfare of the households (Kassam et al., 2016).

Grassland Conversion and Infrastructure Development

Direct conversions into cropland is almost negligible in Mongolia and in TAR, while in Inner Mongolia > 10 per cent of steppes have been sown with crops or planted to create artificial grasslands. Extreme conditions do put severe constraints on further conversion across most grasslands of the two subregions. Nonetheless, conversion remains an issue in oases, especially along inland rivers. Levels of intensification are increasing in Mongolian oases and are already high in neighboring China, especially in the Hexi corridor of Gansu as well as in the Junggar and Tarim basins of Xinjiang. Moist meadows are lost due to salinization or direct conversion. With respect to spatial scales involved, these will, however, remain local phenomena.

Fragmentation is an obvious consequence of sedentarization and fencing schemes as implemented in Chinese grasslands (Squires and Hua, 2017). Development of infrastructure, such as railways and major roads, adds to these and has now even reached the more remote parts of the Qinghai-Tibetan Plateau. A large-scale barrier is formed by the fence that now runs along much of the Chinese Mongolian border (> 4,600 km) that hampers migrations of animals in both countries. Fragmentation is starting to affect Mongolia as well (Batsaikhan et al., 2014), where roads and railways are being built in remote parts of the country with increasing exports to neighboring China.

Large-scale Connections—Eutrophication and Invasive Species

Eutrophication is a common threat for many less productive, high nature value grasslands in North America or Europe (Dengler and Tischew, 2018). In grasslands of the Mongolian and Tibetan subregion, direct nutrient input in terms of fertilizer application is of limited importance and largely confined to some high-input pastures in Inner Mongolia. Fertilizing the remaining grasslands is unfeasible with respect to their sheer size (Table 8.1) and the often low nutrient use efficiency (Wesche and Treiber, 2012). Air-borne nitrogen inputs are more relevant in terms of spatial extent. Inputs are low on an average due to the low number of people and industrial facilities over much of the two subregions, yet levels are increasing. Experimental tests imply that detrimental effects on, for instance, community composition are limited and positive effects may prevail across most of the drier grasslands (Kinugasa et al., 2012). In more productive regions with higher inputs from aerial sources, nitrogen deposition probably has the same detrimental effects on biodiversity as described for grasslands of Europe and North America (Zhang et al., 2014). Further research is, however, needed to assess the role of phosphorus, which is also critical for plant growth, but is naturally rare in the region and not deposited with aerial inputs.

The Mongolian and Tibetan subregions are no exception to the general rule that Eurasian grasslands are donors rather than recipients of invasive species. Crested wheat grass, *Agropyron cristatum*, is a striking example of a species that is a key for ecosystem functioning in Mongolian steppes. At the same time, *A. cristatum* is considered a problematic species because of competitive superiority against native herbs and grasses elsewhere (www.invasiveplantatlas.org). Harsh conditions apparently prevent non-native species to naturalize in grasslands of Mongolia and China. In striking contrast to Europe, even croplands within the two subregions hardly support a rich fauna of non-indigenous weeds. Species currently spreading, such as annual *Chenopodiaceae*, on heavily disturbed sites are almost exclusively native to the area.

Formal Conservation of Grasslands

The spatial share of legally protected areas in the Mongolian and Tibetan subregions is higher than in other grassland regions of the world (Wesche et al., 2016). Across the whole region, protected areas account for some 14 per cent of the grassland ecoregions (China circa 15 per cent, Mongolia 10 per cent, Russia 15 per cent). As before, care has to be taken with respect to regional differences. A single complex of 11 reserves on the QTP accounts for 0.8 million km^2, thus heavily influencing total numbers (Fig. 8.5). In other parts, the share of protected areas typically is between 5–10 per cent of a given grassland region.

In 2016, protected areas in Mongolia comprised 17 per cent of the country (www.protectedplanet.net). The largest reserves are situated in the semi-(deserts) on the southern border and the forest steppes of the north. In Inner Mongolia, 12 per cent of the country is conserved, with steppes and desert steppes lacking sufficient protection (Ma et al., 2016). The large territory of protected areas in the Tibetan steppe reflects the political aim of the Chinese government to reduce grazing impacts and protect the natural environment. This is rooted in its importance for China as a whole and South Asia: the QTP comprises ice fields that contain the largest reserve of fresh water outside the polar region, while

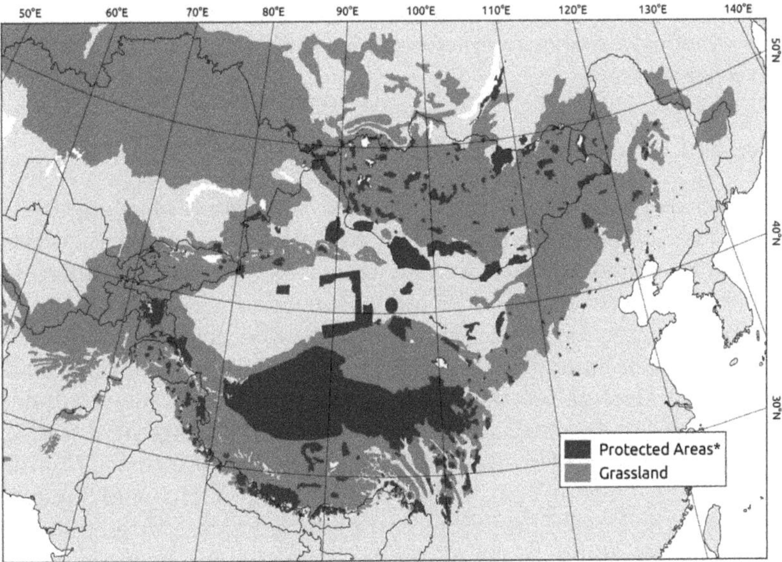

Fig. 8.5 Distribution of main protected areas as registered by IUCN, UNEP-WCMC world database of protected areas (http://www.protectedplanet.net), categories I–IV. (The background shape indicates the extent of grassland ecoregions.)

the Three Rivers Nature Reserve encompasses the headwaters of the Yangtze, Yellow and Mekong rivers. Other important rivers are the Brahmaputra, Indus and the Ganges, all associated with the Tibetan plateau as their headwaters.

Great care is also needed with respect to measures actually implemented on the ground. Outside the core zone, grazing is allowed in almost all reserves of the Tibetan subregion. In Mongolia, Hustai National Park may be the only protected area where livestock access is controlled (except for some winter grazing). Admittedly, zero grazing is not an option if management aims at natural conditions, not least because the grasslands have an evolutionary history of grazing. Wild grazers roam across much of the region and the system may in fact more closely resemble natural conditions if domestic livestock take over that role.

The privatization of rangeland in Inner Mongolia and the QTP has facilitated the introduction of the Payment for Ecosystem Services Scheme as an additional formal instrument of grassland conservation (Tang et al., 2012). It has become a beneficial tool for developing Chinese environmental policy in rural areas. Mandatory grazing bans are enforced and compensated, e.g., in Xilingol payments are about 95 CNY (circa 15.35 US$, July 2015) per hectare and year for grazing prohibition (Addison and Greiner, 2016). Novel approaches like this may play an important role in the future for management of vast grasslands systems like those of Central Asia, if they are to sustain the livelihood of the human population.

Management and Restoration of High Nature Value Grasslands

The Mongolian and Tibetan subregions are different from most other grassland regions in the world because of its pronounced seasonality of climate conditions and also because of the tremendous diversity in environmental conditions associated with the vast size of its territories. Obviously, management recommendations will also differ from those for grasslands in other areas.

Acknowledge and Promote Flexibility

While application of grazing bans and ecological restoration of overgrazed grasslands in Mongolia is just in the beginning, China has already started large-scale ecological conservation and restoration projects to fight against soil erosion and desertification (Piao et al., 2015). However, recent studies from Inner Mongolia suggest that sedentarization and individual land use rights reduce flexibility and force herders to perform unsustainable grazing practices on their limited grassland (Li and Huntsinger, 2011; Conte and Tilt, 2014). The Tibetan grazing systems are now undergoing a similar change as Inner Mongolia experienced decades ago. In a striking parallel, grassland management strategies such as fencing, grazing bans or rearing livestock in stables have been questioned from an ecological and social point of view (Shang et al., 2014; Wang et al., 2014; Hua and Squires, 2015). The fixed boundaries of the private properties and protected areas caused a loss of flexibility and increased the vulnerability of the pastoral system to global changes (Ptackova, 2011). Although herders are compensated, it is questioned whether payment is sufficient and social impacts can be stabilized (Yeh, 2009). Thus government-induced sedentarization programmes tend also to increase the financial vulnerability of pastoralist societies and remove socio-cultural and ecological diversity (Kassam et al., 2016). Restoration of livestock mobility could be a key to sustainable pastoralism, but this would imply drastic changes in existing land use policies towards a comprehensive sustainability science framework concerning both social and ecological needs (Wu et al., 2015).

Wherever coefficients of interannual variability in precipitation are high (typically above 30 per) or snow disasters introduce environmental stochasticity, temporal variability must not be neglected in devising management policies. In Mongolia, herders are still partly mobile and are capable of flexible responses to drought or extremely cold winters. It is perhaps not just a coincidence that Mongolian rangelands are still considered relatively intact. In that perspective, great care should be applied when implementing revised land tenure and user rights in Mongolia. Along the same lines, the suitability of sedentarization, fencing of open range and land tenure schemes should be thoroughly tested and treated with great caution in some of the more environmentally-variable grassland regions of China.

Concentrate Restoration Efforts on Key Resource Areas

In China, steppe degradation triggered large-scale restoration efforts. Many of today's sand fields and semi-deserts occur under typical grassland climates and may have once been grasslands, partly even with scattered trees along water surplus sites (e.g., *Ulmus pumila*, *Populus euphratica*). Efforts to stabilize these sand fields have been massive, including mechanical stabilization (checkerboards), sowing of selected species and plantation of trees (Heshmati and Squires, 2013). Success was, however, limited and in many cases early successional stages rather than properly recovered grasslands were established. In other cases, even undesirable effects, such as soil desiccation, occurred (Deng et al., 2016). In view of the associated costs and the spatial extent, large-scale restoration clearly is not a universally applicable solution.

Restoration may nonetheless work in the much smaller oases regions. These offer key resources for the surrounding rangelands such as forage in times of drought or in unfavorable seasons and they hold a large share of the local population. Concentrating restoration efforts there would have disproportionally greater effects than concentrating on the zonal rangelands.

Learn from Comparison with Other Rangelands but Do Not Simply Copy Established Strategies

Much of the basic and applied scientific literature does not take a proper synoptic perspective. Fifteen years ago, lack of data from certain regions forced researchers to discuss their results in view of the literature from broadly similar environments. Largely owing to massive investments in research by concerned governments (with China in a lead role), studies are now available from the same biogeographic realm, allowing for targeted comparisons and locally adopted planning. What may work in the relatively moist northern parts of Inner Mongolia may not work in the dry south nor in the arid western parts and will almost certainly fail in neighboring Gobi of Mongolia. There also is no need to directly import concepts initially developed in very different contexts, such as the 'tragedy of the commons' and the call for private land tenure in response mentioned above. Instead, we urgently need locally adopted schemes that not only consider the conditions in the given country of interest, but more specifically even in the specific grassland region within that country.

Résumé and Future Prospects

Mongolia, China and neighboring Russia comprise the largest almost continuous grassland regions of the world. Our chapter highlights the unique biodiversity of the region as well as the human use of its biological richness and the threats to the ecosystem. The area is characterized by a high heterogeneity of landscape and climate as well as of cultural traditions and land use systems. Nomadic herders have shaped these grasslands over thousands of years with a sustainable use of biological resources. Yet, now livestock numbers have reached unprecedented densities. At the same time regional climatic warming is extreme and the precipitation pattern may change as well. These factors call for adapted rangeland management decisions that take heterogeneous habitats and resilience capacities into consideration, while protecting biological diversity and cultural identity of the herders. Food security for the growing human population in these regions can only be assured with a sustainable use of these fragile grassland ecosystems.

Abbreviations

BP = Years Before Present; ES = Ecosystem Service(s); ESV = Ecosystem Service Value; QTP = Qinghai-Tibetan Plateau; TAR = Tibet Autonomous Region

References

Addison, J., M. Friedel, C. Brown, J. Davies and S. Waldron. 2012. A critical review of degradation assumptions applied to Mongolia's Gobi Desert. Rangel. J. 34: 125–137.

Addison, J. and R. Greiner. 2016. Applying the social-ecological systems framework to the evaluation and design of payment for ecosystem service schemes in the Eurasian steppe. Biodivers. Conserv. 25: 2421–2440.

Babel, W., T. Biermann, H. Coners, E. Falge, E. Seeber, J. Ingrisch, P.M. Schleuß, T. Gerken, J. Leonbacher, (...) and T. Foken. 2014. Pasture degradation modifies the water and carbon cycles of the Tibetan highlands. Biogeosciences 11: 6633–6656.

Batsaikhan, N., B. Buuveibaatar, B. Chimed, O. Enkhtuya, D. Galbrakh, O. Ganbaatar, B. Lkhagvasuren, D. Nandintsetseg, J. Berger, (...) and T. Whitten. 2014. Conserving the world's finest grassland amidst ambitious national development. Conserv. Biol. 28: 1736–1739.

Bayartogtokh, B., U. Aibek, S. Yamane and M. Pfeiffer. 2014. Diversity and biogeography of ants in Mongolia (*Hymenoptera*: *Formicidae*). Asian Myrmecol. 6: 63–82.

Bazha, S.N., P.D. Gunin, E.V. Danzhalova and T.I. Kazantseva. 2008. Diagnostical parameters of pastural digression of steppe plant communities of the Mongolian biogeographical province of Palearctica [in Russian]. Volga Ecol. J. 4: 251–263.

Bazha, S.N., P.D. Gunin, E.V. Danzhalova, Y.I. Drobyshev and A.V. Prishcepa. 2012. Pastoral degradation of steppe ecosystems in Central Mongolia. pp. 289–319. *In*: M.J.A. Werger and M.A. van Staalduinen (eds.). Eurasian Steppes: Ecological Problems and Livelihoods in a Changing World. Springer, Dordrecht.

Berger, J., B. Buuveibaatar and C. Mishra. 2013. Globalization of the cashmere market and the decline of large mammals in Central Asia. Conserv. Biol. 27: 679–689.

Bragina, T.M., K.A. Vanselow, A. Nowak and V. Wagner. 2018. Grasslands of Kazakhstan and Middle Asia: The ecology, conservation and use of a vast and globally important area (pp. 141–169 this volume).

Briske, D.D., M.L. Zhao, G.D. Han, C.B. Xiu, D.R. Kemp, W. Willms, K. Havstad, L. Kang, Z.W. Wang, (...) and Y.F. Bai. 2015. Strategies to alleviate poverty and grassland degradation in Inner Mongolia: Intensification vs production efficiency of livestock systems. J. Environ. Manag. 152: 177–182.

Buuveibaatar, B., S. Strindberg, P. Kaczensky, J. Payne, B. Chimeddorj, G. Naranbaatar, S. Amarsaikhan, B. Dashnyam, T. Munkhzul, (...) and T.K. Fuller. 2017. Mongolian Gobi supports the world's largest populations of khulan *Equus hemionus* and goitered gazelles *Gazella subgutturosa*. Oryx 51: 639–647.

Chen, F.H., G.H. Dong, D.J. Zhang, X.Y. Liu, X. Jia, C.B. An, M.M. Ma, Y.W. Xie, L. Barton, (...) and M.K. Jones. 2015a. Agriculture facilitated permanent human occupation of the Tibetan Plateau after 3600 B.P. Science 347: 248–250.

Chen, H., Q.A. Zhu, C.H. Peng, N. Wu, Y.F. Wang, X.Q. Fang, Y.H. Gao, D. Zhu, G. Yang, (...) and J.H. Wu. 2013. The impacts of climate change and human activities on biogeochemical cycles on the Qinghai-Tibetan Plateau. Global Change Biol. 19: 2940–2955.

Chen, X.Q., S. An, D.W. Inouye and M.D. Schwartz. 2015b. Temperature and snowfall trigger alpine vegetation green-up on the world's roof. Global Change Biol. 21: 3635–3646.

Conte, T.J. and B. Tilt. 2014. The effects of China's grassland contract policy on pastoralists' attitudes towards cooperation in an Inner Mongolian banner. Hum. Ecol. 42: 837–846.

Dagvadorj, D., B. Khuldorj and R.Z. Aldover. 2009. Mongolia Assessment Report on Climate Change 2009. Ministry of Nature, Environment and Tourism, Ulaanbaatar, 228 pp.

Deng, L., W.M. Yan, Y.W. Zhang and Z.P. Shangguan. 2016. Severe depletion of soil moisture following land-use changes for ecological restoration: Evidence from northern China. For. Ecol. Manag. 366: 1–10.

Dengler, J. and S. Tischew. 2018. Grasslands of Western and Northern Europe—between intensification and abandonment (pp. 27–63 this volume).

Dinesman, L., N. Kiseleva and A. Knyazev. 1989. History of the steppe ecosystems of the Mongolian People's Republic [in Russian]. Nauka, Moscow, 212 pp.

Dong, X.B., W.K. Yang, S. Ulgiati, M.C. Yan and X.S. Zhang. 2012. The impact of human activities on natural capital and ecosystem services of natural pastures in North Xinjiang, China. Ecol. Model. 225: 28–39.

Dorligsuren, D. and D. Dorjgotov. 2017. Rangelands and pasturelands in Mongolia: Land and people in transition. pp. 201–213. *In*: V.R. Squires, Z. Shang and A. Ariapour (eds.). Rangelands along the Silk Road: Transformative Adaptation under Climate and Global Change. Nova Publishers, New York.

Dorji, T., Ø. Totland, S.R. Moe, K.A. Hopping, J. Pan and J.A. Klein. 2013. Plant functional traits mediate reproductive phenology and success in response to experimental warming and snow addition in Tibet. Global Change Biol. 19: 459–472.

Du, F. 2012. Ecological resettlement of Tibetan herders in the Sanjiangyuan: A case study in Madoi county of Qinghai. Nomadic Peoples 16: 116–133.

Dulamsuren, C. 2004. Floristische Diversität, Vegetation und Standortbedingungen in der Gebirgstaiga des Westkhentej, Nordmongolei. Ber. Forschungszentrums Waldökosyst. A 191: 1–290.

Enkhtur, K., M. Pfeiffer, A. Lkhagva and B. Boldgiv. 2017. Response of moths (*Lepidoptera*: *Heterocera*) to livestock grazing in Mongolian rangelands. Ecol. Indic. 72: 667–674.

Fernández-Giménez, M.E. 1999. Sustaining the steppes: A geographical history of pastoral land use in Mongolia. Geograph. Rev. 89: 315–342.

Fernández-Giménez, M.E. and B. Allen-Diaz. 1999. Testing a non-equilibrium model of rangeland vegetation dynamics in Mongolia. J. Appl. Ecol. 36: 871–885.

Ge, X.Y., Y. Yu, N.X. Zhao, H.S. Chen and W.Q. Qi. 2003. Genetic variation in the endangered Inner Mongolia endemic shrub *Tetraena mongolica* Maxim (*Zygophyllaceae*). Biol. Conserv. 111: 427–434.

Gunin, P.D., E.A. Vostokova and N.I. Dorofeyuk. 1999. Vegetation Dynamics of Mongolia. Kluwer Academic Publishers, Dordrecht, 238 pp.

Harris, R.B. 2010. Rangeland degradation on the Qinghai-Tibetan plateau: A review of the evidence of its magnitude and causes. J. Arid Environ. 74: 1–12.

Heshmati, G.A. and V.R. Squires (eds.). 2013. Combating Desertification in Asia, Africa and the Middle East: Proven Practices. Springer, Dordrecht, 368 pp.

Hijmans, R.J., S.E. Cameron, J.L. Parra, P.G. Jones and A. Jarvis. 2005. Very high resolution interpolated climate surfaces for global land areas. Int. J. Climatol. 25: 1965–1978.

Hilbig, W. 1995. The Vegetation of Mongolia. SPB Academic Publishing, Amsterdam, 258 pp.

Hua, L.M. and V.R. Squires. 2015. Managing China's pastoral lands: Current problems and future prospects. Land Use Policy 43: 129–137.

Hua, L.M., S.W. Yang, V.R. Squires and G.Z. Wang. 2015. An alternative rangeland management strategy in an agro-pastoral area in western China. Rangel. Ecol. Manag. 68: 1009–1118.

Hua, L.M., Y. Niu and V.R. Squires. 2018. Climatic change on grassland regions and its impact on grassland-based livelihoods in China (pp. 355–368 this volume).

IPCC. 2014. Climate Change 2014: Impacts, Adaptation, and Vulnerability. Part B: Regional Aspects. Contribution of Working Group II to the Fifth Assessment, Report of the Intergovernmental Panel on Climate Change. Cambridge University Press, Cambridge, 688 pp.

IUCN. 2016. The IUCN Red List of Threatened Species. Version 2016-3. URL: http://www.iucnredlist.org/.

Jamsran, U. 2010. Involvement of local communities in restoration of ecosystem services in Mongolian rangeland. Global Environ. Res. 14: 79–86.

John, R., J.Q. Chen, N. Lu and B. Wilske. 2009. Land cover/land use change in semi-arid Inner Mongolia: 1992–2004. Environ. Res. Lett. 4: Article 045010.

Kang, L., X.G. Han, Z.B. Zhang and O.J. Sun. 2007. Grassland ecosystems in China: Review of current knowledge and research advancement. Phil. Trans. R. Soc. B 362: 997–1008.

Karamysheva, Z.V. and V.N. Khramtsov. 1995. The steppes of Mongolia. Braun-Blanquetia 17: 1–70.

Kassam, K.-A.S., C. Liao and S.K. Dong. 2016. Socio-cultural and ecological systems of pastoralism in inner Asia: Cases from Xinjiang and inner Mongolia in China and the Pamirs of Badakhshan, Afghanistan. pp. 137–175. *In*: S.K. Dong, K.-A.S. Kassam, J.F. Tourrand and R.B. Boone (eds.). Building Resilience of Human-Natural Systems of Pastoralism in the Developing World: Interdisciplinary Perspectives. Springer, Cham.

Khurelpurev, O. and M. Pfeiffer. 2017. *Coleoptera* in the Altai Mountains (Mongolia): Species richness and community patterns along an ecological gradient. J. Asia-Pac. Biodivers. 10: 362–370.

Kinugasa, T., A. Tsunekawa and M. Shinoda. 2012. Increasing nitrogen deposition enhances post-drought recovery of grassland productivity in the Mongolian steppe. Oecologia 170: 857–865.

Klein, J., J. Harte and X.Q. Zhao. 2004. Experimental warming causes large and rapid species loss, dampened by simulated grazing, on the Tibetan Plateau. Ecol. Lett. 7: 1170–1179.

Klein, J.A., J. Harte and X.Q. Zhao. 2007. Experimental warming, not grazing, decreases rangeland quality on the Tibetan Plateau. Ecol. Appl. 17: 541–557.

Knapp, A.K., C. Beier, D.D. Briske, A. Classen, Y.Q. Luo, M. Reichstein, M.D. Smith, S.D. Smith, J.E. Bell, (...) and E.S. Weng. 2008. Consequences of more extreme precipitation regimes for terrestrial ecosystems. BioScience 58: 811–821.

Kreutzmann, H. 2012. Pastoral practices in transition: Animal husbandry in High Asian Contexts. pp. 1–30. *In*: H. Kreutzmann (ed.). Pastoral Practices in High Asia. Agency of 'Development' Affected by Modernisation, Resettlement and Transformation. Springer, Dordrecht.

Lai, C.H. and A.T. Smith. 2003. Keystone status of plateau pikas (*Ochotona curzoniae*): Effect of control on biodiversity of native birds. Biodivers. Conserv. 12: 1901–1912.

Lavrenko, E.M. and Z.V. Karamysheva. 1993. Steppes of the former Soviet Union and Mongolia. pp. 3–59. *In*: R.T. Coupland (ed.). Natural Grasslands: Eastern Hemisphere & Resume. Elsevier Science, Amsterdam.

Lehmkuhl, F., A. Zander and M. Frechen. 2007. Luminescence chronology of fluvial and aeolian deposits in the Russian Altai (Southern Siberia). Quat. Geochronol. 2: 195–201.

Lehnert, L.W., K. Wesche, K. Trachte, C. Reudenbach and J. Bendix. 2016. Climate variability rather than overstocking causes recent large-scale cover changes of Tibetan pastures. Sci. Rep. 6: Article 24367.

Leisher, C., S. Hess, T.M. Boucher, P. van Beukering and M. Sanjayan. 2012. Measuring the impacts of community-based grasslands management in Mongolia's gobi. PLOS One 7: Article e30991.

Li, J.W., Z.F. Liu, C.Y. He, W. Tu and Z.X. Sun. 2016. Are the drylands in northern China sustainable? A perspective from ecological footprint dynamics from 1990 to 2010. Sci. Total Environ. 553: 223–231.

Li, W.J. and L. Huntsinger. 2011. China's grassland contract policy and its impacts on herder ability to benefit in Inner Mongolia: Tragic feedbacks. Ecol. Soc. 16(2): Article 1.

Li, W.J. and Y.B. Li. 2012. Managing rangeland as a complex system: How government interventions decouple social systems from ecological systems. Ecol. Soc. 17(1): Article 9.

Li, Z.G., G.D. Han, M.L. Zhao, J. Wang, Z.W. Wang, D.R. Kemp, D.L. Michalk, A. Wilkes, K. Behrendt, H. Wang and C. Langford. 2015. Identifying management strategies to improve sustainability and household income for herders on the desert steppe in Inner Mongolia, China. Agric. Syst. 132: 62–72.

Liancourt, P., L.A. Spence, B. Boldgiv, A. Lkhagva, B.R. Helliker, B.B. Casper and P.S. Petraitis. 2012. Vulnerability of the northern Mongolian steppe to climate change: Insights from flower production and phenology. Ecology 93: 815–824.

Liao, C., C. Barrett and K.A. Kassam. 2015. Does diversification improve livelihoods? Pastoral households in Xinjiang, China. Dev. Change 46: 1302–1330.

Liu, G.F., X.F. Xie, D. Ye, X.H. Ye, I. Tuvshintogtokh, B. Mandakh, Z.Y. Huang and M. Dong. 2013. Plant functional diversity and species diversity in the Mongolian steppe. PLOS One 8: Article e77565.

Liu, J.Y., X.Z. Deng, X.B. Yu and W.H. Kuang. 2014. Nearly 30 years of terrestrial ecosystem change in China. pp. 292–330. *In*: M. Dunford and L. Weidong (eds.). The Geographical Transformation of China. Routledge, Oxford.

Liu, N., Y.J. Zhang, S.J. Chang, H.M. Kan and L.J. Lin. 2012. Impact of grazing on soil carbon and microbial biomass in typical steppe and desert steppe of inner Mongolia. PLOS One 7: Article e36434.

Lkhagvadorj, D., M. Hauck, C. Dulamsuren and J. Tsogtbaatar. 2013a. Pastoral nomadism in the forest-steppe of the Mongolian Altai under a changing economy and a warming climate. J. Arid Environ. 88: 82–89.

Lkhagvadorj, D., M. Hauck, C. Dulamsuren and J. Tsogtbaatar. 2013b. Twenty years after decollectivization: Mobile livestock husbandry and its ecological impact in the Mongolian forest-steppe. Hum. Ecol. 41: 725–735.

López-Pujol Jordi, F.M. Zhang, H.Q. Sun, T.S. Ying and S. Ge. 2011. Centres of plant endemism in China: places for survival or for speciation? J. Biogeogr. 38: 1267–1280.

Ma, W.J., G. Feng and Q. Zhang. 2016. Status of nature reserves in Inner Mongolia, China. Sustainability 8(9): Article 889.

Mallon, D.P. and Z.G. Jiang. 2009. Grazers on the plains: Challenges and prospects for large herbivores in Central Asia. J. Appl. Ecol. 46: 516–519.

Manafzadeh, S., Y.M. Staedler and E. Conti. 2016. Visions of the past and dreams of the future in the Orient: The Irano-Turanian region from classical botany to evolutionary studies. Biol. Rev. 92: 1365–1388.

Meyer, M.C., M.S. Aldenderfer, Z. Wang, D.L. Hoffmann, J.A. Dahl, D. Degering, W.R. Haas and F. Schlütz. 2017. Permanent human occupation of the central Tibetan Plateau in the early Holocene. Science 355: 64–67.

Middleton, N., H. Rueff, T. Sternberg, B. Batbuyan and D. Thomas. 2015. Explaining spatial variations in climate hazard impacts in western Mongolia. Landsc. Ecol. 30: 91–107.

Miehe, G., S. Miehe, J. Böhner, K. Kaiser, I. Hensen, D. Madsen, J.Q. Liu and L. Opgenoorth. 2014. How old is the human footprint in the world's largest alpine ecosystem? A review of multiproxy records from the Tibetan Plateau from the ecologists' viewpoint. Quat. Sci. Rev. 86: 190–209.

Miller, D.J. 2005. The Tibetan steppe. pp. 305–342. *In*: J.M. Suttie, S.G. Reynolds and C. Batello (eds.). Grasslands of the World. FAO, Rome.

National Bureau of Statistics of the People's Republic of China 1996–2016. China Statistical Year Book. URL: http://data.stats.gov.cn.

Olson, D.M., E. Dinerstein, E.D. Wikramanayake, N.D. Burgess, G.V.N. Powell, E.C. Underwood, J.A. D'Amico, I. Itoua, H.E. Strand, (...) and K.R. Kassem. 2001. Terrestrial ecoregions of the world: A new map of life on Earth. BioScience 51: 933–938.

Paknia, O., M. Grundler and M. Pfeiffer. 2013. Species richness and niche differentiation of darkling beetles (*Coleoptera*: *Tenebrionidae*) in Mongolian steppe ecosystems. pp. 47–72. *In*: M.B.M. Prieto and J.T. Diaz (eds.). Steppe Ecosystems: Biological Diversity, Management and Restoration. Nova Science Publishers, Hauppauge, NY.

Peng, S.S., S.L. Piao, P. Ciais, J.Y. Fang and X.H. Wang. 2010. Change in winter snow depth and its impacts on vegetation in China. Global Change Biol. 16: 3004–3013.

Piao, S.L., G.D. Yin, J.G. Tan, L. Cheng, M.T. Huang, Y. Li, R.G. Liu, J.F. Mao, R.B. Myneni, (...) and Y.P. Wang. 2015. Detection and attribution of vegetation greening trend in China over the last 30 years. Global Change Biol. 21: 1601–1609.

Ptackova, J. 2011. Sedentarisation of Tibetan nomads in China: Implementation of the Nomadic settlement project in the Tibetan Amdo area, Qinghai and Sichuan Provinces. Pastoralism Res. Policy Pract. 1: Article 4.

Retzer, V. and C. Reudenbach. 2005. Modelling the carrying capacity and coexistence of pika and livestock in the mountain steppe of the South Gobi, Mongolia. Ecol. Model. 189: 89–104.

Rumbaur, C., N. Thevs, M. Disse, M. Ahlheim, A. Brieden, B. Cyffka, D. Duethmann, T. Feike, O. Fror, (...) and C. Zhao. 2015. Sustainable management of river oases along the Tarim River (SuMaRio) in Northwest China under conditions of climate change. Earth Syst. Dyn. 6: 83–107.

Samiya, R. and M. Mühlenberg. 2006. Environmental Conservation in Mongolia. Munhiin useg, Ulaanbaatar, 324 pp.

Sasaki, T., T. Okayasu, U. Jamsran and K. Takeuchi. 2008. Threshold changes in vegetation along a grazing gradient in Mongolian rangelands. J. Ecol. 96: 145–154.

Schaller, G.B. 1998. Wildlife of the Tibetan Steppe. University of Chicago Press, Chicago, 374 pp.

Shagdarsuren, O. 2005. Biology of the Livestock of Mongolia and Special Characters of Pastoralism. MUIS-iim Hevleh uildver, Ulaanbaatar, 302 pp.

Shang, Z.H., M.J. Gibb, F. Leiber, M. Ismail, L.M. Ding, X.S. Guo and R.J. Long. 2014. The sustainable development of grassland-livestock systems on the Tibetan plateau: problems, strategies and prospects. Rangel. J. 36: 267–296.

Squires, V.R., X.S. Lu, Q. Lu, T. Wang and Y.L. Yang. 2010. Rangeland Degradation and Recovery in China's Pastoral Lands. CABI, Wallingford, 264 pp.

Squires, V.R. (ed.). 2012. Rangeland Stewardship in Central Asia: Balancing Livelihoods, Biodiversity Conservation and Land Protection. Springer, Dordrecht, 460 pp.

Squires, V.R. and M. Hua. 2017. Land fragmentation: a scourge in China's pastoral areas. Livest. Res. Rural Dev. 29: 32–46.

Squires, V.R., Z.H. Shang and A. Ariapour (eds.). 2017. Rangelands along the Silk Road: Transformative Adaptation under Climate and Global Change. Nova Publishers, New York, 360 pp.

Starrs, P.F., L. Huntsinger and S. Spiegal. 2018. North American grasslands and biogeographic Regions (pp. 369–389 this volume).

Tang, Z., Y.L. Shi, Z.B. Nan and Z.M. Xu. 2012. The economic potential of payments for ecosystem services in water conservation: A case study in the upper reaches of Shiyang River basin, northwest China. Environ. Dev. Econ. 17: 445–460.

Taylor, J.L. 2006. Negotiating the grassland: The policy of pasture enclosures and contested resource use in Inner Mongolia. Hum. Organ. 65: 374–386.

Tuvshintogtokh, I. 2014. The Steppe Vegetation of Mongolia [in Mongolian]. Bembi san, Ulaanbaatar, 610 pp.

Urgamal, M., B. Oyuntsetseg, D. Nyambayar and C. Dulamsuren. 2014. Conspectus of the Vascular Plants of Mongolia. Admon Printing Press, Ulaanbaatar.

Vandandorj, S., E. Munkhjargal, B. Boldgiv and B. Gantsetseg. 2017. Changes in event number and duration of rain types over Mongolia from 1981 to 2014. Environ. Earth Sci. 76: Article 70.

Volkov, V.V. 1967. Bronze and Early Iron Age of Northern Mongolia [in Russian]. Shinzhleh Ukhaa Akademiiyn Hevlel, Ulaanbaatar, 146 pp.

Wan, Y.F., Q.Z. Gao, Y. Li, X.B. Qin Ganjurjav, W.N. Zhang, X. Ma and S. Liu. 2014. Change of snow cover and its impact on alpine vegetation in the source regions of large rivers on the Qinghai-Tibetan Plateau, China. Arct. Antarct. Alp. Res. 46: 632–644.

Wang, H., S.L. Zhou, X.B. Li, H.H. Liu, D.K. Chi and K.K. Xu. 2016. The influence of climate change and human activities on ecosystem service value. Ecol. Eng. 87: 224–239.

Wang, Y., J. Wang, S.C. Li and D.H. Qin. 2014. Vulnerability of the Tibetan pastoral systems to climate and global change. Ecol. Soc. 19(4): Article 8.

Wang, Y. 2016. The Shaping of Tibetan Grasslands: Combined Effects of Livestock-grazing and Environment: A Multi-site, Interdisciplinary Study. Ph.D. Thesis Naturwissenschaftliche Fakultät I. Martin-Luther-Universität Halle-Wittenberg, Halle (Saale), 143 pp.

Wang, Y. and K. Wesche. 2016. Vegetation and soil responses to livestock grazing in Central Asian grasslands: a review of Chinese literature. Biodivers. Conserv. 25: 2401–2420.

Wen, L., S.K. Dong, Y.Y. Li, X.Y. Li, J.J. Shi, Y.L. Wang, D.M. Liu and Y.S. Ma. 2013. Effect of degradation intensity on grassland ecosystem services in the alpine region of Qinghai-Tibetan Plateau, China. PLOS One 8: Article e58432.

Wesche, K., K. Nadrowski and V. Retzer. 2007. Habitat engineering under dry conditions: The impact of pikas (*Ochotona pallasi*) on southern Mongolian mountain steppes. J. Veg. Sci. 18: 665–674.

Wesche, K. and J. Treiber. 2012. Abiotic and biotic determinants of steppe productivity and performance—a view from Central Asia. pp. 3–43. *In*: M.J.A. Werger and M. van Staalduinen (eds.). Eurasian Steppes: Ecological Problems and Livelihoods in a Changing World. Springer, Dordrecht.
Wesche, K., D. Ambarlı, J. Kamp, P. Török, J. Treiber and J. Dengler. 2016. The Palaearctic steppe biome: a new synthesis. Biodivers. Conserv. 25: 2197–2231.
Wingard, J., P. Zahler, R. Victurine, O. Bayasgalan and B. Buuveibaatar. 2014. Guidelines for addressing the impact of linear infrastructure on large migratory mammals in Central Asia. pp. Annex 1–139. *In*: UNEP (ed.). Convention on Migratory Species. UNEP, Quito.
Wu, J.G., Q. Zhang, A. Li and C.Z. Liang. 2015. Historical landscape dynamics of Inner Mongolia: Patterns, drivers, and impacts. Landsc. Ecol. 30: 1579–1598.
Yeh, E.T. 2009. Greening western China: A critical view. Geoforum 40: 884–894.
Yu, H.Y., E. Luedeling and J.C. Xu. 2010. Winter and spring warming result in delayed spring phenology on the Tibetan Plateau. Proc. Natl. Acad. Sci. USA 107: 22151–22156.
Yue, T.X., N. Zhao, R.D. Ramsey, C.L. Wang, Z.M. Fan, C.F. Chen, Y.M. Lu and B.L. Li. 2013. Climate change trend in China, with improved accuracy. Clim. Change 120: 137–151.
Yunatov, A.A. 1950. The Main Features of the Vegetation Cover of the Mongolian People's Republic [in Russian]. Akadademia Nauk SSSR, Moscow, 224 pp.
Zhang, Y.H., X.T. Lü, F. Isbell, C. Stevens, X. Han, N.P. He, G.M. Zhang, Q. Yu, J.H. Huang and X.G. Han. 2014. Rapid plant species loss at high rates and at low frequency of N addition in temperate steppe. Global Change Biol. 20: 3520–3529.
Zhang, Z.B., R. Pech, S. Davis, D.Z. Shi, X.R. Wan and W.Q. Zhong. 2003. Extrinsic and intrinsic factors determine the eruptive dynamics of Brandt's voles *Microtus brandtii* in Inner Mongolia, China. Oikos 100: 299–310.
Zhu, T.C. 1993. Grasslands of China. pp. 61–82. *In*: R.T. Coupland (ed.). Ecosystems of the World 8b: Natural Grasslands. Elsevier, Amsterdam.

9

Grassland Biodiversity in Japan: Threats, Management and Conservation

Atushi Ushimaru,[1,*] *Kei Uchida*[2] and *Takeshi Suka*[3]

Introduction

The Japanese archipelago consists of approximately 6,800 islands, which are located on the east end of the Eurasian continent. The Japanese islands stretch from the subtropical zone in the south (e.g., Naha: mean temperature 17.0°C in January, 28.9°C in July) to the subarctic zone in the north (e.g., Utoro: mean temperature –5.6°C in January, 19.0°C in August) although most areas of the islands belong to the temperate zone (e.g., Tokyo: mean temperature 5.2°C in January, 26.4°C in August) and the total land area is circa 378,000 km^2 (20°25′–45°31′ N, 122°56′–153°59′ E) and the highest elevation is 3,776 m a.s.l. Four main continental islands, Hokkaido, Honshu, Shikoku and Kyushu, cover more than 97 per cent of total land area (Fig. 9.1). The climate is influenced by the monsoon and characterized by high annual precipitation (circa 1,000–3,500 mm per year) throughout the country (Numata, 1961).

Various types of forest are climatic vegetation in most terrestrial areas of the country, whereas natural grasslands exist in very limited areas, such as alpine zones, riparian areas and coasts. Forest vegetation covers up to 67 per cent of the total land surface (Ministry of the Environment of Japan, 2016), whereas total grassland area was less than 5.5 per cent at the end of the 20th century (Fig. 9.1). Today, in Japan, there is a diverse range of natural and semi-natural forest vegetation. This includes subtropical and warm temperate broadleaved evergreen forests in the southwestern parts (Ryukyu and Satsunan islands, Kyushu, Shikoku and the western Honshu), cool temperate beech forests and secondary (*satoyama*) forests in montane and lowland zones of Honshu, Shikoku and Kyushu and

[1] Graduate School of Human Development and Environment, Kobe University, 3-11 Tsurukabuto, Kobe 657-8501, Japan.

[2] Department of General Systems Studies, University of Tokyo, 3-8-1 Komaba, Tokyo 153-8902, Japan.

[3] Nagano Environmental Conservation Research Institute, 2054-120 Kitago, Nagano 381-0075, Japan.
Emails: k.uchida023@gmail.com; suka-takeshi@pref.nagano.lg.jp

* Corresponding author: ushimaru@kobe-u.ac.jp

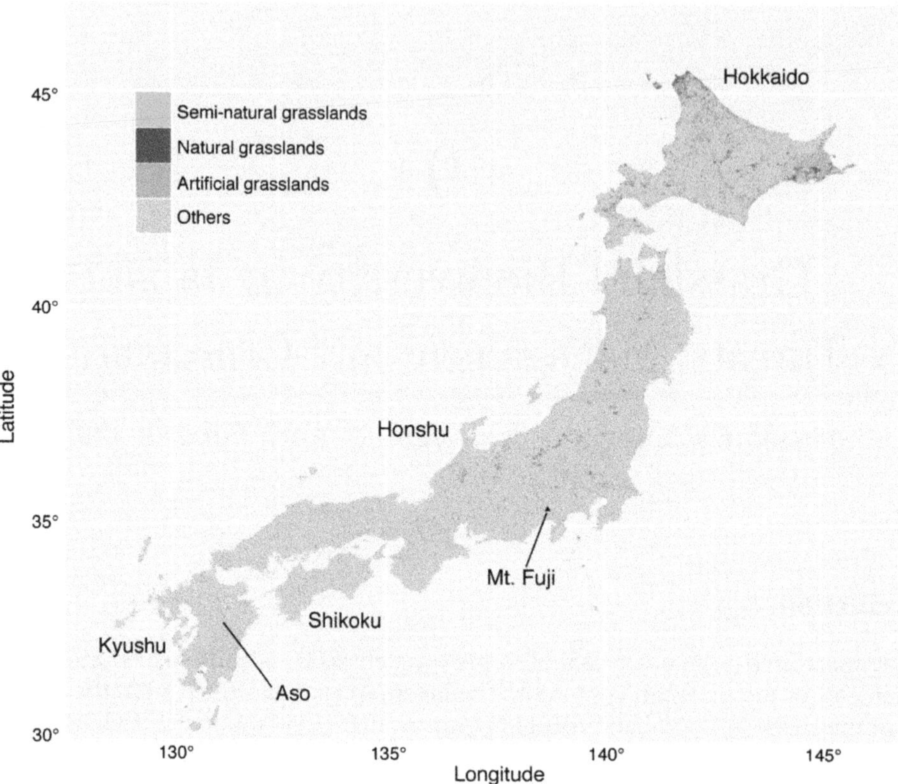

Fig. 9.1 Distribution map of Japanese grassland vegetation. Relatively small islands in the south and south-west regions are not indicated. (The map was drawn based on the vegetation survey (1988–1996) by Japanese Ministry of the Environment.)

boreal coniferous-broadleaved mixed forests and alpine vegetation in Hokkaido and at higher elevations of Honshu (Ministry of the Environment of Japan, 2016).

Grasslands, especially semi-natural grasslands have decreased in area during the last century, mainly due to land-use changes driven by socio-economic changes in Japan (see Threats to Biodiversity). After World War II, Japan has experienced drastic population migration from rural to urban (industrial) areas within the country under very rapid economic growth, based on the shift in industrial structure from agriculture to manufacturing (mainly the heavy chemical industry) during the late 1950s to the early 1970s, which resulted in a decrease in economic values of semi-natural grasslands (Ministry of the Environment of Japan, 2016). In this period, the situation caused land-use changes in semi-natural grasslands, which in turn resulted in very rapid loss and degradation of Japanese semi-natural grasslands after this period and until the 2000s. The rapid loss, fragmentation and degradation in semi-natural grasslands have caused significant grassland biodiversity declines (Ministry of the Environment of Japan, 2012) as in European countries (e.g., Krauss et al., 2004; Cousins et al., 2007), although studies that quantitatively describe grassland diversity loss are still limited in Japan (see Koyanagi and Furukawa, 2013; Nagata and Ushimaru, 2016; Uchida et al., 2016a; Koyama et al., 2017).

Here, we review studies on Japanese grasslands and overview the historical and current status of grasslands (especially of semi-natural grasslands) and the biodiversity crisis. Numata (1961, 1969) conducted intensive reviews of old Japanese literature and established a categorization and within-country distribution of grassland vegetation from ecological and husbandry points-of-view. Since then, however, no such reviews have been published in English. Moreover, in Japan the grassland biodiversity crisis, which has become obvious during the last two decades, has never been reviewed. The present paper is a review about the Japanese grasslands, reflecting recent progress toward quantifying biodiversity change in semi-natural grasslands.

Origin and Types of Grasslands

During the Holocene (the postglacial period), the temperate and humid climate in Japan would have facilitated succession of most vegetation to forests in the absence of continuous natural and/or anthropogenic disturbances. Various grassland types, however, still exist in the present land surface of Japan. In some alpine, lowland floodplain and coastal areas, harsh climates and frequent natural disturbances have maintained natural grasslands. Meanwhile, mainly in hilly and slope montane areas (and sometimes in lowland floodplain and coastal areas), human activities, such as burning, grazing and mowing, have produced and maintained several types of semi-natural grasslands (Fig. 9.2).

We refer to alpine meadows and also riparian and coastal grasslands maintained by natural disturbances as 'natural grasslands' (Fig. 9.1). Alpine meadows are distributed at higher altitude of central Honshu (> circa 2,500 m a.s.l.) and Hokkaido (> circa 500–1,500 m a.s.l.). In alpine meadows (Fig. 9.2a), glacial relict (Palaearctic and tundra) forb, grass and dwarf shrub species are dominant under very cold weather, heavy snow and wind disturbance (Kudo et al., 2010; Ministry of the Environment of Japan, 2012). In riparian zones, many wet and semi-wet grassland types, such as *Phragmites-*, *Miscanthus-* and annual plant-dominated grasslands, have established in response to frequent flooding-induced disturbances (Washitani, 2001; Osawa et al., 2010). Coastal grasslands (Fig. 9.2b) are maintained by high salinity and frequent wave and wind disturbances (Ishikawa et al., 1995) and are influenced by very infrequent but strong tsunami disturbances (Hayasaka et al., 2012). Although some riparian and coastal grasslands have been maintained by burning, grazing and mowing as semi-natural grasslands (Washitani, 2001; Tsuda et al., 2002; Obata et al., 2012), we could not distinguish them in the map (Fig. 9.1) due to limited nation-wide information. Today, natural grasslands cover only circa 6,000 km^2 (alpine meadow: circa 380 km^2; riparian grasslands: circa 1,500 km^2; coastal grasslands: circa 700 km^2; and others), that is circa 1.6 per cent of the total land area. Thus, in Japan, the natural grasslands are very narrowly distributed within the county unlike, for example, the Central Asian regions (Pfeiffer et al., 2018).

Historically, semi-natural grasslands used to be the most-widely-distributed grassland type in Japan. The Japanese black soil *kurobokudo* (Andosols in FAO Soil classification), which contains many grass phytolith and charcoal particles, indicates long historical use of semi-natural grasslands associated with fire (Kawano et al., 2012; Miyabuchi et al., 2012). *Kurobokudo* is found from southern Kyushu to eastern Hokkaido and covers circa 17 per cent of the Japanese land surface. *Kurobokudo* is broadly distributed in montane areas of Kyushu, Shikoku and Honshu (including the foot of Mt. Fuji) and lowlands of Kanto plain and the south-eastern Hokkaido (Fig. 9.3). This wide distribution in Japan contrasts to the general rarity of Andosols, the grassland soil group to which *kurobokudo* belongs and which cover less than one per cent of that in the world (IUSS Working Group WRB, 2015).

Fig. 9.2 Grassland types in Japan. (a) alpine meadow, (b) coastal grassland, (c) semi-natural pasture, (d) sown pasture, (e) *saisochi* meadow, (f) *satokusachi* meadow, (g) semi-natural grassland on ski slope, (h) a training ground of the Japan Self-Defense Force. (Photos by A. Ushimaru (all except b and f) and K. Uchida (b and f).)

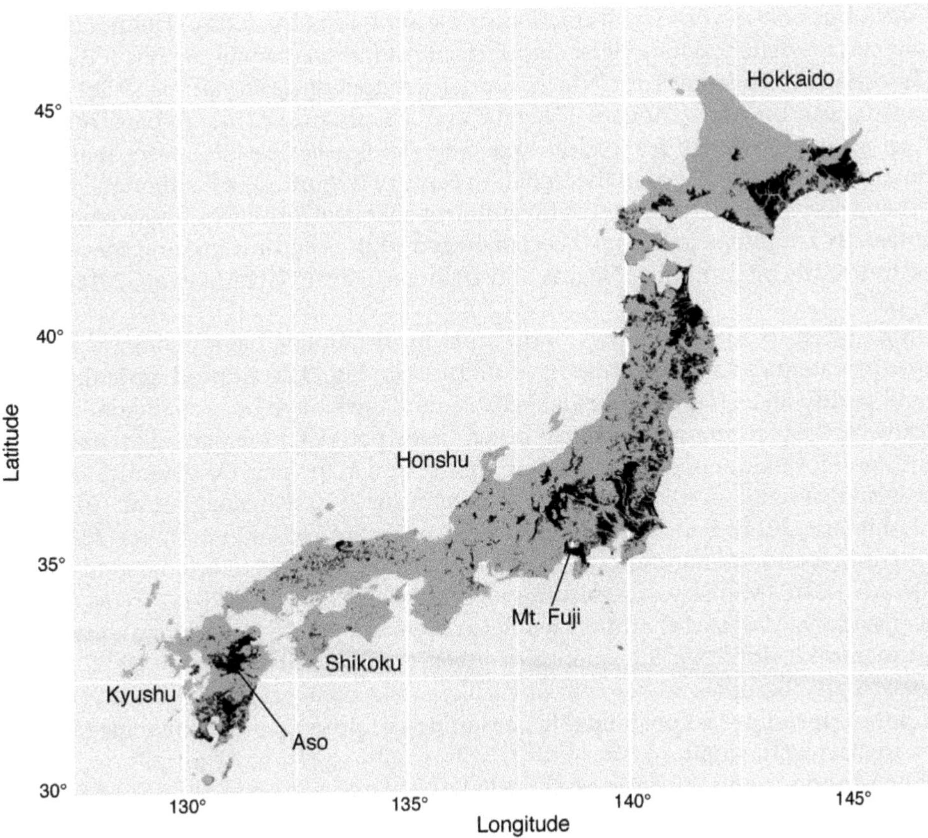

Fig. 9.3 Distribution map of the Japanese black soil *kurobokudo* at the soil surface. Relatively small islands in the south and south-west regions are not indicated. (The map was drawn, based on base soil classification by the Japanese Ministry of Land, Infrastructure and Transport website.)

At present, however, only circa one per cent (circa 3,600 km^2) of the land supports semi-natural grasslands, which can largely be categorized into pastures and meadows. The semi-natural grasslands are concentrated in the Aso region of Kyushu, around Mt. Fuji, the center and the northeastern areas of Honshu (Fig. 9.1). The distribution of past pastures and current pastures and meadows overlaps with the *kurobokudo* distribution (Watanabe, 1992; Nagata and Ushimaru, 2016; Figs. 9.1 and 9.3). Today, semi-natural pastures (*hobokuchi* or *maki*, Fig. 9.2c) are distributed narrowly within the country (such as in Toimisaki, Aso, Sanbe, Oki and Shiriya-zaki). Combing, grazing and burning are a traditional management practice for semi-natural pastures. Here, we use the term 'semi-natural pastures' to discriminate them from sown pastures, where alien grasses are anthropogenically introduced.

Japanese traditional livelihoods have maintained several meadow types for different uses (Figs. 9.2e, f). Meadows exclusively dominated by *Miscanthus sinensis* and *Phragmites australis* (*kayaba*) used to be widespread in hilly and floodplain areas, respectively. The grasses were used for thatch roofing of traditional houses and buildings (Stewart et al., 2009), but only a very small fraction of the type remains today. The honored *kayaba* maintained for thatch roofing of Ise Jingu, is a most famous remaining one today.

Traditionally, meadows for winter fodder production (*saisochi*; Fig. 9.2e) have been managed near breeders' houses (Nagata and Ushimaru, 2016). Expansive meadows producing green manure for paddy and crop fields (*shibakusachi*) were maintained in mountain sides near villages in the mid-17th century (Ogura, 2006), although almost all of these have been abandoned after the invention of chemical fertilizers. These meadow types were usually commons and have been managed with both burning and mowing by the respective local communities (Nagata and Ushimaru, 2016; Uchida et al., 2016a; Koyama et al., 2017).

In addition to such meadows, individual farm families have maintained relatively narrow private meadows (*satokusachi* or *keihansochi*; Fig. 9.2f) around agricultural fields. Around paddy and other crop fields, belts of meadows have been maintained on levees typically by frequent mowing (one to seven times per year) for pathogen/insect control recently and for fodder and green manure production in the past (Matsumura and Takeda, 2010; Uematsu et al., 2010; Uematsu and Ushimaru, 2013; Koyanagi et al., 2014; Uchida and Ushimaru, 2014). Paddy and other fields cover circa 68,000 km^2 (circa 18.4 per cent; Fig. 9.4) and total semi-natural grassland areas around them were estimated to cover circa 1,850 km^2 in 2010 (Ministry of Agriculture, Forestry and Fisheries, 2016). In tea production areas, meadows (*chagusaba*) around tea fields are managed by annual mowing to supply green manure to improve the quality of green tea (Kusumoto and Inagaki, 2016). The chagusaba area occupies 39 per cent of each tea field on average. Thus, the accumulated area of these meadows is considered to be significant, although individual meadows of the types are narrow or small.

Furthermore, semi-natural grasslands have been maintained as ski and golf areas since the early 20th century. Ski areas are usually maintained by annual mowing like meadows (Fig. 9.2g). Some ski meadows have been established on previously pasture-use lands (Y. Yaida, A. Ushimaru, K. Uchida et al., unpublished data) while others have been newly created by forest clear-cutting during the mid-20th (Tsuyuzaki, 1995). Although golf courses have a low plant diversity due to the artificial introduction of grass species (e.g., *Zoysia* spp.), the oldest golf course of the country, Kobe Golf Club in Kobe City, Hyogo Prefecture, has maintained diverse semi-natural grassland plant species just besides the courses due to the mowing management (A. Ushimaru, unpubl. observation).

Japanese semi-natural pastures and meadows used to harbor many plant and animal species, which are distributed or whose close relatives live in the Korean peninsula and meadow steppe of northeastern China (Kadota, 2011; Uchida et al., 2016b). These species are therefore called 'continental-grassland relicts', which are suggested to have expanded their distribution during the glacial periods likely via the Korean peninsula (Murata, 1988). Although the warm and wet conditions of the Holocene have been suitable for forest expansion in Japan, expansive semi-natural grassland use by humans may likely have provided grassland habitat conditions for them like in East-European semi-natural grasslands (Suka et al., 2012; Török et al., 2018). Dominant species are known to change between pastures and meadows as well as with climatic conditions (altitude): *Miscanthus sinensis* and *Pleioblastus* spp. dominate in warm-temperate (southern) pastures and *Zoysia japonica* dominates in cool-temperate (central) pastures. *Pleioblastus* spp., *Miscanthus sinensis* and *Sasa* spp. are dominant in southern, central and northern meadows, respectively (Numata, 1969).

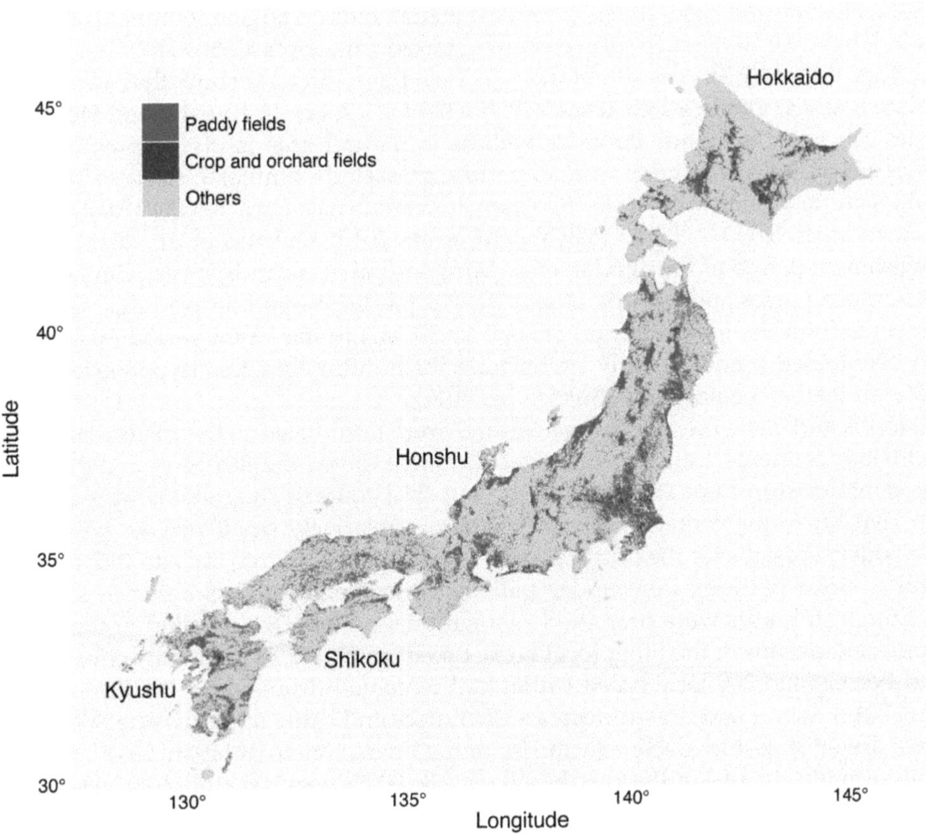

Fig. 9.4 Distribution map of agricultural lands in Japan. Relatively small islands in the south and south-west regions are not indicated. (The map was drawn based on the vegetation survey (1988–1996) by the Japanese Ministry of the Environment.)

Sown pastures (Fig. 9.2d), where alien plants have been planted or seeded, are the most frequent grassland type nowadays (circa 10,500 km^2, circa 2.9 per cent of the land surface) and distributed predominately in Hokkaido, but rarely in other areas (Fig. 9.1). Competition against native species and climatic and soil conditions might have prevented the establishment and spread of introduced forage crops from Europe and North America in old pastures and meadows other than those in Hokkaido, whose climate is suitable for these northern-type introduced crops (Numata, 1961). *Poa pratensis* is the dominant species in sown pastures in Hokkaido (Numata, 1969). The contribution of native species to biodiversity in sown pastures becomes much lower as management intensity increases (Japanese Society of Grassland Science, 2010).

Historical Transition in Agronomic Use of Semi-natural Grasslands

The origin of semi-natural grasslands dates back to the late Pleistocene or early Holocene in Japan. *Kurobokudo* started to form after this period (Kawano et al., 2012; Miyabuchi et al., 2012; Hosono and Sase, 2015). In the Aso region, today's largest semi-natural grassland area in Japan, the *kurobokudo* formation has started from circa 13,500 BP (Miyabuchi et al., 2012; Figs. 9.1 and 9.3). At a site of the Aso area, phytolith data show that *Miscanthus* has displaced *Sasa* since circa 13,500 BP (Miyabuchi et al., 2012). In Kyushu and Honshu, *Sasa* species can grow on forest floors as well as in natural grasslands, whereas *Miscanthus* species are typically dominant in semi-natural grasslands. Similarly, microscopic charcoal in lake sediments and soils in the Aso area appeared since circa 10,000–13,000 BP and has increased since 5,000–7,000 BP (Miyabuchi et al., 2012; Kawano et al., 2012). Thus, the establishment period of the first Japanese semi-natural grasslands is very similar to that of east European grasslands (Török et al., 2018). It has been hypothesized that semi-natural grasslands from the Jomon period (13,000 BC–300 BC) to the Yayoi period (AD ~250) have been maintained predominantly by burning for hunting, but this hypothesis has rarely been examined archeologically (Suka et al., 2012).

Horses and cattle (i.e., pasturage culture) were introduced to the main islands during the 3rd–5th centuries from the Eurasian continent. From the 8th to mid-19th centuries, large public pastures under control of the Imperial court or shogunates were distributed from Kyushu to the center of Honshu, whereas relatively small private pastures were more widely spread over the country, apart from Hokkaido. From the late 19th to mid-20th century, public pastures were under military control and hybrids between local breeds and European horses were promoted as a national policy. This resulted in the extinction of some and decline of the other local horse breeds in Japan. In 2006, only circa 2,000 local breed horses (and 200 local breed cattle) had remained (Ministry of the Environment of Japan, 2016). Rapid and drastic decreases in horses and cattle on farms after World War II were a driver of pasture decline (Ministry of the Environment of Japan, 2016). Husbandry of the currently circa 4.4 million cattle and circa 86,000 horses is mainly conducted on sown and partly on semi-natural pastures; Hokkaido, northeastern Honshu and Aso region are the main areas of current husbandry.

Although historical transition in distributions of *saisochi* (Fig. 9.2e) and *shibakusachi* meadows is largely unknown, it should be associated with the development of cultivated fields. Fertilizing rice paddy field with grass-compost (green manure), *karishiki*, was carried out already in the 8th century (Arioka, 2004). The meadows had expanded around the 17th century when the paddy field area and national population rapidly increased (Ogura, 2006; Mizumoto, 2015). Utilizing *karishiki* (green manure) was a major manuring method in that century. In the 18th century, continuous development of new paddy fields encroached even on the meadows and the conversions and overuse of the meadows caused land erosion, frequent floods and landslide disasters (Mizumoto, 2015). At the end of the 19th century, enormous areas of meadow were still maintained for producing fodder and green manure (Ogura, 2006).

After the late 19th century, however, mainly due to land abandonment and conversion to crop lands and plantations, the semi-natural grasslands have decreased to only one to three per cent of the national land area and were heavily fragmented (Ogura, 2006; Suka et al., 2012; Fig. 9.5). Some pastures and meadows are still active for husbandry or for sightseeing and tourism (Japanese Society of Grassland Science, 2010), but many remaining semi-natural grasslands will be abandoned (Nagata and Ushimaru, 2016; Uchida et al., 2016a; Koyama et al., 2017). Only *satokusachi* and *chagusaba* have continued to be maintained when paddy and tea fields are active, although total *satokusachi* are continuously decreasing, in consequence of agricultural field declines (Fig. 9.6).

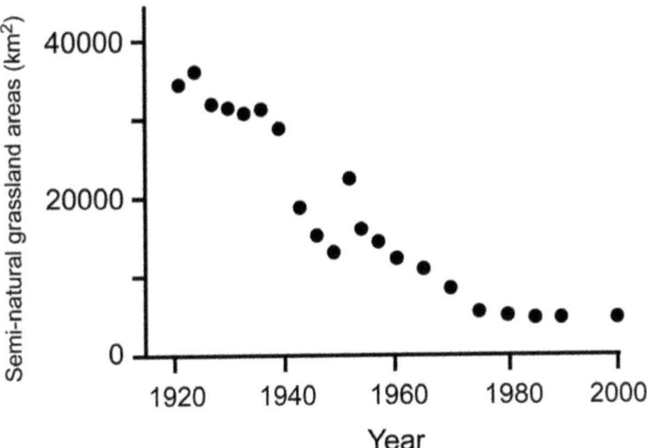

Fig. 9.5 Change in total semi-natural grassland area during the last century. (The figure was modified from Ogura, 2006.)

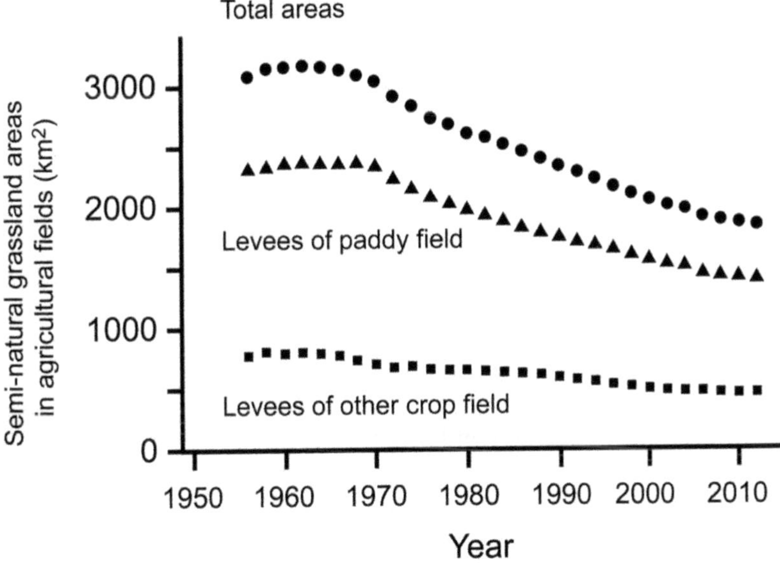

Fig. 9.6 Changes in total area of semi-natural grasslands around agricultural fields during the past 60 years. (This figure was drawn based on data of the Ministry of Agriculture, Forestry and Fisheries data, 2013.)

Value of Grasslands

Direct Economic Value

Even today, materials from semi-natural grasslands in Japan are producing provisioning services (one form of ecosystem services) for some economic utilizations, i.e., meat (mainly beef cattle), green manure for farm products and construction materials for thatched roofs. Although these products have only insignificant importance for the whole Japanese economy at present, utilizing the traditional materials adds high value to the products. In

Fig. 9.7 Provisioning service of Japanese semi-natural grasslands: Thatched roof of Tsuchi-no-miya, one of 125 Shinto shrines of Ise Jingu. *Miscanthus sinensis* plants from the honored *kayaba* privately managed by Ise Jingu were used for the roof. (Photo provided by Ise-Jingu.)

the Aso area, beef cattle pastured in semi-natural grasslands as well as some vegetables fertilized with green manure are certificated by local communities and are welcomed by supporting consumers (Japanese Society of Grassland Science, 2010). In Shizuoka prefecture of Honshu, tea fertilized with green manure from *chagusaba* have a reputation for high quality. Agro-cultural activities in both the area were identified as practices in the Globally Important Agricultural Heritage systems (GIAHS) by FAO in 2013 (Kusumoto and Inagaki, 2016). Thatched roofs are one of the typical components of the Japanese traditional architecture, e.g., the historic villages of Shirakawa-go and Gokayama, a World Heritage site, and Ise Jingu (Fig. 9.7). Builders holding traditional skills utilize plants from *kayaba* for thatching.

Regulating and Cultural Services

The semi-natural grasslands in Japan produce regulating and cultural services (other forms of ecosystem services) as well. The *kurobokudo* holds the highest percentage of carbon storage among various soil types, implying a high efficiency of carbon dioxide fixation that contributes to the climate regulation (Toma et al., 2013). In addition, sceneries and flowers of semi-natural grasslands have fostered a great variety of the Japanese traditional culture, e.g., old literature, Japanese traditional events, *ikebana* (the traditional flower arrangement), *sado* (the tea ceremony), *ukiyoe* (the traditional wood block printing) and traditional landscapes (as depicted in *ukiyoe*) (Fig. 9.8). The plants, collectively called *aki no nana-kusa* (the seven flowers of autumn, *Lespedeza* spp., *Miscanthus sinensis, Pueraria lobata, Dianthus superbus* var. *longicalycinus, Patrinia scabiosifolia, Eupatorium japonicum, Platycodon grandiflorus*), have received popular appreciation in classic literature and traditional

Fig. 9.8 Cultural services of Japanese semi-natural grasslands: (a) semi-natural pasture and (b) meadow in *ukiyoe* (*Fuji-sanjurokkei*, 36 views of Mt. Fuji) by Hiroshige Utagawa and (c) a flower arrangement (*bonbana*) using grassland herbs at the Bon Festival (the traditional Buddhist festival). Native horses and orange flowers of their unfavorite (toxic) grassland shrub species, *Rhododendron molle* subsp. *japonicum* were depicted. *Aki no nana-kusa*, *Miscanthus sinensis* (tall grass), *Patrinia scabiosifolia* (yellow flowers), *P. grandiflorus* (blue flowers) and/or *P. lobata* (vine) are found in both *ukiyoe* (b) and *bonbana* (c). (Utagawa Hiroshige's *Ukiyoes* kindly provided by Hiroshige Museum of Art, Ena; photo by Y. Urayama.)

events (Fig. 9.8). They all together appear in a famous poem of the 7th–8th century in *Manyoshu*, the oldest anthology (collection) of the Japanese traditional poems. They are typical species of semi-natural grasslands in Japan. The people have used these plants in traditional ceremonies and medicines. Grassland flowers, including *aki no nana-kusa*, have been arranged in *ikebana* and have been displayed in scenes of the tea ceremony and the Bon Festival (Fig. 9.8). Nowadays, graceful sceneries of vast semi-natural grasslands, e.g., Aso, Akiyoshi-dai, Kirigamine, as well as natural alpine meadows on high mountains in Chubu and Hokkaido, attract many tourists and mountaineers.

Grasslands as Biodiversity Hotspots in Japan

Both natural and semi-natural grasslands are important ecosystems for biodiversity conservation, but in different aspects. Japan harbors more than 1,900 endemic vascular plant species and is listed as a global biodiversity hotspot (Mittermeier et al., 2005). Within the country, several alpine meadows of Honshu and Hokkaido (Mt. Apoi, Mt. Yubari, South Alps and Mt. Yatsugatake) are regional hotspots of endemic alpine (glacial relict) herb species, which have been differentiated from their continental ancestors (Ebihara, 2011).

In contrast, Japanese semi-natural grasslands exhibit a lower endemism in plant species composition because they share the same species with grasslands of the Korean Peninsula and meadow steppes of northeastern China (Kadota, 2011). However, semi-natural grasslands harbor many plant and animal species, which are now at risk of extinction (Koyanagi and Furukawa, 2013; Uematsu and Ushimaru, 2013; Nakahama et al., 2016; Uchida et al., 2016a,b,c; Koyama et al., 2017). Aso grasslands are a regional hotspot in terms of endangered species distribution in Japan (Fig. 1 in Kadoya et al., 2014). Semi-natural grasslands around Mt. Fuji and Nagano prefecture are also local biodiversity hotspots, which harbor numerous endangered species (Kitahara and Sei, 2001; Kubo et al., 2009; Ministry of the Environment of Japan, 2014; Nagata and Ushimaru, 2016; Uchida et al., 2016a). *Satokusachi* and *chagusaba* meadows have maintained a high diversity of endangered species, too (Kusumoto and Inagaki, 2016; Uchida et al., 2016c). Riparian gravel bar grasslands have many river-side specialist and endangered species, such as *Potentilla chinensis*, the endemic *Anaphalis margaritacea* subsp. *yedoensis*, *Artemisia capillaris*, *Eusphingonotus japonicus* and *Omophron aequalis* (Washitani, 2001; Yoshioka et al., 2010; Ministry of the Environment of Japan, 2014). Moreover, coastal grassland vegetation also exhibits low endemism but high endangered plant diversity: 23 per cent of the specialist species of these vegetation types are listed in the national red list (Sawada et al., 2006). Thus, the Japanese grasslands are regional (or local) hotspots of endemic or endangered species.

Threats to Grasslands

Decrease in Grassland Area during the Last Century

Over the past 80 years (circa 1920–2000), semi-natural grasslands have rapidly declined throughout Japan (Ogura, 2006; Fig. 9.5), severely limiting the habitats of many grassland species. Approximately 38,000 km^2 of semi-natural grasslands existed at the beginning of the 20th century (circa 10.1 per cent of Japan's total lands), but only circa 4,300 km^2 grassland (circa 1.1 per cent of total land area) remained in 2000 (Ogura, 2006; Fig. 9.5). By 2010, this number has declined to 3,800 km^2 (Fig. 9.5; Ministry of Agriculture, Forestry and

Fisheries, 2013). Based on our own estimation using vegetation survey data, the total area of semi-natural grasslands is circa 3,600 km^2 (circa one per cent of the total land surface) (Fig. 9.1). Although estimates differ between data sources, there is agreement that only a small fraction of the grasslands has remained till today. Additionally, after World War II, meadows around paddy and crop fields in Japan rapidly declined. According to Ministry of Agriculture, Forestry and Fisheries of Japan (2016), *satokusachi* meadows of 3,090 km^2 were maintained around the agricultural fields in 1956, but they decreased to 1,850 km^2 in 2010 (Fig. 9.6).

The flora of alpine grasslands is mostly protected within national parks under the law administrated by the Ministry of the Environment of Japan. Although these grassland areas have been protected from anthropogenic impacts in general, the expected climate changes may decrease the grasslands, as well as other alpine vegetation in the future (Hotta et al., 2014). In Mt. Apoi-Dake, a recent decrease in snowfall is reported to cause a decline in alpine meadow area by promoting a rapid expansion of Japanese stone pine (*Pinus pumila*) to higher altitude (Ministry of the Environment of Japan, 2012). Reduced snowfall is also a potential driver of the degradation of alpine meadows by facilitating a Shika deer (*Cervus nippon*) invasion to the vegetation (Takatsuki, 2009; Ministry of the Environment of Japan, 2012). These factors will limit the distribution of alpine meadows in the future. Predicted significant rise of annual mean temperature and average annual precipitation until the end of the 21th century in Japan may raise the velocity of succession towards woody vegetation in both natural and semi-natural grasslands further.

Riparian grasslands have undergone rapid changes to other vegetation types, such as riparian forests in recent decades (Nakamura et al., 2016). The Japanese government has constructed large dams and consolidated river banks because the society often faces natural disasters by flooding after typhoon events (cf. Osawa et al., 2011). These artificial constructions heavily reduce river disturbance events, which prevented successions from grasslands to forest vegetation in riparian zones in the past (Washitani, 2001; Nakamura et al., 2016). Although coastal grasslands harbor many endangered species that rapidly declined during recent decades (Sawada et al., 2006), 41 per cent of Japan's coastlines have been 'developed' artificially with concrete to counteract tsunamis or big waves associated with typhoon-induced surges.

Overall, areas of natural and semi-natural grasslands are rapidly lost due to human impact, except for those of protected alpine meadows. As a consequence of the widespread loss and fragmentation of natural and semi-natural grasslands, grassland plant and herbivorous insect diversity has declined drastically (Suka et al., 2012; Koyanagi and Furukawa, 2013).

Effects of Land Use Changes in Semi-natural Grasslands

Conversion of semi-natural ecosystems to other land uses is a crucial issue for maintaining future biodiversity worldwide (Foley et al., 2005). Semi-natural grassland landscapes have experienced major land-use changes in recent decades (Suka et al., 2012; Uchida and Ushimaru, 2014; Nagata and Ushimaru, 2016; Normile, 2016) as in European (Dengler et al., 2014; Török et al., 2018) and American countries (Queiroz et al., 2014). Major drivers of semi-natural grassland area loss in Japan have been abandonment, intensified use, afforestation, conversion to arable fields and urban development. After World War II, the energy and green revolutions in rural areas promoted abandonment of husbandry and green manure production, subsequently leading to conversion of grasslands into other land uses (Ministry of the Environment of Japan, 2012). Depopulation and aging farmers in rural areas have further accelerated this trend (Uematsu et al., 2010; Koyanagi and Furukawa, 2013; Normile, 2016; Osawa et al., 2016).

Land abandonment, which promotes subsequent vegetation succession from grasslands and fields to secondary forests, is the most influential driver of decline in grassland specialist species in Japan (Suka et al., 2012; Nagata and Ushimaru, 2016; Uchida et al., 2016a). Huge semi-natural grassland areas have changed to secondary forests, which are often dominated by *Betula platyphylla* and/or shrubs (Osumi, 2005), although no reliable statistics are available on how much grasslands have been abandoned to date (Ministry of Environment of Japan, 2016). Moreover, abandonment has been increasing in paddy fields with low accessibility and productivity (Uematsu et al., 2010; Osawa et al., 2016). *Satokusachi* meadows on such fields are usually oligotrophic and harbor many endangered species, which have high conservation values (Uematsu et al., 2010; Uematsu and Ushimaru, 2013; Osawa et al., 2016; Uchida et al., 2016c). Paddy field abandonment has caused a decline not only in endangered species but also in whole grassland species diversity around paddy fields (Fukamachi et al., 2005; Uematsu et al., 2010; Uchida and Ushimaru, 2014). Although numerous studies on biodiversity conservation in abandoned semi-natural grasslands have been conducted in Europe and North America, similar research in east Asia is very scarce (see Queiroz et al., 2014; Normile, 2016). In European countries, rewilding of abandoned grasslands and agricultural lands is sometimes considered to restore forest ecosystems (Navarro and Pereira, 2012). However, in Japan with its already large forest fraction, a further increase would not increase total species richness (Normile, 2016). Instead one should study abandoned east Asian semi-natural grasslands, which have received little attention, as the patterns and consequences of biodiversity losses may differ from region to region (Uematsu and Ushimaru, 2013; Queiroz et al., 2014; Uchida and Ushimaru, 2014; Normile, 2016).

The effects of intensified use (i.e., land consolidation and nitrogen input) are conspicuous in *satokusachi* meadows around paddy fields in Japan (Matsumura and Takeda, 2010; Uematsu et al., 2010; Uchida and Ushimaru, 2014). More than 75 per cent of paddy fields are now consolidated (Uematsu and Ushimaru, 2013), which largely reduced plant and herbivorous insect diversity via changing mowing management frequency and characteristics of the surrounding landscape (Uchida and Ushimaru, 2014). Paddy consolidation, which converts small, irregular and poorly drained paddy fields into large, quadrangular, well-drained fields to improve productivity and to allow mechanized farming, rarely leads to grasslands that resemble the original ones in composition and structure (Matsumura and Takeda, 2010; Uematsu et al., 2010). Nutrient input and neutralization of the formerly acidic soil in consolidated paddy fields can cause declines in rare plant species (Uematsu and Ushimaru, 2013) and increase of invasive alien plants, such as *Andropogon virginicus*, *Erigeron annuus* and *Solidago altissima* (Hiradate et al., 2008). These alien species are listed in the 100 worst invasive alien species in Japan (Ecological Society of Japan, 2002) and/or in the invasive alien species watch (Ministry of the Environment of Japan, 2009) because of their negative impacts on ecosystems and human activities.

Although nitrogen input is known to cause biodiversity loss in European semi-natural grasslands (Benton et al., 2003; Kleijn et al., 2009; Dengler and Tischew, 2018; Török et al., 2018), the effects on grassland biodiversity have not been conspicuous in Japanese semi-natural grasslands. Semi-natural pastures and meadows are usually established on *kurobokudo*, which typically consists of volcanic ash and humus and is very acidic ($pH < 5.7$) and oligotrophic (< 200 mg P_2O_5/kg) (Hiradate et al., 2008; Nagata and Ushimaru, 2016;

Koyama et al., 2017). Numata (1961) reported that manuring had always failed to effectively improve productivity of Japanese pastures due to the oligotrophic soil conditions; it has been gradually discouraged.

For areas promoting intensified livestock farming, farmers improved pastures by introducing alien grass and herb species (Numata, 1961). They eliminated dominant native plant species (e.g., *Pleioblastus* spp., *Pteridium aquilinum* and *Miscanthus sinensis*) and introduced non-native species (e.g., *Trifolium* spp., *Dactylis glomerata*, *Phleum pratense* and *Poa pratensis*) for increasing annual productivity (plant biomass) for livestock growing during the 1970s to 2000s (Ohara, 1969). For example, large semi-natural grassland areas in Aso were converted to sown pastures by the national grassland improvement project. Until the 1960s, huge semi-natural grassland areas of eastern Hokkaido were 'improved' into sown pastures, dominated by non-native grass *P. pratensis* from Europe (Numata, 1961; Figs. 9.1 and 9.2). K. Uchida et al. (unpublished data) recently examined plant and butterfly diversity in sown pastures and found the diversity was much lower than other semi-natural grassland types. Alien crops have often naturalized in semi-natural grasslands nearby sown pastures (Tsuda et al., 2002), although abundance and richness of these plants are generally low in unimproved semi-natural grasslands other than those on Hokkaido due to acidic and oligotrophic soils and warm climates (Hiradate et al., 2008; Nagata and Ushimaru, 2016). In some sown pastures, farmers repeatedly apply fertilizers to improve growth of alien plant biomass (unpubl. interview with a farmer in Gifu Senmachi).

As stated earlier, the Tokugawa shogunate and the Meiji governments promoted new paddy and vegetable field development projects on semi-natural and riparian grasslands during the 17th–19th centuries (Mizumoto, 2015). Figures 9.3 and 9.4 show that large areas of agricultural fields have been established on *kurobokudo* in southern Kyushu, central Honshu and eastern Hokkaido, supporting the historical land use changes. Some grassland species have stayed in *satokusachi* meadows around the developed paddy fields (Suka et al., 2012), but this large habitat loss likely has diminished grassland biodiversity.

Moreover, many semi-natural grassland areas have been converted to conifer plantations for post-World War II rebuilding since the 1950s. Total plantation area drastically increased from circa 49,700 km^2 in 1952 to circa 1,02,900 km^2 in 2012 according to various data from the Ministry of Agriculture, Forestry and Fisheries. However, there are no reliable data on how much semi-natural grasslands have been converted to plantations in Japan. This is because plantations have increased also via cutting natural and secondary forests.

Rapid urbanization since the 1960s have caused drastic decreases in lowland paddy and vegetable fields and riparian floodplain grasslands (Washitani, 2001; Tsuji et al., 2011), while it had almost no impact on semi-natural grasslands in montane zones. Current assessments of the effects of urbanization on biodiversity of *satokusachi* meadows indicate significant negative impacts on grassland plants and animals via habitat loss and degradation (Tsuji et al., 2011; H. Fujimoto, K. Uchida and A. Ushimaru, unpubl. data). Due to fixed river channels, flood-defence constructions and frequent introduction of invasive alien plant species, riparian grasslands, which used to be widely distributed near urban areas, have decreased in area and become degraded, leading to loss of riparian grassland species (Washitani, 2001; Osawa et al., 2011).

Management Type Changes

Management type changes in remaining semi-natural grasslands have often occurred during recent decades due to labor-saving and/or loss of local traditional knowledge, causing grassland biodiversity loss (Suka et al., 2012; Nagata and Ushimaru, 2016; Uchida et al., 2016a). For example, in Kiso area, a central part of Honshu, the meadows for fodder production had been maintained sustainably by the traditional practice of combining both burning and mowing every second year for at least 300 years. However, the traditional practice has largely been abandoned and replaced by a new management practice. Annual burning is currently prevalent, resulting in negative impacts on biodiversity (Nagata and Ushimaru, 2016; Uchida et al., 2016a). Nakahama et al. (2016) suggested that recent changes in management (mowing) timing were responsible for decline in reproductive success and genetic diversity of the endangered grassland plant species, *Vincetoxicum pycnostelma* in *saisochi* and *satokusachi* meadows.

Conservation of Grasslands

Biodiversity of natural and semi-natural grasslands is now threatened by different drivers as described above. Accordingly, there are a variety of conservation activities to meet differentiated necessities for biodiversity maintenance. Most alpine meadows are currently protected in national parks, so the national (Ministry of the Environment of Japan) and local governments control their conservation planning, while researchers contribute to it. As for riparian and coastal grasslands, the national (Ministry of Land, Infrastructure, Transport and Tourism of Japan and Ministry of the Environment of Japan) and local governments manage their conservation with consultation of academic societies as well. In contrast, most semi-natural grasslands and agricultural fields are private lands or local commons, so their conservation has been conducted by many different stakeholders, including governments, citizens, local committees, non-profit organizations (NPOs) and scientists (for details, see below).

Legal Conservation Status

The principal parts of the alpine natural meadows and several expanses of semi-natural grasslands, such as Aso, Akiyoshi-dai, Hiruzen, Kirigamine and Koshimizu, are conserved as parts of national and quasi-national parks in Japan. Particularly, the core areas of the alpine meadows are protected strictly in most cases. Feeding damage on rare plant species by sika deer, however, has become serious recently in some areas, including the alpine meadows in the south, Alps and the semi-natural grasslands in Kirigamine. Therefore, deer fences have been established to protect the vegetation in these areas. Besides, several endangered species of semi-natural grassland are protected under the Law for the Conservation of Endangered Species of Wild Fauna and Flora and similar local regulations. For example, a small area of semi-natural grassland, which was maintained by a traditional agrarian activity on the Kaida Plateau, Nagano prefecture, is conserved as a reserve for some endangered species of plants and insects by the local government of the prefecture (Nagata and Ushimaru, 2016; Uchida et al., 2016c).

Other Formal Instruments of Grassland Conservation

For the conservation of semi-natural grasslands, not only regulatory measures but also support measures should play an important role under the present socio-economic situation. Some formal instruments are functioning, although they are insufficient for implementation of grassland conservation required. As for an international approach, identification of activities in the Aso area and in the tea plantations with *chagusba* as the practices in GIAHS by FAO encourages the commitments of local communities. As for a national and local approach, ASO Grassland Restoration Committee (AGRC) has been established under the Law for the Promotion of Nature Restoration and is improving local activities for grassland conservation and restoration. The Committee is engaging in promotions of sustainable use of semi-natural grasslands, biodiversity conservation, environmental education, adding of values to grassland resources, ecotourism and sustainable land use and is soliciting contributions to their support projects. Payment schemes in general for conservation of natural environments to farmers by national and local governments exist, although they are not focused on grasslands and are operating insufficiently for grassland conservation. Apart from the conservation purpose, several training grounds of the Japan Self-Defence Forces retain vast areas of semi-natural grasslands that are known as endangered species habitats.

Management and Restoration of High Nature Value Grasslands

Recently, people have realized the value of semi-natural grasslands in the light of low-cost resources for modern husbandry and biodiversity, landscape and traditional culture conservation (Japanese Society of Grassland Science, 2010). This promotes many public and private activities for management and restoration of semi-natural grasslands within the country (Japanese Society of Grassland Science, 2010). For example, the Law for the Promotion of Nature Restoration was enacted in 2002, the AGRC, which is the largest grassland restoration committee in Japan established by numerous stakeholders (i.e., local landowner associations, citizens, NPOs, non-governmental organizations [NGOs], companies, the national and local governments and researchers) for sustainable maintenance of Aso grasslands in 2005. The AGRC not only promotes maintenance and reintroduction of traditional management practices by volunteers in wider areas within the Aso region but also conducts research for adaptive managements, builds new agro-economic systems and education methods for younger generation and establishes funds for grassland conservation and AGRC activities (AGRC, 2017). Although such huge committees are rare, there are many local grassland restoration activities throughout the country (Japanese Society of Grassland Science, 2010).

Abandoned or unmanaged semi-natural grasslands have usually been restored by reintroduction of burning, grazing and/or mowing. In abandoned Aso grasslands, reintroduction of traditional management that combines spring burning and autumn mowing has successfully restored plant diversity (Koyama et al., 2017). The endangered grassland butterfly species, *Shijimiaeoides divinus asonis*, inhabits only the areas of Aso grasslands burned in every spring (Murata et al., 1998). Reintroduction of another subspecies of the same butterfly, *S. d. barine*, to a semi-natural grassland in Nagano prefecture, central Japan, has succeeded by adopting spring burning because it reduces the generalist egg parasitoid, *Trichogramma chilonis* (Koda and Nakamura, 2010). Because managers should be skilled and experienced to protect people and ecosystems from uncontrolled fire, burning is often not easy to apply to grassland restoration (Japanese Society of Grassland Science, 2010; Valkó et al., 2014). Thus, burning is an unfamiliar restoration measure in European countries, but it is effective together with use of mowing for grassland managements in Japan and North America (Valkó et al., 2014; Nagata and Ushimaru, 2016; Koyama et al., 2017).

In abandoned pastures of Mt. Sanbe, reintroduced extensive cattle grazing enhanced plant-height heterogeneity within the pasture due to the animal preference, leading to increase in autumn-blooming plants and in population size of an endangered plant species, *Pulsatilla cernua* (Japanese Society of Grassland Science, 2010). In contrast, over-grazing caused population decline of this and another endangered species, *Swertia pseudochinensis*. Thus, controlled grazing intensity is essential for grassland restoration. With the recent development of a mobile electronic fence and rental cattle system in Japan, restoration by grazing will become easier (Japanese Society of Grassland Science, 2010).

Reintroduction of mowing is also effective in restoration of Japanese semi-natural grasslands. In the protected natural monument area of Nishiura, Hofu City, Yamaguchi prefecture, removal of the tall grass species, *Miscanthus sinensis* and *Pleioblastus chino* var. *viridis* by reintroduced mowing increased the population of endangered *Iris rossii* (Japanese Society of Grassland Science, 2010). Mowing twice a year has most effectively facilitated the population growth (Japanese Society of Grassland Science, 2010). Restoration by mowing would be suitable for management of narrow semi-natural grasslands, such as *satokusachi* meadows because mowing is more labor-intensive than burning and grazing (Uchida and Ushimaru, 2014; Nagata and Ushimaru, 2016; Nakahama et al., 2016). In Kobe City, mowing has been reintroduced by the Kobe City goverment and researchers on abandoned *satokusachi* for restoration of grassland plant species including a highly endangered species, *Syneilesis tagawae* (cf. Normile, 2016). The project revealed that grassland plant richness increased through reintroduced mowing for two years but only reached half the level of traditionally managed *satokusachi*, indicating that complete restoration will take more time (T. Nagai, K. Uchida and A. Ushimaru, unpubl. data).

Yet, only a handful of researchers have examined the effects of reintroduction of burning, grazing and/mowing on restoration of abandoned or degraded semi-natural grasslands. Therefore, no clear recommendations can be given currently. For example, in Koshimizu Nature Reserve, Hokkaido an experimentally introduced combination of burning and grazing management enhanced the dominance of some alien plants, suggesting the importance of continuous vegetation monitoring and adaptive management strategy (Tsuda et al., 2002). Thus, more studies on grassland restoration are needed to accumulate knowledge of grassland management in the modern ways.

Résumé and Future Prospects

Under the temperate humid climate of the Holocene in Japan, most vegetation potentially would change into forests by ecological succession. Harsh environments, natural disturbances and or human activities, however, have sustained natural and semi-natural grasslands until today. The grassland types vary, depending on locations and management forms, and the aims of using semi-natural grasslands have changed throughout history. Although semi-natural grasslands used to be the most-widely-distributed grassland type, these expanses have shrunk to circa one per cent of the land of Japan during the 20th century following the drastic socio-economic change, particularly owing to the reduction of the grassland resource uses. The internal migration from rural to urban areas under the rapid economic growth and the shift from agriculture to manufacturing (late 1950s to early 1970s) have largely caused this, while the predicted population decline as well as the climate change in the near future may exacerbate the situation. The appreciation of the value of Japanese grasslands, however, as the refuge of a large number of endemic and endangered species and the source of a great variety of traditional cultures has grown in the last decade. Besides, the inclination to rural return among urban residents is rising in recent years and a consumption trend change from 'owing products' to 'using services' can be noted. Taking into account these recent socio-cultural changes, in the effort to achieve realization of the sustainable use of ecosystem services, versatile practices for conservation and management of high value grasslands must be operationalized in Japan. Additionally, flora and fauna of Japanese semi-natural grasslands are similar to those of the Korean peninsula and meadow steppes of the northeastern China and the Russian Far East. In order to elucidate the biogeographical relationship and to compare the historical transition between grasslands of Japan and of the neighboring countries and to conserve these grassland species within the region effectively, intensive exchanges of experience, information and research within the region should be reinforced.

Acknowledgement

We thank Jürgen Dengler for inviting us to write the review and giving us constructive advice and Makihiko Ikegami for helping draw the maps. This work was supported by grants-in-aid for A. Ushimaru (Nos. 20770015, 23570024, 15K12257, 17K07557 and 16H02993) from the Japan Science Society for the Promotion of Science.

Glossary of Japanese Grassland Terms

Chagusaba = Meadows maintained by annual mowing, specially for green tea production.

Houbokuchi (or *maki*) = Pastures that have been maintained by grazing and annual burning.

Karishiki = Green manure produced in *shibakusachi. Kari* and *shiki* mean 'mowing plants' and 'spreading to fields', respectively.

Kayaba = *Miscanthus* or *Phragmites* dominated meadows that have been maintained by mowing and annual burning for thatch roofing.

Kurobokudo = Japanese black soil, which indicates a long history of semi-natural grassland vegetation.

Saisochi = Meadows managed by mowing and burning for winter fodder production.

Satokusachi (or *keihansochi*) = Meadow strips around paddy and vegetable fields. They have usually been managed by frequent mowing.

Shibakusachi = Meadows producing green manure for paddy and crop fields.

Abbreviations

AGRC = ASO Grassland Restoration Committee; BP = Years Before Present; FAO = Food and Agriculture Organization; GHIAS = Globally Important Agricultural Heritage Systems; NGO = Non-governmental Organization; NPO = Non-profit Organization

References

[AGRC] ASO Grassland Restoration Committee. 2017. Aso Grassland Restoration Report: Activity Report 2016 [in Japanese]. URL: http://www.aso-sougen.com/kyougikai/restoration/report.html.

Arioka, T. 2004. Satoyama I [in Japanese]. Housei University Press. Tokyo, Japan.

Benton, T.G., J.A. Vickery and J.D. Wilson. 2003. Farmland biodiversity: Is habitat heterogeneity the key? Trends Ecol. Evol. 18: 182–188.

Cousins, S.A.O., H. Ohlson and O. Eriksson. 2007. Effects of historical and present fragmentation on plant species diversity in semi-natural grasslands in Swedish rural landscapes. Landsc. Ecol. 22: 723–730.

Dengler, J., M. Janišová, P. Török and C. Wellstein. 2014. Biodiversity of Palaearctic grasslands: A synthesis. Agric. Ecosyst. Environ. 182: 1–14.

Dengler, J. and S. Tischew. 2018. Grasslands of Western and Northern Europe—between intensification and abandonment (pp. 27–63 this volume).

Ebihara, A. 2011. Hotspots of endemic plants in Japan [in Japanese]. pp. 29–34. *In*: M. Kato and A. Ebihara (eds.). Endemic Plants of Japan. Tokai University Press, Tokyo.

Ecological Society of Japan. 2002. Handbook on Alien Species in Japan [in Japanese]. Chijin Shokan, Tokyo.

Foley, J.A., R. DeFries, G.P. Asner, C. Barford, G. Bonan, S.R. Carpenter, F.S. Chapin, M.T. Coe, G.C. Daily, (...) and P.K. Snyder. 2005. Global consequences of land use. Science 309: 570–574.

Fukamachi, K., H. Oku and A. Miyake. 2005. The relationships between the structure of paddy levees and the plant species diversity in cultural landscapes on the west side of Lake Biwa, Shiga, Japan. Landsc. Ecol. Eng. 1: 191–199.

Hayasaka, D., N. Shimada, H. Konno, H. Sudayama, M. Kawanishi, T. Uchida and K. Goka. 2012. Floristic variation of beach vegetation caused by the 2011 Tohoku-oki tsunami in northern Tohoku, Japan. Ecol. Eng. 44: 227–232.

Hiradate, S., S. Morita and Y. Kusumoto. 2008. Effects of soil chemical properties on the habitats of alien and endemic plants [in Japanese]. Nougyougijutsu 63: 39–44.

Hosono, M. and T. Sase. 2015. The historical development of 'Kurobokudo' layer: Preliminary discussion from the perspective of man-made ecosystems [in Japanese with English summary]. Quat. Res. 54: 323–329.

Hotta, M., I. Tsuyama, K. Nakao, M. Ozeki, M. Higa, Y. Kominami, T. Matsui and N. Tanaka. 2014. Impacts of climate change on the alpine habitat of the rock ptarmigan in the Hida Mountains, central Japan. Ornithol. Sci. 13(Suppl.): 215–215.

Ishikawa, S., A. Furukawa and T. Oikawa. 1995. Zonal plant distribution and edaphic and micrometeorological conditions on a coastal sand dune. Ecol. Res. 10: 259–266.

IUSS Working Group WRB. 2015. World Reference Base for Soil Resources 2014, Update 2015. International Soil Classification System for Naming Soils and Creating Legends for Soil Maps. FAO [World Soil Resources Report No. 106], Rome.

Japanese Society of Grassland Science. 2010. Ecology and Conservation of Grasslands: For Harmony between Husbandry and Biodiversity [in Japanese]. Gakkai Shuppan Center, Tokyo.

Kadota, Y. 2011. Habitat environments of endemic plants [in Japanese]. pp. 36–38. *In*: M. Kato and A. Ebihara (eds.). Endemic Plants of Japan. Tokai University Press, Tokyo.

Kadoya, T., A. Takenaka, F. Ishihama, T. Fujita, M. Ogawa, T. Katsuyama, Y. Kadono, N. Kawakubo, S. Serizawa, (...) and T. Yahara. 2014. Crisis of Japanese vascular flora shown by quantifying extinction risks for 1618 taxa. PLOS One 9: Article e98954.

Kawano, T., N. Sakaki, T. Hayashi and H. Takahara. 2012. Grassland and fire history since the late-glacial in northern part of Aso Caldera, central Kyusyu, Japan, inferred from phytolith and charcoal records. Quat. Int. 254: 18–27.

Kitahara, M. and K. Sei. 2001. A comparison of the diversity and structure of butterfly communities in semi-natural and human-modified grassland habitats at the foot of Mt. Fuji, central Japan. Biodivers. Conserv. 10: 331–351.

Kleijn, D., F. Kohler, A. Báldi, P. Batáry, E.D. Concepción, Y. Clough, M. Díaz, D. Gabriel, A. Holzschuh, (...) and J. Verhulst. 2009. On the relationship between farmland biodiversity and land-use intensity in Europe. Proc. R. Soc. Lond. B 276: 903–909.

Koda, K. and H. Nakamura. 2010. The effect of bush burning on parasitism by the egg parasitoid, *Trichogramma chilonis* Ishii (*Hymenoptera*: *Trichogrammatidae*) on *Shijimiaeoides divinus barine* (Leech) (*Lepidoptera*: *Lycaenidae*) eggs in Azumino, Nagano Prefecture [in Japanese, with English abstract]. Jpn. J. Environ. Entom. Zool. 21: 93–98.

Koyanagi, T.F. and T. Furukawa. 2013. Nation-wide agrarian depopulation threatens semi-natural grassland species in Japan: Sub-national application of the Red List Index. Biol. Conserv. 167: 1–8.

Koyanagi, T., S. Yamada, K. Yonezawa, Y. Kitagawa and K. Ichikawa. 2014. Plant species richness and composition under different disturbance regimes in marginal grasslands of a Japanese terraced paddy field landscape. Appl. Veg. Sci. 17: 636–644.

Koyama, A., T.F. Koyanagi, M. Akasaka, M. Takeda and K. Okabe. 2017. Combined burning and mowing for restoration of abandoned semi-natural grasslands. Appl. Veg. Sci. 20: 40–49.

Krauss, J., A.M. Klein, I. Steffan-Dewenter and T. Tscharntke. 2004. Effects of habitat area, isolation, and landscape diversity on plant species richness of calcareous grasslands. Biodivers. Conserv. 13: 1427–1439.

Kubo, M., T. Kobayashi, M. Kitahara and A. Hayashi. 2009. Seasonal fluctuations in butterflies and nectar resources in a semi-natural grassland near Mt. Fuji, central Japan. Biodivers. Conserv. 18: 229–246.

Kudo, G., M. Kimura, T. Kasagi, Y. Kawai and A.S. Hirao. 2010. Habitat-specific responses of alpine plants to climatic amelioration: Comparison of fell-field to snowbed communities. Arct. Antarct. Alp. Res. 42: 438–448.

Kusumoto, Y. and H. Inagaki. 2016. Symbiosis of biodiversity and tea production through Chagusaba. J. Resour. Ecol. 7: 151–154.

Matsumura, T. and Y. Takeda. 2010. Relationship between species richness and spatial and temporal distance from seed source in seminatural grassland. Appl. Veg. Sci. 13: 336–345.

Ministry of Agriculture, Forestry and Fisheries of Japan. 2013. The 87th Statistical Yearbook of Ministry of Agriculture, Forestry and Fisheries [in Japanese]. Ministry of Agriculture, Forestry and Fisheries, Tokyo.

Ministry of Agriculture, Forestry and Fisheries of Japan. 2016. Data from Statistics Department [in Japanese]. URL: http://www.maff.go.jp/j/tokei/index.html.

Ministry of the Environment of Japan. 2009. The Invasive Alien Species Act [in Japanese]. URL: http://www.env.go.jp/nature/intro/.

Ministry of the Environment of Japan. 2012. The National Biodiversity Strategy of Japan 2010–2022 [in Japanese]. Nature Conservation Bureau, Ministry of the Environment, Government of Japan, Tokyo.

Ministry of the Environment of Japan. 2014. The 4th Version of the Japanese Red List on 9 Taxonomic Groups [in Japanese]. Ministry of the Environment Government of Japan. Kasumigaseki, Tokyo.

Ministry of the Environment of Japan. 2016. Japan Biodiversity Outlook 2: Report of Comprehensive Assessment of Biodiversity and Ecosystem Services in Japan [in Japanese]. Nature Conservation Bureau, Ministry of the Environment, Government of Japan, Tokyo.

Mittermeier, R.A., P.R. Gil, M. Hoffman, J. Pilgrim, T. Brooks, C.G. Mittermeier, J. Lamoureux and G.A.B. da Foncesca. 2005. Hotspots Revisited: Earth's Biologically Richest and Most Endangered Terrestrial Ecoregions. Conservation International. Washington, DC.

Miyabuchi, Y., S. Sugiyama and Y. Nagaoka. 2012. Vegetation and fire history during the last 30,000 years based on phytolith and macroscopic charcoal records in the eastern and western areas of Aso Volcano, Japan. Quat. Int. 254: 28–35.

Mizumoto, K. 2015. Mura: Hyakushou-tachi no kinsei [in Japanese]. Iwanami Shoten, Publishers. Tokyo.

Murata, K., K. Nohara and M. Abe. 1998. Effect of routine fire-burning of the habitat on the emergence of the butterfly, *Shijimiaeoides divinus asonis* (Matsumura) [in Japanese, with English abstract]. Japanese J. Entomol. New Ser. 1: 21–33.

Nagata, Y.K. and A. Ushimaru. 2016. Traditional burning and mowing practices support high grassland plant diversity by providing intermediate levels of vegetation height and soil pH. Appl. Veg. Sci. 19: 567–577.

Nakahama, N., K. Uchida, A. Ushimaru and Y. Isagi. 2016. Timing of mowing influences genetic diversity and reproductive success in endangered semi-natural grassland plants. Agric. Ecosyst. Environ. 221: 20–27.

Nakamura, F., J. Seo, T. Akasaka and F.J. Swanson. 2016. Large wood, sediment and flow regimes: Their interactions and temporal changes caused by human impacts in Japan. Geomorphology 279: 176–187.

Navarro, L.M. and H.M. Pereira. 2012. Rewilding abandoned landscapes in Europe. Ecosystems 15: 900–912.

Normile, D. 2016. Nature from nurture. Science 351: 908–910.

Numata, M. 1961. Ecology of grasslands in Japan. J. Coll. Arts Sci. Chiba Univ. 3: 327–345.

Numata, M. 1969. Progressive and retrogressive gradient of grassland vegetation measured by degree of succession—Ecological judgment of grassland condition and trend IV. Vegetatio 19: 96–127.

Obata, T., J. Ishii, T. Kadoya and I. Washitani. 2012. Effect of past topsoil removal on the current distribution of threatened plant species in a moist tall grassland of the Watarase wetland, Japan: Mapping of selected sites for wetland restoration by topsoil removal [in Japanese, with English abstract]. Jpn. J. Conserv. Ecol. 17: 221–233.

Ogura, J. 2006. The transition of grassland area in Japan [in Japanese]. J. Kyoto Seika Univ. 30: 160–172.

Ohara, H. 1969. Grassland: Cultivation, Management and Utilization [in Japanese]. Meinun Shobo, Tokyo.

Osawa, T., H. Mitsuhashi, H. Niwa and A. Ushimaru. 2010. Enhanced diversity at network nodes: River confluences increase vegetation-patch diversity. Open Ecol. J. 3: 48–58.

Osawa, T., H. Mitsuhashi, H. Niwa and A. Ushimaru. 2011. The role of river confluences and meanderings in preserving local hotspots for threatened plant species in riparian ecosystems. Aquat. Conserv. Mar. Freshw. Ecosyst. 21: 358–363.

Osawa, T., K. Kohyama and H. Mitsuhashi. 2016. Multiple factors drive regional agricultural abandonment. Sci. Total Environ. 542: 478–483.

Osumi, K. 2005. Reciprocal distribution of two congeneric trees, *Betula platyphylla* var. *japonica* and *Betula maximowicziana*, in a landscape dominated by anthropogenic disturbances in northeastern Japan. J. Biogeogr. 32: 2057–2068.

Pfeiffer, M., C. Dulamsuren, Y. Jäschke and K. Wesche. 2018. Grasslands of China and Mongolia: spatial extent, land use and conservation (pp. 168–198 this volume).

Queiroz, C., R. Beilin, C. Folke and R. Lindborg. 2014. Farmland abandonment: Threat or opportunity for biodiversity conservation? A global review. Front. Ecol. Environ. 12: 288–296.

Sawada, Y., T. Hattori and K. Uchida. 2006. The Endangerment of coastal plants of Honshu, Shikoku and Kyushu Islands in Japan—Based on national and prefectural red data books [in Japanese]. Environ. Inf. Sci. 20: 71–76.

Stewart, J.R., Y. Toma, F.G. Fernandez, A. Nishiwaki, T. Yamada and G. Bollero. 2009. The ecology and agronomy of *Miscanthus sinensis*, a species important to bioenergy crop development, in its native range in Japan: a review. GCB Bioenergy 1: 126–153.

Suka, T., T. Okamoto and A. Ushimaru. 2012. Souchi to Nihon-jin: Nihon-rettou sougen 1 man-nen no tabi [in Japanese]. Tsukiji Shokan Publishing, Tokyo.

Takatsuki, S. 2009. Effects of sika deer on vegetation in Japan: A review. Biol. Conserv. 142: 1922–1929.

Toma, Y., J. Clifton-Brown, S. Sugiyama, M. Nakaboh, R. Hatano, F.N.G. Fernández, R. Stewart, A. Nishiwaki and T. Yamada. 2013. Soil carbon stocks and carbon sequestration rates in seminatural grassland in Aso region, Kumamoto, Southern Japan. Global Change Biol. 19: 1676–1687.

Török, P., M. Janišová, A. Kuzemko, S. Rūsiņa and Z. Dajić Stevanović. 2018. Grasslands, their threats and management in Eastern Europe (pp. 64–88 this volume).

Tsuda, S., H. Fujita, M. Ajima, K. Nishisaka and T. Tsujii. 2002. Conservation and management of grassland vegetation on coastal sand dune in the Koshimizu Nature Reserve, Hokkaido, Japan [in Japanese, with English abstract]. Grassl. Sci. 48: 283–289.

Tsuji, M., A. Ushimaru, T. Osawa and H. Mitsuhashi. 2011. Paddy-associated frog declines via urbanization: a test of the dispersal-dependent-decline hypothesis. Landsc. Urban Plan 103: 318–325.

Tsuyuzaki, S. 1995. Ski slope vegetation in central Honshu, Japan. Environ. Manag. 19: 773–777.

Uchida, K. and A. Ushimaru. 2014. Biodiversity declines due to abandonment and intensification of agricultural lands: patterns and mechanisms. Ecol. Monogr. 84: 637–658.

Uchida, K., S. Takahashi, T. Shinohara and A. Ushimaru. 2016a. Threatened herbivorous insects maintained by long-term traditional management practices in semi-natural grasslands. Agric. Ecosyst. Environ. 221: 156–162.

Uchida, K., T. Shinohara, S. Takahashi, N. Nakahama, Y. Takami and A. Ushimaru. 2016b. Rediscovery of *Celes akitanus* (*Orthoptera, Acrididae*) from semi-natural grasslands in Japan. Entomol. Sci. 19: 89–96.

Uchida, K., M.K. Hiraiwa and A. Ushimaru. 2016c. Plant and herbivorous insect diversity loss are greater than null model expectations due to land-use changes in agro-ecosystems. Biol. Conserv. 201: 270–276.

Uematsu, Y., T. Koga, H. Mitsuhashi and A. Ushimaru. 2010. Abandonment and intensified use of agricultural land decrease habitats of rare herbs in semi-natural grasslands. Agric. Ecosyst. Environ. 135: 304–309.

Uematsu, Y. and A. Ushimaru. 2013. Topography- and management-mediated resource gradients maintain rare and common plant diversity around paddy terraces. Ecol. Appl. 23: 1357–1366.

Valkó, O., P. Török, B. Deák and B. Tóthmérész. 2014. Review: Prospects and limitations of prescribed burning as a management tool in European grasslands. Basic Appl. Ecol. 15: 26–33.

Washitani, I. 2001. Plant conservation ecology for management and restoration of riparian habitats of lowland Japan. Popul. Ecol. 43: 189–195.

Watanabe, M. 1992. A comparative study of culture and the value of Kuroboku-soil resources [in Japanese]. Bull. Chuo-Gakuin Univ. Inst. Comp. Stud. Cult. 6: 189–210.

Yoshioka, A., T. Kadoya, S. Suda and I. Washitani. 2010. Invasion of weeping lovegrass reduces native food and habitat resource of *Eusphingonotus japonicus* (Saussure). Biol. Invasions 12: 2789–2796.

PART 3
Grasslands of Other Biogeographic Regions
Problems and Prospects

10

Rangelands/Grasslands of India
Current Status and Future Prospects

Devendra R. Malaviya,[1,*] *Ajoy Kumar Roy*[2] and *Pankaj Kaushal*[3]

Introduction

In India, it is estimated that over 100 million ha is presently underutilized which includes over 25–30 million ha of degraded forest lands, 45–50 million ha agricultural lands unsuitable for crop production, 9–10 million ha of sodic wastelands and the rest are ravines, pasture lands and 'wastelands'. Livestock and grazing-based livestock husbandry continues to play an important role in the rural economy of the country. For increasing milk production from the current level of 128 million tons (mt) to 160 mt by 2020, Indian needs 494 mt of dry fodder, 825 mt of green fodder and 54 mt concentrates. At present India is facing 36 per cent green fodder deficit for the existing livestock. Fodder is being cultivated only in 4 per cent of agricultural land, hence, fodder availability from non-arable lands needs to be significantly increased.

Many of the natural grasslands have degraded due to overgrazing, in addition to large areas being converted to plantations/protected areas/industrial establishment. Under the British, nomadic pastoralists were sedentarized and the grasslands they depended on were converted to agriculture, leading to salinization of the soil and rendering once productive grasslands to wastelands. This ignores the fact that grasslands in India have existed as natural ecosystems as far as 50 million years ago as evidenced by fossil records (Vanak et al., 2015). In India, the agricultural scenario is characterized by the traditional predominance of mixed farming system. Livestock rearing is a major source of income that provides employment and livelihood to the rural families. The grasslands and pastures not only form a major source of forage for the livestock, but also provide habitat to a large

[1] ICAR-Indian Institute of Sugarcane Research, Lucknow, India.
[2] ICAR-Indian Grassland and Fodder Research Institute, Jhansi, India.
[3] ICAR-National Institute of Biotic Stress Management, Raipur, Chhattishgarh, India.
Emails: royak333@rediffmail.com; pkaushal@rediffmail.com
* Corresponding author: drmalaviya47@rediffmail.com

variety of wild animals and birds. There is an immediate need to map the grazing lands in the country, demarcate these on the ground and initiate policy steps to implement/maintain sustainable land use.

The livestock population, over the years, has shown a steady growth on broadly two counts: (i) increase in the number of stall-feeding-based bovine livestock, viz., buffaloes and hybrid cattle, and (ii) increase in the number of free-grazing-based livestock like goats and sheep which pertain to resource-poor households, landless pastoralists, nomadic and semi-nomadic tribes and marginal farmers. Livestock production is the backbone of Indian agriculture contributing 4 per cent to national GDP and source of employment and ultimate livelihood for 70 per cent population in rural areas. Livestock population is around 629.7 million and is expected to grow at the rate of 0.55 per cent in the coming years. It has 56.7 per cent of world's buffaloes, 12.5 per cent cattle, 20.4 per cent small ruminants, 2.4 per cent camel, 1.4 per cent equine, 1.5 per cent pigs and 3.1 per cent poultry. There is a shift in composition of livestock towards small ruminants due to natural resource degradation in arid and semiarid regions and an increasing demand for meat.

Status of Indian Grasslands/Rangelands

India has a geographical area of 3,287,263 sq km^2 with total land of 3,090,846 sq km^2 of which grassland area is 5,35,441 sq km^2 (17.32 per cent) and forest area 7,68,436 sq km^2. With only 2 per cent of the world's geographical area, India supports 20 per cent of the world's livestock. Only 12.15 million ha of land in the country is officially classified as permanent pastures/grazing lands, but grazing is estimated to occur on about 40 per cent of the land area in the country. It holds worlds' 16 per cent cattle, 55 per cent buffalo (20 per cent), and sheep (5 per cent). The share of forages in cultivated land remained < 5 per cent in the country for many years. The fodder resource from grasslands is fast deteriorating owing to heavy grazing pressure of 3.42 adult cattle unit (ACU) per ha. The country's pastures witnessed severe decline from about 70 million ha to 38 million ha during the 50 years post-independence. The fast degrading grazing lands have carrying capacity of less than one ACU. The area of various categories of degraded land/wasteland is presented in Table 10.1.

Table 10.1 Harmonized area statistics of degraded lands/wastelands of India.

Type of Degradation	Arable Land (m ha)	Open Forest (< 40 per cent Canopy) (m ha)
Water erosion (> 10 t/ha/yr)	73.27	9.30
Wind erosion (Aeolian)**	12.40	–
Exclusively salt affected soils	5.44	–
Salt-affected and water eroded soils	1.20	0.10
Exclusively acidic soils (pH < 5.5)	5.09	–
Acidic (pH < 5.5) and water eroded soils	5.72	7.13
Mining and industrial waste	0.19	–
Water logging (permanent) (water table within 2 mts depth)*	0.88	–
Total	**104.19**	**16.53**

Source: ICAR-NAAS, 2010.

Temperate region has up to 70 per cent area under pasture as compared to quite less area in the tropical region. Distribution of pasture lands varies in Himachal Pradesh (36.4 per cent), Sikkim (13.3 per cent), Karnataka (6.5 per cent), Madhya Pradesh (6.3 per cent), Rajasthan (5.4 per cent), Maharashtra (5.1 per cent) and Gujarat (4.5 per cent). Common Property resources (CPRs) have decreased from 70 million ha in 1947 to 38 m ha in 1997, thus, decreasing availability of CPR per head of livestock from 0.25 ha to 0.08 ha.

Nomads travel from place to place for their livelihood. Pastoral nomads in India also specialize in animal breeding. In the mountains, Gujjars, Bakarwals, Gaddi, Bhotiyas, Sherpas and Kinnauris are common whereas on the plateaus, plains and deserts Dhangars, Raikas and Banjaras are important pastoral communities.

Type of Grassland/Rangeland

The Indian sub-continent represents a wide spectrum of eco-climates, ranging from humid tropical to semi-arid, temperate to alpine. Agro-bio-diversity in India is distributed in eight very diverse phytogeographical and 15 agroecological regions. Because of wide variation in climate, various types of grasslands with different species composition and productivity are available in these regions, details of which are described here. Five hundred fifty tribal communities of 227 ethnic groups are spread over 5,000 forested villages. During 1954 and 1962, the Indian Council of Agricultural Research conducted grassland surveys and classified the grass cover of India into following major types (Dabadghao and Shankarnarayan, 1973):

***Sehima-Dichanthium* type**: Spreads over whole of peninsular India including central India plateau and Aravali ranges and coastal region covering an area of about 17,40,000 km^2 at an elevation between 300 and 1,200 m. Dominant grasses are *Dichanthium annulatum, Sehima nervosam, Chrysopogon fulvus, Heteropogon contortus, Iseilema laxum, Themeda, Bothriochola pertusa, Cynodon dactylon*, etc., although 24 species of perennial grasses, 89 species of annual grasses and 129 species of dicots, including 56 legumes, are reported from these grasslands.

***Dichanthium-Cenchrus-Lasiurus* type**: Spread over an area of about 4,36,000 km^2 in sub-tropical and semi-arid regions of Gujarat, Rajasthan except the Aravali ranges, western Uttar Pradesh, Delhi, Punjab and Haryana at an elevation of 150 to 300 m. Dominant grass species of the grasslands are *Cenchrus ciliaris, C. setigerus, Dichanthium annulatum, Cynodon dactylon* and *Lasiurus*.

***Phragmites-Saccharum-Imperata* type**: Covering about 2,800,000 km^2 in the Gangetic Plain (Uttar Pradesh, Haryana, Bihar and West Bengal) and Brahamaputra valley in north-eastern states, these grasslands are existing at an elevation between 300 to 500 m. There are 10 perennial grasses, 26 annual grasses and 56 herbaceous species, including 16 legumes. The Gangetic Plain is one of the most thickly populated regions in the world and so the original grassland type is almost gone. Some of the wet grasslands harbor many globally threatened wildlife species.

***Themeda-Arundinella* type**: Covers entire sub-mountain tract of north-western part of Uttar Pradesh, Punjab, Himachal Pradesh, Jammu & Kashmir and Haryana. Principal species are *Themeda, Anathera, Arundinella benghalensis, Chrysopogon, Cynodon dactylon, Heteropogon contortus*. These grasslands cover about 2,30,000 km^2 at elevations between

350 and 1,200 m. There are 37 major perennial grass species, 32 annual grass species and 34 dicots including 9 legumes.

***Temperate Alpine* type**: The Hindu Kush Himalayan (HKH) region is the largest mountain system in the world, spanning over 4.3 million km^2 and covering 3,500 km-long fragile environment. Grasslands in the HKH region are the source of livelihood for approximately 25 to 30 million pastoralists and agro-pastoralists. Over 1,80,000 km^2 are under grasslands in HKH region. There is an urgent need for managing grassland ecosystems and pastoral livelihoods in the region (Sharma and Shrestha, 2015). Temperate and alpine pastures in India are spread over an approximate area of 74,809 sq. km in the high hills of Uttarakhand, Jammu & Kashmir, Himachal Pradesh, North Bengal and the north-east region at > 2,000 m elevation. A large part of temperate grasslands in India is represented in the Himalayan region where due to variations in altitude and other agro-climatic parameters, such as rainfall and temperature, a wide variation in type of vegetation is observed. Thus, 47 perennial grasses, five annual grasses and 68 dicots, including six legumes are represented by this region. With an average rainfall of 100 cm, the cold desert is present in Leh, Kargil, Lahaul, Spiti, Jammu & Kashmir and some parts of Chamba and Kinnaur districts in Himachal Pradesh. The fast blowing winds in the region erode the immature sandy soil. Natural vegetation comprises of a few tree species and shrubs. The Alpine arid vegetation is dominated by the *Artemisia-Caragana, Hippophae- Myricaria,* and *Ephedra gerardiana* communities. The characteristic species in the Trans-Himalayas are *Saussurea, Potentilla, Corydalis, Astragalus* and *Oxytropis*. The land use system is horti-agri-pastoral. The grasslands here are heavily grazed both by draught animals and by migratory graziers. The grass cover comprises mainly of *Themeda, Arunidnella, Hamertheria, Heteropogon,* etc., livestock composition is dominated by small ruminants.

Arid and Semi-arid grasslands: Located mainly in Rajasthan and Gujarat and occupies nearly 16.6 per cent of India's geographical area (i.e., 5,48,850 sq km). This semi-arid region receives 400 to 1,000 mm rainfall and is dominated by grass and shrub species. It has dry deciduous forest, but extensive tracts of grasslands are seen in the Deccan plateau of central India, the Malwa plateau in western India, and in Saurashtra and Kutch of Gujarat. These are the best places for foraging and/or nesting for more than 100 bird species. In the zone, there are eight national parks, totalling 1,319 sq km or 0.24 per cent, and 83 wildlife sanctuaries, covering nearly 14,000 sq km or 2.56 per cent of surface.

Ethnic Grasslands

The country has some old and natural grasslands with ethnic value. Understanding the history of these can give lot of information on how to develop new grasslands. These pastures, with unique floristic compositions, have evolved to climax/sub-climax stages over hundreds of years of ecological succession and it may not be possible to bring these back once these are destroyed.

Banni Grasslands: Made up land formed by the detritus brought down and deposited predominantly by the Indus river, which was reported to flow through the Great Rann in the past. The great and the little Ranns of Kachchh were the old arms of the sea in the old geological period. Due to the eruption and formation of the Allah Bund near Kori Creek, the land got blocked and was filled up by deposits brought down by the Indus river. Banni, once considered the largest grassland in Asia, has drastically degraded due to salinity ingress, lack of knowledge, increased human and livestock population and invasion of *Prosopis juliflora*. Various species found in Banni are *Dichanthium-annulatum, Sporobolus helvolus, Chloris barbate, Cenchrus biflorus, Eleusine bianata, Elysecarpus rugosus, Heylandis latebrosa, Digitarea sanguinalis, Var Ciliaris, Crotolaria medicaginea, Indigofera* spp., *Sida* spp., *Malanocenchrus jacquemontii, Sporobolus diander, Cenchrus setigerus, Aristida adscensionis, Aristida funiculata, Setaria rhachitricho, Eragrostis minor and major, Eragrostis trimula, Desmostachya bipinnata, Cyperus rotundus, Cressa cretica, Eragrostis bulbosa, Kochia* spp., *Suaeda fruticosa*, of which the first 12 are palatable while the remaining are highly salt-tolerant grasses.

Shola Grassland of the Western Ghats: The Western Ghats on the west coast of the Indian peninsula occupy only circa 5 per cent of India's land area (about 1,32,606 sq km), yet they harbor nearly 27 per cent of its total flora. The forests confined to the windward side of this area are the best representatives of non-equatorial tropical forests in the world. These are spread at an high altitude (> 1,700 m) and interspersed with tropical forests. They are maintained by fire and frost and appear to be climax vegetation as an ancient and geographic relict species of ungulate (Niligir Tahr) is found in the Shola grasslands and nowhere else in the world.

Thar Deserts: This is the smallest desert in the world but has great biodiversity (Ajai and Dhinwa, 2018) with especially high avian diversity (~ 300 species) represented by the Great Indian bustard (*Ardeotis nigriceps*), Houbara bustard (*Chlamydotis undulate*), Cream-coloured courser (*Cursorius cursor*), Hoopoe lark (*Alaemon alaudipes*), various species of sandgrouse, raptors, wheatears, larks, pipits and munias. The Desert National Park or wildlife sanctuary (3,162 sq km) in addition to five more wildlife sanctuaries of 12,914 sq km lie in this zone. Besides over-grazing, expansion of agriculture, salinization due to wrong irrigation practices, the desert ecosystem is also being altered due to invasive species, such as *Prosopis chilensis* (Ajai and Dhinwa, 2018). Tremendous changes in the avifaunal structure of the Thar desert are taking place due to the canal (Indira Gandhi Nahar Project) and species never seen earlier are now regularly seen here (Rahmani, 1997a, 1997b; Rahmani and Soni, 1997). However, this project is changing the desert ecosystem by changing the crop pattern tradition. Due to easy availability of water everywhere, unsustainable livestock grazing is taking place and the famous Sewan grasslands which have survived for hundreds of years with low grazing pressure are now under tremendous pressure.

Terai Grasslands: The Terai-Duar savannas and grasslands in India form an ecoregion that stretches across the middle of the Terai belt is about 25 km wide in an area of 3,54,800 sq km in the state of Uttarakhand, through southern Nepal to the northern part of West Bengal. These are actually a mosaic of tall grasslands, savannas and evergreen deciduous forests. The grasslands are counted amongst the tallest in the world and are well maintained by silt deposited by the yearly monsoon floods. The area is dominated by Kans and Baruwa grass. The ecoregion is home to several endangered species like the Indian rhinoceros, elephant, tiger, bear, leopard and also some other wild animals. These grasslands are interspersed with sal (*Shorea robusta*) forest which contains endangered bird species, such as Swamp francolin, Bengal florican and Finn's weaver (*Ploceus megarhynchus*).

Kangeyam Grassland: 'Korangadu' is a traditional pasture land farming system existing in the semiarid tract of Tamil Nadu state of South India, comprising Dharapuram, Kangeyam, Palladam, Moolanur and Allimanthayam areas with an annual rainfall of 600–675 mm and laterite red soil or gravel type. The region lies in rain shadow region of Western ghats. The majority of rural population depends on livestock rearing. Korangadu has predominantly three major plant species spatially in three tiers. The lower tier is characterized by *Cenchrus* grass, with an overstorey of trees, including *Acacia leucophloea*. The land is fenced with a thorny shrub, *Commiphora berryii* as a live fence. Korangadu is owned privately by individual farmers and more than 50,000 individuals keep their own paddocks of about 1–2 ha size of land. Approximately 50,000 ha of Korangadu pasture land is distributed among 500 villages in Erode, Karur, Dindigul and Coimbatore districts of Tamil Nadu state.

Grassland Productivity

Grasslands do not form a prominent feature of vegetation in the tropical part of India because in the moist lowlands, grass faces very tough competition from trees and shrubs. Studies on grass communities, including their succession and trends and the level of productivity, show that *Sehima-Dichanthium* cover is highly productive at its climax stage, whereas *Themeda-Arundinella* yields are the lowest. Most of these communities are in their last stage of retrogression and thus represent poor condition grasslands, except *Phragmitis-Saccharum-Imperata*, which is predominantly unpalatable. Their actual harvestable biomass is between 0.2 and 3.5 t/ha as against the potentials of 4–6 t/ha (Singh, 2015). In the protected area, the tropical grasslands are evolved under a system of grazing, drought and periodic fire whereas in unprotected areas, these are maintained by livestock grazing and other biotic factors. Hence, the majority of grasslands in arid and semiarid region of the country belong to the 'poor' category with productivity ranging from 0.5 to 1.0 t/ha. The reason for low productivity is that the grasslands are 'common' lands of the community and are the responsibility of none, in spite of having the potential to be more productive ecosystems. All types of grassland ecosystems are under tremendous grazing pressure. For example, in the semi-arid grasslands, the carrying capacity is one Adult Cattle Unit (ACU) per ha (Shankar and Gupta, 1992), but the stocking rates are as high as 51 ACU per ha, while in the arid areas, the carrying capacity is 0.2–0.5 ACU per ha but the stocking rates are one to 4 ACU per ha (Raheja, 1966). Grazing pressure was more pronounced in states like Punjab, Uttar Pradesh and Tamil Nadu. The carrying capacity of permanent pastures and grazing lands declined by 15 per cent during 1980–81 to 2007–08. Per unit of pastures and grazing lands do not even sustain single adult cattle unit of 270 kg body weight with 3 per cent of DM requirement for its maintenance (Dixit et al., 2015).

Further, due to poor management or community interest, many of the grazing lands have been invaded by invasive alien species, like *Lantana, Eupatorium, Parthenium, Prosopis juliflora, Leucaena*, etc., thus, severely impacting productivity. Thus, a total of 120.7 m ha in the country is constituted by degraded or wastelands. The present day rangelands/ grasslands are no longer providing the much needed forage for livestock; hence, the opportunity lies in growing fodder trees, bushes and grasses in wastelands to augment and supplement fodder and feed requirements (Singh, 2015).

Factors Affecting Sustainability

Degradation of Indian grasslands is a serious concern in multiplicity due to low fodder availability, environmental degradation (soil erosion, increased water run-off, poor carbon sequestration), decreased livestock production and diminishing livelihood for poor people. Social, legal and climatic factors are affecting these grasslands in different ways.

Social Factors: Grasslands are the 'common' lands of the community and while there have been robust traditional institutions ensuring their sustainable management in the past, today, due to take-over by government or breakdown of traditional institutions, they are the responsibility of none. Common property land resources constitute about 15 per cent of the total geographical area of India, of which 23 per cent is community pasture and grazing lands and 16 per cent have been classified as village forests and woodlots. However, post-independence focus on agricultural productivity and land reform was one of the factors for encroaching pastoral land which further compounded with industrialization. Historically also, the British and the Mughals shared a preference for agricultural production systems and introduced taxes and collected revenue, staking ownership over the land and resources. In rural Punjab, they set about acquiring 20 lakh acres of common grazing lands, expropriating land formerly used pastoralists, setting up agricultural colonies, constructing canals and granting blocks of land to peasants (Arnold and Guha, 1952).

With minor exceptions, the grasslands of India and Sri Lanka are considered to be anthropic, the early consequence of habitat conversion (Misra, 1983; Yadava, 1990). It is thought that the whole of the subcontinent was formerly wooded; in reality, up to half the land classified as 'forest' may be grasslands (Pemadasa, 1990). Some of the factors which affected growth of grazing lands are absence of any nodal agency to coordinate and steer grassland and fodder development programme, gradual erosion of the traditional agro-forestry/silvi-pastoral systems; lack of fodder banks and value addition facilities; lack of field level research on management protocol in respect of ecologically sensitive grasslands; un-organized use of grazing lands and lack of inter-sectoral dialogue between the key departments.

Social issues do get in the way of implementing programmes. For example, in a livelihood improvement programme in semi-arid India, due to misunderstanding about bans on grazing, the outcome was unsatisfactory as the whole community migrated out of the watershed the project had planned to rehabilitate (Puskur et al., 2004). Community response, however, in Ahmednagar district of Maharashtra was different and the people followed the practices as suggested.

In arid regions of the country, in the states of Rajasthan and Gujarat, the fear of rapid desertification propelled the forest department and other agencies to plant *Prosopis juliflora* (Ghotge, 2004) on a considerable portion of grazing lands, destroying local varieties and species on which animals used to graze. *Prosopis* is now considered to be a pest species and has colonized vast areas.

Policy Issues for Protection of Grasslands: It is long felt that the lack of comprehensive grazing-cum-fodder and pasture management policies at national and state levels is the major cause of degradation and diversion of grazing lands. Currently 'The Cattle Trespassers Act', formulated in 1871, is the only Act applicable to regulate grazing in public and forest land. Most of the states have not identified grasslands 'deemed forest', hence, need to be notified as Protected Areas under the Wild Life (Protection) Act, 1972 or notified as Protected or Reserve Forest under the Indian Forest Act, 1927. Although not covered in the National Environment Policy 2006 (NEP), many of the grasslands in the country, which are sensitive to climate change, developmental pressures and invasion by alien invasive plants need due attention. This also got the focus of attention during the 23rd International Grassland Congress held in India in 2015. The Forest Policy of 1954, critical of unrestricted and uncontrolled grazing faced difficulties in its implementation due to uncontrolled heavy and continuous grazing pressure. Scheduled Areas and Scheduled Tribes Commission, 1966 recommended that the Forest Department should promote growth of improved varieties of grasses in forest areas and grazing fees should be regulated. The National Commission on Agriculture (NCA, 1976) recommended strict control on grazing and regulation on grazing. National Forest Policy 1988, and its implementation also recommended that "a National Grazing Policy should come into effect at the earliest." Regrettably, few of these measures have been implemented.

Many of the ecologically sensitive pasture lands, viz. Shola grasslands of Nilgiris; Sewan grasslands of Bikaner; Jodhpur and Jaisalmer; semi-arid grasslands of Deccan; Rollapadu grasslands in the semi-arid tracts of Andhra Pradesh; Banni grasslands of Gujarat and Alpine grasslands of Sikkim and western Himalaya are already on the verge of no return. Some of the recommendations of the Task Force recently constituted by Government of India include recognition to certain grasslands, viz., Shola, Sewan, semi-arid grasslands of Deccan, Rollapadu grasslands, Banni grasslands and Alpine grasslands as ecologically sensitive ecosystems; conservation and management of Alpine meadows (Bugyals) and framing of national grazing policy.

Government of India has formulated 'Draft Grazing and Livestock Management Policy, 1994', and 'Draft National Policy for Common Property Resource Lands (CPRLs)', which need effective implementation. These policies emphasize to develop large blocks of grass reserves away from human habitation (in arid and semi-arid regions) and as fodder banks for drought years. Similarly, traditional rights to graze have been recognized to those who primarily reside in a forest or on forest lands and is dependent for 75 years through Forest Rights Act, 2006.

Climate Change and Grazing Lands: Environment is concern for one and all on this earth. Rising temperature due to various reasons and its impact is being studied by a large number of scientific organizations. In this context, the vast grazing lands, if managed properly, can have a significantly positive impact. Growing concern for environment and pollution is forcing a switch from fossil energy source to renewable energy. This affects grasslands in two ways: (i) more of biomass will be digested in biomethanation plants to produce 'clean' energy and (ii) a large area will be required for huge structures of solar and wind power generation.

Eighty per cent of India's pasture/rangelands are already degraded (Anonymous, 1999). The 73 per cent of the 34.5 (excluding 0.5 of feed crops) million Km^2 of the pastures and rangelands in dry areas is fast degrading causing a threat to 456 Gt of carbon being stored in the top 30 cm soil on earth (Sharma and Khadak, 2015). Overgrazing has also affected 23 of the 35 global hotspots for biodiversity loss (Mittermeier et al., 2004).

The concentration of Green House Gases (GHG) in the atmosphere has been on the rise since the beginning of the industrial revolution which was further accentuated by massive deforestation. There is an increase of average global temperatures by 0.6°C since year 1800. Most GHG are contributed into the atmosphere by soil organic matter (SOM) oxidation/ erosion and respiration by all living organisms (Table 10.2), however, most of it is taken out of the atmosphere by plant photosynthesis followed by diffusion into the oceans, leaving net annual increase of about 4.5–6.5 billion (giga) tons (Gt) of carbon (C) (or 16.5–23.8 Gt of CO_2eq/year) (Steinfeld et al., 2006).

The agricultural areas (including grasslands and wastelands) have a potential to sequester 26–31 per cent (FAO, 2014) of the 4.5–6.5 Gt C eq/year net GHG emissions (Steinfeld et al., 2006). Emissions by various livestock production cycles on grasslands also prove negative for environmental protection although sources of GHG emission in the production cycle are 34 per cent from deforestation for creating grasslands, 25 per cent from enteric fermentation and 30.5 per cent from manure as both methane and nitrous

Table 10.2 Atmospheric carbon sources and sinks (Steinfeld et al., 2006).

Source/Sink of the GHG	Estimated Carbon (C) eq in Billion (or giga) Tons (Gt)/Year	
	In to Atmosphere	Out of Atmosphere
Fossil fuel burning	4–5	
SOM oxidation/erosion	61–62	
Respiration by all organisms	50	
Deforestation	2	
Photo-syntheses		110
Diffusion into oceans		2.5
Total	117–119	112.5
Global net increase into atmosphere (1 CO_2 = 3.67 C)	+4.5 to 6.5 or average ≈ 5.5 (20 Gt CO_2eq), but may be up to +9 to 11 (33 to 40 Gt CO_2eq)	

oxide. Hence, we must concentrate on limiting enteric fermentation in animals and managing manures and fertilizer. Free animal grazing on rangeland (extensive systems) and manure and fertilizer management should receive priority for GHG reduction from animal activities on grasslands. Response of grassland ecosystem to climate change is critical in grazing animal production efficiency as climate change will impact forage quality (nutritive value). Foliage quality decreases at elevated CO_2 because of higher C:N ratios. Further, with doubling of CO_2 climate change, a 40 per cent increase in warm grasslands and a 50 per cent decrease in cool grasslands are predicted (Babu et al., 2015).

Drivers of Grassland/Rangeland Restoration

When it is well accepted that grasslands have a great role in livestock production, livelihood support and environmental protection together with the realization that these grazing lands are degrading fast, it is time to concentrate on ways and means to revegetate them as they need technological input as well as strong socio-political will.

Recently Developed Technological Options

There has been significant technological development in grassland science which includes varietal development, agronomical practices and integrated resource management. If incorporated properly, these technologies are likely to boost the productivity in addition to contributing to forage availability and environmental protection. Hereunder are described some innovative technologies which are proved to enhance fodder production from these lands.

Alternate Land Use Systems: Recent technological developments have established popularizing of alternate systems, such as horti/silvi/agro-forestry systems, which are potential means for increasing forage resource. The silvi-pastoral system under marginal and sub-marginal conditions has the potential of 10 t/ha green fodder against only 2–4 t/ha without tree component. The system serves the twin purpose of forage and firewood production and ecosystem conservation. Even under semi-arid marginal soil conditions the system gives 5–7 t/ha/yr. With some irrigation resource the agri-horti-silvicultural system comprising fruit trees, fodder crops, fast growing NFTs produces tree-lopping fodder

3.0 t/ha) and fuelwood (1.8–2.5 t/ha). In temperate conditions, introduction of Fescue in apple orchard gives 83.5 per cent higher fodder yield over local grasses in Himachal Pradesh. Productivity of cold climate grasslands and meadows of central Himalaya is degrading due to various reasons, such as soil compaction from vehicles and livestock; soil erosion; nutrient enrichment favoring weed growth; and depleting of soil biota. There has been good success in developing grasslands in Himalayas of Nepal wherein during the 1980s. The Department of Livestock Services (DLS) used risers and bunds on private sloping terraces for planting napier grass (*Pennisetum purpureum*) together with sowing of degraded forest-land with *Stylosanthes guainansis* cv, cook and molasses grass (*Melins minutiflora*).

Introducing Tree Component: Fodder tree species, such as *Prosopis cineraria, Acacia nilotica, Albizia lebbek, Azadirachta indica*, and *Dalbergia sissoo*, etc., are native to the Indian sub-continent. Taking a lesson from earlier experience, *Prosopis juliflora*, considered at one time as boon for desert/dry area, may be discouraged from planting on large scale as it is now becoming a problem because of its prolific growth. On an average, tree leaf fodder production of 0.2 to 2.0 ton/ha/year can be obtained up to 50 per cent by pruning height of the trees every year from the various agroforestry systems. The tree leaves contain 8–33 per cent crude protein. Tree leaves, like babul (*Acacia nilotica*), jackfruit (*Artocarpus heterophyllus*), pipal (*Ficus religiosa*), plantain (*Musa paradisica*), siris (*Albizia lebbek*) and subabul (*Leucaena leucocephala*) contained higher levels of Ca, Cu, Zn and Fe (Rai et al., 2007). The important top feed shrubs and trees used during the drought period include *Ailanthus excelsa, Acacia catechu, A. leucophloea, A. tortilis, Balanites roxburghii, Prosopis cineraria, P. juliflora, Azadirachta indica, Albizia lebbeck, Melia azadirach, Hardwickiabinata, Grewia ovate, Ficus bengalensis, F. religiosa, Anogeissus pendula, Bauhinia variegate, B. racemosa, Butea monosperma, Cordia dichotoma, Flacourtia indica, Moringa oleifera, Dichros tachys nutans, Morus alba, Ziziphus mauritiana*, and *Z. nummularia*. Additionally, considerable success has been achieved in the country in cultivating halophytic forages, such as chenopods, especially *Atriplex* in areas subject to total summer drought or on badly salt-affected lands.

Forest Grazing Land: For increasing fodder availability, forests are also being looked as a resource because socially 100–250 m people are intertwined directly or indirectly with forests. However, there is need to focus on decentralized forest management for profitable economic returns for communities because most have long traditions of forest use and sense of customary rights. Policies for CPLRs need to consider diversity for types of CPLRs, spatial distribution and changes over time due to physical degradation, conversion of CPLRs and declining local interest.

Revegetating Ravines: A considerable area in the country is ravine which is unstable, faces heavy soil erosion and degradation in addition to social problems. These need to be revegetated and made productive and ecologically stable. The forage trees, which are suitable for checking the erosion of the land, include *Acacia eburnea, A. nilotica, A. leucophloea, Balanities roxburghii, Cordarothii, Azadirachta indica, Pongamia pinnata, Dichrostachis cineria, Leucaena leucocephala, Prosopis juliflora, Butea monosperma*; and the grasses like *Dichanthium annulatum, D. caricosum, Bothriochloa pertusa* and *Cynodon dactylon*.

Cactus as Fodder: Global livestock population is increasing steadily. Water is an important limiting factor in dry lands. In this scenario, spineless cactus, more specifically *Opuntia* and *Nopalea*, becomes one of the most prominent crops of the 21st century. Cactus is the major feed commodity for livestock feeding in the semiarid region of Brazil. Cactus represents 75 per cent of maize grain energy, but produces at least 20x more in harsh semiarid environments. Cactus has potential to produce > 20 t DM ha^{-1} yr^{-1} and provide 180 t ha^{-1}

yr^{-1} of fresh good-quality water stored in the cladodes for the livestock (Dubeux et al., 2015). *Opuntia ficusindica* has been exploited as human food, forage, medicinal and energy crop in countries like Mexico, USA, Brazil, South Africa, Israel, Italy, Morroco, etc.

Socio-political Issues and Public-private Partnership

Grasslands are spread in the land of common, hence, it is property of all but responsibility of none. This establishes that for improvement of this ecosysytem there is a need for initiative from the government with involvement of all stakeholders. As forest and forest fringe area are also grazing resource for marginal farmers as well as the poor nomadic population, the conflicts between grazing lands and forests also need to be suitably resolved. The Wildlife Protection Act of 1972 and Project Tiger placed the need for conservation of India's wildlife. Forests, where animals used to graze, were closed to grazing. Grazing areas for domestic animals shrank further and poor people, who were dependent on these lands, were pushed further and further to the periphery. For improving grasslands, the Task Force for Desert and Grasslands has emphasized the need of a National Grazing Policy; grasslands as a place for biodiversity conservation; modification required in environment impact assessment guidelines by including ecologically fragile and environmentally sensitive areas; network of grassland ecologists; to include grasslands and desert ecosystems in Protected Area system; park of Jaisalmer and Barmer be declared as a Biosphere Reserve. In addition to these points favouring grassland development, the Sub Group III on Fodder and Pasture Management of the Working Group on Forestry and Sustainable Natural Resource Management constituted by Planning Commission of India, 2011 has given importance to certain points in favour of improving pasture sustainability and fodder availability: creation of nodal agencies on CPRs to coordinate and steer various research; educational and extension programmes; mapping of ecologically sensitive grasslands; rehabilitation and productivity enhancement of degraded forests through silvi-pastoral practices; develop fodder blocks in forest fringe villages; development of seed/germplasm banks and nurseries in every state for pasture development.

Involving societies through formation of grass growers co-operatives and ensuring benefit to the local people by providing the fodder produced from reserved *vidis* (as in Gujarat) to them may prove to be successful in attracting their interest. Similarly, incentives to farmers and pastoralists to continue traditional practices could be beneficial for wildlife and help in sustainable use of grasslands and deserts. People's participation is the crux of getting successful implementation of the programme because farmers need to be educated on the importance of rotational or seasonal grazing, control on free ranging animals, grassland plots to serve as nucleus for seed bank and improvement of livestock breed. People's participation in grassland management has given encouraging experience with *patels* in western India where shepherds herding groups negotiate access to harvested fields and sort out disputes that arise during migration.

Making Rangeland/Grassland an Economically Viable Option of Livelihood

Potential of Grasslands in Terms of Productivity, Water Harvesting, Soil Conservation and Economic Returns

The multiple dimensions of grasslands, rangelands and the forest cover include their role in maintaining the water regime and hydrological cycle; ecological balance; livelihood of millions of livestock and rural people; and conservation of soil, rare wildlife and biodiversity and every dimension is a strength to make these lands economically viable.

Tropical perennial grasses offer high water-use, high production and improved sustainability outcomes. This is due to high green leaf area and root density. Mostly, these grasses grow on soils with at least 150–200 mm of plant available water capacity (PAWC) and can develop root depths up to 1.6 metres. In Australia tropical grasses show high WUE, with values ranging from 13.7 (Swann forest bluegrass) to 28.8 kg dry matter/ha/mm (premier digit) with annual dry matter (DM) production up to 14.7 t DM/ha (Anonymous, 2014).

Generally, about 5 per cent to 10 per cent of the land area in a village is reserved for community pastures. Even after encroachment or diversion for other purposes, a significant portion is still available for common grazing which can be brought under silvipasture development by involving the local people. In a project initiated by BAIF in Asind taluka of Bhilwara district in Rajasthan, as a part of community land development under silvipasture activities, such as establishment of live hedges, gully plugging, contour bunding, sowing of forage seeds such as *Cenchrus setigerus* and *Stylosanthus hamata* (Stylo) and planting of saplings of *Acacia* and *Prosopis cinereria* (*Khejdi*) with education for primarily cut-and-carry fodder and to let animals for grazing for a period of 15–20 days after cutting, resulted in 3.5–4.0 tons of dry fodder per ha in subsequent years. In semi-arid India, particularly in Andhra Pradesh and Gujarat, common pool resources (CPRs) are often significant for poor people's livelihoods. They provide sustenance and income for household survival, opportunities for risk sharing and coping with seasonal crises or unusual shocks (e.g., sickness, drought) (Anonymous, 2002).

Worldwide land grazed by animals covers 32 million km^2 and pastoralism is the most widespread human land-use system on earth. However, with increasing population growth, it is apparent that biomass availability will control milk production in rangeland systems (Wright et al., 2015). In West Africa alone, 32 per cent of the livestock produced in the region is from pastoral (rangeland) systems. In Kenya, 80 per cent of all red meat produced in the country is raised in rangelands. The situation in India is also similar and a large population is dependent on rangelands for nutritional security in the form of livestock and milk production. Hence, the National Wasteland Development Board was established in 1985 under the Ministry of Forests and Environment mainly to tackle the problem of degradation of land, restoration of ecology and to meet the growing demands of fuelwood and fodder at the national level. The Government of India is taking up this colossal task of improving wastelands through its IWDP by revitalizing and reviving village-level institutions in order to harness the actual productivity of 'marginal areas' and enormous output of livestock products from these areas. Even under degraded conditions the livestock production from these lands by poor farmers is quite significant (up to 50 per cent of total), as evidenced by the fact that India is the largest exporter of sheep and goat meat in the world (23 Mt of sheep and goat meat in 2013–14); the world's largest milk producer and the world's biggest exporter of beef. The sheep and goat meat has been produced practically in its entirety in India's marginal rain-fed areas (Köhler-Rollefson, 2015). The quality of meat and ghee (clarified butter) produced from animals raised in the Thar Desert has also been found to be superior (Kamal Kishore, pers. comm.). Thus, the potential of grasslands can be exploited. In an estimate, even if 75 per cent area under *Sehima-dichanthium* cover is improved in terms of productivity increase from the present 0.65 t/ha to 1.75 t/ha, the fodder availability is likely to increase from the present 83

Mt to 224 Mt. In the watershed development programmes implemented by BAIF in the Saurashtra region of Gujarat and several districts in Karnataka, the immediate impact was the regeneration of various native grass species on field bunds and borders.

Grazing Lands as Global Resource for Biodiversity Conservation, Production of Medicinal Plants and Carbon Sequestration

Agro-bio-diversity provides the foundation for food production and grasslands are important for *in situ* conservation of genetic resources. Of the total of 10,000 species, only 100 to 150 forage species are in cultivation whereas a majority of the rest are found in the vast grazing lands. Despite the importance of grasslands and deserts for biodiversity conservation because of livestock dependency and emphasis on poverty alleviation programmes, the species diversity is under great threat. It is notable that some of the most threatened species of wildlife are found in the grasslands and deserts (e.g., Great Indian bustard, Lesser florican, Indian rhinoceros, Snow leopard, Nilgiri tahr, Wild buffalo, etc.).

Indian grasslands are very rich in biodiversity and represent the world's mega centers of crop origin and plant diversity in Western Ghats, Deccan Plateau, Central India, north-western Himalayas and the Northeastern hills. This diversity includes about 141 genera belonging to 47 families of higher plants which are endemic. Approx, 1,256 species of Gramineae belonging to 245 genera are represented in Indian grasslands and of these, about 21 genera and 139 species are endemic. These grasslands are maintained/conserved by 550 tribal communities of 227 ethnic groups. About 50 wild relatives of grass crops in the Southern Western Ghats are documented as a habitat for several endangered species. Additionally, Kerala state, the land of backwaters, also possesses a rich diversity in its wetland ecosystem. High species diversity and endemism among grasses in the form of 230 species of 22 genera is seen. Important genera conserved in the region are *Aeluropus, Echinochloa, Elytrophorus, Hygroryza, Hymenachne, Leersia, Leptochloa, Phragmitis, Pseudoraphis, Sacciolepis, Acroceras, Spinifex*, etc. In five conservation and community reserves, covering more than 80,000 hectares, the number of endangered/critically endangered species of mammals, birds, fishes and plants are being protected by local people under the guidance of ATREE, an NGO. Some of the grasslands with high ethinic value are also conserving vide diversity, e.g., Terai-Duar savanna and grasslands have characteristic species, such as *Saccharum spontaneum, S. benghalensis, Phragmitis kharka, Arundo donax, Narenga porphyracoma, Themeda villosa, T. arundinacea,* and *Erianthus ravennae* and shorter species such as *Imperata cylindrica, Andropogon* spp., and *Aristida ascensionis*. Kangayam grassland possesses eight perennial and six annual grass species with nine legumes and 16 forbs. Some of the novel germplasm in grasses developed in India, such as *Pennisetum squamulatum* (2n = 56); *P. pedicellatum* (2n = 72); *Panicum maximum* (3x, 4x, 5x, 6x, 7x, 8x, 9x, 10x, 11x cytotypes) are relevant to grassland diversity enrichment (Malaviya et al., 2015).

Peninsular India forms an important genetic resource centre for many grasses. About 50 wild relatives of relevant grass crops in southern Western Ghats are documented from here (Raj and Sivadasan, 2006). Grasses like *Arundo donax, Cynodon dactylon, Desmostachya bipinnata, Heteropogon contortus, Chrysopogon zizanioides, C. aciculatus, Saccharum spontaneum*, etc., are widely used as traditional medicines in the Western Ghats. There are seven species belonging to two genera (*Chrysopogon* and *Cymbopogon*) from which aromatic grass oils—Vetiver oil, Palmerosa oil, Citronella oil, lemon grass oil, ginger grass oil—are extracted and used (Raj and Sivadasan, 2007a). Dry grassland biomes of peninsular India are the prime habitat for several endangered species. Conservation of dry grasslands is a global challenge (Vanak, 2013).

Similarly, in Kerala state, the land of backwaters is endowed with rich plant diversity. Out of the approximately 400 grass species recorded from Kerala, about 230 species belonging to 22 genera are inhabited in both uplands and wetland areas. Among those, about 140 species are exclusively found in the wetlands. The genera such as *Aeluropus, Echinochloa, Elytrophorus, Hygroryza, Hymenachne, Leersia, Leptochloa, Phragmitis, Pseudoraphis, Sacciolepis, Acroceras, Spinifex*, etc., are found only in wetland or coastal regions of the state (Raj and Sivadasan, 2007b).

Thus, grasslands serve as best hotspots for *in situ* biodiversity conservation with people's participation which is nowadays considered to be economical and have become order of the day in several CGIAR centres also. Although, the practice of community-based conservation remains problematic because of its high dependence on centralized bureaucratic organizations for planning and implementation (Pimbert and Pretty, 1997), the community-based conservation is likely to be more cost effective and sustainable when national regulatory frameworks are left flexible enough to accommodate local peculiarities. Through 'UN Convention on Biological Diversity (CBD)-Article 8j' and 'Nagoya Protocol on Access and Benefit-Sharing 2014', it is mandated that the establishment of community protocols in which communities document their role in biodiversity conservation, articulate their demands for access and benefit-sharing (Köhler-Rollefson and Meyer, 2014). A number of livestock-keeping communities from India, Pakistan and Kenya have developed such community protocols, also referred to as Biocultural Community Protocols or BCPs (Köhler-Rollefson et al., 2012).

In India, there are nearly 95 national parks and 500 wildlife sanctuaries. Most of these are in the forest ecosystems. According to the report of the Forestry Commission (2006), nearly 40 per cent of these Protected areas (PAs) suffer from livestock grazing and fodder extraction. There are only a handful of PAs having grasslands. Notable ones are Velavador National Park (34 km^2) in Gujarat, Desert National Park (3,162 km^2 but less than 100 km^2 really protected), Kaziranga National Park (> 500 km^2, 60 per cent wet grassland), Manas Tiger Reserve (> 500 km^2, 40 per cent under wet grassland), Sailana Florican Sanctuary (2.50 km^2 grassland) in Gujarat. Some of the rarest species, viz., Bengal florican, One-horned rhinoceros, Pygmy hog, Hispid hare, Wild buffalo, Hog deer, Swamp deer in *terai* grassland, the Great Indian bustard are also found in the grasslands whereas according to reports of the Wildlife Institute of India (WII), less than one per cent of the grasslands come under the Protected Area Network. Hence, there is need to bring grasslands under the protected area network. The very presence of bustard species can be considered as an indicator of grassland ecosystem and by conserving the bustards and their habitats, a very large number of species dependent on healthy grasslands will also be protected.

Constraints and Pathways for Rejuvenating Tropical Grasslands/ Rangelands

Beyond doubt and delay, there is an urgent need to revegetate vast Indian grazing lands. However, several factors pose as constraints towards restoration. The most important constraint is funds. This is a huge area and mostly a common property resource. Hence, funds need to be allocated by the government although it needs people's participation. The social values and people's participation in both planning and implementation of these schemes is also of great importance. An environment is to be created in such a way that the communities feel responsible for restoration of these grasslands. Further, the grassland resources and its beneficiaries make up a complex environmental and institutional context in terms of varying agro-ecological conditions, rampant poverty among pastoralists, high-altitude climatic harness, environmental fragility, inaccessibility, conflicts on tenure issues of common pool resources and accelerated climate change in recent times. Specifically, the high-altitude grasslands and the people dependent on them deserve more investments and regional cooperation for continued supply of ecosystem services that have global goods and services value for humankind (Sharma and Shrestha, 2015).

Even if there is a plan/scheme for restoration of these grasslands, the availability of a large quantity of seeds of range grasses and legumes is likely to come in the way because of it being of low priority and low cost at the moment. To overcome the seed-related issues, research initiative has been taken up in India, however, more needs to be done. The innovative defluffing technique, using common cotton batting machine, helps in extracting the caryopsis from the fluffy seed material of *Deenanath* grass and significantly reducing the volume. The naked seed so produced is good for pelleting and precise sowing. Similarly, the *in vitro* maturation of panicles and application of IAA during pre-anthesis stage in Guinea grass has resulted in significantly increased seed-set and germination respectively. The technique helps in almost nil field loss and quality seed production. In the case of cultivated grass, the *in vitro* rooting of stem cuttings and rapid multiplication of *Bajra* napier hybrid rooted slips, using high density stem cutting nursery were found to reduce the time and space constraints in multiplication and transport of rooted slips (Malaviya et al., 2013). All these innovative technologies are cost effective, simple and can boost the grass seed and planting material production which otherwise is miniscule as compared to the demand.

Hence, to restore the grasslands in India there is a need of mission mode projects to revegetate the denuded grazing land. This will have long-term positive environmental impact also. Hence, considering these vast lands as a global resource, the programme should have backup support of international agencies. More to mention is that internationally it may be realized that tropical grasses which are C4 or intermediate have greater potential of C sequestration. Programmes to be implemented need careful planning by involving communities and taking lessons from other such programmes attempted in various parts of the country over the last few decades. Thoughtful interaction between the researcher and the developmental agencies is required as voluminous work on ecology of grasslands can be the key to success of revegetating projects. It is important to know whether the grassland ecosystems are in equilibrium systems and whether they will return if stresses are removed. Thus, ecological succession knowledge is important.

In this context, some of the recommendations of 23rd International Grassland Congress (recently held in Delhi, Nov 20–24, 2015) are quite relevant and may be considered:

- Savanna and old grasslands, with high ethnic values are a source of subsistence to millions; hence need to be conserved
- Constitute Grassland Authority of India
- Real-time remote sensing data
- Subsistence to market oriented and business plans involving communities
- Rural assessment of goals, skills, resource, climate change and socio-cultural modifiers.

Conclusion

In order to manage these grasslands for multiple use, we need to develop a comprehensive and integrated plan. A systematic approach following GIS-based inventory development of degraded rangelands, policy support for grazing and utilization of CPRs, incorporation of suitable recently developed technologies, community awakening and education, involvement of shepherds and the landless, rejuvenation of water bodies, quality seeds availability and market linkage for sustainability are likely to show positive results. Since there are interactions between different CPRs, and between CPRs and farmland, development of one CPR alone is not enough. Hence, livelihood approach to work in a more integrated way is to be followed. Since, one of the important output of grasslands is fodder availability round the year, various sources should also be looked into. Since India is characterized by tropical monsoon climate, active growth in grazing lands occurs only during the monsoon months with surplus fodder during rainy months and deficits of various levels in other months. Thus, there is already growing emphasis on animal feed security systems and fodder banks to overcome such problems.

References and Further Readings

Ajai and P.S. Dhinwa. 2018. Desertification and land degradation in Indian sub-continent: Issues, present status and future challenges. pp. 181–201. *In*: M.K. Gaur and V.R. Squires (eds.). Climate Variability Impacts on Land and Livelihoods, Springer, N.Y.

Anonymous. 1999. CSE (Centre for Science and Env.). Milk that ate the Grass. Down to Earth. April 15, pp. 1–5.

Anonymous. 2002. Common Pool Resources in Semi-arid India—Dynamics, Management, and Livelihood Contributions. Summary Findings of NRSP Project R7877. March 2002. http://www.nri.org/IndianCPRs.

Anonymous. 2014. Root Depths, Growth and Water-use Efficiency. Future Farm Industries CRC (FFI CRC) New South Wales. Tropical Perennial Grasses, 10.

Arnold, A. and R. Guha. 1952. Nature and Culture and Imperialism: Essays on the Environmental History of South Asia. OUP, 1995.

Babu, C., R. Kumar, V. Choudhary and V. Kumar. 2015. Tropical Grassland Ecosystems and Climate Change.

Bhattacharya, N. 1998. Pastoralists in a Colonial World. NSSO 54th Round, January 1998–June 1998, pp. 49–85.

Blasing, T.J. 2014. Recent Greenhouse Gas Concentrations. CDIAC, US Dept. of Energy, North Ridge National Lab. http://cdiac.ornl.gov/pns/current_ghg.html.

Dabadghao, P.M. and K.A. Shankarnarayan. 1973. The Grass Cover of India. Indian Council of Agriculture Research, New Delhi.

Dixit, A.K., M.K. Singh, A.K. Roy, B.S. Reddy and N. Singh. 2015. Trends and contribution of grazing resources to livestock in different states of India. Range Mgmt. & Agroforestry 36(2): 204–210.

Dubeux Jose, C.B. Jr. 2015. Cactus, a crop for dry areas. *In*: *Broadening Horizons*. Feedipedia. http://www.feedipedia.org/. N°28 April 2016.

Ghotge, N.S. 2004. Livestock and Livelihoods: The Indian Context. CEE India and Foundation Books, New Delhi.

Huntsinger, L. and P. Hopkinson. 1996. Viewpoint: Sustaining rangeland landscapes: A social and ecological process. Journal of Range Management 49: 167–173.

ICAR-NAAS. 2010. Degraded and Wastelands of India: Status and Spatial Distribution. Indian Council of Agricultural Research, p. 158.

Köhler-Rollefson, I., A.R. Kakar, E. Mathias, H.S. Rathore and J. Wanyama. 2012. Biocultural community protocols: Tools for securing the assets of livestock keepers. pp. 109–118. *In*: Biodiversity and Culture: Exploring Community Protocols, Rights and Consent, PLA 65.

Köhler-Rollefson, I. and H. Meyer. 2014. Access and Benefit-sharing of Animal Genetic Resources. Using the Nagoya Protocol as a Framework for the Conservation and Sustainable Use of Locally Adapted Livestock Breeds. GIZ, Eschborn.

Köhler-Rollefson, I. 2015. Ecologically and socially sustainable livestock development in marginal areas. pp. 283–288. *In*: D. Vijay, M.K. Srivastava, C.K. Gupta, D.R. Malaviya, M.M. Roy, S.K. Mahanta, J.B. Singh, A. Maity and P.K. Ghosh (eds.). Sustainable Use of Grassland Resources for Forage Production, Biodiversity and Environmental Protection: Proceedings of 23rd International Grassland Congress. November 20–24, 2015. New Delhi, India. Range Management Society of India, Indian Grassland and Fodder Research Institute.

Malaviya, D.R., D. Vijay and C.K. Gupta. 2013. Forage seed research—innovations at IGFRI. pp. 2.1–2.4. *In*: D.R. Malaviya, D. Vijay, D. Bahukhandi, C.K. Gupta, Vikas Kumar and H.C. Pandey (eds.). Quality Seed Production and Seed Standards in Forage Crops and Range Grasses: Challenges, Advances and Innovations. Indian Grassland and Fodder Research Institute, Jhansi, India.

Malaviya, D.R., A.K. Roy and P. Kaushal. 2015. Conservation of grassland plant genetic resources through people participation. pp. 318–326. *In*: D. Vijay, M.K. Srivastava, C.K. Gupta, D.R. Malaviya, M.M. Roy, S.K. Mahanta, J.B. Singh, A. Maity and P.K. Ghosh (eds.). Sustainable Use of Grassland Resources for Forage Production, Biodiversity and Environmental Protection: Proceedings of 23rd International Grassland Congress. November 20–24, 2015. New Delhi, India. Range Management Society of India, Indian Grassland and Fodder Research Institute.

Misra, R. 1983. Indian savanna. pp. 151–166. *In*: F. Bourlière (ed.). Tropical Savannas. Ecosystems of the World 13. Amsterdam/Oxford/New York: Elsevier.

Mittermeier, R.A., P. R-Gil, M. Hoffmann, J.D. Pilgrim, T.B. Brooks, C.G. Mittermeier, J.L. Lamoreux and G.A.B. Fonseca. 2004. Hotspots Revised: Earths Biologically Richest and Most Endangered Eco-regions, Mexico City, CEME, pp. 390.

Nalule, A.S. 2010. Social Management of Rangelands and Settlement in Karamoja Subregion. European Commission, Humanitarian Aid and FAO.

Pemadasa, M.A. 1990. Tropical grasslands of Sri Lanka and India. Journal of Biogeography 17: 395–400.

Pimbert, M.P. and J.N. Pretty. 1997. Diversity and Sustainability in Community-based Conservation. Paper Presented at the UNESCO-IIPA Regional Workshop on Community-based Conservation, February 9–12, 1997, India.

Puskur, R., J. Bouma and C. Scott. 2004. Sustainable livestock production in semi-arid watersheds. Economic and Political Weekly. July 31, 2004, pp. 3, 447.

Rahmani, A.R. 1997a. Wildlife in Thar. WWF. New Delhi, India, pp 1–100.

Rahmani, A.R. 1997b. The effect of Indira Gandhi Nahar Project on the avifauna of the Thar Desert. J. Bomb. Nat. Hist. Soc. 94: 233–266.

Rahmani, A. R. and Soni R.G. 1997. Avifaunal Changes in Indian Thar Desert. Journal of Arid Environments 36: 687–703.

Rai, P., Ajit and A.K. Samanta. 2007. Tree leaves, their production and nutritive value for ruminants: A review. Animal Nutrition and Feed Technology 7: 135–159.

Raj, M.S.K. and M. Sivadasan. 2006. Proc. National Conference on Forest Biodiversity Resources: Exploitation, Conservation and Management. March 21–22, 2006. Madurai Kamaraj University, Madurai, India, p. 12.

Raj, M.S.K. and M. Sivadasan. 2007a. Proc. National Conference on Medicinal and Aromatic Plants. December 10–12, 2007. Gulbarga University, Karnataka, p. 52.

Raj, M.S.K. and M. Sivadasan. 2007b. Proc. International Seminar on Present Trends and Future Prospectus of Angiosperm Taxonomy on October 4–6, 2007. Agharkar Research Institute, Pune.

Shankar, V. and J.N. Gupta. 1992. Restoration of degraded rangelands. In: Restoration of degraded lands - Concepts and Strategies. (Ed. Singh J S) Rastogi Publication. Meerut, India. pp. 115–155.

Sharma, E. and R.M. Shrestha. 2015. High-Altitude Grassland Management and Improvement of Pastoral Livelihoods in the Hindu Kush Himalayan Region. ICIMOD, Kathmandu.

Sharma, P.N. and S. Khadka. 2015. Emission of green house gases from grasslands and their mitigation. pp. 374–383. *In*: D. Vijay, M.K. Srivastava, C.K. Gupta, D.R. Malaviya, M.M. Roy, S.K. Mahanta, J.B. Singh, A. Maity and P.K. Ghosh (eds.). Sustainable Use of Grassland Resources for Forage Production, Biodiversity and Environmental Protection: Proceedings of 23rd International Grassland Congress. November 20–24, 2015. New Delhi, India. Range Management Society of India, Indian Grassland and Fodder Research Institute.

Singh, G. 2015a. Exploitation of wastelands for fodder production and agroforestry. pp. 198–207. *In*: D. Vijay, M.K. Srivastava, C.K. Gupta, D.R. Malaviya, M.M. Roy, S.K. Mahanta, J.B. Singh, A. Maity and P.K. Ghosh (eds.). Sustainable Use of Grassland Resources for Forage Production, Biodiversity and Environmental Protection: Proceedings of 23rd International Grassland Congress. November 20–24, 2015. New Delhi, India. Range Management Society of India, Indian Grassland and Fodder Research Institute.

Singh, P. 2015b. Tropical grasslands—trends, perspectives and future prospects. pp. 56–69. *In*: M.M. Roy, D.R. Malaviya, V.K. Yadav, T. Singh, R.P. Sah, D. Vijay and A. Radhakrishna (eds.). Sustainable Use of Grassland Resources for Forage Production, Biodiversity and Environmental Protection: Souvenir 23rd International Grassland Congress, November 20–24, 2015. New Delhi, India. Range Management Society of India, Indian Grassland and Fodder Research Institute.

Singh, J.S., Y. Hanxi and P.E. Sajise. 1985. Structural and functional aspects of Indian and southeast Asian savanna ecosystems. pp. 34–51. *In*: J.C. Tothill and J.C. Mott (eds.). International Savanna Symposium 1984. Australian Academy of Science, Canberra.

Steinfeld, H., P. Gerber, T. Wassenaar, V. Castel, M. Rosales and C. De. Haan. 2006. Livestock's Long Shadows-Environmental Issues and Options. FAO, Rome. p. 390.

Vanak, A.T. 2013. Conservation & Sustainable Use of the Dry Grassland Ecosystem in Peninsular India: A Quantitative Framework for Conservation Landscape Planning Report submitted to the Ministry of Environment and Forests, Government of India April 1st 2012–March 31st 2013.

Vanak, A.T., A. Hiremath and N. Rai. 2015. Wastelands of the mind: The identity crisis of India's savanna grasslands. pp. 33–34. *In*: M.M. Roy, D.R. Malaviya, V.K. Yadav, T. Singh, R.P. Sah, D. Vijay and A. Radhakrishna (eds.). Sustainable Use of Grassland Resources for Forage Production, Biodiversity and Environmental Protection: Souvenir 23rd International Grassland Congress, November 20–24, 2015. New Delhi, India. Range Management Society of India, Indian Grassland and Fodder Research Institute.

Wright, I.A., P. Ericksen, A. Mude, L.W. Robinson and J. Sircely. 2015. Importance of livestock production from grasslands for national and local food and nutritional security in developing countries. pp. 48–55. *In*: M.M. Roy, D.R. Malaviya, V.K. Yadav, T. Singh, R.P. Sah, D. Vijay and A. Radhakrishna (eds.). Sustainable Use of Grassland Resources for Forage Production, Biodiversity and Environmental Protection: Souvenir 23rd International Grassland Congress, November 20–24, 2015. New Delhi, India. Range Management Society of India, Indian Grassland and Fodder Research Institute.

Yadava, P.S. 1990. Savannas of north-east India. Journal of Biogeography 17: 385–394.

Zhaoli, Y. 2004. Co-Management of Rangelands: An Approach for Enhanced Livelihoods and Conservation. ICIMOD Newsletter, No. 45. Kathmandu: ICIMOD.

11

North American Grasslands and Biogeographic Regions

Paul F. Starrs,[1,*] *Lynn Huntsinger*[2] and *Sheri Spiegal*[3]

Introduction

North American grasslands are the product of a long interaction between people, land, and animals. While grasslands may form because of aridity, cold, or soil limitations, many have been created or expanded by human activity, most often by burning. Deliberately setting fires is a common part of the indigenous and traditional management portfolio in North America. It aims at reduction of trees and shrubs, manipulation of species composition, opening of areas for game, hunting, plant gathering, human habitation and diverse other purposes. The grazing of domestic livestock was not introduced to the continent until the 16th century and remained mostly a subsistence pastoralism until it became a widespread commercial activity in the 19th century. Instead, the grasslands found by Euroamerican colonists had already been shaped and sometimes created by the great hunting economies of native people and developed in the late Pleistocene as the last Ice Age drew to a close (Krech, 1999; Merchant, 2007).

Overall, the grasslands of North America include more than one billion hectares (Fig. 11.1). Working with the ecological qualities of life form and climate, the North American Commission on Environmental Cooperation has divided the continent into 15 Ecological Regions (CEC, 1997). Adopting that generalized view of North American grasslands with modifications to emphasize biogeographic regions and to distinguish cold and hot deserts, we recognize four generalized types: Mediterranean Grasslands, Temperate Grasslands, Desert Grasslands, and Tropical Dry Forest. Temperate Grasslands can be further broken down to include Tallgrass, Mixed Grass, and Shortgrass Prairies and the Southern Coastal Plains. Desert Grasslands include the Hot Desert, Cold Desert,

[1] Department of Geography, University of Nevada, Reno.
[2] Department of Environmental Science, Policy, and Management, University of California, Berkeley, California.
[3] Jornada Experimental Range, United States Department of Agriculture – Agricultural Research Service, Las Cruces, New Mexico.
* Corresponding author: starrs@unr.edu

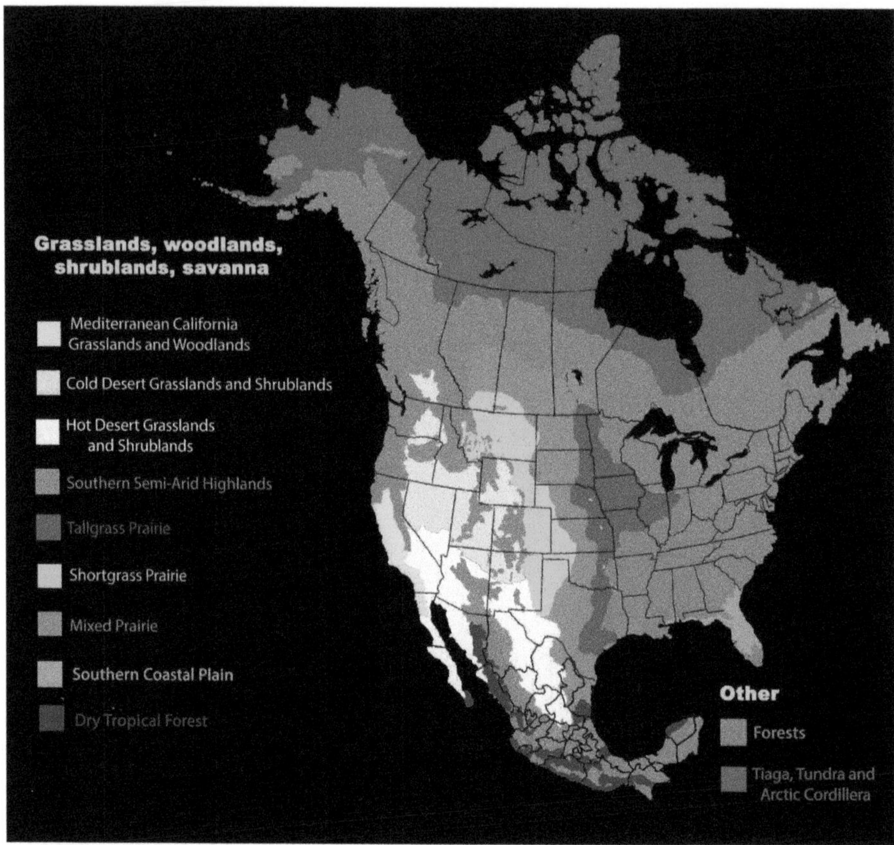

Fig. 11.1 Grassland regions of North America. (Modified from CEC, 1997.)

and Semi-arid Highlands (Table 11.1). All of these grasslands may also be referred to as rangelands, defined as "lands on which the native vegetation…is predominantly grasses, grass-like plants, forbs, or shrubs suitable for grazing or browsing use" (US-EPA, 2017).

Internal variation in physiography is pronounced in this landscape: elevations ascend from a low of 86 m below sea level (Death Valley, California, USA) to alpine-zone summits and volcanic peaks of some 4,000 m (Mt. Whitney, 4,418 m, lower 48 United States; Mt. Robson, 3,954 m, Canada; Pico de Orizaba, 5,610 m, Mexico). A defining trait of the realm is its vast surface area.

The Great Plains support an extensive, historically unbroken grassland in the continent's temperate interior, extending from the prairies of Canada to Mexico (green areas in Fig. 11.1). While now interrupted for grain production and irrigated agriculture, the Plains include the prairie types and, to the southwest, grade into Desert Grasslands. These prairies provided forage for the vast herds of bison—numbering into tens of millions—that once roamed in the 19th century and earlier them (Isenberg, 2000; Lott, 2003). Bison were the backbone of the economy of indigenous tribes, and a combination of herbivory, burning and environmental factors kept the prairies an open sea of grass. These grasslands grow on rich, wind-deposited loess soils and in the most fertile areas to the east, they have long been converted for crop production in the United States and Canada. With aridity increasing from east to west, the grassland grades from the Tallgrass Prairie, through the

Table 11.1 Major grassland types in North America: Status and threats to sustainability.

Generalized Grassland Type	Subtypes	Characteristics	Current Status	Threats
Mediterranean Rangelands	Mediterranean California Grasslands and Woodlands	Annual species; oak woodland and savanna with scattered trees; shrublands or chapparal particularly on shallow soils. Remnant native bunchgrasses, along the coast or often in moister areas. High biodiversity.	Native herbaceous species largely replaced by non-native annual grasses and forbs; grazing land; irrigated crops occupy good soil and flat areas.	Residential and urban development; changes in fire frequency; invasive annuals; cultivation.
Temperate Grasslands	Shortgrass and Mixed Grass Prairies	Mid sized bunch grasses and sod-forming shortgrasses, adapted to intensive grazing.	Grazing land and cropland, rainfed crops. Bison reserves.	Climate change, inappropriate cultivation; mining, energy production, and related infrastructure; drought and fire.
	Tallgrass Prairie	Large, very dense bunchgrasses, extremely productive and adapted to intensive grazing.	Mostly replaced by crops, notably rain-fed corn and soybeans.	Climate change, mining and related infrastructure; drought and fire.
	Southern Coastal Plain	Subtropical bunchgrass grassland and savannah interspersed with woodland, lakes, and wetlands.	Extensive timber plantations and urban expansion.	Climate change, urban expansion and population growth, non-native invasives.
Desert Grasslands	Cold Desert Grasslands and Woodlands	Sagebrush (*Artemisia*) steppe at elevations of 1,000 m and more, native bunchgrasses, non-native invasives most notably cheatgrass, Pinyon-juniper woodlands.	Grazing land and mining, irrigated crops, wild horses, huge expansion of cheatgrass.	Invasives esp. cheatgrass, increased fire frequency; mining; loss of sagebrush to fire.
	Hot Desert Grasslands and Shrublands	Widely spaced bunchgrasses in dry areas with, higher density in moist swales, cacti, woody plant invasion as well as pasture grass invasion in some areas.	Grazing land and mining, irrigated crops, spreading irrigated agriculture.	Thickening of woody plants, drought, increasing aridity, soil loss; loss of water table due to pumping for irrigation; energy production.
	Southern Semi-arid Highlands	Prairies, small trees, parkland; grassland interspersed with desert scrub at low elevations, oak and juniper at intermediate elevations, and coniferous forests at high elevations.	High biodiversity; small farms; urbanization.	Increasing shrub densities, invasive annual grasses, nitrogen deposition from power plants; urban sprawl, overgrazing, inappropriate cultivation; wealth differentials.
Tropical Grasslands	Dry Tropical Forest	Forests lose most of their leaves in the dry season, supporting the development of a forage base of shrubs and warm-season grasses.		Significant conversion to agricultural use; grazing may limit tree regeneration.

Mixed Grass Prairie, to the Shortgrass Prairie at the foot of the Rocky Mountains (Fig. 11.1). Inappropriate plowing of the arid Shortgrass Prairie resulted in the greatest environmental disaster in American history, the Dust Bowl (Worster, 1979) from 1934 to 1940, as prolonged but not untypical drought and the breakup of the sod-forming grasses by the plowing of eager homesteaders and absentee speculative farmers (Hewes, 1974) caused much of the topsoil to blow away. The eastern and northern reaches of the prairies are largely in corn, soybean, and other crop production today, while to the west, grazing is the common activity on non-irrigated areas.

The Southern Coastal Plains of northern and central Florida are typically savanna important for livestock production (Swain et al., 2013). Undeveloped areas of Florida are woodland, savanna, and a vast grassland, including the Everglades wetland, but sizable population growth has expanded urban and developed land in this region. Between 1970 and 2000, the population increased by more than 140 per cent, and large urban areas are now present on the Florida peninsula. Aside from agriculture and the extensive pine plantations, tourism and associated service and recreational industries are important economically.

The grasslands and oak woodlands of drought-prone Mediterranean California fed the elk, antelope, and deer that sustained Californian tribes, providing diet staples, such as acorns and grass seed. Frequent intentional burning by indigenous groups is unanimously attested to by native Californians and well-studied using tree rings, lightning records, and other forms of historical analysis. In many areas, tree and shrub density has increased dramatically, causing a precipitous rise in wildfire risk and changing the species composition of woodlands and savanna, for example, with invasion of Douglas fir (*Pseudotsuga menziesii* (var. *menziesii*)) and coyote brush (*Baccharis pilularis*) in central and northern coastal California. The grasslands of the state have been profoundly altered by introductions of annual grasses from other Mediterranean regions and for the most part, the introduced grasses have naturalized and become the major forage base. Among the plant functional groups supplanted by these new species are native bunchgrasses and native perennial and annual forbs. New invasive species, some considered pests, continue to arrive. The grasslands with soils level enough for the plow are largely converted to crops, with livestock grazing remaining the major land use in the hills. New technologies allow grapes, almonds and other specialty crops to move into hill areas. Residential and urban sprawl have consumed vast acreages in California's mild, attractive climate.

In the deserts of the western United States and northern Mexico, invasive species, native and non-native, are changing fire frequencies and the agricultural productivity of the grasslands. In these arid and semi-arid lands, bunchgrasses with deep roots once shared the land with shrubs, sparse trees, and in the hotter zones, they grew alongside cacti, including the giant saguaro. In the Cold Desert, a steppe with snowy winters, bunchgrasses typically share the range with sagebrush or shadscale, giving the land a silvery aspect with high albedo. Here, fire supported by invasion of the invasive annual cheatgrass (*Bromus tectorum*) threatens sagebrush and other native species. Indeed, invasive annual and perennial grasses and forbs invading and expanding beyond the disturbed soils left by abandoned farmland, excessive livestock and horse grazing, mining waste and road construction, is a major issue in the Cold Desert and portions of the Hot Desert. Grasses, colonizing the gaps between bunchgrasses and shrubs, help increase ignition and spread of fires, creating burnt-over lands easily colonized by more invasives, many from the Eurasian steppe. In other parts of the Hot Desert, mesquite (*Prosopis* spp.), a largely unpalatable thorny shrub-tree, readily expands its range, for reasons not fully understood but which may include climate change, seed distribution by grazing, and fire suppression.

In Mexico, the Tropical Dry Forest loses most of it leaves in the dry season, supporting the development of a forage base of shrubs and warm-season grasses. Mexico's grasslands and semi-arid highlands were grazed by domesticated animals beginning not long after the Columbian arrival in 1492 (Sluyter, 1996, 2012). The conversion of native shrublands and dry forests to non-native pasture plants such as buffelgrass (*Pennisetum ciliare*) and the spreading and thickening of mesquite and acacia shrubs and trees (*Acacia* spp.), have severely diminished the extent of some native grassland types. A significant portion of the rural population lives in poverty and/or on a subsistence basis. Smallholders, using degraded rangelands and growing subsistence crops, or feed-grains already in surplus, remain trapped in a cycle of producing little cash income and depleting their natural resource base year to year. About 40 per cent of Mexico's Tropical Dry Forest has been converted to agriculture in the last few years and it is considered an endangered type.

Table 11.1 gives the different types derived from the North American Commission on Environmental Cooperation (CEC, 1997). Further and different divisions of our generalized grassland types are common and many more fine-grained divisions can be made, but they do not serve the purpose of this chapter and for the most part are not discussed here. In addition, there are meadows and clearings and alpine grasslands, occupying much smaller areas but providing significant habitat and forage, especially in montane areas, where they offer green forage in the summer.

Following a long history of grazing by large ruminant herbivores (Mack and Thomspon, 1982) and indigenous management through fire and pruning, North American grasslands are now the domain of livestock graziers, miners, and recreationists. In some rural areas, mining and energy production cause boom and bust economic dynamics, and with that arises a measure of popular protest against such development. Rangelands that once went virtually uninterrupted are now crossed east-west by interstate highways and railroads, and sliced-and-diced north-south by equivalent thoroughfares. Recent additions include electrical transmission lines of up to a 1,000 kV and international pipelines created to move slurries, tar-sands, and precursor petrochemical products to refineries and ports along the Gulf of Mexico.

Economies associated with livestock ownership grade from indigenous and subsistence to fully globalized operations drawing on multinational capital reserves; the land owned and controlled can vary from none at all to hundreds of thousands of hectares and the numbers of livestock included in an operation range from one or a handful to tens of thousands. Today the grasslands of the continent support some 42 million beef cows, 5.5 million ewes, and 11 million goats. The semi-arid (precipitation 300–500 mm) U.S. Great Plains and Canadian prairie and plains are home to much of the livestock—16 per cent of the North American (= North American area) area contains 57 per cent of beef cows. Texas alone has almost 50 per cent of U.S. goats (many of them mohair goats, bred for soft hair). Unintended, but un-eradicable users of grasslands include feral pigs and feral burros and horses (mustangs, wild horses) that are part of the human heritage and cannot be easily removed for management or unwieldy political reasons (NRC, 2013; Nordrum, 2014; Snow et al., 2017). The role of grazing livestock, their owners and their tenders—variously the cowhand, buckaroo, charro or vaquero—constitute in each country a national legacy. Yet the grasslands are increasingly an amenity-rich magnet drawing industrial and residential development, as once-small rural outposts become home to companies seeking low-cost space and amenities that will attract and keep workers. Competition for the grasslands is having an impact on land use and conversion (see Chap. 17, this volume).

Physiography and Morphology of North American Grasslands

The morphology and physiography of western North America ranges from the geologically ancient and glacially-eroded continental shield in central Canada to the slowly westward upward-ramping Great Plains of the United States, to a massive and still-forming cordillera that extends from mid-Mexico through the Rocky Mountains of the United States into the Canadian Rockies (Fig. 11.2). Along the western edge of North America runs a moving system of tectonic plates and terranes with active faults that produce igneous features of various ages, ranging from the volcanoes of Mexico and the Sierra-Cascade region of the Pacific Coast to the vast batholiths of the Sierra Nevada of California and significant sections of Canada. The Pacific Slope is especially diverse: wet and volcanic in the Canadian north and Pacific northwest, increasingly arid moving south through California, and truly dry with distinctive endemic vegetation down the Baja California peninsula. The entire 5,000 km is subject to irregular earthquakes. While the effects of geologic change are gradual on the eastern edge of the arid zone from Canada to Mexico, these shifts are day-to-day visitors to the west, and microclimates and endemism make the plant biogeography

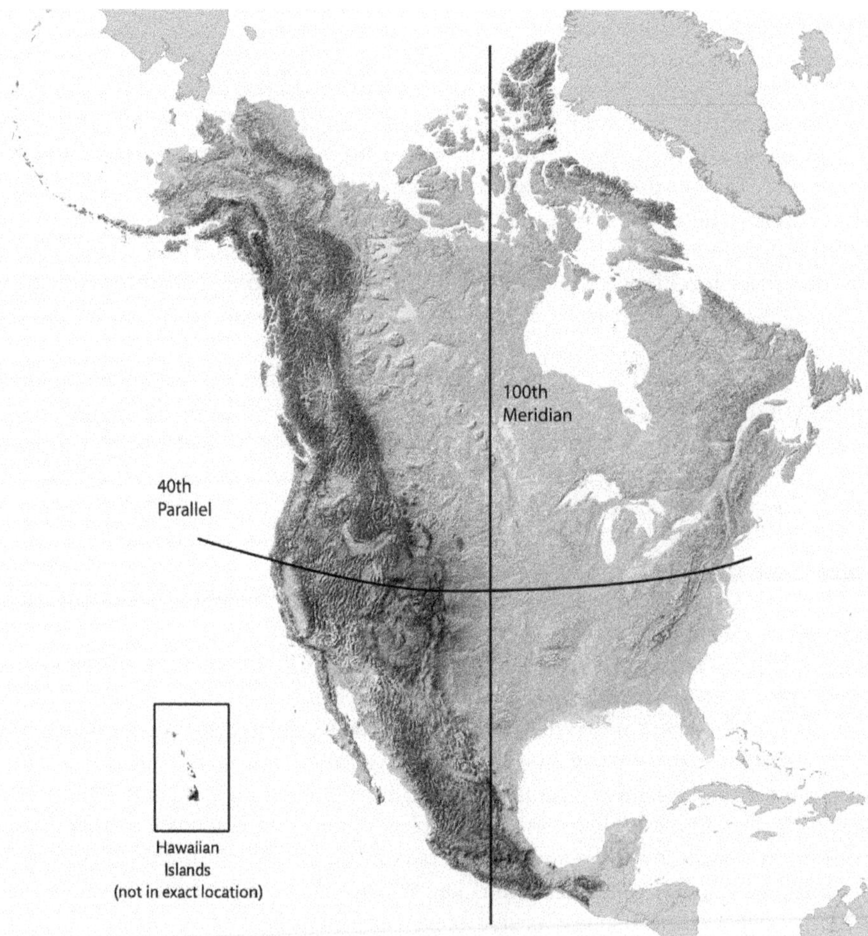

Fig. 11.2 Shaded relief map showing continental terrain. (Modified from CEC, 2007.)

something that seemingly varies every few dozen kilometers traveled and hundreds of meters ascended.

While variation is paramount, one generalization is fundamental—the land west of the 100th meridian (which runs largely through the Mixed Grass Prairie) is distinguished by fault-shaped landforms, often-dramatic relief and a pattern of aridity (Fig. 11.2). That meridional divide traditionally defined the 20-inch (500 mm) isohyet—more rain falls east of that line; as a rule, less to the west, except in the higher mountains. In Mexico and Canada, the same distinctions broadly apply: northern Mexico is far more arid than subtropical southern Mexico and western Canada far drier than the Canadian east. To the east of the 100th meridian, entering the climate and soils of the Prairie Grasslands, rain-fed agriculture becomes possible and potentially profitable.

A Transect

One way to visualize the topography and vegetation of the continent is to follow a transect from west to east, as marked differences in rangeland and habitat are discernable along this path (Huntsinger and Starrs, 2006). Traveling from the Pacific Coast and Mediterranean California to the Midwestern-Great Plains involves a journey through distinctive precipitation regimes, ecological types and recognizable regions (Fig. 11.3).

At the Pacific Coast, grasslands, woodlands and chaparral are situated in a Mediterranean climate on a highly heterogeneous area of sandstone formation. Toward the north, these grade into volcanics and toward the east, they venture into the granites of the uplifted Sierra Nevada. The Mediterranean climate, with dry, hot summers and mild yet wet winters, spans from the coast into the Sacramento Valley and even east into the Sierra Nevada foothills. Most of the rangeland in the lower elevations is under private ownership (Huntsinger and Bartolome, 2015), with mountain areas either private land, or managed in some form by governmental entities, such as the Forest Service or National Park Service.

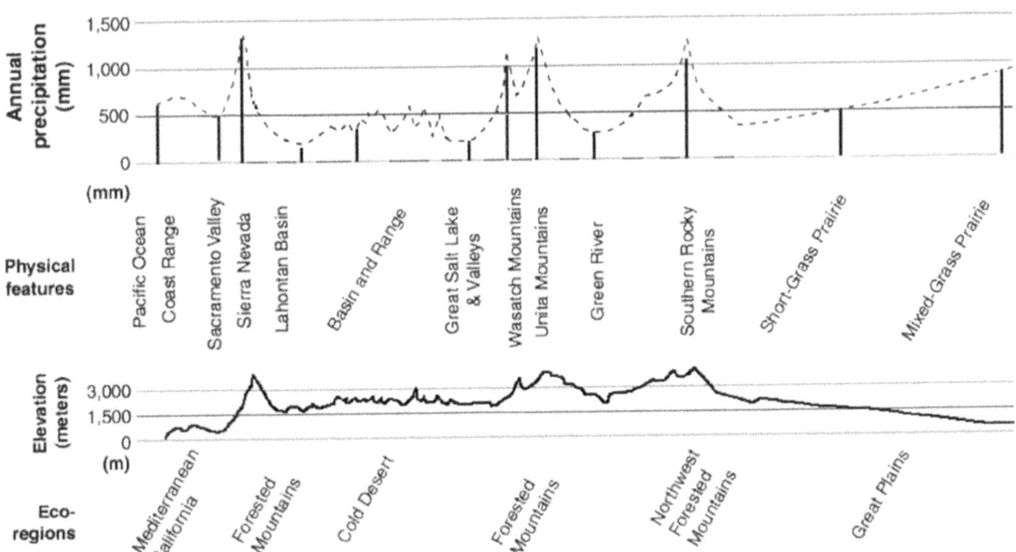

Fig. 11.3 A transect across the continent, west-east on the 40th parallel. (Modified from USDA, 1941 and Huntsinger and Starrs, 2006; for location, see Fig. 11.2.)

The hot dry summers, combined with a one-hundred-year legacy of fire suppression, have resulted in frequent wildfires that have grown in extent and severity over time, especially in the mountains. Soil depth, fire, and climate result in a dynamic relationship between the woodlands, savanna, and grasslands, all of which share a herbaceous layer of annual grassland. In this western region of the transect, water is highly limited and irregularly distributed. Most arises from snowmelt off the many mountain ranges and is either diverted from mountain streams and rivers or pumped from subsurface flows on alluvial fans.

The forests of the Sierra Nevada provide summer range used in transhumance historically and even today. Mountain meadows are green and nourishing when Mediterranean grasslands and the Cold Desert grasslands to the east are dry. Use by livestock, however, has markedly declined in recent decades as recreational use increased and fire suppression produced a thickening of woody vegetation. The Sierra also provides most of California's and northern Nevada's water supply. As lowland watered areas were converted to intensive crop production by the 20th century, water is today distributed through complex and sophisticated irrigation systems and waterworks developed to support the extremely valuable cropland agriculture of California (Starrs and Goin, 2010).

Eastward on the transect, in the rain shadow cast by the Sierra Nevada, lie Cold Desert grasslands where elevations are 1,000–2,500 m (Fig. 11.3). The region continues east for a thousand kilometers, broken by successive north-south mountain ranges with their own supply of snow-derived water and grassy meadows. While a physiographer might prefer the term 'basin and range' province, the area is variously called the Intermountain West, the 'Great Basin,' or described as Cold Desert Steppe.

The rolling topography of the *Artemisia*-dominated lowlands looks like a silver sea of sagebrush. The vegetation is a shrub-bunchgrass rangeland, with shrubs spaced 0.5–2 m apart and about 60 per cent bare soil. Typically, the basins of the region have playas, or salt pans, at their lowest points, sometimes with yearlong water—a pattern evident in some isolated parts of Canada, but quite common through the United States and extending southward into arid Mexico. Salt-tolerant grasses and shrubs grow on the clayey soils.

Sometimes set aside as military or native American Indian reservations, the lowlands in this part of the transect are otherwise mostly under jurisdiction of the U.S. Bureau of Land Management (BLM). Montane regions are more than 90 per cent administered by the U.S. Forest Service (USFS). Private land is scarce; 93 per cent of Nevada, California's neighbor to the east, is public land. Invasion of non-native species, particularly cheatgrass and Russian thistle (*Salsola* spp.), have increased fire frequencies, putting at risk sagebrush and the animal species dependent on it. In the Cold Desert, the alluvial fans exiting the mountain ranges are typically below or near the headquarters of range-based livestock operations—alluvium at the outwash routinely provides the best soil for cultivation and the water is close to or at the surface through the summer, sometimes allowing artificial or natural flood irrigation of meadows. Much of the surface water and groundwater recharge in the region is produced by orographic precipitation that concentrates in the mountains, and then emerges by runoff through rivers, streams, springs and aquifer-tapping to solve human supply needs. Runoff is strongly seasonal, often episodic, and sometimes ephemeral. Transhumance is a common rangeland use in this region, drawing on transient water with a contemporary and historic importance that varies from ecological regions.

Mountain ranges alternate with the many basins of the Cold Desert, each terminating in a playa or salt pan, until finally the Great Salt Lake is reached, nestled in the Wasatch Front—a dramatic active-fault mountain range marking the boundary between the Cold Desert and the Rocky Mountains to the east. The Rockies form a cordillera that runs from

Mexico all the way north through Canada. Along the 40th parallel, they are at their most formidable: fifty-four peaks in Colorado rise above 4,267 m. A comparatively mesic climate and the high mean elevation make this an area rich in meadow, lake, and conifer lands suited to transhumance. From the early 20th century onward, a significant part of this region was under the control of the U.S. Forest Service or, for less remunerative, more decorative, or recreation-appropriate lands, under the administration of the National Park Service.

Once the Rockies are crossed to the east, a steep drop in elevation leads to the shortgrass prairies of the western Great Plains Region (Fig. 11.3). These rangelands are dominated by short sod-forming grasses, less than 25 cm tall, growing on loess soil blown from the north. These were once home to large herds of migratory bison and antelope. The shortgrass prairie grades into tallgrass prairie to the east—but the tallgrass prairie, aside from some 'National Grassland' and other reserves, is today the major corn and soybean growing area of the United States, and is almost fully cultivated and privately owned. Western ranches that graze cattle on rangeland give way to intensive livestock operations—backgrounding pastures and confinement feedlots (CAFOs, concentrated animal feeding operations) where cattle and pigs are held to feed on corn, soybeans, wheat and other grains produced where the tallgrass and shortgrass prairies once grew. While the relatively fewer feedlots elsewhere—northern Mexico, California, the plains of western Canada—draw on agricultural byproducts and aftermath, considerable numbers of livestock are shipped from all over the western U.S. for feeding on the prairies.

Dry Tropical Forest, Southern Coastal Plains, Hot Deserts, and the Southern Semi-arid Highlands are missed by the transect (Table 11.1; Figs. 11.2 and 11.3). In Mexico, land redistribution has shaped today's land tenure patterns. Extensive, low productivity rangelands of the Hot Desert region of Northern Mexico are mostly in large ranches and community landholdings. Where agriculture is possible, small holders with subsistence goals are common, though in the last decade, irrigated agriculture is making inroads into the grasslands, despite the specter of groundwater depletion. Rural land in Mexico is mostly private or a part of the community-owned *ejidos,* though there are a growing number of nature and recreation reserves and Mennonite colonies centered on irrigated agriculture.

In the United States the Hot and Cold Deserts (Fig. 11.1), 19th-century land distribution policies resulted in large areas of land-uncultivatable due to aridity of topography remaining under diverse forms of government control. Government ownerships are substantially more complex now than a hundred years ago, with areas increasingly designated for special protection, such as wildlife reserves, wilderness areas and parks, endangered species habitat, or reserved for activities like 'off-highway vehicle use' cultural activities. While deserts have by far the largest proportion of public or government lands, lands in the prairies where rain-fed agriculture is possible have higher rates of private landownership than the other regions discussed herein.

Retirees and second-home owners from cooler regions often seek out southwestern locations in the United States (U.S.) or northern Mexico, while tourism spreads people throughout the West (Starrs and Wright, 1995). Northern Mexico, the source of so much of the 19th century livestock culture of what would become the United States and Canada, has long attracted residents interested in warm and remote areas. Except for border zones where urban populations are quite large and linked to high-intensity assembly and manufacturing industries, there remain extensive sparsely populated areas still used for livestock grazing. Much the same is true in Canada, where the westernmost plains are dotted with cities that function as supply centers for petroleum and mineral booms and where exploited resource products are conveyed to international markets (sometimes via pipelines that cross from Canada to the United States). Only tourist venues or isolated service-oriented livestock towns exist across much of the Canadian plains and the Northern Rocky Mountains.

Major Grassland Types: Species, Climate and Productivity

In this section, we describe the forage characteristics of the plant communities of the major grassland types of North America (Table 11.1; Fig. 11.1). The Society for Range Management's 'range cover type' designations (Shiflet, 1994) and other sources are used to provide information about grassland productivity and stocking rates (Huntsinger and Starrs, 2006).

Mediterranean Grasslands

This grassland type, relegated mainly to the U.S. state of California and Baja California Norte, is called Mediterranean because the climate is similar to that of Europe's Mediterranean Basin (Table 11.2). Peak productivity is in late spring. Forage quality peaks in early spring and is at its lowest in fall after rains hit summer's dry forage. The woodlands, shrublands and grasslands of this region are used year-round or as the winter part of transhumance, with most calving in fall. The forage base is cool season grasses that can provide nutritional value into the warm season. Gradients of increasing rainfall trend south to north, and east to west, while temperature increases north to south. The climate is moderated by Pacific and coastal fog.

Soils are highly diverse and heterogeneous. Understory grasses are mostly introduced through annual species from other Mediterranean regions, intermixed with native bunchgrasses, annuals, and forbs. *Quercus* spp., are a widespread component, with more than 14 species. Where the canopy is less than 50 per cent, understory annual grassland is usually well developed and highly productive. In deciduous types, the understory develops at evenly high densities. Because the forage base is annual grasses, the most usual management is based on the residue management concept. Enough dry matter is left behind after grazing to protect soil from erosion and to influence the seed bank in a positive way. A high annual (and seasonal) variability in rainfall and subsequent wide swings in forage production are evident, as is a typical eight-month annual drought. These pose a significant challenge to livestock producers. Buying weanlings in the fall and selling them in the spring are ways of coping, as are transhumance, irrigated pasture and the feeding of supplements. Agricultural by-products of a highly productive irrigated agriculture offer an abundant aftermath that can be fed to livestock coming off grass as it dries in late spring. Once plentiful, irrigated pastures are being converted rapidly to higher value crops, such as wine grapes.

Cow-calf production is the common livestock system in the Mediterranean grassland, with yearling grazing operations and some sheep grazing. When used, transhumance often requires a combination of private land in grasslands and public land in the montane regions. Ranching is often confined to hilly areas as lowlands are largely cultivated or developed for urban-residential use. Urbanization is spreading throughout the region and is a significant threat to rangelands and rangeland. The native vegetation is fire-adapted and indigenous Californians commonly use fire as a management tool. A history of post-contact fire suppression, however, has fed a huge buildup of fuels in shrublands, woodlands, and

Table 11.2 Mediterranean grassland characteristics (CEC, 1997; Barbour and Billings, 1988), including selected Rangeland Cover Types, after Shiflet, 1994; Huntsinger and Starrs, 2006).

Generalized Grassland Type	Precipitation and Elevation	Characteristic Species	Forage Production in kg/ha and local recommendations for allowable use[2]
Mediterranean Rangelands **168,523 km²** Koppen: Cs	Elevation: 0 to 800 m. Precipitation: 200–1,000 mm, falling Nov–April, with summer drought. Frost-free period 250–350 days.	Oak woodland and savanna, chaparral, and grassland occur throughout ecoregion and intermix, constrained by soils, precipitation, fire, drought, and human activity. Areas with dense tree or shrub cover are unsuitable for grazing.	500–3,000 kg/ha (Holechek et al., 2001). Minimum residual dry matter (RDM) guidelines with modifications for slope and woody plant cover (Bartolome et al., 2002). Productive coastal prairie, averages 3,000 kg/ha in coastal prairie, can exceed 10,000 kg/ha (Cooper and Heady, 1964).
Typical Rangeland Cover Types (Shiflet 1994)			
blue oak woodland SRM 201	Elevation: 90 to 1,800 m.	Trees: (10–20 per cent cc[1] in savanna; 21–100 per cent in woodland) Dominant species: *Quercus douglasii* Hook & Arn., but mixes with *Q. agrifolia* Née, *lobata* Née, *wislizenii* A. D.C. and *kelloggii* Newb. (Griffin, 1977). *Umbellularia californica* (Hook. & Arn.) Nutt. Average of 70–130 trees/acre. Herbaceous (80–100 per cent cc[1]) annual grassland understory. 250 different species at one site (Shiflet). Scattered shrubs, shrub understory in some areas.	700–3,000 kg/ha (Holechek et al., 2001; UC-DANR, 1996).
chamise chaparral SRM 206	Elevation: 150 to 1,000 m, sandy to rocky well-drained soils.	Shrubs (30–90 per cent cc[1]): *Adenostoma fasciculatum* Hook. & Arn., *Arctostaphyloss* spp., *Ceanothus* spp., *Quercus dumosa*, *Cercocarpus montanus* var. *glaber* (S.Watson) F.L. Martin, *Artemisia* spp. Herbaceous (0 per cent cc[1] where shrub canopy is closed, to 100 per cent cc[1] after fire) native and non-native grasses of annual grassland, *Elymus* spp., *Erigonum* spp.	0–1,500 kg/ha
valley grassland SRM 215	Elevation: 0 to 500 m.	*Herbaceous* (90–100 per cent cc[1]) Introduced annual grasses *Avena* spp., *Bromus* spp., *Festuca* spp., *Hordeum* spp., *Lolium* spp., *Erodium* spp., *Trifolium* spp., *Madia* spp., *common*, components of native grassland include *Nassella* spp., *Leymus* spp., *Melica* and *Danthonia* spp.	400–3,500 kg/ha (Holechek et al., 2001). Production can exceed 4,000 kg/ha (Bartolome et al., 1980).

[1] Canopy cover.
[2] **Used to determine the number of animals to be placed on an area with the goal of sustainable use..**

forests. Californians have attached multiple goals to the rangelands of California: livestock production, natural resource conservation, soil carbon sequestration, recreation and more. To increase chances for optimal management of these cherished lands, in recent decades diverse stakeholders have collaborated to form coalitions to transfer knowledge and share ideas.

The most widespread woodland in the region is the Rangeland Cover Type 'blue oak woodland' (SRM 201, in Shifflet, 1994). It grows on relatively dry sites, typically in hilly areas and with other oaks, and is drought tolerant. Major controls on the number and structure of oaks is the north-south gradient, aspect, elevation, and fire history. The blue oak (*Quercus douglasii* Hook. & Arn.) is winter-deciduous. Regeneration of some species of oaks is considered inadequate in some areas, possibly due to massive species change in understory plants, fire regimes, grazing, changes in wildlife populations and other factors. It is argued that grazing in the understory reduces the high fire hazard in summer that threatens the woodlands; it is also argued that grazing reduces oak seedling survival and growth in some cases. About 80 per cent of this type is in private ownership.

A widespread shrubland is the Rangeland Cover Type 'chamise chaparral'; SRM 206. Growing on steep hillsides and shallow soils, dense chamise stands—sometimes described by using the Spanish term 'chamissal'—are unsuitable for grazing. However, grazing can be good where the brush is patchy, or after fire or clearing. Fire hazard is usually severe, with an abundance of fine fuels and shrubs containing flammable oils. Grazing can be used to reduce and control shrubs following burning or cutting, and using goats for vegetation management is a growing industry. Conversion to grassland is sometimes undertaken on deeper soils.

'Valley grassland' (SRM 215, in Shifflet, 1994), is a widespread Rangeland Cover Type characterized by annual grasses. The grassland on good soils has largely been converted to crops, and only 6 per cent is in public ownership; otherwise, grassland intermixes into woodland and chaparral, and occupies harvested and burned sites. Production is highly seasonal, peaking in spring and drying in summer and fall. On this forage base, calving typically takes place in fall and lambing is pushed back to winter. Most of the grasses are from other Mediterranean regions and invasion by exotics, some unpalatable, is a regular occurrence. Yellow star thistle, medusahead (*Taeniatherum caput-medusae*), and goat grass (*Aegilops cylindrica*) are recently-arrived invasive species that have proved difficult to control and reduce rangeland value.

Mediterranean Grassland Community Profile: Oakdale (Stanislaus County), California, USA

Oakdale (Fig. 11.4), at just 48 m (157 foot) elevation, is one of a chain of livestock service towns in the San Joaquin Valley that have for a hundred years found steady business providing goods and shipment for ranchers and farmers. The valley runs north-south through central California and bears the name of the San Joaquin river, its main drainage source. These communities are mostly at the foot of the Sierra Nevada foothill grazing zone on the Stanislaus river and near deep-valley agricultural alluvial sandy loam soils. While the established business of Oakdale in the 19th century was cattle-raising, for a time sheep herds coming down from the Sierra on a counter-clockwise circle-migration route to graze grasslands and crop stubble added an important element of seasonal use (Igler, 2001). Now, Stanislaus County is a sizable agricultural contributor in California, producing tree crops (almonds and walnuts are prominent), hay for forage and irrigated pasture,

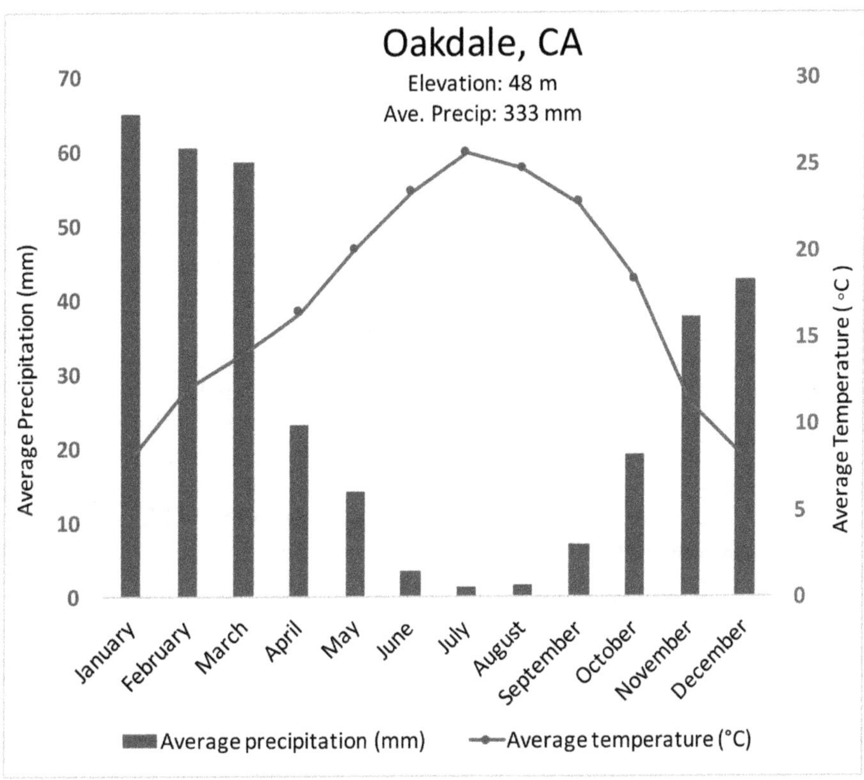

Fig. 11.4 Oakdale, California, in the Mediterranean Grassland region.

kiwifruit, chickens and turkeys, eggs, apricots, and milk and dairy products (Starrs and Goin, 2010).

It is, however, the Sierra Nevada foothill rangelands north and south of Oakdale that have long produced a significant and stable livestock industry. Directed principally now at cattle, grasslands and oak woodlands with a pasturable understory include some sheep-raising operations and replacement dairy herds, but once historically were home to goats and horses and a robust wildlife population including deer and such keystone predators as wolves and grizzly bears. The foothills, rising towards forest land in the Sierra Nevada, are vastly changed (like much of California) from the original species mix: exotic Mediterranean species introduced in the mid- to late-18th century are now established and provide a visual dominant (Alagona et al., 2013). With a pronounced summer drought, livestock are on spare rations from May or June onward, and requires movement of livestock to wetter upland areas, shipping to other locations, or feeding on agricultural aftermath.

Livestock ranchers, especially cattle raisers, are the functional aristocracy of Stanislaus County and the clientele of a broad range of leather- and silver-workers, feed stores and saddle makers. Many artisans cater to a so-called 'Californio' style, better known as the buckaroo tradition (from *vaquero*)—a mode of handling livestock brought from peninsular Spain and in marked contrast to the so-called 'cowboy' practices of Texas and the Great Plains (Starrs and Huntsinger, 1998).

Cold Desert Steppe

The vegetation of the Cold Desert Steppe (Fig. 11.1; Table 11.3) is mostly sagebrush spaced one-half to 2 m apart with bunchgrasses in between—it has been described as a silver 'sea' of sagebrush (*Artemisia* spp.), as the surface is typically rolling or flat. It is found at elevations of more than 1,000 m in the rain shadow of mountain ranges to the west. The cool season grasses that provide the forage are bunchgrasses that usually 'cure well' and can be grazed in winter when there is snow on the ground. Most precipitation falls as snow and productivity peaks in spring, as the snow melts. Supplemental forages, such as alfalfa, often grown on the ranches where livestock are produced, are used in winter. The Cold Desert type extends east into Utah, north into Idaho, and northeast as far as parts of Wyoming, Montana, and Alberta, where Calgary is a northern expression of the ecoregion.

Cow-calf production is by far the most common form of ranching. Range is used year-round or as winter range for transhumance in the many north-south-trending mountain ranges. Some 77 per cent of the region is in public ownership (Holechek et al., 2001), with lower elevation areas controlled by the Bureau of Land Management (BLM) and uplands managed by the U.S. Forest Service. Calving typically is in spring, with weanlings shipped to Mediterranean California in the fall for backgrounding on annual grass range, or to backgrounding pastures and feedlots in the Great Plains.

This type has been widely invaded by and seriously influenced by cheatgrass. It covers the land in fine dry fuel during the summer, creating a fire cycle that eliminates the shrubs, mostly sagebrush. Traditional range improvement for lands that have lost most of the native bunchgrasses is re-seeding with non-native bunchgrasses from the Asian steppe, like wheatgrass, *Agropyron* spp. A healthy stand of introduced or native bunchgrass is sometimes resistant to the invasion of cheatgrass. In native stands that have not been invaded, soil areas between the sagebrush and bunchgrass may be covered by cryptobiotic soil crusts. Some argue that these are an important feature of an healthy ecosystem and must be protected from trampling by livestock, while others claim crusts prevent regeneration of shrubs and grasses and should be broken up by livestock hooves.

'Basin bigbrush,' SRM 401, is a Rangeland cover type typical of the region's lowlands. Found on deeper soils, it is predominantly sagebrush and perennial bunchgrass. It grades with and is sometimes considered invaded by Pinyon-Juniper (*Pinus* spp., *Juniperus* spp.) and is mostly managed by the BLM. Traditional range improvements include removal of sagebrush and seeding with exotic wheatgrass. However, in the last decade with the invasion by cheatgrass and the resulting frequent fire, the emphasis has shifted to attempting to protect and restore sagebrush.

'Pinyon-juniper woodlands,' SRM 412, are typically found at mid-elevations on slopes grading into sagebrush type, and are about 65 per cent in public ownership (Holechek et al., 2001). Understory grazing is usually poor. Traditional improvements are removal of the tree layer and seeding, but this is now relatively rare because of the cost, and controversies about tree removal. Grazing and fire suppression have led to thickening and spreading stands (West, 1984) and the resulting canopy closure reduces the understory and accelerates erosion. Fire feeds into a cycle of invasion and replacement by cheatgrass.

The 'blue-bunch wheatgrass' Rangeland Cover Type, SRM 101, is part of the 'Palouse Prairie' at the northern extent of the Cold Desert Steppe. It is found on the volcanic soils of northeastern Oregon and eastern Washington, western Montana, and interior British Columbia and grades into sagebrush types on drier sites. Shrubs are relatively rare and the native grassland is rich grazing land. Unfortunately, reproduction of the native

Table 11.3 Characteristics of the Cold Desert Steppe (CEC, 1997; Barbour and Billings, 1988), including selected Rangeland Cover Types (after Shiflet, 1994).

Generalized Grassland Type	**Precipitation and Elevation**	**Characteristic Species**	**Forage Production in kg/ha and local recommendations for allowable use**[2]
Cold Desert Steppe **983,104** km^2 Koppen: Bsk	Elevation: 1,000 to 3,000 m. Precipitation: 150 to 500 mm. Undrained basins alternate with mountain ranges running north to south.	Characterized by *Artemisia* spp., commonly *Artemisia tridentata* Nutt., with chenopod shrubs, *Atriplex* spp. on saline soils. *Amelanchier* spp., *Prunus* spp., *Cercocarpus* spp., *Purshia* spp., and *Rosa* common at higher elevations. Herbaceous: bunchgrasses *Agropyron* spp., *Oryzopsis* spp., *Stipa* spp., *Koeleria cristata*, *Poa* spp. Palouse prairie, *Agropyron* spp. and *Festuca ~~i~~idahoensis~~s~~* Elmer, is found on volcanics in northern extent of this zone; *Distichl~~u~~is spicata*, on alkali. Trees: Pinyon-Juniper woodland grades with sagebrush types at higher elevations and may be constrained by fire, elevation, precip and human activity. *Pinus* spp., *Juniperus* spp.	100–500 kg/ha (Holechek et al., 2001). Moderate levels of grazing with periods of rest and/or growing season deferment to maintain sagebrush-bunchgrass plant communities (Davies et al. 2011).
Typical Rangeland Cover Types (Shiflet 1994)			
basin bigbrush SRM 401	Elevation: 1,000 to 2,300 m. Precipitation: 200–350 mm. Deep permeable soils.	Shrub (10–30 per cent cc[1], 40cm tall): *Artemisia tri~~t~~dentata* subsp., *Tridentata* spp., *Chrysothamnus* spp., *Purshia tridentata* (Pursh) D.C., *Tetradymia canescens* DC. Herbaceous (10–20 per cent cc[1]): *Agropyron spicatum* (Pursh) Gould, *Poa secunda*. Presi, *Festuca Idahoensis* Elmer, *Sitanion hystrix* (Nutt.) J.G. Sm. More than 50 per cent of ground is usually bare, with a 10 per cent cryptogam crust.	456–1,260 kg/ha average (adapted from Shiflet, 1994).
pinyon-juniper woodland SRM 412	Throughout cold desert at higher elevations, intermixes with sagebrush types.	Trees: *Pinus monophylla, edulis* spp., others; *Juniperus* spp. About nine species of junipers and four species of pinyon pines. Herbaceous: diverse assemblage of cold desert species. Shrubs: *Artemisia* spp.	100–600 kg/ha (Holechek et al., 2001).
bluebunch wheatgrass SRM 101	Elevation: 1,700–2,300 m. Precipitation: 200–500 mm.	Herbaceous: *Agropyron spicatum* (Pursh) Gould, with *Poa secund~~a~~a*. Presi, *Sporobolus cryptandrus* (Torr.) A. Gray, *Aristida longiseta* Steud., *Stipa comata* Trin. & Rupr., *Koeleria pyramidata* (Lam.) P. Beauv., diverse forbs and annuals. Shrubs (rare): *Chrysothamnus nauseosus*.	445–1,335 kg/ha

[1] Canopy cover.
[2] Used to determine the number of animals to be placed on an area with the goal of sustainable use.

bunchgrass is rare and overgrazing has severely reduced its presence. The type has been heavily invaded by cheatgrass, and other *Bromus* species and medusahead. Large areas are now dominated by introduced annual grasses and forbs, and significant areas have been converted to crops. Unusual in this region, public ownership of the type is only 15 per cent (Holechek et al., 2001) in the US. The Palouse Prairie extends into British Columbia, where it is a valued grazing land.

Cold Desert Steppe Community Profile: Elko, Nevada, USA

Elko in the northeastern corner of Nevada (Fig. 11.5) shares the buckaroo livestock culture that once dominated Oakdale, California to the west. Sheep herds trailed from southern Nevada and the Owens Valley and California once ended their market drives at the stockyards in Elko. That stopped in the 1940s, as cattle gained ascendancy—and control over grazing resources—from the roaming sheep drovers. Long a relatively small city, Elko and its surrounding area has seen a boom that quadrupled population in the great movement of urban people eager to join the rural parts of the American West (Starrs and Wright, 1995). The surge is emblematic of the pulse of amenity-seekers who are finding attractive sites to settle, creating a mix of urban and rural life far more characteristic of the United States than of Canada or northern Mexico, where the traditional segregation of city and countryside is maintained.

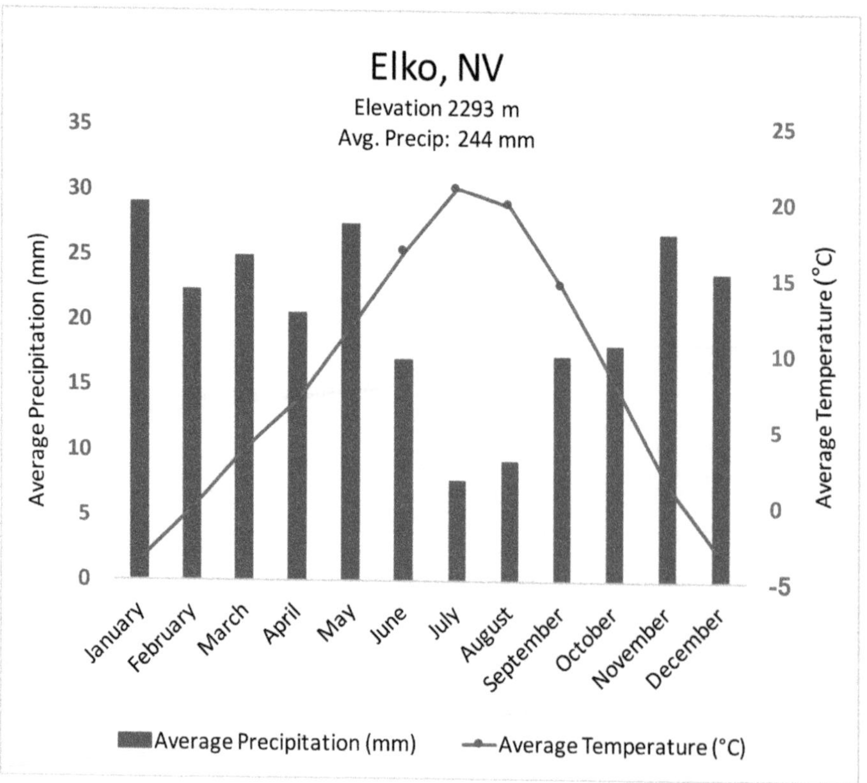

Fig. 11.5 Elko, Nevada, in the Cold Desert Steppe.

A changed economy is evident in Elko, but by no means is it complete. Ranching and associated crops (alfalfa hay, pasture crops) remain important and in two distinct forms that are also characteristic of the Cold Desert Steppe region. The first distinct form comprises small holdings in land and livestock, cultivated and grazed with the intensity of homesteads. The bulk of the region, however, is taken up with the second form—a few vast holdings dependent on seasonal (and sometimes year-round) access to federal land, be it BLM acreage in the relatively level lowlands, or uplands under management of the U.S. Forest Service, which in Elko means the Humboldt National Forest. The large ranches in Elko include a relatively small nucleus of private land, usually irrigated cropland with hay harvested into bales or stacks as winter-spring forage, and large areas (as much as 3,00,000 hectares) of BLM land and Forest Service land, which can rise above 3,500 m.

The sole variation in this pattern of land use can be found along the Humboldt river, where the historic transcontinental railroad tracks run almost parallel to the waterway. Since these lands were granted to the Central Pacific Railroad in the 19th century for resale, much of this acreage went to private hands. In the 1980s, microscopically disseminated gold deposits were discovered in these grazing lands, and many an area ranch was purchased by multinational gold producers. Mine companies now manage relict cattle operations in close proximity to their heap-leach gold mines. Mining executives can fly directly from San Francisco, Salt Lake and Reno to inspect their holdings and sample rural fare. For all the excitement of a mineral boom, Elko seeks to retain much of its cattle-culture roots, hosting the annual Cowboy Poetry Gathering and the Western Folklife Center, and is home to the buckaroo-oriented Capriola Saddlery and Garcia Bit & Spur.

Hot Desert

The Hot Desert grassland type (Fig. 11.1; Table 11.4) spans over 1 million km^2 and contains the Chihuahuan, Sonoran, and Mojave deserts. The Mojave is in the northwest grading into Mediterranean California, the Sonoran lies below in, and western Mexico, southeastern California, and Arizona, and the Chihuahuan lies to the east in Mexico, New Mexico, and Texas, grading into the Great Plains to the east.

Shrubs and grasses are interspersed with cacti and low trees in the hot deserts. Peak productivity may follow irregular rainfall events, or, in the southwest, summer monsoon rains. The forage base is shrubs and warm season grasses, with some cool season annuals. There are scattered perennial bunchgrasses and a lot of bare ground. About 55 per cent of the region is in public ownership in the United States, mostly as BLM land, but also in military reservations (Holechek et al., 2001).

Compared with other North American grassland types, sheep and goats are relatively common in the Hot Desert type, particularly in Texas and Mexico. Typically, grazing, calving, kidding, and lambing are year-round. The predominant form of cattle production is cow-calf, with some backgrounding on irrigated or sown pastures.

Ruminant ungulates were rare in the Chihuahuan Desert during the Holocene until sheep, goats, and cattle were introduced late in the 16th century (Havstad et al., 2006). These animals grazed at low densities from the late 1500s through the late 1800s, when large herds were imported via railroad. The spike in livestock numbers was dramatic—in New Mexico, numbers increased from fewer than 200,000 cattle and 1.6 million sheep in 1870, to 1.4 million cattle and 4.6 million sheep in the late 1880s (Schikedanz, 1980). Along with fire suppression and drought, this novel land use contributed to a critical transition from bunchgrass dominance to mesquite dominance (*Prosopis* spp.) over large expanses

Table 11.4 Characteristics of the Hot Desert (CEC, 1997; Barbour and Billings, 1988), including selected Rangeland Cover Types (after Shiflet, 1994).

Generalized Grassland Type	Precipitation and Elevation	Characteristic Species	Forage Production in kg/ha and local recommendations for allowable use[2]
Hot Desert **1,022,152 km²** Koppen: Bwh	Elevation: –86 m to 2,500 m. Precipitation: 50–380 mm. Proportion of summer rainfall increases to the south. Arid with seasonal extremes.	Low growing shrubs and grasses predominate. Shrubs (0–20 per cent cc[1]): *Larrea tridentata* Creosote bush, common throughout, *Flourensia cernua*, *Prosopis juliflora* (and other *Prosopis* spp.), *Sarcobatus vermiculatus*, *Atriplex* spp. on saline soils. Diverse cacti and succulents. Herbaceous (15–40 per cent cc): *Hilaria* spp., *Oryzopsis hymenoides*, *Sporobolus* spp., *Muhlenbergia* spp., *Bouteloua* spp. common. *Atriplex* spp., *disticlus* spp. on saline areas.	150–500 kg/ha (Holechek et al., 2001). Chihuahuan Desert: Take 30% of the perennial grass production each year, on average (Holechek, 1991).
Typical Rangeland Cover Type (Shiflet, 1994)			
grama-tobosa shrub SRM 505	Elevation 1,100 m to 1,700 m. Precipitation: 220 to 450 mm. Summer monsoonal rains important, summer growing season, but some precipitation and growth in winter. Also in Southern Semi-arid Highlands type.	*Herbaceous: Bouteloua eriopoda* and *Hilaria mutica* Includes a variety of *Bouteloua species*, *Aristida* spp., *Sporobolus* spp., diverse summer forbs. *Opuntia* spp., *Yucca* spp., *Agave* spp. and succulents are common. Shrubs: *Prosopis* spp., *Larrea* spp., *Flourensia* spp., *Gutierrezia* spp., *Acacia* spp., *Koeberlinia* spp., *Atriplex* spp., *Ephedra* spp., *Rhus* spp. May become dominant. In poorer soil and moisture conditions, dominated by *Bouteloua hirsuta* or *Bouteloua curtipendula* and in conditions of high pH or salinity by *Distichlis spicata*, *Sporobolus airoides* and *Hilaria mutica* (Miranda and Hernández, 1985).	0.14 AU/ha when in excellent condition range in plains, lowlands and valleys and 0.03 AU/ha in poor condition range on hillsides (FIRA, 1986).

[1] Canopy cover.
[2] AU = Animal Unit = a year's air-dry forage for a 450 kg cow, approximately 4,271 kg.

(Grover and Musick, 1990). While livestock grazing is recognized as only one of multiple interacting factors influencing ecosystem processes in this post-transition landscape (Herrick et al., 2006), grazing is known to contribute to irreversible grass loss, soil erosion, and dust emissions (Nash et al., 2003). Further, the region's unpalatable shrubs, low and variable primary production, widely spaced watering points, multiyear drought, and rough terrain make livestock production perennially challenging. Developing new, sustainable options for livestock producers is a major focus of rangeland scientists in the region (Spiegal et al. 2018).

The Mojave Desert, a rain-shadow desert due to the orographic effect of the Sierra Nevada covers parts of Nevada, Utah, Arizona, and California in the U.S. Alluvial deposits are the dominant landform, with mountains, hills, playas and pediments composing the rest of the landscape. Geologists separate the Mojave into eastern and western sections along a sinuous vertical line on the low-elevation ground of US Route 66, from Death Valley National Monument through Boron, California and down toward Amboy, California. Rugged mountain ranges rise steeply from valley floors east of the central trough. Mojavean plant communities are distributed by elevation, with alkali sink and saltbush scrub at lowest elevations and Bristlecone-limber pine and white fir (*Abies concolor*) forests at the highest elevations. Middle elevations support a mix of shrubs and bunchgrasses. *Larrea tridentata* (creosotebush) is the most abundant species in the Mojave Desert, found at 0–1,200 m.

The urban population increased by 350 per cent between 1970 and 1990 in the Mojave Desert (Hunter et al., 2003), and with that population growth came intensification and expansion of exurbanization, off-highway vehicle use wind and solar energy development, livestock grazing, and crop agriculture. As in many parts of California, annual grasses and forbs from the Mediterranean Basin and northern Africa (largely *Bromus rubens*, *Schismus* spp. and *Erodium cicutarium*) entered Mojave grasslands in the 19th century. Today their extensive cover is increasing the frequency and intensity of grass fires (Brooks and Berry, 2006).

The Navajo Nation is a U.S. territory covering 71,000 km^2 that includes extensive areas of Hot Desert. The Navajo are known as a sheep grazing culture, after sheep were introduced by Spanish colonists in the 16th century. Navajo weavers produce highly valued blankets from wool, and some still live a subsistence lifestyle in Arizona and New Mexico. Numerous Hispanic communities are also found in close proximity, and grazing is often an important part of local culture and lifestyle. Disputes over land rights stemming from the acquisition of Mexican territory in 1848 by the United States persist.

Native grasslands in Mexico prevail in areas of higher rainfall on the plains, smooth hill country with relatively deep soils (Miranda and Hernández, 1985). Widespread cacti of the genus *Opuntia*, or prickly pear, may be used as supplemental livestock forage in Mexico. There is some transhumance from the Mexican Hot Deserts to the temperate Mexican Sierra mountains. Sheep or goats may be brought in to graze on 'ephemeral range' that comes when summer monsoon or winter rains stimulate the growth of annual herbaceous plants. Cattle from northern Mexico are mainly exported to feedlots in the United States, to the large CAFOs in southern California's Imperial Valley, and to Texas.

The 'grama-tobosa shrub' Rangeland Cover Type, SRM 505, is typical of the Hot Desert. Although widespread, it is a discontinuous type found on the floors and bajada slopes of basins in the northwest through central to southeast Arizona, southern New Mexico, and the Trans-Pecos of Texas. It is well represented in the northern states of Chihuahua and Coahuila in Mexico, and is found in the Southern Semi-arid Highlands type as well. At its upper elevations, it may grade into the temperate Sierra and at low elevations grade into Hot Desert types with creosote bush, *Larrea tridentata*.

Hot Desert Community Profile: Terlingua, Texas, United States

'Terlingua' identifies a tiny town that has a special significance for residents of the American southwest and Mexico (Fig. 11.6). By no means atypically, it has an identity split between notoriety as a quicksilver (mercury) mining community and recent importance as a gateway for Big Bend National Park. It has a long history of ranching, mostly involving

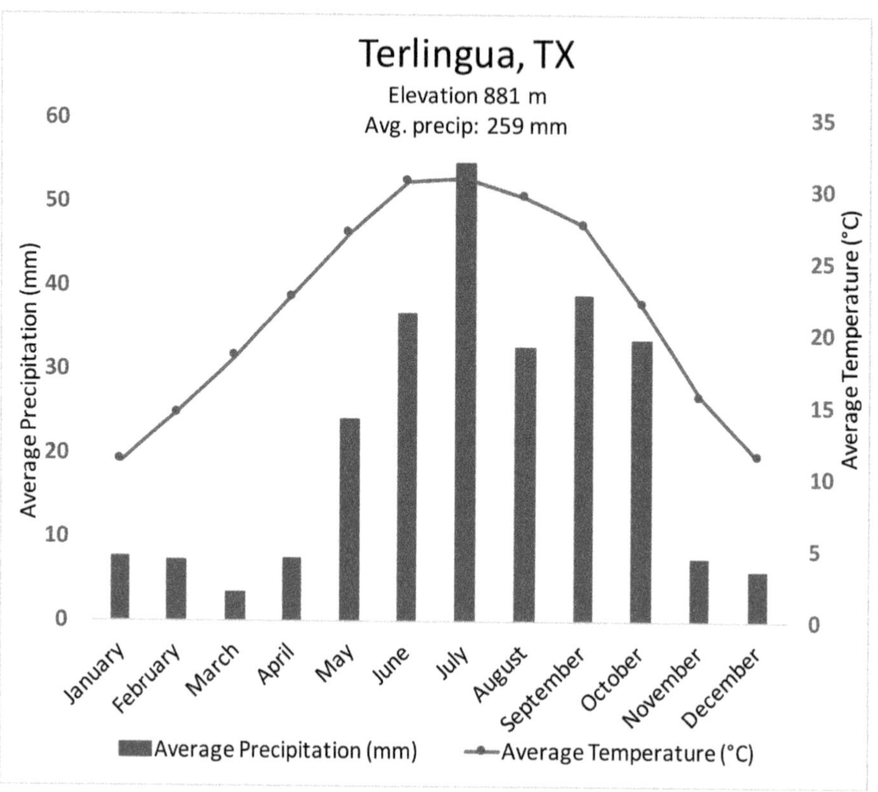

Fig. 11.6 Terlingua, Texas, in the Hot Desert.

cattle. It is country dotted with cacti, small-leaved often high-albedo arid-land shrubs, and is of low net productivity.

Profit from livestock operations is limited and in the last several decades an increase in guest ranching and other visitor-based residential tourism has added income to the communities in and around the Big Bend, the sole US National Park in Texas. In the post-World War II years, this was country better known for smuggling, an illicit trade in cross-border cattle movement, and a population so sparse that only parts of Nevada could compete for the title of 'least populated place.' Terlingua came to fame almost by accident, with country singer Jerry Jeff Walker's *Viva Terlingua!* record album. A modest festival used to be part of Terlingua activities and the cattle trade mixed well with country-music lyrics.

Southern Semi-arid Highlands

Southern Semi-arid Highlands (Fig. 11.7; Table 11.5) are upland, higher-rainfall areas of northern Mexico. They transition to Hot Desert to the east. In general, the vegetation is dominated by grasslands, and in the transition zones, by various scrublands and forests. The flatlands are commonly used for irrigated agriculture. Rainfall increases to the south and the grass is warm season grass. In some areas, overgrazing has shifted the original plant and wildlife communities, with a reduction in plant cover and species composition along

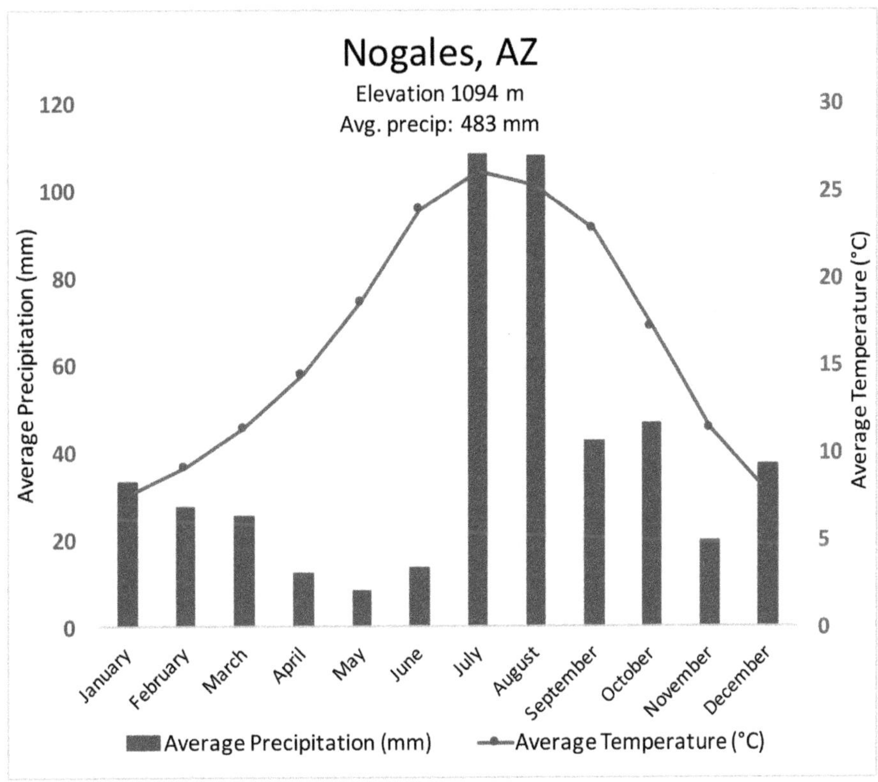

Fig. 11.7 Nogales, Arizona, in the Southern Semi-arid Highlands.

Table 11.5 Characteristics of the Southern Semi-arid Highlands (CEC, 1997; Barbour and Billings, 1988), including selected Rangeland Cover Types (after Shiflet, 1994).

Generalized Grassland Type	Precipitation and Elevation	Characteristic Species	Forage Production in kg/ha and Grazing Capacity[1] in AU/ha
Southern Semi-arid Highlands **269,996 km²** Koppen: Bsk	Elevation: 1,100 to 2,500 m. Precipitation: 300–600 mm. Two major types of soils, relatively dry and moderately deep, and shallow, clay soils. Summer rainfall increases to south. Mean temperatures ranging from 12 to 20°C. In winter, frosts are common, as are periodic droughts.	Herbaceous: *Agropyron* spp., *Aristida* spp., *Hilaria* spp., *Muhlenbergia* spp., *Bouteloua* spp., *Bouteloua gracilis*, foot of western Sierra Madre. Trees: *Prosopis* spp. on deep clay soils. *Quercus* spp., *Juniperus* spp. (western juniper), *Acacia* spp., *Mimosa biuncifera* in warmer regions. Cacti: *Opuntia* spp. in warmer areas. SRM 505 also found here.	Between 0.14 and 0.03 AU/ha (FIRA, 1986).

[1] AU = Animal Unit = a year's air-dry forage for a 450 kg cow, approximately 4,271 kg.

with changes in structure, mainly through shrub species invasion and soil erosion. A 2000 FAO report states that around 39 per cent of the semi-arid lands of northern Mexico (more than 350 mm precipitation) show severe loss of top soil due to water and wind erosion (INEGI, 2000); however, for some specific sites up to 43 per cent of the land has been shown to have a serious loss of top soil (Royo et al., 2004). Many observers also mention on-going problems with over-harvest of timber. The primary agents of natural resource degradation are described as demographic pressure, land-use change, fire, logging, reduction of soil fertility and an asymmetrical economic relationship between Indians and Mestizos, as in general exists between the rich and poor (Ochoa-Gaona and Gonzalez-Espinosa, 2000). The Mexican government has created biotic reserves in response to concerns about inappropriate use of natural resources.

This region contains forage resources particularly important in the Mexican states of Chihuahua, Coahuila and Durango. Cattle, sheep and goats are all produced. It is also important for the cultivation of beans and corn. The area includes SRM Rangeland Cover Type 505, 'grama-tobosa shrub' (Table 11.5) (Shiflet, 1994; McClaran, 1995).

Southern Semi-arid Highlands Community Profile: Nogales, Arizona, USA

The desert town of Nogales (Fig. 11.7) in southeastern Arizona shares a trait common to many a city in the highlands of the semi-arid zone—from the start, it was a kind of oasis settlement on the Sonoran desert borderland. The name itself, from the Spanish, means 'walnut trees' and was a reflection of the local vegetation and spring-based water supply that made a permanent presence possible. Pima Native Americans established settlements first. By the 1690s, Spanish missions secured the imperial claim of Spain to land and authority over its residents. Nogales lay within the small strip of land that was the last-added part of the lower 48 United States, leaving Mexican control only in 1854, when the Gadsen Purchase moved the border with Mexico south. The border now marks the southern edge of the city limits.

Inserted along with the tracks of the railroad in 1882, the border briefly created a separate town, called 'Line City,' later united with Nogales. Now a city with some 20,000 residents, Nogales is parched in the spring months until monsoonal rains arrive in July and August, the two months that provide more than half of the annual precipitation in a concentrated burst. The flavor of the city is unmistakably international, with a porous flow back and forth across the border and a significant Native American presence.

Tropical Dry Forests

About 13 per cent of Mexico is Tropical Dry Forest (Fig. 11.1; Table 11.6), and 10 per cent of cattle production takes place in this area. In wetter areas there are more evergreen trees. Goats are grazed extensively. Cattle production systems are based on grazing of native vegetation or pastures sown to *Hyparrhenia rufa* (in the drier areas, or pastures sown to *Panicum maximum* or *Cynodon plectostachyus* and more recently with *Andropogon gayanus*). Cattle are usually finished within the region on pasture (Améndola et al., 2005).

Though 73 per cent of Tropical Dry Forest has been disturbed (Trejo and Dirzo, 2000), it is considered more resilient than more mesic tropical types (Murphy and Lugo, 1986). From the 1970s–1990s, vast tracts of forest between the cities of Navojoa and Alamos were cleared, burned and replaced with non-native pasture. Twelve to 20 years later, the majority of original woody species were again evident (Bowden, 1993), although grazing

Table 11.6 Characteristics of Tropical Dry Forests (CEC, 1997; Barbour and Billings, 1988), including selected Rangeland Cover Types (after Shiflet, 1994).

Generalized Grassland Type	Precipitation and Elevation	Characteristic Species	Forage Production kg/ha and Grazing Capacity[1] in AU/ha
Tropical Dry Forests **266,079 km²** Koppen: Aw	Elevation: 200 to 1,000 m. Precipitation: 600–1,600 mm falling mostly summer, 5–8 mo dry period. Fine textured soils of plains, shallow hillside soils. Rugged topography. Average temp. 20–29°C.	Low deciduous and sub-deciduous forests; 4–15 m tall with three distinct strata. Trees and shrubs: In areas above 1,000 metres. *Pinus Quercus, Abies, Lysiloma, Leucaena, Acacia, Pithecellobium* spp., *Bursera simaruba, Tabebuia rosea, Enterolobium cyclocarpum, Ipomoea intrapilosa* and *Prosopis juliflora* (Jaramillo, 1994). Herbaceous: Legumes such as *Leucaena, Desmodium, Macroptilium* and *Centrosema* spp. (De Alba, 1976). *Muhlenbergia, Festuca, Piptochaetium, Bromus, Poa, Aristida* spp. and others (Hernández, 1987). Cacti, inc. *Opuntia* spp.	Carrying capacity of sown pastures is between 0.3 and 1 AU/ha while that of native vegetation is as low as 0.08 AU/ha (Améndola et al., 2005). The national rangeland office recommends a stocking rate of 0.03 to 0.04 AU/ha for similar areas (Martínez-Balboa, 1981).

[1] AU = Animal Unit = a year's air-dry forage for a 450 kg cow, approximately 4,271 kg.

may reduce tree regeneration. About 0.07 per cent is in protected status. The majority of Tropical Dry Forests are owned and managed by rural and mostly indigenous communities that are profoundly dependent on them (Valero et al., 2017).

Tropical Dry Forest Community Profile: Cabo San Lucas, Mexico

Now a resort town with some 68,000 year-round residents (2010), Cabo San Lucas (Fig. 11.8) sits at the southern tip of Baja California Sur (south), where for hundreds of years after its founding, it was oriented more towards the sea than towards land. Its coves and fresh water supply made it a favored stopping point that hid English pirates who would launch attacks on Manila galleons. Even today, deep-sea fishing for marlin and other great game fish is among the most popular tourist undertakings, drawing not just on Pacific Ocean waters, but also the deep chasms that mark the southern Gulf of California. But with a separation of more than 200 km from the Mexican mainland, Cabo was nothing if not isolated and as late as the 1930s, had a population of only several hundred people. Water supply was a crucial question. To the north, it is 1,300 km to Tijuana and the US border and that distance was not easily traversed.

By the 1970s, the Baja Highway had improved and cruise ships and airplane flights were landing in southern Baja. With a reliable fresh water supply, an environment considered ideal for tourism was discovered. Tropical Dry Forest can be considered perhaps the world's singlemost sought-after ecoregion as a setting for seaside tourism. While well adapted to aridity, the vegetation is highly susceptible to type conversion, especially when vegetation is introduced in golf courses and resort landscaping. And those changeovers are common enough to pose a risk throughout the southern Baja peninsula. Tourist journals describe coming across widely dispersed cattle along the dirt roads of the countryside. A rangeland-

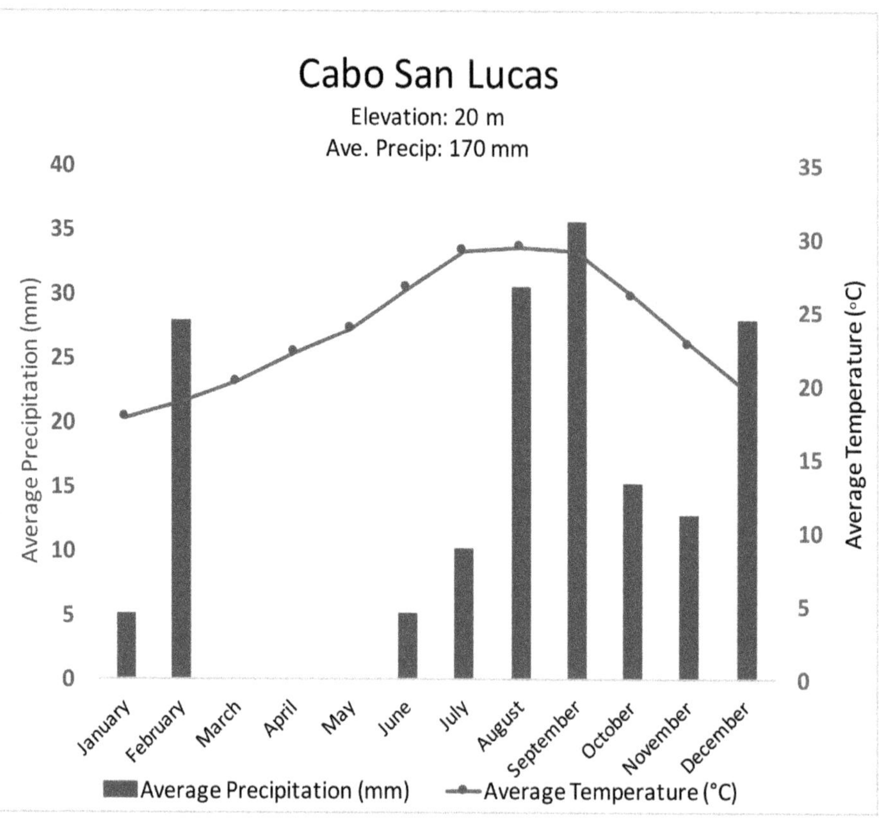

Fig. 11.8 Cabo San Lucas, Baja California Sur, Mexico.

based livestock herding culture that dates back to Jesuit and Dominican settlements in Baja (Aschmann, 1959) has seen its isolation broken and the risks to the marginal economy of cattle, sheep and goat ranching in Baja increase with human arrivals (Crosby, 1981).

Temperate Grasslands of the Great Plains: Tallgrass, Mixed Grass and Shortgrass Prairie

There is currently an agreement among ecologists that the vast grasslands of the Great Plains (Fig. 11.1; Table 11.7) are maintained by climate, fire and grazing. Native American and lightning-lit fires, and grazing of the infamous once-vast herds of bison and antelope, have been largely supplanted by livestock grazing and plowing. Season-long grazing is typical, but in the northern portions, including Canada, the season can be very short with only 200 livestock feeding days not unusual. Stored forages, agricultural by-products and feeds from nearby and intermixed farmlands are used. Range productivity peaks as warm season grasses grow in response to summer rains and soil moisture from winter snowmelt. Spring calving is the norm. The region contains the exceptionally productive tallgrass prairies, whose deep soils can produce 11,000 kg/ha (Shiflet, 1994). These have long been converted to rain-fed corn and soybeans. Cattle are finished on local backgrounding pastures and then large feedlots supplied with soybeans, grains, and agricultural byproducts grown in the region, and many are shipped in from other regions for finishing.

Table 11.7 Characteristics of the Great Plains Tallgrass, Mixed Grass and Shortgrass prairies (CEC, 1997; Barbour and Billings, 1988), with select Rangeland Cover Types (after Shiflet, 1994).

Generalized Grassland Type	Precipitation and Elevation	Characteristic Species	Forage Production (kg/ha) and local recommendations for allowable use[2]
Temperate Grasslands: Tallgrass, Mixed Grass, Shortgrass prairies of the Great Plains **3,406,536 km^2**	Elevation: 0 to 2,000 m. Precipitation: 250 to 1,000 mm. Winters in the northern Great Plains are exceptionally cold.	Herbaceous (80–100 per cent cc[1]) Shortgrass prairie from the west grades through mixed grass prairie to tallgrass in the east. In the north, boreal plains of Canada intergrade with boreal forest to the north; to the south, scrubby vegetation grades into desert. Tallgrass dominants include big and little bluestem *Andropogon gerardii* Vitman, *Schizachyrium scoparium* (Michx.) Nash, and *Panicum virgatum* L., *Sorghastrum nutans* (L.) Nash, see below for others. Shrubs and Trees: rare.	300–3,500 kg/ha (Holechek et al., 2001). Take 40–50% of the perennial grass production each year, on average (George et al., 2013). May reach 11,000 kg/ha
Typical Rangeland Cover Types (Shiflet, 1994)			
blue grama-buffalo grass Shortgrass SRM 611 Koppen: Bsk-Dfa	Elevation: 800 to 1,700 m. Precipitation: 270–500 mm. Western northern and central plains. Undulating plains, medium to fine textured soils.	Herbaceous layer (80–100 per cent cc[1]) Shortgrass prairie: *Bouteloua gracilis* (Willd. Ex Kunth) Lag. ex Griffiths, *Bouteloua dactyloides* (Nutt.) J.T. Columbus; with *Bouteloua hirsuta* Lag., *Agropyron smithii* Rydb., *Bouteloua curtipendula* (Michx.) Torr., *Aristida purpurea* Nutt. Diverse forbs, including *Opuntia polyacantha* Haw.	600–1,000 kg/ha
bluestem-grama prairie Mixed Grass SRM 604 Koppen: Bsk	Elevation: 330 to 1,000 m. Precipitation: 550–700 mm. Medium to moderately fine-textured soils with good drainage, rolling upland plains bisected by breaks.	Herbaceous layer (90–100 per cent cc[1]) Mixed grass prairie: *Schizachyrium scoparium* (Michx.) Nash, *Andropogon gerardii* Vitman, *Bouteloua curtipendula* (Michx.) Torr., *Bouteloua gracilis* (Willd. ex Kunth) Lag. ex Griffiths, *Bouteloua hirsuta* Lag., *Bouteloua dactyloides* (Nutt.) J.T. Columbus important in understory on dry sites. Diverse forbs. Distinct two-layered appearance.	1,000–2,500 kg/ha (Holechek et al., 2001).
bluestem-prairie sandreed Tallgrass SRM 602 Koppen: Dfa	Elevation: 700 to 1,300 m. Precipitation: 400–580 mm. Sand hills of Nebraska, other sand hills areas in the Great Plains. Deep sandy soils.	Western displacement of tallgrass prairie. Herbaceous (> 80 per cent cc[1]): *Andropogon gerardii* Vitman, *Schizachyrium scoparium* (Michx.) Nash, *Calamovilfa longifolia* (Hook.) Scribn., *Panicum virgatum* L., *Stipa comata* Trin. & Rupr., *Sorghastrum nutans* (L.) Nash. *Bouteloua gracilis* (Kunth) Lag. ex Griffiths and other shortgrass species often form an understory.	1,000–3,000 kg/ha (adapted from Holechek et al., 2001).

[1] Canopy cover.
[2] Used to determine the number of animals to be placed on an area with the goal of sustainable use..

The rainfall gradient is from west to east, and from south to north, with sod-forming shortgrasses to the west until an increasing proportion of tallgrasses to the east culminates in tallgrass prairie. In short and mixed grass areas, cool season grass may be interplanted to extend forage quality and production. Wheat cultivation is important in mixed grass areas, but shortgrass areas are used primarily for grazing.

'Blue grama-buffalo grass,' SRM 611, is a Rangeland Cover Type of the shortgrass prairie. Predominantly used for livestock grazing, grass is less than 30 cm tall. It extends to high elevations on the eastern slope of the Rocky Mountains. Cattle and sheep are commonly produced. About 5 per cent is in public ownership (Holechek et al., 2001).

Rangeland Cover Type 'bluestem-grama prairie', SRM 604, is a mixed grass prairie type found in south-central Nebraska, west central Kansas, western Oklahoma, extreme northern Texas and eastern Colorado. A mixed grass prairie, with both tall and shortgrasses, overgrazing may remove tallgrass species, converting it to shortgrass (Bose, 1977). As with shortgrass areas, about 5 per cent is in public ownership (Holechek et al., 2001).

Great Plains Community Profiles: Valentine, Nebraska, USA

Billing itself as 'the heart city', Valentine (Fig. 11.8), in the core of the expansive sandhill country of Nebraska, can just barely be described as a city: its 2010 population was 2,737 residents. The town is both the county seat of government and the center for agricultural machinery dealerships, government services, and the largest urban area along the Niobrara river, which meanders along the town's northern edge. The town sits on the edge of 52,000 square kilometers (20,000 sq miles) of relict Pleistocene sand dunes. Soils were blown in and the area became an isolated westernmost tallgrass prairie remnant of spectacular stands of grass with interstitial cacti, and an unparalleled richness in fauna and insects. But though enormous in extent—the sandhills are 20 per cent larger than all of Switzerland—this is, in fact, a relict vegetation displaced considerably west of its normal range and made possible by the piezometric draw of sand dunes on the perched Ogallala Aquifer, a great lens of what is widely considered 'fossil' water from geological times past. The sandhills are dramatic visually and in animal stocking rates, since the carrying capacity of the tallgrass prairie is three to five times that of a comparable area of shortgrass rangeland. And the climatic cycles of great spring rains, which usually extend into the first month or two of summer, produce a growth of grass and other fodder that is almost unique, now, in the United States and a relic of what once was far more widespread. Yet it remains at risk, with deep wells pushed down into the Ogallala Aquifer initiated in the 1960s, but reaching levels high enough to draw down fossil water levels, causing a range of ecological issues (Wise, 2015; Parker and Wilson, 2016).

A contrasting side of the Great Plains is found farther less than 720 km (430 mi) west in Sheridan, Wyoming, a wholly different view historically, culturally and economically. There the landscape is shortgrass prairie, soaked by spring rain but with little summer rainfall until the months of September and October, when thunderstorms can bring both relief from the heat and an enhanced likelihood of wildland fires (climate graph not pictured). Sheridan is a bustling service center, historically important for livestock growers, dude ranch clients, and cultivators of cropland plowed from the westernmost reach of the shortgrass prairie. Since the late 1970s, a variety of energy-based industries, especially those involved in coal and petroleum derivatives, have arisen. But the 19th-century legacy of Sheridan was its role as a railhead for lines running up the Big Horn Range and other mountains that herald arrival at the Rocky Mountains for westbound travelers. Just west of Sheridan is Eaton's, the oldest dude (guest) ranch in the western states still in operation.

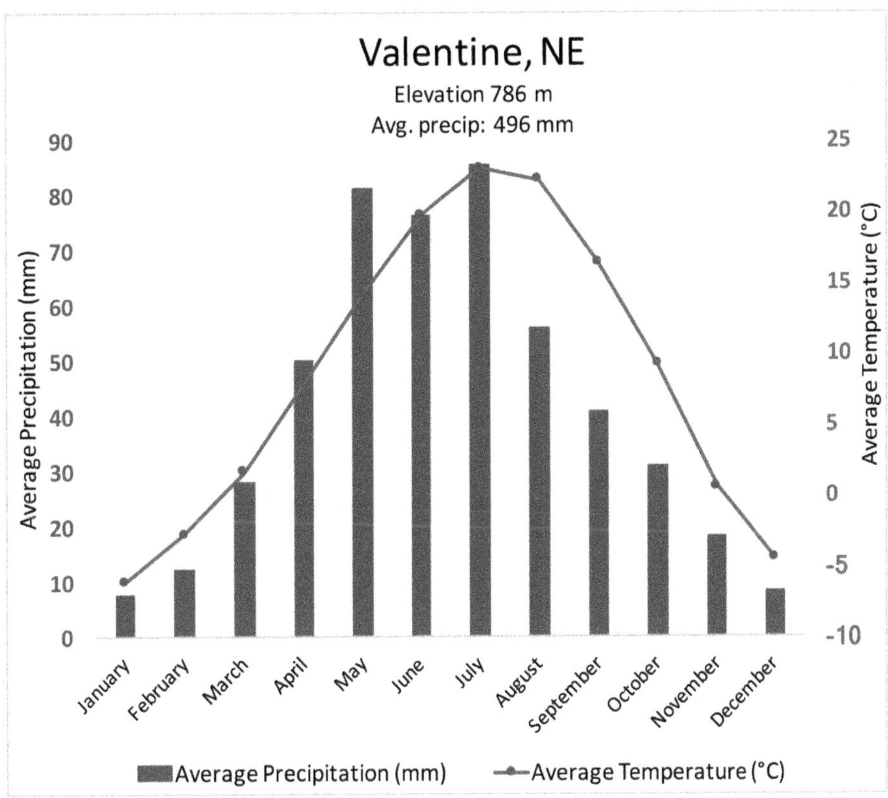

Fig. 11.9 Valentine, Nebraska in the mixed grass Prairie.

The smaller landholdings just to the south in Big Horn have an intricate ownership pattern that owe much to claims for land made by the second sons and daughters of European aristocracy, who came to the western United States to attempt to gain their own fortune and recognition (Gressley, 1966; Starrs, 1998).

Temperate Grasslands of the Southern Coastal Plains

The Southern Coastal Plain of Florida and Georgia, USA, is rich in biodiversity, containing forests, woodlands, grasslands, marshes, and savannahs (Table 11.8). The region has a mild, mid-latitude, humid subtropical climate, characterized by hot humid summers and warm to mild winters. The mean annual temperature is approximately 19° to 22°C. The frost-free period ranges from 280 to 360 days. The mean annual precipitation is 1,338 mm, ranging from 1,170 mm to 1,650 mm. Originally, pine and mixed hardwood forests covered much of the ecoregion. Southern floodplain forests have been cleared for lumber and converted to pine plantations that favor the faster-growing slash (*Pinus elliottii* Engelm) and loblolly pine (*Pinus taeda* L.) species. Longleaf pine and other forests have also been converted to cropland, pasture, mining and urban uses. The native longleaf pine (*Pinus palustris*) was once the dominant tree species; however, its current extent has been reduced by as much as 98 per cent (Wear and Greis, 2002).

Numerous low-gradient perennial streams and large rivers, wetlands, and lakes are found here. In Florida, an area of more rolling discontinuous highlands that contains

Table 11.8 Characteristics of the Southern Coastal Plains (CEC, 1997; Barbour and Billings, 1988; Drummond, 2016; and selected Rangeland Cover Types after Shiflet, 1994).

Generalized Grassland Type	Precipitation and Elevation	Characteristic Species	Forage Production (kg/ha) and/or local recommendations for allowable use[2]
Temperate Grasslands: Southern Coastal Plain SRM 800 Koppen: Cfa **139,217 km²**	Elevation: 0 to 900 m. Precipitation: 1,100 to 1,500 inches.	Herbaceous (80–100 per cent cc[1]).	1,727 to 11,500 kg/ha (Holechek et al., 2001). Take 45–60% of the perennial grass production each year, on average (George et al., 2013).
Typical Rangeland Cover Types (Shiflet, 1994)			
South Florida Flatwoods SRM 811 Koppen: Cfa	Elevation: 5 to 15 ft. Precipitation: 1,200–1,500 mm. Nearly level, deep, acid, sandy Spodosols, seasonally poorly to excessively drained. Coarse textured throughout or course textured in the upper part and moderately coarse textured or moderately fine textured in the lower part.	Typically savanna, with scattered pine trees and an understory of sawpalmetto and grasses. Herbaceous layer (55–85 per cent cc[1]). Some areas in extreme south Florida have few, if any, trees. *Trees–Live oak, Quercus virginiana* Mill*; Slash pine, Pinus elliottii* Engelm*; South Florida slash pine, Pinus elliottii var. densa.* *Shrubs–Ground blueberry, Vaccinium myrsinites* Lam.*; Gallberry, Ilex glabra* (L.) A. Gray*; Saw palmetto, Serenoa repens* (W. Bartman) Small*; Tarflower, Bejaria racemosa* Vent.*; Shining sumac, Rhus copallina* L.*; Wax myrtle, Myrica cerifera* (L.) Small*; Blackberry, Rubus* spp. *Grasses–Herbaceous cover 55 to 85* per cent. *Chalky bluestem, Andropogon capillipes* Nash*; Creeping bluestem, Schizachyrium.* *Grasses–Herbaceous cover 55 to 85* per cent. *Chalky bluestem, Andropogon capillipes* Nash*; Creeping bluestem, Schizachyrium stoloniferum* (Michx.) Nash var. *stoloniferum* (Nash) Wipff*; Lopsided indiangrass, Sorghastrum secundum* (Elliott) Nash*; South Florida bluestem, Schizachyrium rhizomatum* (Swallen) Gould*; Low panicum, Panicum* spp.*, Pineland threeawn, Aristida stricta* Michx. Per cent annual vegetative production by weight is 75 per cent grasses and grasslike plants, 15 per cent trees and shrubs and 10 per cent herbaceous plants.	3,500 to 5,200 kg/ha

[1] Canopy cover.
[2] Used to determine the number of animals to be placed on an area with the goal of sustainable use.

numerous lakes, Ultisols, Spodosols, and Entisols are common, with thermic and hyperthermic soil temperature regimes and aquic and some udic soil moisture regimes. The region is largely used for extensive pine plantations and forestry, pasture for beef cattle, citrus groves, tourism and recreation, and fish and shellfish production. Some large areas of urban, suburban, and industrial uses coexist. Fire suppression causes the scrub of the region's pine savannas to become too dense, resulting in habitat degradation.

Some 26 per cent of Florida, the most characteristic part of the Southern Coastal Plains zone, is public land, but as is common throughout the Mexico, the United States, and Canada, the control of stockwater supplies and much of the best grazing land is in private hands. What appears in inventory maps is a telling mosaic of developed cities and residential areas, with significant areas still in agriculture, including commercial citrus groves that historically have thrived in areas somewhat elevated about the coastal zones (McPhee, 1967).

Grazing is a longterm fact of Florida life, dating back to the establishment in 1565 of St. Augustine by the Spanish, a founding that makes it the oldest continually occupied Euroamerican city in the continental United States. The cattle industry of the Southeast United States owes much to early Spanish implantation in the second expedition of Juan Ponce de Leon, who in 1521 brought a small herd of Andalusian cattle with him. This marked the start of a formal livestock-raising culture, and included significant innovations spilling forward from the Guadalquivir Basin in southern Spain. Prominently, it gave birth in the next 200 years to the southern 'cracker,' a poor white herder with minimal property holdings, living off squatting and grazing otherwise unwanted grassland resources. Those small holders and herders would be pushed away in the 1700s and 1800s as plantation agriculture displaced livestock herding, and grazing herds had to move inland and up into the southern Appalachian mountains (Owsley, 1949; Jordan, 1981; Jordan, 1993). Cowhands in Florida and the Southern Coastal Plains were indeed a rough if accomplished lot, and diverse to a remarkable degree, with African-American, Native American, and white hands working with catahoula cow dogs, snubbing posts, bullwhips for moving and controlling animals, and a way of life and doing business that was notable for its discomforts.

Historically, the grasslands of the Southern Coastal Plains were marginal lands, and for several hundred years a livestock-based economy extended into Georgia, the Carolinas, and, in part, west into Alabama and Mississippi. The great green stretches were sufficient that Marjory Stoneman Douglas in 1947 would publish an account of south-central Florida titled *Everglades: The River of Grass,* with a title that suggests nearly all. But grasslands are in this day and age defined by recreation, urbanization, commercial fishing, and some notable industries. Use of the grasslands, which are made up of heat and rainfall-seeking plant species, is sporadic. Nonetheless, the many introduced pasture grasses of west African and Latin American origin can handle a high stocking rate, and the livestock-per-hectare loading is considerable and far higher than in less humid areas to the west in Mexico and Canada. What makes the Southern Coastal Plains an especially distinctive grassland type is the periodic inundation of areas of grass in hurricanes driving up the Caribbean and Gulf of Mexico, with epic rainfall totals. Gainesville, Florida, for example, has a mean annual precipitation of over 1,200 mm, with a pronounced summer maximum.

Southern Coastal Plains Community Profiles: Gainesville, Florida and Environs

Best known as the home of the University of Florida, Gainesville sits in a zone of hardwoods, but surrounded by the typical piney flatwoods and scrub grasslands that are typical on the limestone bedrock that makes up much of peninsular Florida. As a whole, the state in 2016 is the ninth largest cattle producer in the United States, with over a million animals and 15,000 beef producers, and most of the animals are grazing on rangelands (Fig. 11.10).

The culture and life of 19th-21st-century livestock ranchers is a narrative of note (Akerman, 1976). Sixteenth-century Florida was the site of the oldest deliberate implantation of cattle in what was then Spanish territory, but is now the United States. Of late, there is increased recognition of what is described as the cracker cattle breed—a recognition and restoration of a long-traditional feature of the so-called Florida cracker-cattle culture (St. Clair, 2006; FCM, n.d.). With animals introduced so long ago from Old World to New, it is perhaps unsurprising that the Florida cracker cattle breed bears a striking physical resemblance to the morucha breed, a rather independent roan-hued breed considered characteristic in Salamanca Province, in western Spain. Cattle breeds typically used in the region tend to stay cool by wallowing in seasonal wetlands, influencing water quality (Shukla et al., 2011) and species composition (Tweel and Bohlen, 2008). Florida Cracker cattle are better adapted to the Florida heat than are the conventional breeds (Sponenberg and Olson, 1992). Florida Cracker cattle are genetically related to the heritage Raramuri Criollo cattle breed under investigation for sustainable livestock production in the Hot Chihuahuan Desert (Anderson et al., 2015) and there is concurrent interest in evaluating possible gains in environmental and economic outcomes through using cracker cattle instead of conventional breeds in the Southern Coastal Plains.

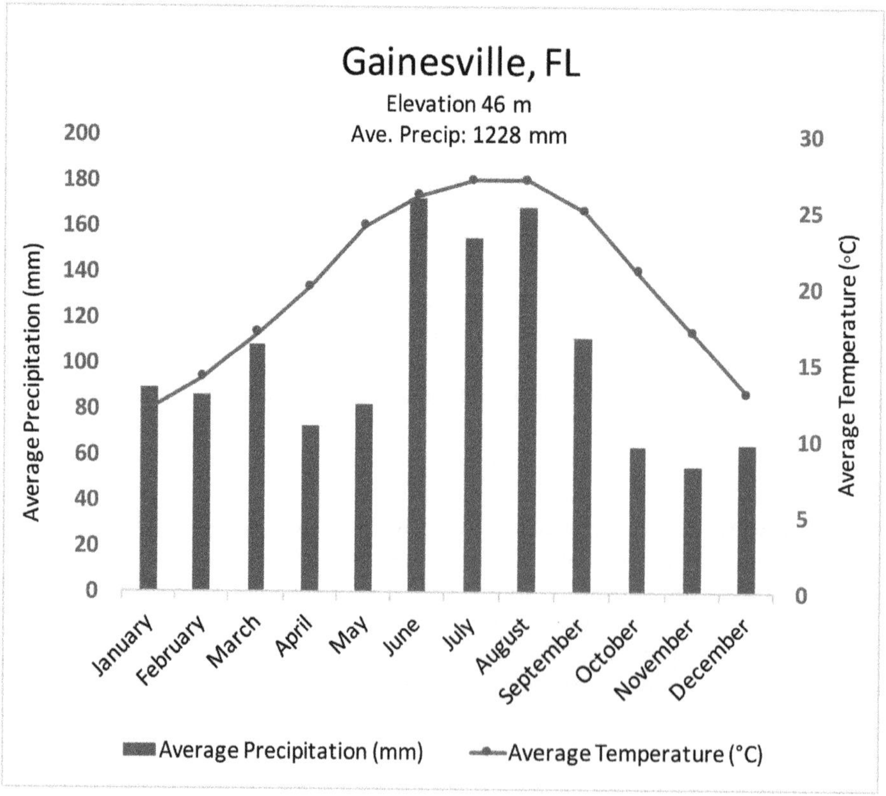

Fig. 11.10 Gainesville, Florida in the Southern coastal plains.

Conclusion

North America has diverse and plentiful grasslands. The wide climatic and topographical variation, the recent introduction of the plow and domestic livestock, has left a legacy of rich biodiversity and complex ecosystems. The grasslands used for grazing livestock today were until very recently the domain of indigenous economies as varied as the North American environment, cultures that shaped and were shaped by the ecosystem. The subjugation of Native people has left us ill-informed about many aspects of the history of human-ecosystem interaction in North America and led to many false assumptions about the 'pristine,' a mystique about a supposed wilderness that is at the heart of much of the North American ideology of frontier conquest and environmentalism. Ecologists are trying still to disentangle human and natural influences, the better to understand how the ecological cornucopia found by European colonists might be restored. Unfortunately, they are still highly reluctant to listen to the very people who created that cornucopia. In North America, grasslands were often manipulated, if not created, by native groups seeking game, grain, fiber and goods they provide. It has been estimated that, for example, in the Sierra Nevada of California, there are now six times as many trees as there were in the mid-19th century.

Perevolotsky and Seligman discuss how views of ecological history affect our interpretation of current activities and their outcomes on rangelands (1998). For Europe, the domestication of large livestock took place 9,000 years ago and led to profound change in the environment, which then achieved a kind of bounded rebalancing with human communities and that has persisted to the current day. In contrast, North America is at that much earlier point, with ecosystems isolated from domestic livestock and the plow until a few hundred years ago and now undergoing a major change (Huntsinger, 2016). Indeed, the entire grassland flora of California has shifted to dominance by non-native species, and in many areas, from a perennial dominated system to a system dominated by annual plants. In the Intermountain West, sagebrush country has been altered on a broad scale by an invasive grass from the Asian steppe. A scenario driven by species introductions and human practices is now made even more complex by climate change. North American grasslands are changing and this state of change is likely to persist for a long time. Natural resource managers, pastoralists and other grassland users face a future of constant adaptation and adjustment in their use and management, while grassland species face an unknown and unpredictable future.

References and Further Reading

Akerman, J.A. 1976. Florida Cowman: A History of Florida Cattle Raising. Kissimmee: Florida Cattlemen's Association, 280 p.

Alagona, P.S., A. Linares, P. Campos and L. Huntsinger. 2013. History and recent trends. pp. 25–60. *In*: P. Campos, L. Huntsinger, J.L. Oviedo, P.F. Starrs, M. Díaz, R.B. Standiford and G. Montero (eds.). Mediterranean Oak Woodland Working Landscapes: Dehesas of Spain and Ranchlands of California. Dordrecht, Heidelberg, New York, London: Springer, 508 p.

Améndola, R. and P.A. Epigmenio Castillo. 2005. Country Pasture/Forage Resource Profile: Mexico. Agriculture Department, Crop and Grassland Service, FAO Country Pasture and Forage Resource Profiles. www.fao.org/ag/AGP/AGPC/doc/Counprof/mexico/mexico.htm <accessed June 2017>.

Anderson, D.M., R.E. Estell, A.L. Gonzalez, A.F. Cibils and L.A. Torell. 2015. Criollo cattle: Heritage genetics for arid landscapes. Rangelands 37: 62–67.

Aschmann, H.H. 1959. The Central Desert of Baja California: Demography and Ecology. Ibero-Americana, Number 42. Berkeley and Los Angeles: University of California Press, 315 p.

Barbour, M.G. and W.D. Billings. 1988. North American Terrestrial Vegetation. New York: Cambridge University Press, 708 p.

Bartolome, J.W., M.C. Stroud and H.F. Heady. 1980. Influence of natural mulch on forage production on differing California annual range sites. J. Range Mgmt. 33: 4–8.

Bartolome, J.W., W.E. Frost, N.K. McDougald and J.M. Connor. 2002. California guidelines for Residual Dry Matter (RDM) management on coastal and foothill annual rangelands. University of California - Division of Agriculture and Natural Resources, Rangeland Management Series Publication 8092, Oakland, CA.

Bose, D.R. 1997. Rangeland Resources of Nebraska. Society for Range Management—Old West Regional Commission Joint Publication, Society for Range Management, Denver, CO.

Brooks, M.L. and K.H. Berry. 2006. Dominance and environmental correlates of alien annual plants in the Mojave Desert, USA. Journal of Arid Environments 67: 100–124.

Bowden, C. 1993. The Secret Forest. Albuquerque: University of New Mexico Press, 141 p.

CEC [Commission for Environmental Cooperation]. 1997. Ecological Regions of North America: Toward a Common Perspective. Communications and Public Outreach Department of the CEC Secretariat. Number 393, Rue St. Jacques Ouest, Bureau 200 Montréal (Québec) Canada H2Y1N9.

Cooper, D.W. and H.R. Heady. 1964. Soil analysis aids grazing management in Humboldt County. Calif. Ag. 18: 4–5.

Crosby, H.W. 1981. Last of the Californios. LaJolla, California: Copley Books, 196 pp.

Davies, K.W., C.S. Boyd, J.L. Beck, J.D. Bates, T.J. Svejcar and M.A. Gregg. 2011. Saving the sagebrush sea: An ecosystem conservation plan for big sagebrush plant communities. Biol. Cons. 144(11): 2573–2584.

Davies, K.W., C.S. Boyd, J.L. Beck, J.D. Bates, T.J. Svejcar and M.A. Gregg. 2011. Saving the sagebrush sea: An ecosystem conservation plan for big sagebrush plant communities. Biol. Cons. 144(11): 2573–2584.

Drummond, M.A. 2016. Southern Coastal Plain. United States Geological Survey Land Cover Trends Project. Accessed 3/26/18 at [https://landcovertrends.usgs.gov/east/eco75Report.html].

FCM [First Coast Magazine]. n.d. The Science behind Florida Cracker Cattle. http://firstcoastmagazine.com/news/the-science-behind-florida-cracker-cattle/ <accessed 25 June 2017>.

FIRA. 1986. Instructivos técnicos de apoyo para la formulación de proyectos de financiamiento y asistencia técnica. Serie Ganadería. Forrajes. FIRA-Banco de México, Mexico DF, 256 p.

George, M.R., W. Frost and Neil McDougald. 2013. Grazing Management. In George, M.R. (ed.), Ecology and Management of Annual Rangelands. University of California Division of Agriculture and Natural Resources, Oakland, California.

Gressley, G. 1966. Bankers and Cattlemen. New York: Alfred A. Knopf, 320 p.

Griffin, J.R. 1977. Oak woodland. pp. 383–415. *In*: M.G. Barbour and J. Major (eds.). Terrestrial Vegetation of California, Sacramento: California Native Plant Society, 1020 p.

Grover, H.D. and H.B. Musick. 1990. Shrubland encroachment in southern New Mexico, USA: An analysis of desertification in the American Southwest. Climatic Change 17: 305–330.

Havstad, K.M., E. Fredrickson and L.F. Huenneke. 2006. Grazing livestock management in an arid ecosystem. pp. 278–304. *In*: K.M. Havstad, L.F. Huenneke and W.H. Schlesinger (eds.). Structure and Function of a Chihuahuan Desert Ecosystem: The Jornada Basin Long-Term Ecological Research Site, Oxford University Press, New York.

Herrick, J.E., K.M. Havstad and A. Rango. 2006. Remediation research in the Jornada Basin: Past and future. pp. 278–304. *In*: K.M. Havstad, L.F. Huenneke and W.H. Schlesinger (eds.). Structure and Function of a Chihuahuan Desert Ecosystem: The Jornada Basin Long-term Ecological Research Site, Oxford University Press, New York.

Hewes, L. 1974. The Suitcase Farming Frontier: A Study in the Historical Geography of the Central Great Plains. Lincoln: University of Nebraska Press, 281 p.

Holechek, J.L. 1991. Chihuahuan Desert rangeland, livestock grazing, and sustainability. Rangelands 13(3): 115–120.

Holechek, J.L., R.D. Pieper and C.H. Herbel. 2001. Range Management: Principles and Practices. 4th ed. Upper Saddle River, NJ: Prentice Hall, 587 p.

Hunter, L.M., M.D.J. Gonzalez G., Stevenson, M. et al. 2003. Population and land use change in the California Mojave: Natural habitat implications of alternative futures. Population Research and Policy Review 22: 373. https://doi.org/10.1023/A:1027311225410.

Huntsinger, L. and P.F. Starrs. 2006. Livestock production in North America: A biogeographical approach. Sécheresse 17: 219–233.

Huntsinger, L. and J.W. Bartolome. 2015. Cows? In California? Rangelands and livestock in the Golden State. Rangelands 36: 4–10.

Huntsinger, L. 2016. The tragedy of the common narrative: Re-telling degradation in the American West. pp. 293–326. *In*: R.H. Behnke and M. Mortimore (eds.). The End of Desertification? Disputing Environmental Change in the Drylands, Berlin: Springer Earth Systems Sciences, Springer-Verlag, 560 p. DOI 10.1007/978-3-642-16014-1 <accessed July 2017>.

Igler, D. 2001. Industrial Cowboys: Miller & Lux and the Transformation of the Far West, 1850–1920. Berkeley and Los Angeles: University of California Press, 267 p.

INEGI [Instituto Nacional de Estadística, Geografía e Informática]. 2000. Indicadores de desarollo sustentable en México. Sistemas Nacionales de Estadístico y de Información Geográfica. México DF: INEGI. www.inegi.gob.mx/prod_serv/contenidos/español/biblioteca/default.asp <accessed June 2017>.

Isenberg, A.C. 2000. The Destruction of the Bison: An Environmental History, 1750–1920. Cambridge: Cambridge University Press, 206 p.

Jordan, T.G. 1981. Trails to Texas: Southern Roots of Western Cattle Ranching. Lincoln: University of Nebraska Press, 220 p.

Jordan, T.G. 1993. North American Cattle Ranching Frontiers: Origins, Diffusion, and Differentiation. Albuquerque: University of New Mexico Press, 439 p.

Krech (III), S. 1999. The Ecological Indian: Myth and History. New York: W.W. Norton & Company, 318 p.

Lott, D.F. 2003. American Bison: A Natural History. Berkeley and Los Angeles: University of California Press, 237 p.

Mack, R.N. and J.N. Thompson. 1982. Evolution in steppe with few large, hooved mammals. Amer Naturalist 119: 757–773.

Martínez-Balboa, A. 1981. La Ganadería en Baja California Sur. Vol. 1: La Paz, B.C.S., México: Edit. J.B, 229 p.

McClaran, M.P. 1995. Desert grasslands and grasses. pp. 1–30. *In*: M.P. McClaran and T.R. Van Devender (eds.). The Desert Grassland. Tucson, University of Arizona Press.

McPhee, J. 1967. Oranges. New York: Farrar Straus Giroux, 176 p.

Merchant, C. 2007. American Environmental History: An Introduction. New York: Columbia University Press, 504 p.

Miranda, F. and X.E. Hernández. 1985. Fisiografía y vegetación en las zonas áridas del centro y noreste de México. pp. 255–272. *In*: Xolocotzia Tomo I. Chapingo, Mexico: Universidad Autónoma Chapingo.

Murphy, P.G. and A.E. Lugo. 1986. Ecology of tropical dry forest. Ann. Rev. Ecol. Systematics 17: 67–88.

Nash, M.S., E. Jackson and W.G. Whitford. 2003. Soil microtopography on grazing gradients in Chihuahuan desert grasslands. J. Arid. Envi. 55: 181–192.

Nordrum, A. 2014. Can Wild Pigs Ravaging the US Be Stopped? The USDA is spending $20 million to solve a pig problem that has spread to 39 states and counting. Scientific Amer October. https://www.scientificamerican.com/article/can-wild-pigs-ravaging-the-u-s-be-stopped/ <accessed 21 June 2017>.

NRC [Committee to Review the Bureau of Land Management Wild Horse and Burro Management Program, National Research Council, National Academy of Sciences]. 2013. Using Science to Improve the BLM Wild Horse and Burro Program: A Way Forward. Washington DC: The National Academies Press, 450 p. http://www.nap.edu/catalog.php?record_id=13511 DOI: 10.17226/13511 <accessed July 2017>.

Ochoa-Gaona, S. and M. Gonzalez-Espinosa. 2000. Land use and deforestation in the highlands of Chiapas, Mexico. Appl. Geog. 20: 17–42.

Owsley, F.L. 1949. Plain Folk of the Old South. Baton Rouge: Louisiana State University Press, 235 p.

Parker, L. and R. Wilson. 2016. What happens to the U.S. Midwest when the water's gone? National Geographic (August). http://www.nationalgeographic.com/magazine/2016/08/vanishing-midwest-ogallala-aquifer-drought/ <accessed 26 June 2017>.

Perevolotsky, A. and N. Seligman. 1998. Role of grazing in Mediterranean rangeland ecosystems: Inversion of a paradigm. BioScience 48: 1007–1017. http://www.jstor.org/stable/1313457 <accessed July 2017>.

Royo, M.M., T.J. Sierra, C.A. Melgoza and R.R. Carrillo. 2004. Situación actual de los pastizales del noroeste de Chihuahua. XL Reunión Nacional de Investigación Pecuaria, 203 p.

Schickedanz, J.G. 1980. History of grazing in the Southwest. pp. 1–9. *In*: K.G. McDaniel and C. Allison (eds.). Proceedings: Grazing Management Systems for Southwest Rangelands: A Symposium, Las Cruces: New Mexico State University.

Shiflet, T.N. 1994. Rangeland Cover Types of the United States. Denver, Colorado: Society for Range Management, 152 p.

Shukla, S., D. Goswami, W.D. Graham, A.W. Hodges, M.C. Christman and J.M. Knowles. 2011. Water quality effectiveness of ditch fencing and culvert crossing in the Lake Okeechobee basin, southern Florida, USA. Ecol. Engineering 37: 1158–1163.

Snow, N.P., M.A. Jarzyna and K.C. VerCauteren. 2017. Interpreting and predicting the spread of invasive wild pigs. J. Appl. Ecol. DOI:10.1111/1365-2664.12866.

Spiegal, S., B.T. Bestelmeyer, D.W. Archer, D.J. Augustine, E.H. Boughton, R.K. Boughton, M.A. Cavigelli, P.E. Clark, J.D. Derner, E.W. Duncan, C.J. Hapeman, R.D. Harmel, P. Heilman, M.A. Holly, D.A. Huggins, K. King, P.J.A. Kleinman, M.A. Liebig, M.A. Locke, G.W. McCarty, N. Millar, S.B. Mirsky, T.B. Moorman, F.B. Pierson, J.R. Rigby, G.P. Robertson, J.L. Steiner, T.C. Strickland, H.M. Swain, B.J. Wienhold, J.D. Wulfhorst, M.A. Yost and C.L. Walthall. 2018. Evaluating strategies for sustainable intensification of US agriculture through the Long-Term Agroecosystem Research network. Envir. Res. Lett. 13(3): 034031.

Sponenberg, D. and T. Olson. 1992. Colonial Spanish cattle in the USA: History and present status. Archivos de Zootecnia 41: 401–414.

Starrs, P.F. 1998. Let the Cowboy Ride: Cattle Ranching in the American West. Baltimore, Maryland: Johns Hopkins University Press, 396 p.

Starrs, P.F. and J.B. Wright. 1995. Great basin growth and the withering of California's Pacific idyll. Geog. Rev. 85: 224–244.

Starrs, P.F. and L. Huntsinger. 1998. The cowboy & buckaroo in American ranch hand styles. Rangelands 20: 36–40.

Starrs, P.F. and P. Goin. 2010. A Field Guide to California Agriculture. Berkeley and Los Angeles: University of California Press, 503 pp.

St. Clair, D. 2006. Cracker: The Cracker Culture in Florida History. Gainesville: University of Florida Press, 256 p.

Swain, H.M., E.H. Broughton, P.J. Bohlen and L.O. Lollis. 2013. Trade-offs among ecosystem services and disservices on a Florida ranch. Rangelands 35: 75–87.

Trejo, I. and R. Dirzo. 2000. Deforestation of seasonally dry tropical forest: A national and local analysis in Mexico. Biol. Conser. 94: 133–142. Doi: 10.1016/S0006-3207(99)00188-4.

Tweel, A.W. and P.J. Bohlen. 2008. Influence of soft rush (*Juncus effusus*) on phosphorus flux in grazed seasonal wetlands. Ecol. Engineering 33: 242–251.

UC–DANR [University of California Division of Agriculture and Natural Resources]. 1996. Guidelines for Managing California's Hardwood Rangelands. Integrated Hardwood Range Management Program. Communications Services Publication, 3368.

US–EPA [United States Environmental Protection Agency]. 2017. Agriculture: Pasture, Rangeland and Grazing Operations. U.S. Environmental Protection Agency. www.epa.gov/agriculture/agriculture-pasture-rangeland-and-grazing <accessed 27 June 2017>.

Valero, A., J. Schipper and T. Allnut. 2017. Jalisco [Mexico] Tropical Dry Forests, Tropical and Subtropical Dry Broadleaf Forests, WWF-UK. www.worldwildlife.org/ecoregions/nt0217 <accessed July 2017>.

Wear, D.N. and J.G. Greis. (eds.). 2002. Southern Forest Resource Assessment, General Technical Report SRS-53: Asheville, N.C., U.S. Department of Agriculture, Forest Service, Southern Research Station, 635 p.

West, N.E. 1984. Successional patterns and productivity potentials of pinyon-juniper ecosystems. pp. 1301–1332. *In*: Developing Strategies for Rangeland Management, Prepared by Committee on Developing Strategies for Rangeland Management, National Research Council, National Academy of Sciences. Boulder, Colorado: Westview Press, 2022 p.

Wise, L. 2015. A drying shame: With the Ogallala Aquifer in peril, the days of irrigation for western Kansas seem numbered. Kansas City [Kansas] Star, 24 July. www.kansascity.com/news/state/kansas/article28640722.html <accessed July 2017>.

Worster, D. 1979. Dust Bowl: The Southern Plains in the 1930s. New York: Oxford University Press, 305 p.

12

Southern African Grassland in an Era of Global Change

Graham Paul von Maltitz

Introduction

Grasslands, either as pure grasslands (i.e., grasslands without a tree component and which will be referred to simply as grasslands), or savanna grasslands (i.e., a mix of grass and trees, which will be referred to simply as savanna) constitute the predominant vegetation type of southern Africa. It is only South Africa and Lesotho that have extensive grasslands and these contribute about 28 per cent of the South African landscape. In other southern African countries, grasslands are limited to high mountainous areas or hydromorphic grasslands in seasonally-flooded areas. With the exception of Lesotho which is almost exclusively grassland, savanna vegetation is the dominant vegetation of all Southern African countries, contributing about 33 per cent of South Africa's land area, but the major proportion of the land area is in the remaining countries.

For this chapter southern Africa will be considered as those countries south of 10° and include South Africa, Botswana, Namibia, Swaziland, Lesotho, Angola, Zambia, Zimbabwe, Malawi and Mozambique (Fig. 12.1).

A Grassland Typology

There are many ways in which the grassland and savanna areas of southern Africa can be classified. This chapter will use as its basis a simplified classification based in part on structure (trees/no trees), but mostly on functional attributes of the plants, based on the logic of Huntley (1982). Since the theme of this book is on grasslands, the typology is based primarily on functions of the grass layer, using a grass and grazer centric lens of analysis.

A two by two classification is proposed (Fig. 12.2). The X axis is the presence or absence of a tree layer, i.e., distinguishing between pure grasslands and savanna woodlands. The Y

Natural Resources and the Environment, CSIR, PO Box 395 Pretoria 0001, South Africa.
Email: gvmalt@csir.co.za

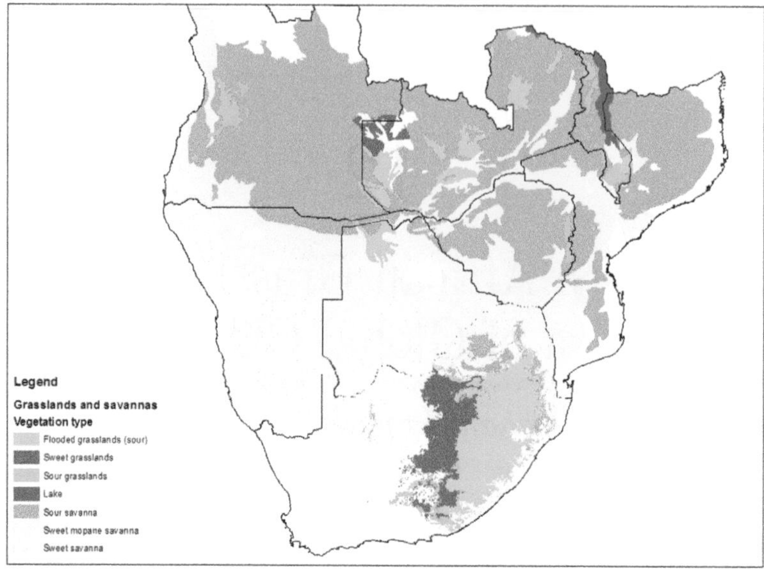

Fig. 12.1 Simplified map of major grasslands and savannas in southern Africa. (For South Africa, this is based on Musina and Rutherford, 2010; for the remainder of southern Africa Olsen et al., 2001 is used as the basis for the mapping units.)

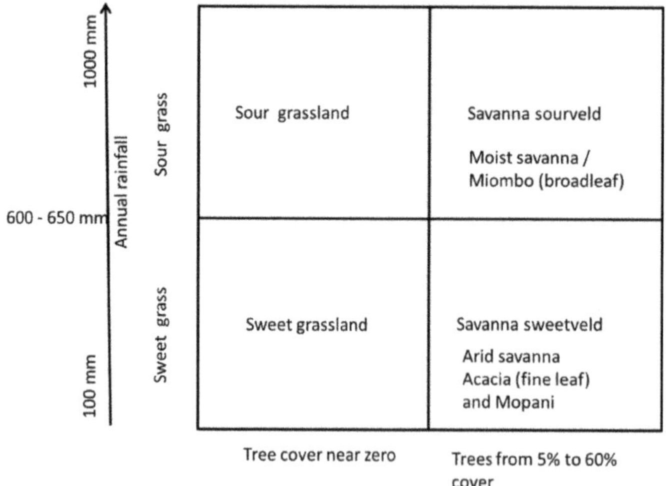

Fig. 12.2 Simple typology of grasslands in southern Africa.

axis is based on palatability of the grass, what is locally termed 'sweet' and 'sour'. Sour grass gets its name because it is unpalatable to livestock in winter. In simple terms, there is a high ratio of carbon to nitrogen in the grass which results in crude protein decreasing from about 10 per cent in summer to 3 per cent during winter, with an accompanying increase in fibre (Weinmann, 1948; Plowes, 1957; Elliot and Folkertsen, 1961). Andropogonoid grasses are very common within the temperate sourveld. These grasses tend to be tall and tufted and are typically found on highly leached soils of low nutrient status in areas with over 600 mm of annual rainfall (Mucina and Rutherford, 2006). Due to the fact that these grasses provide poor grazing, high biomass tends to accumulate and fires are a very common feature of this vegetation (Huntley, 1982; Frost, 1996; Mucina and Rutherford, 2006). In fact, these grasses become moribund without fire and fire is an important component of maintaining these grasslands in a healthy state (Huntley, 1982). The sweet grasses remain palatable in winter and tend to be finer in structure with a higher nitrogen ratio. Because these sweet grasses remain palatable, they are heavily grazed, even in winter, and hence there is far less standing biomass to carry a fire. Further these grasses tend to be in more arid areas, thus limiting the build up of biomass. This sweet/sour divide is based on the attributes of individual grass species and in essence it is the predominance of either the sweet or sour species that leads to the classification of the grassland as either sweet or sour (Ellery, 1995). Figure 12.1 gives a spatial division of sweet and sour grasslands in southern Africa.

The division between sweet and sour grass is closely linked to aridity with the changeover taking place at about 600 to 650 mm in savannas (Huntley, 1982) (Fig. 12.3a). Ellery et al. (1995), looking specifically at grasslands, agrees that moisture is a key determinant of the divide between sweet and sour grasslands, but found here is a complex interplay with temperature, soils and nutrient availability. Ellery (1995) uses the growth day concept rather than precipitation in his models. Mucina and Rutherford (2006) suggest that change can be anywhere between 500 and 700 mm precipitation. In the southern African region, the African term 'veld' is used to denote rangelands, so locally the term 'sweetveld' and 'sourveld' is commonly used to distinguish between the different levels of palatability. The divide between sweetveld and sourveld can be abrupt when there is a sharp change in geology or soils. However, in areas close to the critical moisture cut-offs there can be a landscape with mixed characteristics linked to different soils (Huntley, 1982). A common pattern in the region is that the top of the slope has sandy soils with low nutrients and a sourveld nature and bottom slope has heavier soils, higher nutrients and sweetveld characteristics. In South Africa, areas with both sweetveld and sourveld components are referred to as mixed veld. In the savannas, tree species and their associated attributes change with tops of slopes having sourveld trees and bottom slopes, sweetveld trees. Within the savanna regions, it is also quite common for a mid catenal area to have a hydromorphic grassland zone (seep zone) where water is forced to the surface. This, where it occurs, is seen as a narrow band of grassland which tends to follow the contour and divides the upslope vegetation types from the valley bottom vegetation.

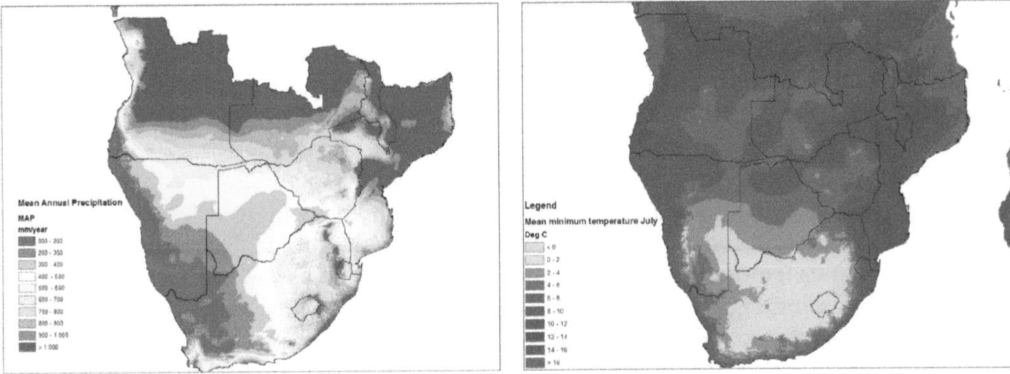

Fig. 12.3 (a) Mean annual precipitation and (b) mean minimum temperature of the coldest month. Derived from the World Climate Data database.

At Nylsvley Nature Reserve, there are patches of *Senegalia* dominated savanna (typical sweetveld) imbedded within a broadleaf savanna dominated by *Burkea* (a common broadleaf, indicative of sour savanna). Within these patches the grass layer shows distinct differences between sweetveld species within the *Senegalia* patches whilst the surrounding area is dominated by sourveld species. Blackmore et al. (1999) showed that although there were no soil texture differences, there was a marked increase in nutrients within the *Senegalia* area. It is speculated that the origin of the higher nutrients was anthropogenic, but that the system has maintained this difference over possibly hundreds of years.

At a local scale three additional factors can affect the occurrence of more palatable grass in an otherwise sourveld area. Firstly termiteria, which are very common, especially in the sourveld savanna, can concentrate nutrients and result in more palatable species, such as *Cynodon dactylon, Urochloa mosambicensis, Panicam maximum* (Davies et al., 2014). Secondly, trees tend to concentrate nutrients, through for instance leaf fall, and it is common that under larger trees, there is a concentration of the palatable grass, such as *Panicum maximum* (Ludwig et al., 2008). This, despite the fact that the trees have a negative overall impact on grass production due to competition for water and the suppression of sunlight (Belsky, 1994; Scholes and Archer, 1997). Thirdly, in some instances, there are grazing lawns, where high densities of grazers concentrate. In these areas it would appear that the high grazing pressure maintains an area of higher nutrition grasses that have adapted to high grazing (Hempson et al., 2015).

At the extremes the distinction between grassland and savanna is obvious, but there are intermediate areas where a relatively high number of trees or bushes may be found in areas classified as grassland. This potential encroachment of trees into the grasslands will be discussed in detail in a later section. For practical purposes a cut-off of about 5 per cent tree cover is suggested as a divide (Scholes and Walker, 1993). It should be pointed out that in many instances when tree elements start appearing within the grasslands, these are often more shrubs rather than large trees. This occurs both along the grassland—savanna divide, but also on the arid western boundary of the grasslands where the adjacent vegetation type is dwarf scrubland vegetation of the nama-karoo (Mucina and Rutherford, 2006). Encroachment of these shrub elements into the grassland has been a longstanding element of concern (Acocks, 1966; Cowling et al., 2004; Hoffman and Ashwell, 2002). Though recent data shows that there is an increase of grass into some areas of the nama-karoo, possibly as a consequence of changed rainfall (Stevens et al., 2015).

Understanding the factors leading to the differentiation between grasslands and savanna, or the different savanna and grassland functional types, is important when attempting to understand possible impacts of global change. Fire is clearly an important determinant of the differentiation between grasslands and forests. Most of the moister grasslands could climatically favour forests, and in fact, forests are often embedded within these grasslands (Bond et al., 2003; Mucina and Rutherford, 2006). However, forests tend to be restricted to fire refugia (von Maltitz et al., 2003; Mucina and Rutherford, 2006). Although fire plays an important role in determining the degree of woodiness in savannas, on its own it cannot fully explain the differentiation between savanna and grassland. Ellery et al. (1991) have shown that the savanna/grassland divide is largely temperature dependent, due to cold winter temperatures and the occurrence of frost as a main factor. As can be seen from comparing Figs. 12.1 and 12.3, there is a strong commonality between area of low mean minimum temperature of July (mid-winter) and the occurrence of grasslands in areas with sufficient rain to support grasslands.

Within savannas there is a divide between savanna sweetveld, functionally similar to what Huntley (1982) terms as arid savannas (also referred to as thornveld, acacia veld, bushveld, Kalahari bushveld, mopane veld (Mucina and Rutherford, 2006)), and the sourveld savanna aligned with Huntley's moist savanna (also referred to as miombo, dry deciduous forest, *Brachystegia, Burkea, Isoberlinia or Jubernaria* forest (Frost, 1996)). It is important to emphasise that the savanna sourveld (and the imbedded hydromorphic sourveld grasslands) are distinctly tropical in nature and therefore floristically very different from the grassland sourveld which is considered as temperate in nature. This differs from the grassland and savanna sweetveld where many of the same grass, forb and tree species occur in both the grassland and savanna, and are tropical in nature.

Where there is a strong geological divide, the distinction between the savanna sweet and sourveld types can be very distinct, but on a common substrate such as the Kalahari sands this can be a slow gradient. As Huntley (1982) points out, in addition to the sweet/sour nature of the grass vegetation, there are a large range of both functional and floristic attributes that make the functioning of these different savanna types very different. This then impacts on the ecosystem services emerging from these systems and the management practices available. Over large areas, especially in the large dry river valleys, a unique vegetation type dominated by the tree species *Colophospermum mopane* (mopane veld) occurs. In the mopane veld, the mopani trees dominate the tree layer. Mopane is an exception to the rule in the sense that they are a broadleaf and thornless species within the arid sweetveld savannas, which are more typically dominated by fine-leaved and spinescent species.

Overview of the Southern Africa Grasslands and Savanna

A strong seasonality in rainfall is a common feature of the African grasslands and savannas. Rainfall is predominantly or exclusively in summer and as one moves to the more arid areas, the duration of the rainy season tends to reduce to about four months, often with within-season drought periods. As can be seen from Fig. 12.3a, there is a broad gradient in rainfall from the arid south-west to a more mesic east and northern region. High rainfall also occurs along some of the large mountain ranges, such as the Drakensberg, resulting in the high altitude cool and moist habitats favored by the sour grasslands. A further feature of the region is that there tends to be an inverse relationship between rainfall and the variance in rainfall (Tyson et al., 1975). The arid areas have high between-season variance in rainfall, with drought periods or periods with far higher than mean rainfall being common. This variance in rainfall is strongly linked to El-Niño Southern Oscillation (ENSO) cycles with strong El-Niño years typically resulting in severe droughts in the region. Multiple wet or drought years in a row are common (Reason et al., 2000; Rouault and Richard, 2005).

Several workers (Coe et al., 1976; Dye and Spear, 1982; Lamprey, 1983) have shown that there is a linear relation between herbage production and annual rainfall. The relation is modified by soil type and the ratio of trees to grasses. However this appears to apply only in the more arid areas and above 1,000 mm, this relationship is weak or absent (Scholes and Walker, 1995). In addition, in the arid areas (less than about 650 mm on heavy soil or 600 mm on sand), the grass in general tends to be sweet (Huntley, 1982; Scholes and Walker, 1996; Ellery et al., 1995). As a consequence of the above and given the high between-season variance in rainfall in arid areas, the sweetveld areas are well known for their high variance in forage productivity between the years. As such, the carrying capacity of livestock or game during any one year is strongly related to that year's rainfall.

The South African grasslands are dominated on a biomass basis by grass (Poaceae), although from a species perspective (Mucina and Rutherford, 2006) the grass only accounts for about one-sixth of the high plant biodiversity (about 3,370 plant species).

These grasslands occur predominantly on the high inland plateau of South Africa, Lesotho and to a small extent, Swaziland, and have mostly perennial C4 grasses in the warmer area, though C3 grasses become dominant in the higher altitude cooler area. Mucina and Rutherford (2006) suggest that higher altitude, cooler areas should be considered temperate rather than tropical in nature. Since the South African grassland is relatively isolated from other large areas of grassland is not surprising that the area has a high occurrence of endemics, including 15 endemic mammals. These are predominantly found within the sourveld, temperate climatic regions (Mucina and Rutherford, 2006). A number of important wetland areas are found imbedded in the grasslands, and these are very important for bird biodiversity. Further, small patches of indigenous forest are often found embedded within the moist sour grasslands, proving that climatically these areas could support forest (von Maltitz et al., 2003). The grassland can therefore be divided into two categories—climatic climax (sweet) and fire climax (sour) grasslands (Acocks, 1988; Tainton, 1999). When sourveld grasslands are discussed in this chapter it will refer exclusively to the temperate South African grassland.

The large hydromorphic grasslands as found in Botswana and Zambia are clearly tropical in nature and dominated by C4 grasses of the genera *Andropogon, Eragrostis, Aristida, Elionurus, Rhynchelytrum,* and *Tristachia.* These include the world-famous Okavango Delta in Botswana as well as the Barotose Floodplains in Zambia. Small wooded 'islands' are found on higher ground, such as caused by termiteria. In addition to these large floodplains, small low-lying areas, referred to as 'dombos', form hydromorphic grasslands that are a common feature throughout the 'miombos'. These grasslands are regarded as an integral component of the miombo (Frost, 1996) and will be discussed jointly with them.

Southern African savanna are characterised as having a continuous layer of grass (hemicryptophytes) and a discontinuous layer of trees (phanaerophytes) (Scholes and Archer, 1997). The savannas are considered tropical in nature and areas of temperate grassland with trees (such as the *Protea* trees in the sourveld grasslands) are not considered as part of the savanna (Mucina and Rutherford, 2006). Savanna have a high alpha but low beta and gamma (Scholes, 1997). As such, similar species are found over literally thousands of km^2 in Southern Africa. Key distinctions between the sweet (arid) savanna and the sour (moist savanna) is given in Table 12.1.

Table 12.1 Key distinctions between sweetveld and sourveld savanna. (Based on data from Huntley (1982); Frost, 1996; Owen-Smith and Kooper, 1987.)

Sweet (arid) Savanna	**Sour (moist) Savanna**
In mean rainfall typically less than 650 mm per year.	Mean rainfall typically over 600 mm.
Extent of tree cover limited by climate. Fires are relatively infrequent, especially in the more arid extremes.	Extent of tree cover controlled by fire—these areas could climatically be forest. Fire is very frequent.
Typically on eutrophic soils (though also on the extensive Kalahari sands).	Typically on dystrophic and highly leached soils.
Trees typically fine leaved, relatively palatable to browsers, but defended by thorns. An exception being mopane. Leguminous trees are mostly from the Fabaceae. Other common genera include *Commiphera, Colophospermum, Adensonia.* Trees are strongly deciduous.	Trees typically have larger leaflets, high tannins in the leaves limit browsing. Thorns are rare. Leguminous trees are mostly from the Caesalpinioidea and weekly deciduous.
Good grazing from largely palatable genera.	Most grass is poor grazing especially during winter. Genera include Hyparrhenia, Andropogon, Loudetia and Digitaria.

Use of the Grasslands and Savannas

At a macro level the grasslands and savannas of southern Africa fall into three main land tenure arrangements:

1. Private large land holdings that can be either freehold or leasehold. This is the predominant land tenure in South Africa and to a lesser extent in Namibia, but less common in the other African countries. Within the South African grasslands, this is by far the most common tenure arrangement, with communal land being more prevalent in the South African savanna regions. Private farms are typically managed at moderate grazing pressure, using some form of rotational management when livestock is involved (Palmer and Ainslie, 2006).
2. Communal land is managed and used by a group of people and often incorporates customary land management practices. This is the common tenure arrangement in all the countries other than South Africa, and even within South Africa about 14 per cent of the land area is under communal tenure. These lands typically have continuous grazing at high stocking densities, though in some of the sourveld areas there is a transhumance system with alternate winter grazing (Palmer and Bennett, 2013).
3. State conservation land as in some countries, such as Botswana, are of exceptionally large areas under conservation (30 per cent). Lesotho, Swaziland and South Africa have the smallest area under conservation in the region with one per cent, 4 per cent and 7 per cent respectively.

Vast areas of the grassland, especially grasslands with intermediate rainfall, have been transformed to croplands (Low and Rebelo, 1996; Fairbanks et al., 2000; Mucina and Rutherford, 2006). Since most of the non-hydromorphic grasslands occur in South Africa, most of the debate on grasslands is specifically South African. Despite the grasslands having few natural trees, it is the grasslands where a large proportion of plantation forestry takes place. This is restricted to high rainfall regions, often within the mountains where steep slopes make crop agriculture difficult. In addition, much of South Africa's mineral resources and 40 per cent of the mines are located within the grasslands. In total, an estimated 30 per cent to 32 per cent of the grasslands have been permanently transformed to the above-mentioned activities or for urban expansion (Fairbanks et al., 2000). A small proportion, about 2.2 per cent (Low and Rebelo, 1996), is under conservation; the remaining area of grasslands is largely used for livestock, and in most cases, cattle and sheep ranching. An estimated 6 million cattle and 13 million sheep are farmed in the South African grasslands (Davies et al., 2014). Satellite imagery indicates about 2.5 per cent of the grassland being degraded, though Mucina and Rutherford (2006) suggest that this may be an underestimation and that up to 30 per cent of grasslands may be secondary grasslands on previously transformed land. A recent trend has been the reverting from livestock to game farming with lifestyle residences and tourism being drivers of this change.

The grasslands contain most of the country's important mountain catchment areas and as such are managed for water production. These are largely the same areas historically used for plantation forestry. Plantation forestry has, itself, been found to be a streamflow-reducing activity, because trees transpire more water than the grassland they replaced, and this impacts on streamflow (Gush et al., 2002). An even greater threat to streamflow than plantation forestry has been the widespread expansion of alien invasive plant (AIP) species, many of which grow in the plantation forests. This has had a major impact on streamflow (le Maître, 1996, 2000, 2004). In general, it seems that the grasslands, and in particular the sour grasslands, are more susceptible to alien woody plant invasion than the savannas (Richardson and Brown, 1986; Le Maitre et al., 2000). However, savanna is not immune from AIP, with *Prosopis* being a particular concern in very arid areas (Harding and Bate, 1991).

The savannas, and especially the sweetveld savannas, are largely used for livestock or game management. Their carrying capacity is closely linked to rainfall, at least until the changeover to sourveld occurs. They are also home to a large proportion of Africa's rural population and are widely used for small-scale subsistence agriculture, as well as some large-scale commercial agriculture. The sweetveld savannas are where many of southern Africa's best known and largest conservation areas are situated. There is a growing trend away from livestock to game ranching. In South Africa, literally thousands of cattle ranches have been converted to either wildlife or a mix of wildlife and livestock. Drivers of this are multiple, including poor returns from livestock, a move towards lifestyle residences, stock theft (it is harder to steal wildlife), crime (a lion, or even an ostrich, roaming around your house is quite a strong deterrent to intruders) (ABSA, 2003). Hunting plays an important role as a revenue generator in the savannas and many game farms include this as part of their income stream. However, photographic safaris from luxury lodges are also growing in importance, with many of the historic hunting areas in countries, such as Namibia, Botswana and Mozambique, becoming ecotourism destinations. Namibia has had extensive success in establishing community-run conservancies on communal land. Similar processes have also occurred in Botswana. Historically it was Zimbabwe that initiated the concept of community-based natural resource management (CBNRM) and the linking of this to ecotourism, but political instability and a reluctance to fully devolve management rights has hindered its uptake in Zimbabwe (Bond et al., 2004; Child, 2004; DeGeorges and Reilly, 2009).

The sourveld savannas (miombos) are largely on ancient and highly leached soils (Huntley, 1982; Frost, 1996). Despite relatively high biomass production, livestock farming on these areas is problematic, especially during winter. Grazing by cattle and livestock is also limited in savanna sourveld by the widespread presence of tsetse fly, which prevent livestock farming (Frost, 1996). Despite this the sourveld savannas support large densities of livestock and are often considered as overgrazed (Moyo et al., 1993). The miombos have historically been seen as forest and the colonial and post-colonial management has historically been focused on timber extraction. Though the miombos have a number of high-value hardwood species, the rate of tree growth and the form of the trees limit the timber value of the area for forestry (Grundy, 1990; Dewees et al., 2011). The miombos are, however, used far more extensively than for just timber, and more than any of the other regions considered, this is mostly a communal land. Despite the trees being poor timber, the tree component is still the key driver of ecosystem processes, playing an important role in nutrient cycling as well as a limiting factor to grass production (Frost, 1996; Ribeiro et al., 2015). Traditional slash-and-burn agriculture, termed *Chitemene*, is still widely practiced as a mechanism for maintaining soil fertility (Frost, 1996). Where modern farming practices are employed, including liming and fertilization, high production of crops, such as maize, can be achieved, but careful soil nutrient management is required (Huntley, 1984). For small-scale subsistence farmers, yields are typically very low (Campbell, 1996). These savannas are well known for the importance that non-wood products, such as honey and mushrooms, play in rural diets and livelihoods (Mucina and Rutherford, 1996; Shackleton et al., 2010). Charcoal production from these forests is also a critically important rural livelihood, often bringing greater financial return than agriculture (Zulu and Richardson, 2013). Deforestation as a consequence of agricultural expansion, charcoal production and poor fire management are major concerns in the savanna sourveld (Frost, 1996; Mugo and Ong, 2006; Zulu, 2010; Sepp, 2011; Zulu and Richardson, 2013).

Unlike the grasslands, the savannas tend to be well conserved from a formal conservation perspective. Within South Africa 9 per cent of the savanna is under conservation (Mucina and Rutherford, 2006), with even higher levels being conserved in countries like Namibia, Botswana, Zambia and Zimbabwe. Conservation is not uniform over savanna types and it is the areas that were historically unfavorable for human and livestock habitation that have tended to be the conserved areas. Many of the conserved areas evolved from areas historically used for hunting, and carry a high density of wildlife. Elephants, as a mega-herbivore, play an important role in savanna ecology (Owen Smith, 1988), and can have large impacts on tree dynamics.

Managing Grasslands and Savanna: Fire and Herbivory

How best to manage grasslands and savanna from a livestock or biodiversity management perspective has dominated rangeland literature for years. Although this will be touched on briefly, a large set of macro-level drivers are starting to have far greater importance in relation to grassland and savanna management. These include aspects, such as, land-use-change and impacts from climate change including CO_2 fertilization and IAP and this will be covered in the later section.

From a livestock management perspective, managers have two main options for management intervention—changing the timing and intensity of livestock grazing and changing the nature of fires. Extensive literature exists on the impacts of both these variables and the interplay between them (e.g., Trollope, 1999; van Niekerk, 1984; Peddi et al., 1995; O' Connor and Roux, 1995; Hoffman, 2003; Hardy and Tainton, 1995; Hardy et al., 1999). For instance, Trollope (1999) suggests that sour grassland that is burnt early before spring rains and then grazed heavily and continuously before it can produce substantial re-growth may lead to a substantial decline in plant vigour and changes in species composition. Elaborate rotational grazing arrangements have been proposed and implemented on commercial farms, often with limited differences in results (O'Connor, 1985). In communal areas there is often *de facto* open access management as either there were never effective community rules to deal with livestock numbers, or the rules were largely broken down. Typically this results in year-round grazing except in some sourveld areas where transhumance practices have evolved, such as in the high altitudes of Lesotho (Quinlan and Morris, 1993) and the Barotose plains of Zambia (Turpie et al., 1999). Excluding transhumance, two fundamentally different management practices tend to take place and these are linked to tenure, one with relatively low-stocking densities and planned rotation, and the other with high-stocking densities and continuous grazing (Palmer and Ainslie, 2006). In many instances there are very clear fence-line contrasts between adjacent areas with private or communal livestock management. Further, since each private farmer can choose, within reason, his own livestock management practices, there can be strong fence line contrasts between adjacent private farms (Hoffman et al., 2002).

The South African grasslands are predominantly privately-owned large commercial farms. As such they tend to have large herds of animals that are managed based on economic carrying-capacity principles and for the purpose of maximising economic return, which could be either from meat sales or from milk production. If managing for milk production then improved pasture and irrigation is often a part of the grazing regime. Small areas of grasslands in South Africa, and almost all the grasslands in Lesotho, are managed as communal rangelands. In addition to financial returns, livestock (and especially cattle) are kept for cultural occasions, as an emergency source of finance or for investment purposes. In contrast to commercial farms, the livestock in communal areas is typically managed for multiple products including milk, draft power, meat, use in traditional ceremonies and sales (Shackleton et al., 2000; Palmer and Ainslie, 2006; Vetter, 2013). Livestock-stocking rate tends to approach ecological carrying capacity, with die-offs occurring during drought years and the herds re-building during periods of good rain (Illius and O'Connor, 1999).

The South African sourveld probably had a relatively low density of grazers before modern human settlement as the area has poor palatability in winter (Hardy et al., 1999; O'Connor, 1995). Current livestock practices potentially stock these areas at six to 20 times their historic livestock density (O'Conner, 2005) and this is thought to have had major changes on ecological structure, in many cases leading to a high persistence of unpalatable species (Hardy et al., 1999). The way grazing and fire are managed in combination can have significant impacts on the proportion of palatable versus unpalatable species in these systems (Harding et al., 1999). Sheep, with their more selective grazing appear to have a more detrimental impact than cattle on species composition (Hardy and Tainton, 1993; Hardy et al., 1999). Understocking or too infrequent a burning can also potentially lead to unpalatable species proliferation (Hardy et al., 1999). Rest periods from grazing in addition to rotation of grazing is typically advocated (Hardy and Hurt, 1989; Hardy and Tainton, 1993, 1995; Hardy et al., 1999).

The hydromorphic Zambian grasslands are an important grazing resource for communal livestock management. Cattle are moved into the very low fertility miombo woodlands during summer and grazed in the floodplains in winter where these areas provide better quality winter grazing than the surrounding areas (Turpie et al., 1999).

Fires are routinely used to stimulate new growth, especially in the sourveld. In the communal areas fires are offer initiated early in the dry season and grazed as soon as new sprouts appear (Frost, 1995). In contrast to communal land, the commercial farmers have greater control over fire and stock management and tend to burn later in the dry season, while withholding stocking directly after the grass begins to re-sprout. It is the combination of fire, browsing and grazing that is largely responsible for impacting tree grass ratios and the palatability of the grass layer. In the arid savannas, fires are less frequent due to the reduced fuel load and the fact that the dry grass can still be grazed. Some veld management practices advocate intense grazing as a method of stimulating plant growth rather than the use of fire. An extreme variant of this is the holistic management method advocated by Savory (2013), where he advocates very short periods of extremely high grazing intensity. This system has gained interest in the region but remains contentious (Carter et al., 2014).

Concern for land degradation has been expressed throughout the region, and particularly for communal rangelands. There are a number of studies that suggest that despite the lack of degradation in most communal grazing areas, actual production levels have been maintained over long periods (Scoones, 1993; Tapson, 1993; Shackleton, 1993; Harrison and Shackleton, 1999). There is extensive debate in the literature whether the grasslands and savannas should be considered as equilibrium of disequilibrium systems, and the impact this may have on management choices (Behnke, 1993; O'Connor and Bredenkamp, 1997; Illius and O'Connor, 1999; Vetter, 2005, 2009). There is growing evidence that the more arid (sweet) areas may in best be regarded as disequilibrium that closely tracks annual rainfall. One important consequence of this is that stocking rates may be better determined in a dynamic way based on annual rainfall, rather than being based on a mean stocking rate based on long-term trends. Sour savanna on the other hand may be considered as disequilibrium as in that fire plays an important role in maintaining the system. Scholes and Walker (1995), however, warn that this is the natural dynamic of a system and that the savannas should not be viewed as an arrested stage in a succession to forest.

Long-term cattle numbers from the communal areas suggest that in areas of prolonged drought there are widespread livestock die-offs, but that when the rains return, the grass vegetation can recover quite readily (Vetter, 2009; Scoones, 1993). However, in these systems there is a strong concern that overgrazing may become a factor promoting a trend to increased woody plant dominance (Joubert, 2003; Bester, 1996). In the very arid areas, this is manifested as increased density of karoo bushes. In the slightly moister areas, it is through an increased density of savanna tree species.

In the moist (sour) grassland areas, there is mixed evidence that grazing pressure changes have major impacts on species composition, in particular leading to a shift from palatable to unpalatable species. The interaction between fire and grazing is primarily determined by the influence of fire on large-scale forage patterns and by the influence of grazing on fire fuel loads (Archibald et al., 2005). In the literature the terms 'increaser species' and 'decreaser species' are widely used. Under heavy grazing, many of the palatable perennial species decrease and are replaced by unpalatable perennials and annuals. Managing grazing to ensure that there is not a substantive change in sward composition is a key management objective for many livestock farmers. The timed use of fire as a management tool has also been advocated, with one of the common moist grassland palatable species, *Themeda trianda*, having a well-known positive association with frequent burning.

Fire is also advocated as a management tool to reduce the spread of savanna trees into the grasslands. It was found in the Eastern Cape of South Africa that a combination of heavy browsing using goats, coupled with hot fires, could effectively control *Senegalia (acacia) karoo* expansion in grasslands (Trollope, 1974). Frequent late dry season fires can transform a miombo woodland into open tallgrass savanna with isolated fire-tolerant canopy trees and scattered understorey trees and shrubs (Timberlake et al., 2010). Many studies have reported that once the trees reach a certain height, they are less susceptible to fire (Furley et al., 2008). Surviving fire and herbivory until trees are large enough not to be effected by fire is what Higgins et al., refer to as the 'fire trap'.

Within the savannas, fire and herbivory are also major management tools. In the sweet savannas the combined impact of overgrazing and reduced fires (partly through deliberate fire management, and partly due to the grazing reducing fuel load), has historically been seen as a major driver of increased densities of woody plants (what is locally referred to as bush encroachment). This problem has been especially severe in Namibia where an estimated 26 million hectares have been affected (Bester, 1996, 1999). The impact of bush encroachment will drastically reduce the grass production and hence have major impacts on livestock (and in particular cattle) carrying capacity. In Namibia the national cattle herd size has reduced from 2.1 million in 1958 to 8,00,000 in 2001 (De Klerk, 2004). Some studies have estimated that there can be up to a tenfold reduction in cattle-carrying capacity. It reduces visibility, making the location of livestock difficult and in some cases the bush can be so dense as to retard livestock movement. Namibian and South African ecologists and farmers have done extensive research on mechanisms to cost-effectively reduce bush encroachment and these range from aerial spraying to the use of heavy machinery, to hand harvesting. Given the inherent low stocking rates in arid areas of Namibia, clearing of bush is often not financially viable from the perspective of enhanced grazing capacity. However, there is now a large established industry of charcoal production (CBEND, 2007). Bush clearance has been shown to increase grass yields by up to 400 per cent (Barnes, 1979). However, such increases tend to be associated with adverse changes in herbaceous species composition. For example, bush clearance at a woodland site in Central Zimbabwe led to the replacement of *Panicum maximum*, a palatable grass species, by unpalatable species (O'Connor, 1985).

In moist savanna the use of fire is critical as the grass becomes moribund without burning and mature grass is unpalatable to livestock; so it cannot be grazed away. In these areas, fire is also an important mechanism for stimulating new grass growth which is palatable to livestock. Gammon (1976) found the scope to control defoliation patterns through grazing management practices in miombo as limited. Despite the high biomass production in these systems, they have relatively low livestock-carrying capacity, as only about one per cent of total net primary production is consumed by livestock, with the balance being burned off, decomposed or consumed by termites and other insects (Frost, 1995). The sour savannas mostly have sufficient rainfall to be a closed forest (from a rainfall limiting perspective), and fire is an important component on preventing canopy closure. There is growing evidence that browsing of young trees is also important in this regard as it maintains the trees within the fire 'kill' zone. Most trees are, however, only top killed and can re-sprout rapidly from their root system. Deforestation from human activity, including charcoal production, is a major concern in these areas (Sepp, 2010; Dewees et al., 2011).

Drivers of Change in the Grasslands and Savannas

Despite decades of detailed grassland and savanna research on management techniques, the true threats to the resource base may come from macro factors that go beyond simple fire and grazing management choices open to rangeland managers.

The South African grassland biome is considered as one of the most vulnerable biomes in the country. It is poorly conserved, highly transformed and now faces additional threats from raised CO_2 levels and climate change (Mucina and Rutherford, 2006; Davies et al., 2014; Rutherford et al., 1999). The threat to the savannas is less severe but there is speculation that raised CO_2 may have major impacts on the tree/grass ratio, potentially reducing or eliminating the grass component (Bond, 2008; Bond and Midgely, 2000, 2012; Buitenwerf, 2012).

Climate change is anticipated to have severe impact in the region, with the greatest impacts occurring in the more arid western half of the region. In this region, centred on the southern Kalahari, both observation data and climate models suggest that this already hot area is warming at about twice the global average (Engelbrecht et al., 2015). This means that even if the rise in global temperature is limited to 1.5°C, this region is likely to experience a three degree rise in temperature. Rainfall data trends are less clear and vary between models and scenarios; however there is a growing consensus that it is likely to be the already arid southwestern side of the sub-continent that will experience a decreased rainfall, whilst the moister eastern side may well experience slight increases in rainfall (Hewitson et al., 2005; DEA, 2013). The implication of this is that the nama-karoo may well expand into the western boundary of the grassland and savanna, with the dwarf karoo bushes replacing grass. The already arid Kalahari region may experience long-term reduction in production potential. Further, the savannas are likely to encroach into the grassland along the savanna grassland divide, shrinking the extend of the grasslands (Rutherford et al., 1999). The cooler temperature sour and species-rich grassland will be forced further upslope on the large mountain ranges. Extensive land transformation and fragmentation within the grassland will exacerbate this process.

A factor affecting many savannas, and also potentially affecting grasslands, is the densification of woody species. In the savanna this is largely from species indigenous to the area, but that previously occurred at lower densities. As stated above, this process has historically been ascribed to impacts from livestock management, and though management clearly plays a role, there is growing evidence that CO_2 fertilization may also be having a direct effect on the increased density of woody plants (Bond and Midgely, 2012). The rational underpinning this hypothesis is that raised CO_2 levels may favour C3 plants over C4 plants, but also favour woodiness as a growth strategy (Archer et al., 1995; Fensham, 2005). The increase in CO_2 will have important, but poorly understood interactions with both fire and herbivory (O'Conner, 2014). Higgins et al. (2000) have postulated that woody plants are suppressed by what they term a 'fire trap', that is short trees are consistently top-killed by fires before they get large enough to escape the impact of grass fires. Grazing pressure on the newly re-sprouted trees may increase the strength of this effect. However, increased CO_2 concentrations allow the trees to grow faster and to store substantial reserves in their root systems (Bond and Midgely, 2012; Kgope et al., 2010). This then gives them a greater chance of being able to grow beyond the fire trap height during the period between fires. Open chamber raised CO_2 experiments have shown that some tree species can substantially increase growth rates and root carbon stores under raised CO_2 concentrations (Kgope et al., 2010). This creates what Bond and Midgely (2012) refer to as 'super-seedlings' of savanna trees, which are able to establish at a rate not possible under low atmospheric CO_2. This is supported by widespread evidence that bush encroachment has taken place on areas that have not been subject to livestock grazing (Ward, 2005; Masubelele et al., 2015). A recent study by Stevens et al., 2016 which compared 1940s aerial photographs with modern imagery found increased woody plant density in all areas on both communal, conservation and commercial land, the only exception being the arid savanna areas with high elephant density. Sankaran et al. (2004, 2005) show that most arid savannas have a climate limitation on tree density, whilst for moist savanna, the area can potentially become forest. Bond and Midgely (2012) postulate that increased CO_2 may well change the nature of dynamics towards greater trees, but emphasise that there is still a scarcity of data to understand the dynamics at different locations. This is echoed by O'Connor (2014), who points out that drivers of bush encroachment are complex and situation-dependent.

A CO_2-induced increased woodiness could have profound impact on the savanna. There is a general negative relationship between tree cover and grass cover (Scholes and Walker, 1995). Some authors have suggested that a very low tree density might in fact enhance grass quantity and palatability (Belsky, 1994), but a tree density over about 10 per cent of cover clearly decreases grass, with a tree cover of over 60 per cent potentially leading to the total exclusion of grass. This is particularly likely in the moister savannas where water does not limit tree cover (Scholes and Archer, 1997). Bush encroachment clearly leads to grass reduction and to the overall carrying capacity of the veld, especially for domestic grazers such as cattle or wildlife like zebra, wildebeest and buffalo (Smith, 2001; De Klerk, 2004; Scholes and Archer, 1997). In the wildlife areas there is a secondary impact on predators, such as cheetah that have evolved on open plains (Joubert, 2003). In the arid areas, it is seldom seen that sufficient grass biomass builds up to carry a sufficiently hot fire to top-kill the bush and as the bush becomes denser, the ability to carry a fire decreases. The concern is that the system will enter a new stable state where grass is largely absent. In addition to reduced grazing, there are a number of additional effects, such as changed biodiversity of the herbaceous layer, increased evapotranspiration with trees accessing slightly deeper soil water than the grass they replaced, a possible decreased level of the water table (Bockmühl, 2006), impaired movement and visibility. Long-term impacts on soil carbon are unclear, though there is clearly an increase in above-ground carbon stores, with Namibia declaring itself carbon neutral as a nation based on the carbon sequestered in bush encroachment.

Increased tree cover, especially if this is driven by global change, will have a profound impact on how the areas are managed. Browsers rather than grazers and/or game rather than livestock may perform better in these areas. Also there might be a trend toward biomass farming rather that livestock farming. Already charcoal is an important product from some areas (SEA, 2010; C.A.R.E., n.d.). The use of woody biomass to drive micro electrical power generation stations is been seriously considered (C.A.R.E., n.d.; SEA, 2010). Current biomass accumulation rates are still poorly understood as most studies focus on stem counts per hectare rather than biomass. Further, there is no data on what impact long-term repeated extraction of woody biomass would have on the ecology of the system. The exact factors leading to bush encroachment still remain relatively poorly understood. Clearly a multitude of factors operate in tandem, with the relative importance of individual factors in different locations and under different management regimes being relatively poorly understood (O'Connor, 2014; Bond and Midgely, 2012). The situation is also dynamic as atmospheric CO_2 levels are still likely to increase for many years before stabilising and a very limited set of enhanced CO_2 studies need to be relied on in an attempt to understand the potential future options.

The impact of climate change and raised CO_2 on the sour savanna systems is poorly understood. Pienaar et al. (2015) suggest that climate change may well shrink the distribution of *Brachystigia*. Wigley (2009) warns that moist savannas may in fact be more susceptible to bush encroachment than arid savanna, but this is based on research in KwaZulu Natal, South Africa, and not in the miombo. It is likely that thickening will occur in the sour savanna and that the role of fire will be critically important in determining the extent to which grass is maintained in the system.

Though poorly researched, a change in the spatial divide between sweetveld and sourveld as a consequence of climate change is unlikely in the short term. Although the mean annual rainfall is clearly an important factor in the sweetveld/sourveld divide, the changes in soil nutrient status and the leaching of nutrients in moister conditions will take time. As the Nylsvley example shows, the system can remain stable in a changed state from sourveld to sweetveld over centuries.

Impacts of IAP may well increase in the future (Richardson et al., 2004). Many of these species have weedy characteristics and as such, the disturbance of climate change may well favor them. Although IAP are a threat to all grasslands and savannas, it appears that it is the sour savanna where they pose the greatest threat and also where they have the greatest impact on streamflow (Le Maitre et al., 1996, 2014).

Figure 12.4 gives a graphic representation of the identified major threats. In all cases, raised CO_2 levels and other changes due to climate change may well impact the current dynamics. In the sour grasslands, the biggest climate threat is that raised temperature may well reduce the available habitat. In the sweetveld grasslands and savanna, increased CO_2 may well increase woody density at the expense of grass and grazing. The extent of this threat to sour savanna is less clear, but increased woodiness could convert these areas into forests, and the role of fire will remain a critical factor in determining this. Management practices may also impact all these systems. Fire is of critical importance in the sourveld, but in the grasslands there is a strong interaction with grazing that is largely absent from the sourveld savanna. Human-induced deforestation remains a critical concern in the sour savanna, whilst in many areas of the sweet savanna, it is woody encroachment that remains a problem.

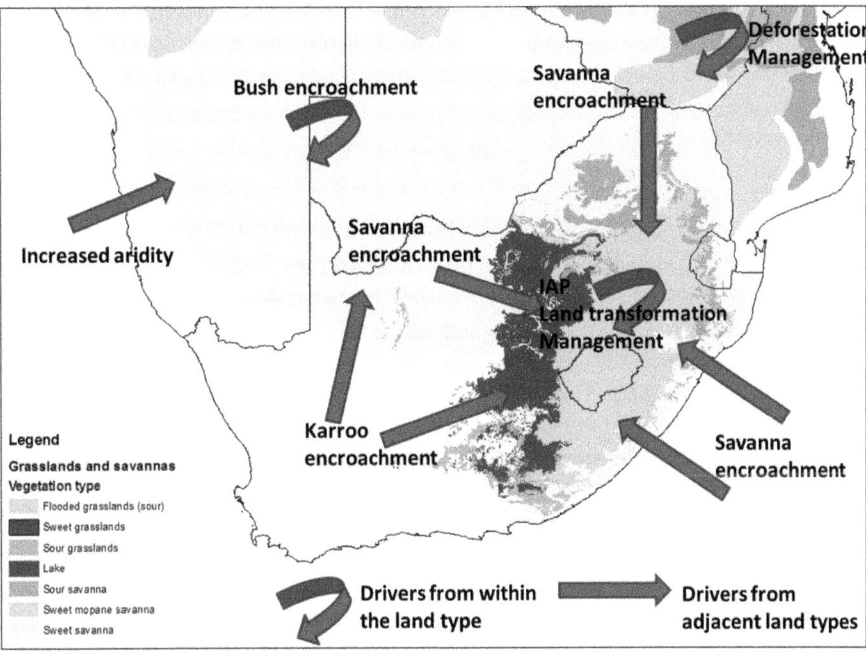

Fig. 12.4 Symbolic representation of some of the key threats to different grassland and savanna types under global change.

Conclusion

From a functional perspective it is useful to divide the southern African grasslands into four groups based on the palatability of the vegetation and on the occurrence of the woody element. This, therefore, separates pure grasslands from savanna grasslands and areas of sweetveld from areas of sourveld. Although this differentiation is relatively simplistic, many functional aspects of the system can be well explained by this division and it has a number of over-arching implications for management. Caution must, however, be maintained that individual areas within any of these broad divisions may have local specific dynamics and implications for management. Sweetveld grassland, although sought after for livestock and wildlife ranching, requires large land holdings, especially at the arid extremes. These areas appear to withstand relatively high grazing pressure but are prone to droughts and bush encroachment. The sourveld areas make winter grazing difficult, but are still widely grazed. Shifts to less palatable grass species or grazing-resistant life forms are common. A careful balance of grazing practices and fire is needed to maintain the veld for optimum grazing.

The management practices and objectives in southern Africa also change substantively between areas managed as large commercial farms, which attempt to optimise financial returns versus areas being managed as communal areas where there might be a variety of management objectives which may include financial returns, but also a multitude of other management objectives relating to culture, multiple use of livestock and subsistence needs. Commercial farmers attempt to maintain herd size to a maximum economic stocking rate, whilst their communal counterparts often maintain herds at an ecological stocking rate. Both practices have had a profound impact on the environment, but these have not always been the same. In some cases, the vegetation has proven to be extremely resilient, though species composition often changes between the different management practices. Further, a growing spatial extent of the region is being managed for wildlife rather than livestock. This is partially for strict conservation measures in national conservation areas, but also on private farms where wildlife may be found for hunting, tourism or a combination of both. Management for diversity of species as well as aesthetics may be more important than animal density in these situations.

Many aspects of global change are posing new threats to the grasslands and savannas. The South African grasslands are one of the most transformed and fragmented habitats in the country and in addition are likely to suffer profound impacts from climate change. A shrinkage in spatial extent of the grasslands is anticipated. For the savannas, a poorly understood potential impact from global CO_2 increase may well lead to significant change in the tree:grass ratio. This could have a profound impact on cattle-carrying capacities and management choices. It might even lead to new and novel use and management practices. Global warming is likely to cause an intrusion of savanna into grassland. Savannas, themselves may become more woody and move towards a forest or thicket type vegetation with a reduced grass component as a consequence of raised CO_2 levels. The sour savannas might move towards forest. The role of fire will remain an important component in these dynamics.

Changes to grasslands and savanna as a consequence of global change will have a direct impact on livelihood, livelihood options and the management practices undertaken. In addition, there will be a substantial impact on plant and animal biodiversity.

References and Further Readings

ABSA. 2003. Game Ranch Profitability in Southern Africa. ABSA Group Economic Research. South Africa Financial Sector Forum, www.finforum.co.za.

Acocks, J.P.H. 1966. Non-selective grazing as a means of veld reclamation. Proceedings of the Grassland Society of Southern Africa 1: 33–39.

Acocks, J.P.H. 1988. Veld Types of South Africa. Memoir of the Botanical Survey of South Africa. No. 40. Government Printer, Pretoria, 128.

Archer, S., D.S. Schimel and E.A. Holland. 1995. Mechanisms of shrub land expansion: Land use, climate or CO_2. Clim. Change 29: 91–99.

Archibald, S., W. Bond, W. Stock and D. Fairbanks. 2005. Shaping the landscape: Fire-grazer interactions in an African savanna. Ecol. Appl. 15: 96. DOI: 10.1890/03-5210.

Behnke, R., I. Scoones and C. Kerven. 1993. Range Ecology at Disequilibrium: New Models of Natural Variability and Pastoral Adaptation in African Savannas. Overseas Development Institute, London.

Belsky, A.J. 1994. Influences of trees on savanna productivity: Test of shade, nutrients, and tree-grass competition. Ecology 75: 922–932.

Bester, F.V. 1996. Bush encroachment, a thorny problem. Namibian Environment 1: 175–177.

Blackmore, A.C., M.T. Mentis and R.J. Scholes. 1990. The origin and extent of nutrient enriched patches within a nutrient-poor savannah in South Africa. J. Biogeogr. 17: 463–70.

Bockmühl, F. 2006. Platveld Aquifer Study. Special Volume: Recharge Considerations (Ref: 12/1/2/16/9). Draft report prepared for Department of Water Affairs, MAWF, Windhoek. May 2006.

Bond, I.B., D. Child, B. de la Harpe, J. Jones, J. Barnes and H. Anderson. 2004. Private land contribution to conservation in South Africa. *In*: B. Child (ed.). Parks in Transition: Biodiversity, 33 Rural Development and the Bottom Line. London, Sterling, VA: IUCN, SASUSG, and Earthscan.

Bond, W.J. and G.F. Midgley. 2000. A proposed CO_2-controlled mechanism of woody plant invasion in grasslands and savannas. Global Change Biology 6: 865–869.

Bond, W.J., G.F. Midgley and F.I. Woodward. 2003. What controls South African vegetation—climate or fire? South African Journal of Botany 69: 79–91.

Bond, W.J. 2008. What limits trees in C_4 grasslands and savannas? Annual Review of Ecology, Evolution, and Systematics 39: 641–659.

Bond, W.J. and G.F. Midgley. 2012. Carbon dioxide and the uneasy interaction of trees and savannah grasses. Philosophical Transactions of the Royal Society B: Biological Sciences 367: 601–612.

Buitenwerf, R., W.J. Bond, N. Stevens and W.S.W. Trollope. 2012. Increased tree densities in South African savannas: 50 years of data suggests CO_2 as a driver. Global Change Biology 18(2): 675–684.

C.A.R.E. Ltd (no date) Cost-benefit Analysis of Encroachment: Bush Removal in Namibia (2008–2009). http://www.care.demon.co.uk.

Campbell, B. 1996. The miombo in transition, woodlands and welfare in Africa. CIFOR, Bogor, Indonesia.

Campbell, B.M., P. Frost and N. Byron. 2006. Miombo woodlands and their use: Overview and key issues. pp. 1–10. *In*: B. Campbell (ed.). The Miombo in Transition: Woodlands and Welfare in Africa. Centre for International Forestry Research, Bogor, Indonesia.

Carter, J., A. Jones, M. O'Brien, J. Ratner and G. Wuerthner. 2014. Holistic management: Misinformation on the Science of Grazed Ecosystems. International Journal of Biodiversity. http://dx.doi.org/10.1155/2014/163431.

CBEND. 2007. Turning Namibian Invader Bush into Electricity: The CBEND Project, Desert Research Foundation of Namibia (DRFN), Windhoek, Namibia.

Child, B. (ed.). 2004. Parks in Transition: Biodiversity, Rural Development and the Bottom Line. London: Earthscan.

Coe, M.J., D.H. Cumming and J. Phillipson. 1976. Biomass and production of large African herbivores in relation to rainfall and primary production. Oecologia (Berlin) 22: 341–354.

Cowling, R.M., D.M. Richardson and S.M. Pierce. 2004. Vegetation of southern Africa, Cambridge University Press.

Davies, A.B., M.P. Robertson, S.R. Levick, G.P. Asner, B.J. van Rensburg and C.L. Parr. 2014. Variable effects of termite mounds on African savanna grass communities across a rainfall gradient. J. Veg. Sci. 25: 1405–1416. DOI: 10.1111/jvs.12200.

De Klerk, J.N. 2004. Bush encroachment in Namibia: Report on Phase 1 of the Bush Encroachment Research, Monitoring, and Management Project, Windhoek, Namibia, Ministry of Environment and Tourism, Directorate of Environmental Affairs, Windhoek.

DEA (Department of Environmental Affairs). 2013. Long-Term Adaptation Scenarios Flagship Research Programme (LTAS) for South Africa. Climate Trends and Scenarios for South Africa. Pretoria, South Africa.

DeGeorges, P.A. and B.K. Reilly. 2009. The realities of community-based natural resource management and biodiversity conservation in sub-saharan Africa. Sustainability 1: 734–788. DOI: 10.3390/su1030734.

Dewees, P., B. Campbell, Y. Katerere, A. Sitoe, A.B. Cunningham, A. Angelsen and S. Wunder. 2011. Managing the Miombo Woodlands of Southern Africa: Policies, Incentives, and Options for the Rural Poor. Washington: Program on Forests (PROFOR).

Dovie, D.B.K., C.M. Shackleton and E.T.F. Witkowski. 2006. Valuation of communal area livestock benefits, rural livelihoods and related policy issues. Land Use Policy 23: 260–271.

Dye, P.J. and P.T. Spear. 1982. The effects of bush clearing and rainfall on grass yield and composition in SW Zimbabwe. Zimb. J. Agric. Res. 20: 103–118.

Ellery, W.N., R.J. Scholes and M.T. Mentis. 1991. Differentiation of the grassland biome of South Africa based on climatic indices. S. Afr. J. Science 87: 499–503.

Ellery, W.N., R.J. Scholes and M.C. Scholes. 1995. The distribution of sweetveld and sourveld in South Africa's grassland biome in relation to environmental factors. African Journal of Range & Forage Science 12(1): 38–45. DOI: 10.1080/10220119.1995.9647860.

Elliot, R.C. and K. Folkertsen. 1961. Seasonal changes in composition and yields of veld grasses. Rhodesia Agricultural Journal 58: 186–197.

Engelbrecht, F., J. Adegoke, M.M. Bopape, M. Naidoo, R. Garland, M. Thatcher, J. McGregor, J. Katzfey, M. Werner, C. Ichoku and C. Gatebe. 2015. Projections of rapidly rising surface temperatures over Africa under low mitigation. Environ. Res. Lett. 10(8): 085004. DOI: 10.1088/1748-9326/10/8/085004.

Fairbanks, D.H.K., M.W. Thompson, D.E. Vink, T.S. Newby, H.M. van den Berg and D.A. Everard. 2000. The South African land-cover characteristics landbase: A synopsis of the landscape. South African Journal of Science 96: 69–82.

Fensham, R.J., R.J. Fairfax and S.R. Archer. 2005. Rainfall, land use and woody vegetation cover change in semi-arid Australian savanna. J. Ecol. 93: 596–606.

Furley, P.A., R.M. Rees, C.M. Ryan and G. Saiz. 2008. Savanna burning and the assessment of long-term fire experiments with particular reference to Zimbabwe. Prog. Phys. Geogr. 32(6): 611–634.

Frost, P. 1996. The ecology of miombo woodlands. pp. 11–55. *In*: B. Campbell (ed.). The Miombo in Transition: Woodlands and Welfare in Africa. Bogor: CIFOR; 1996.

Gammon, D.M. 1976. Studies of the Patterns of Defoliation, Herbage Characteristics and Grazing Behaviour during Continuous and Rotational Grazing of the Matopos sandveld of Rhodesia. D.Sc. Agric. thesis, University of Orange Free State, Bloemfontein.

Grundy, I.M. 1990. The Potential for Management of the Indigenous Woodland in Communal Farming Areas of Zimbabwe with Reference to the Regeneration of Brachystegia spiciformis and Julbernardia globiflora. Unpublished M.Sc. thesis, University of Zimbabwe, Harare.

Gush, M.B., D.F. Scott, G.P.W. Jewitt, R.E. Schulze, L.A. Hallowes and A.H.M. Gorgens. 2002. A new approach to modelling streamflow reductions resulting from commercial afforestation in South Africa. South. Afr. For. J. 196: 27–36.

Harding, G.B. and G.C. Bate. 1991. The occurrence of invasive Prosopis species in the north-western Cape, South Africa. S. Afr. J. Sci. 87: 188–192.

Hardy, M.B. and N.M. Tainton. 1993. Mixed-species grazing in the Highland Sourveld of South Africa: an evaluation of animal production potential. Proceedings XVII International Grassland Congress, Rockhampton, Australia, 2026–2028.

Hardy, M.B. and N.M. Tainton. 1995. The effects of mixed-species grazing on the performance of cattle and sheep in Highland Sourveld. African Journal of Range and Forage Science 12(3): 97–103.

Hardy, M.B., D.L. Barnes, A. Moore and K.P. Kirkman. 1999. The management of different types of veld. *In*: N.M. Tainton (ed.). Veld Management in South Africa. University of Natal Press, Pietermaritzburg.

Harrison, Y.A. and C.H. Shackleton. 1999. Resiliance of south African communal grazing lands after the removal of high grazing pressure. Land Degradation and Development 10: 225–239.

Hempson, G.P., S. Archibald, W.J. Bond, R.P. Ellis, C.C. Grant, F.J. Kruger, L.M. Kruger et al. 2014. Ecology of grazing lawns in Africa. Biological Reviews 90: 979–994. http://dx.doi.org/10.1111/brv.12145.

Hewitson, B., M. Tadross and C. Jack. 2005. Scenarios from the University of Cape Town. *In*: R.E. Schulze (ed.). Climate Change and Water Resources in Southern Africa: Studies on Scenarios, Impacts, Vulnerabilities and Adaptation. WRC Report No. 1430/1/05. Water Research Commission, Pretoria, South Africa.

Higgins, S.I., W.J. Bond and W.S.W. Trollope. 2000. Fire, resprouting and variability: A recipe for tree–grass coexistence in savanna. Journal of Ecology 88: 213–229.

Higgins, S.I. and S. Scheiter. 2012. Atmospheric CO_2 forces abrupt vegetation shifts locally, but not globally. Nature 488: 209–213.

Hoffman, M.T. 2003. Nature's method of grazing: Non-Selective Grazing (NSG) as a means of veld reclamation in South Africa. South African Journal of Botany 69(1): 92–98.

Hoffmann, T. and A. Ashwell. 2002. Nature Divided: Land Degradation in South Africa. University of Cape Town Press, Cape Town.

Huntley, B.J. 1982. Southern African savannas. pp. 101–119. *In*: B.J. Huntley and B.H. Walker (eds.). Ecology of Tropical Savannas, Springer-Verlag, Berlin.

Illius, A.W. and T.G. O'Connor. 1999. On the relevance of non-equilibrium concepts to arid and semi-arid grazing systems. Ecological Applications 9: 798–813.

Joubert, D.F. 2003. The Impact of Bush Control Measures on the Ecological Environment (Biodiversity, Habitat Diversity and Landscape Consideration). Unpublished specialist report prepared for the Ministry of Environment and Tourism, Windhoek.

Kgope, B.S., W.J. Bond and G.F. Midgley. 2010. Growth responses of African savanna trees implicate atmospheric [CO_2] as a driver of past and current changes in savanna tree cover. Austral. Ecol. 35: 451–463. (DOI: 10.1111/j.1442- 9993.2009.02046.x).

Lamprey, H.F. 1983. Pastoralism yesterday and today: the overgrazing problem. pp. 643–666. *In*: F. Bourlière (ed.). Tropical Savannas. Ecosystems of the World 13, Elsevier, Amsterdam.

Le Maitre, D., D.B. Versfeld and R.A. Chapman. 2000. The impact of invading alien plants on surface water resources in South Africa: a preliminary assessment. Water SA 26(3): 397–408.

Le Maitre, D.C., B.W. van Wilgen, R.A. Chapman and D.H. McKelly. 1996. Invasive plants in the Western Cape, South Africa: modelling the consequences of a lack of management. Journal of Applied Ecology 33: 161–172.

Le Maitre, D.C., I. Kotzee and P.J. O'Farrell. 2014. Impacts of land-cover change on the water flow regulation ecosystem service: Invasive alien plants, fire and their policy implications. Land Use Policy 36: 171–181.

Ludwig, F., H. De Kroon and H.H.T. Prins. 2008. Impacts of savanna trees on forage quality for a large African herbivore. Oecologia 155(3): 487–496. DOI: 10.1007/s00442-007-0878-9.

Masubelele, M.L., M.T. Hoffman and W.J. Bond. 2015. A repeat photograph analysis of long-term vegetation change in semi-arid South Africa in response to land use and climate. Journal of Vegetation Science 26(5): 1013–1023. See more at: http://www.pcu.uct.ac.za/pcu/staff/hoffman#sthash.QdwCTICc.dpuf.

Moyo, S., P. O'Keef and M. Sill. 1993. The Southern African Environment: Profiles of the SADC Countries. The ETC Foundation, London, p. 354.

Mucina, L. and M.C. Rutherford. 2006. The Vegetation of South Africa, Lesotho and Swaziland. Strelitzia 19. South African National Biodiversity Institute, Pretoria.

Mugo, F. and C. Ong. 2006. Lesson's from Eastern Africa's Unsustainable Charcoal Trade. ICRAF Working Paper no. 20. Nairobi, Kenya: World Agroforestry Center.
O'Connor, T.G. 1985. A Synthesis of Field Experiments Concerning the Grass Layer in the Savanna Regions of Southern Africa, South African National Scientific Programme Report No. 114, CSIR, Pretoria.
O'Connor, T.G. and P.W. Roux. 1995. Vegetation changes (1949–71) in a semiarid, grassy dwarf shrubland in the Karoo, South Africa: Influence of rainfall variability and grazing by sheep. Journal of Applied Ecology 32: 612–626.
O'Connor, T.G. and G.J. Bredenkamp. 1997. Grassland. *In*: R.M. Cowling, D.M. Richardson and S.M. Pierce (eds.). Vegetation of Southern Africa. Cambridge University Press, United Kingdom.
O'Connor, T.G. 2005. Influence of land use on plant community composition and diversity in Highland Sourveld grassland in the southern Drakensberg, South Africa. Journal of Applied Ecology 42: 975–988.
O'Connor, T.G., J.R. Puttick and M.T. Hoffman. 2014. Bush encroachment in southern Africa: Changes and causes. African Journal of Range & Forage Science 31(2): 67–88.
Owen-Smith, N. 1988. Megaherbivores: The Influence of Very Large Body Size on Ecology. Cambridge University Press, Cambridge, United Kingdom.
Palmer, T. and A. Ainslie. 2006. Country Pasture/Forage Resource Profiles: South Africa. Rome, Italy: FAO.
Palmer, A.R. and J.E. Bennett. 2013. Degradation of communal rangelands in South Africa: Towards an improved understanding to inform policy. African Journal of Range & Forage Science: 1–7.
Peddie, G.M., N.M. Tainton and M.B. Hardy. 1995. The effect of grazing intensity on the vigour of Themeda triandra and Tristachya leucothrix. African Journal of Range and Forage Science 12: 111–115.
Pienaar, B., D.I. Thompson, B.F.N. Erasmus, T.R. Hill and E.T.F. Witkowski. 2015. Evidence for climate-induced range shift in Brachystegia (miombo) woodland. S. Afr. J. Sci. 111(7/8): Art. #2014-0280, 9 pages. http://dx.doi.org/10.17159/sajs.2015/20140280.
Plowes, D.C.H. 1957. The seasonal variation in twenty common veld grasses at Matopos, southern Rhodesia, and related observations. Rhodesia Agricultural Journal 54: 33–55.
Quinlan, T. and C.D. Morris. 1993. Implications of changes to the transhumance system for conservation of the mountain catchments in eastern Lesotho. Journal of Range and Forage Science 11(3): 76–81.
Reason, C.J.C., R.J. Allan, J.A. Lindsey and T.J. Ansell. 2000. ENSO and climatic signals across the Indian Ocean basin in the global context: Part I, Interannual composite patterns. International Journal of Climatology 20: 1285–1327.
Ribeiro, N.S., S. Syampungani, N.M. Matakala, D. Nangoma and A.I. Ribeiro-Barros. 2015. Miombo Woodlands Research Towards the Sustainable Use of Ecosystem Services in Southern Africa.
Richardson, D.M. and P.J. Brown. 1986. Invasion of mesic mountain fynbos by Pinus radiata. S. Afr. For. J. 56: 529–536.
Richardson, D.M. and B.W. van Wilgen. 2004. Invasive alien plants in South Africa: how well do we understand the ecological impacts? South African Journal of Science 100: 45–52.
Rouault, M. and Y. Richard. 2003. Intensity and spatial extension of drought in South Africa at different time scales. [Online] Available at: http://www.wrc.org.za.
Rutherford, M.C., G.F. Midgley, W.J. Bond, L.W. Powrie, R. Roberts and J. Allsopp. 1999. South African country study on climate change. Plant Biodiversity: Vulnerability and Adaptation Assessment. SANBI.
Sankaran, M., R. Jayashree and N.P. Hanan. 2004. Tree–grass coexistence in savannas revisited—insights from an examination of assumptions and mechanisms invoked in existing models. Ecology Letters 7: 480–490.
Sankaran, M., N.P. Hanan, R.J. Scholes, J. Ratnam, D.C. Augustine, B.S. Cade et al. 2005. Determinants of woody cover in African savannas. Nature 438: 846–849.
Savory, A. 2013. How to Green the Desert and Reverse Climate Change, TED: Ideas Worth spreading, filmed February 2013, posted March 2013, http://www.ted.com/talks/allan_savory_how_to_green _the_world_s_deserts_and_reverse_climate_change.html (accessed 3 September 2016).
Scholes, R.J. and B.H. Walker. 1993. An African Savanna: Synthesis of the Nylsvley Study. Cambridge University Press, Cambridge UK.
Scholes, R.J. 1997. Savanna. *In*: R.M. Cowling, D.M. Richardson and S.M. Pierce (eds.). Vegetation of Southern Africa Cambridge, UK: Cambridge University Press.
Scoones, I. 1993. Why are there so many animals? Cattle population dynamics in the communal areas of Zimbabwe. pp. 62–76. *In*: R.H. Behnke, I. Scoones and C. Kerven (eds.). Range Ecology at Disequilibrium, ODI, IIED and CS, London.
SEA. 2010. Strategic Environmental Assessment of replication of the project Combating Bush Encroachment for Namibia's Development (CBEND). Prepared by the Southern African Institute for Environmental Assessment (SAIEA) for the National Planning Commission Secretariat.

Sepp, S. 2010. Wood Energy: Renewable, Profitable and Modern. Postfach, Germany: GTZ (Deutsche GesellschaftfürTechnischeZusammenarbeit); http://www.gtz.de/de/dokumente/.
Shackleton, C.M. 1993. Are the communal grazing lands in need of saving? Dev. South Afr. 10(1): 65–78.
Shackleton, S., C. Shackleton and B. Cousins. 2000. Re-valuing the communal lands of southern Africa: New understandings of rural livelihoods. ODI Natural Resource Perspectives No. 62. Overseas Development Institute, London.
Shackleton, S., M. Cocks, T. Dold, S. Kaschula, K. Mbata, G. Mickels-Kokwe and G.P. von Maltitz. 2010. Non-wood forest products: Description, use and management. *In*: E.N. Chidumayo and D.J. Gumbo (eds.). The Dry Forests and Woodlands of Africa: Managing for Products and Services. Earthscan. London.
Smit, G.N. 2001. The influence of tree thinning on the vegetative growth and browse production of Colophospermum mopane. South African Journal of Wildlife Research 31: 99–114.
Stevens, N., W.J. Bond, M.T. Hoffman and G.F. Midgley. 2015. Change is in the air: Ecological trends and their drivers in South Africa. SAEON, Pretoria. ISBN: 978-0-620-65261-2 http://www.saeon.ac.za/enewsletter/archives/2015/december2015/Change%20is%20in%20the%20air_WEB%20VERSION.pdf.
Stevens, N., B.N.F. Erasmus, S. Archibald and W.J. Bond. 2016. Woody encroachment over 70 years in South African savannahs: overgrazing, global change or extinction aftershock? Phil. Trans. R. Soc. B 371: 20150437. http://dx.doi.org/10.1098/rstb.2015.0437.
Tainton, N.M. 1999. The ecology of the main grazing lands of South Africa. *In*: N.M. Tainton (ed.). Veld Management in South Africa. University of Natal Press, Pietermaritzburg.
Tapson, D. 1993. Biological sustainability in pastoral systems: The Kwazulu case. pp. 118–135. *In*: R.H. Behnke, I. Scoones and C. Kerven (eds.). Range Ecology at Disequilibrium, ODI, IIED and CS, London.
Timberlake, J., E. Chidumayo and L. Sawadogo. 2010. Distribution and characteristics of African dry forests and woodlands. *In*: E. Chidumayo and D.J. Gumbo (eds.). The Dry Forests and Woodlands of Africa: Managing for Products and Services. Earthscan, London, UK.
Trollope, W.S.W. 1974. Role of fire in preventing bush encroachment in the Eastern Cape. African Journal of Range and Forage Science 9: 67–72.
Trollope, W.S.W. 1999. Veld burning. *In*: N.M. Tainton (ed.). Veld Management in South Africa. University of Natal Press, Pietermaritzburg.
Turpie, J., B. Smith, L. Emerton and J. Barnes. 1999. Economic value of the Zambezi Basin Wetlands. Report prepared for IUCN Zambezi Basin Wetlands Conservation and Resource Utilization Project.
Tyson, P.D., G.J. Dyer and M.N. Mametse. 1975. Secular changes in South African rainfall: 1880–1972. Quart. J. Roy. Meteor. Soc. 101: 817–833.
Van Niekerk, A., M.B. Hardy, B.D. Mappeldoram and S.F. Lesch. 1984. The effect of stocking rate and lick supplementation on the performance of lactating beef cows and its impact on highland Sourveld. Journal of the Grassland Society of South Africa 1(2): 18–21.
Vetter, S. 2005. Rangelands at equilibrium and non-equilibrium: recent developments in the debate. Journal of Arid Environments 62: 321–341.
Vetter, S. 2009. Drought, change and resilience in South Africa's arid and semi-arid rangelands. South African Journal of Science 105: 29–33.
Vetter, S. 2013. Development and sustainable management of rangeland commons—Aligning policy with the realities of South Africa's rural landscape. Afr. J. Range Foroage Sci. 30: 1–9.
von Maltitz, G.P., L. Mucina, C. Geldenhuys, M. Lawes, H. Ealey, H. Adie, D. Vink, G. Fleming and C. Bailey. 2003. Classification system for the South African indigenous forests. An objective classification for the Department of Water Affairs and Forestry. Report ENV-P-2003-017. Environmentek. CSIR Pretoria.
Ward, D. 2005. Do we understand the causes of bush encroachment in African savannas? African Journal of Range & Forage Science. 22(2): 101–105.
Zulu, L.C. 2010. The forbidden fuel: charcoal, urban woodfuel demand and supply dynamics, community forest management and woodfuel policy in Malawi. Energy Policy 38: 3717–30. http://dx.doi.org/10.1016/j.enpol.2010.02.050.

13

Grasslands of Eastern Africa
Problems and Prospects

Dennis O. Otieno[1,*] and *Jenesio I. Kinyamario*[2]

Introduction

The landscape of east Africa, which includes Ethiopia, parts of the Horn of Africa, Sudan, Kenya, Uganda, Rwanda, Burundi and Tanzania, has changed dramatically over the last 14–10 million years from a relatively flat, homogenous region covered with mixed tropical forest, to a varied and heterogeneous environment, with mountains over 4,000 m high (Maslin et al., 2014). Volcanic activities which occurred between 14 and 8 Ma during the Miocene resulted in the formation of the Great Rift Valley, greatly altering the relief, hydrology and climate of the landscape. The change in physiography led to the rise of a variety of vegetation types, ranging from forests, savannas, grasslands and deserts. The region sits on an area of 6.2×10^6 sq km, 75 per cent of which is covered by grasslands with varying altitudes (Fig. 13.1), soil types, climates and woody vegetation. They comprise expansive semi-arid to arid grasslands, wooded grasslands, savanna, bushlands and woodlands as well as the extensive highlands, including the high elevation mountain regions (FAO, 2005). Productivity of these grasslands is variable and rainfall is the main determining factor. Generally, the regional rainfall pattern decreases northwards, away from the equator, except in the highlands; hence most grasslands in northern Kenya, most parts of Ethiopia and Sudan and the Horn of Africa are drier compared to the southern parts of Kenya, Uganda, Rwanda, Burundi and Tanzania. Some of the most expansive pure grasslands in the region are found in central and southern parts of former Sudan, extending into some parts of Ethiopia, western Kenya and northern and western Tanzania. At higher rainfall availability, grasslands are likely to be converted into farmlands. Most grasslands here have evolved with grazing by wildlife and livestock as part of their environment. They, however, suffer from a history of economic and political marginalization, demographic, socio-political and ecological challenges.

[1] Department of Biological Sciences, Jaramogi Oginga University of Science and Technology, P.O. Box 210-40601, Bondo, Kenya.

[2] School of Biological Sciences, University of Nairobi, P.O. Box 30197, Nairobi, Kenya.

* Corresponding author: dennis.otieno@uni-bayreuth.de

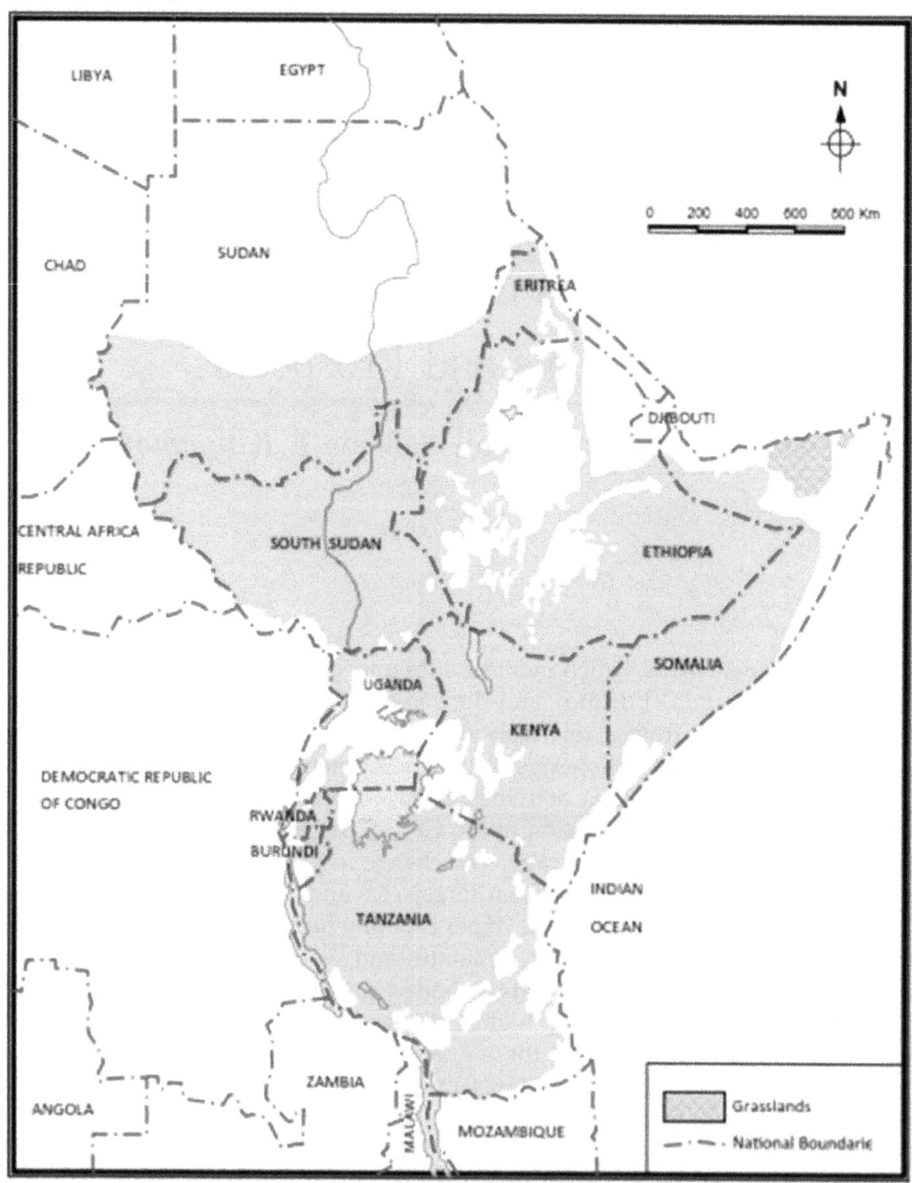

Fig. 13.1 Distribution of grasslands (shaded area) in the east Africa region. The region has a land surface area of about 5.6 M km² out of which 75 per cent (close to 4.2 M km²) is covered by grassland.

Population pressure from within the pastoral system and its surroundings, climate change and drought occurrences, as well as poorly informed government policies have undermined the traditional, customary practices among the grazing communities that were once instrumental in maintaining the ecosystem balance (Little et al., 2001). Due to lack of intervention, large areas of grasslands are currently degraded, have low productivity and are characterized by recurrent famine and conflicts (Opiyo et al., 2011; Reda, 2015).

Majority of the people living in grasslands are pastoralists whose livelihoods depend on livestock (cattle, sheep and goats and camels). Recurrent droughts are a major challenge to these people, who lose large numbers of their herd during the extended periods of drought, exposing these communities to chronic food insecurity (FAO, 2000). Given that the communities have limited alternative income sources and other coping mechanisms, poverty levels are high, with most of the pastoralists living below one US dollar a day. Despite these challenges, the population in and around the grasslands continues to grow and land available for free-ranging animals has continued to shrink due to expanding human settlements and conversion to agriculture. These changing environmental conditions have compelled pastoralists to change their lifestyles, abandon their nomadic way of life and seek alternative sources of livelihoods, settling down and diversifying their income generating activities, with dire consequences on the expanse and quality of the grassland ecosystem.

Origin and Types of Grasslands

East Africa grasslands might have been present since earlier times but their spread was not apparent until Late Pliocene or Early Pleistocene, during the period when large global changes were associated with increased aridity in Africa (Bobe and Behrensmeyer, 2004). Pollen evidence (Shackleton et al., 1984; Kennett, 1995) from northern Kenya shows that grasslands were already present in this part of East Africa as early as 14 Ma (Bonnefille, 1984; Retallack et al., 1990; Retallack, 1992; Magill et al., 2012). The early grasslands were C_3-dominated grasslands as demonstrated by paleosol and dental enamel carbon isotopes from this region, which show predominantly C_3 ecosystem (Cerling et al., 1991). Paleosol carbonates from the Baringo basin of Kenya showed little evidence of C_4 grasslands during the Miocene to Early Pleistocene, but that of a mosaic of C_3-dominated grasslands and wooded vegetation (Kingston et al., 1994). Jacobs and Kabuye (1987) demonstrated that until 12.4 Ma, tropical rainforests were an important component of the East African vegetation. However, based on palynology records, an increase in grasslands occurred during 2.5 Ma (Bonnefille and Letouzey, 1976; Bonnefille and Dechamps, 1983; Bonnefille, 1995). Thus, although C_4 grasses might have been part of the East African vegetation during the Miocene, they were subdued by wooded bushlands or forests and the C_3 grasslands. Evidence of some C_4 domination at around 1.8 Ma is drawn from the Turkana region, which until then was dominated by C_3 plants (Cerling et al., 1988). In Ethiopia, a shift from wooded vegetation to a more expansive, open grassland occurred during the Pliocene (Levin, 2002). These records show that even within the East African landscape, there were temporal variabilities in the spread of grasslands. Sample analyses from the Olduvai Gorge in Tanzania, and also from the Turkana region of northern Kenya, showed two peaks of C_4 domination in 1.7 Ma and also in 1.2 Ma (Cerling, 1992). This spread was associated with increasing aridity in the region until 1.8 Ma, as demonstrated by palynological samples (Bonnefille, 1995). Therefore, the Pleistocene spread of C_4 grasslands occurred at different times in different regions.

Main Grassland Types of the Region

Grassland in the present context includes all the plant formations in which grasses form an important proportion of the vegetation cover (Vesey-Fitzgerald, 1963; Dixon et al., 2006). Species of *Cyperaceae* are often included as 'grasses'. Categorization of grasslands here is predominantly based on the system recommended by Pratt et al. (1966) for the East African rangelands, although with modifications where appropriate. Wooded grasslands, savanna, bushlands and woodlands have also been discussed as grasslands in our context. The herbaceous field layer in woodland, although with a mix of graminoids and other broad leaves, is discussed as if it were grassland, but such deviation from the correct conception of a grassland will be obvious from the context.

Afro-alpine Moorland and Grassland

This grassland type is defined by altitude and not by climate or water availability. The description provided here is biased and predominantly reports on the vegetation characteristics of the Afro-alpine zone of Mt. Elgon, from 3,000 m to 4,100 m above sea level (asl). While this is not a synopsis for grasslands in the Afro-alpine zones of East Africa, the parallelism and similarities in grass types and species are apparent. The description we provide here is also based on the eastern (Koitobos) transect originally described by Beck et al. (1987). This transect is drier as compared to the western transect described by Wesche (2002). In certain instances, we also provide examples from the western transect. In such cases, there is mention to that effect. Above the tree line, at 3,000 m elevation, mostly occupied by Erica, tree density and tree height get drastically reduced and lush expansive dense grassland appears. Given that here the intensity of grazing is also highly reduced, the grass is tall, except in the event of fire, which tends to raze down the accumulated biomass (Wesche, 2002). Vegetation composition here is quite mixed and even though we describe the entire transect as grassland, in some locations the forest canopy (*Erica* and *Dendrosenecio*) is almost closed. In the open locations, pure stands of grasslands occur but this can also be diluted in some locations due to the presence of other herbs and shrubs. On Mt. Elgon, at an elevation between 3,500 and 3,700, *Artemisia* is very conspicuous within the grass vegetation and in some locations, it becomes the dominant herb.

The genera *Festuca* is dominant and most successful on Mt. Elgon's alpine and sub-alpine zones, with a distribution range from 3,000 m to 4,000 m asl. While in some locations it appears as pure stands, its cover in some locations is reduced by the occurrence of other vegetation of the herbaceous community, trees such as *Erica* or other shrubs, such as *Helichrysum*, the giant groundsels, etc. What becomes conspicuous is that in the lower elevation, i.e., 3,000 to 3,600 m asl, tussock formation is reduced because of the dominance of *Sporobolus olivaceus*, but with the increasing dominance of Sporobolus at higher elevations tussocks increase, with increased space size among the single tussocks. On the western side of the mountain, *Exotheca abyssinica, Piloselloides hirsute* and *Polygala steudtneri* are also present in this zone (Wesche, 2002). It is worth noting that at this elevation, the grassland is quite rich in species, both of the grasses, herbs and shrubs. The strong presence of *Artemisia* within the grasslands at this elevation cannot be overlooked. At the higher end of the elevation and extending up to 3,800 m asl, *Sporobolus olivaceus, Dierama cupuliflorum* and *Agrostis gracilifolia* take over as the dominant grasses. Their dominance here is related to fire frequency. Descriptions of the Afro alpine vegetation of Mt. Kilimanjaro (Beck et al., 1983) and Mt. Kenya (Rehder et al., 1988) do not include *S. olivaceus* and *E. abyssinica* as members of this zone, indicating some of the small variations in vegetation distribution patterns of the East African highland vegetation.

The inter-tussock spaces allow for the growth of herbs and shrubs. In the lower elevations of the sub-alpine zone, Wesche (2002) observed that within the elevation range of 3,500 m and 3,800 m asl, *Koelaria gracilis* becomes the dominant grass after fire events. In

well-drained locations, *Koeleria capensis* and *Agrostis gracilifolia* dominate. In locations that are flooded during part of, or throughout the whole year, *Cyperus kerstenii, Juncus dregeanus, Carex runssoroensis, Festuca pilgeri* and sometimes *Panicum snowdenii* form the dominant tussock grasslands. This kind of grassland is found from 3,600 m to 4,100 m elevation, including inside the caldera. The substrate has a soil of organic matter ranging from 38–50 per cent, with a soil pH of 5.6 to 5.9 (Wesche, 2002). On Mt. Elgon however, permanent swamps/bogs are rare, with most flooding periods only associated with extended rainfall events; thus a description here may not appropriately represent the highland swampy vegetation of such ecosystems in other highland grasslands of East Africa, such as the Ruwenzori mountains, which have extensive, permanent high elevation swamps and bogs. Also in such habitats, *Carex monostachya* is found, with *C. kerstenii* dominating. Because of high moisture content in the places they grow, the two sedges are able to survive fires. In better drained locations, tussocks of *Festuca pilgeri* occur, while *Carex runssoroensis* is found at the other end of the drainage scale (van Heist, 1994). On Mt. Elgon, there are areas occasionally dotted with *Dendrocenesio* and *Lobelia.* At similar elevations, on locations with shallow, tussocks of *Andropogon lima* occur. On the Abadares (Schmidt, 1991), described, *A. lima* occupies concave slopes in the moorland. Such a landscape was characterized by interspaces of bare soils among the tussocks. Beck et al. (1987) describe the presence of *A. lima* in a *Carex* bog at an elevation of around 4,100 m asl. From 3,800 m to 3,900 m elevation, the canopy of *Dendrosenecio elegonensis* becomes relatively closed, particularly in the mesic, regularly flooded locations. The understory is, however, covered by a rich mat of grassland that is characterized by a wide range of species, including *Anemone thomsonii, Galium ossirwaense, Festuca abyssinica, Agrostis volkensii, Pentasischtis borussica,* etc.

Woodland

Woodland is an edaphic formation representing an ecotone between woodland and valley grassland or between the prevailingly woody and prevailingly grassy types of vegetation. This was originally one of the dominant grassland types in the East Africa region, with tree canopy cover exceeding 20 per cent (Pratt et al., 1966). Everywhere in the region, this plant formation has been much modified by fires. Often, reference is made to the dominant canopy and understory species for the purpose of distinguishing the several sub-types. Trees can sometimes grow up to 18 m in height with different levels of canopy closure, while grasses and herbs dominate the understory. In south-east Ethiopia, where this type of grassland formation is dominant, woody vegetation accounts for 75 per cent, while the understory comprises 25 per cent (Phaulos et al., 1999). Within the understory (herbaceous layer), grass species make up to 79 per cent of the total herbaceous vegetation, with 56.8 per cent being annuals and 43.2 per cent perennials. Like in other grasslands, grazing intensity influences the distribution of species and hence community composition. In the heavily grazed lands, *Aristida adscension, Sporobolus panicoides, Sporobolus pyramidalis* and *Tragus berteronianus* are dominant, whereas *Aristida vestita, Cenchrus ciliaris, S. pyramidalis* and *Tetrapogon cencriformis* were frequent and/or most frequent species in the medium grazing sites. On light grazing sites, *Bothriochloa radicans, Cenchrus ciliaris, Cynodon dactylon, Eragrostis tenuifolia, Panicum maximum* and *S. pyramidalis* were frequent and/or most frequent species (Table 13.1). Desirable perennial grasses on the heavily grazed sites are replaced with annual grasses and forbs. Many of the grass species identified in the heavily grazed sites were annuals and undesirable species. Thus heavy grazing pressure and low rainfall promote the growth of annual grasses and alter community composition.

Table 13.1 Community composition of wooded grassland in Bale, South-east Ethiopia.

Grass	Other Herbaceous Plants	Sedge	Woody Vegetation
Aristida vestita	*Crotolaria incana*	*Commolina benghalensis*	*Acacia brevispica*
Aristida adscensionis	*Indigofera volkensii*	*Cyprus obtusiflorus*	*Acacia bussie*
Bothriochloa insculpta	*Achyranthes aspera*	Bare ground	*Acacia mellifera*
Bothriochloa radicans	*Belpharis persica*		*Acacia nilotica*
Brachiaria brizantha	*Bidens biternata*		*Acacia oerfota*
Brachiaria dictyoneura	*Hibiscus aponeurus*		*Acacia Senegal*
Brachiaria xantholeuca	*Ocimum basilicum*		*Acacia seyal*
Cenchrus ciliaris	*Sida ovata*		*Acacia tortilis*
Chloris pycnothrix	*Tephrosia ogelii*		*Acokanthera schimperi*
Chloris roxburghina	*Tribulus terrestris*		*Asparagus falcatus*
Chloris virgata			*Balanites aegyptiaca*
Coelachyrum poaflorum			*Becium filamentesum*
Chrysopogon serrulatus			*Boscia mossambicensis*
Cynodon dactylon			*Boswellia neglecta*
Cynodon plectostachyus			*Calotropis procera*
Dactyloctinium aegyptica			*Cassia singueana*
Digitaria senegalensis			*Chenopodium opulifolium*
Digitaria ternate			*Chionothrix tomentosa*
Digitaria velutina			*Combretum collinum*
Eragrostis cilianensis			*Combretum molle*
Eragrostis cylindiriflora			*Combretum hereroense*
Eragrostis superba			*Commiphora kua*
Eragrostis ciliaris			*Commiphora boranensis*
Eragrostis tenuifolia			*Commiphora erythraea*
Heteropogon contortus			*Commiphora myrrha*
Lepthotherium senegalense			*Commiphora schimperi*
Lintonia nutans			*Delonix elata*
Microchloa caffra			*Dichrostachys cinerea*
Microchloa kunthii			*Erythrina abyssinica*
Panicum coloratum			*Euclea divinorum*
Panicum deustum			*Grewia arborea*
Panicum maximum			*Grewia flavescens*
Pennisetum stramineum			*Grewia lilicina*
Paspalum dilatatum			*Grewia penicillata*
Setaria incrassate			*Grewia tembensis*
Setaria verticillata			*Grewia tenax*
Sorghum bicolor			*Grewia velutina*
Sporobolus panicoides			*Phyllanthus sepialis*
Sporobulus pyramidalis			*Psiadia incana*
Tetrapogon cenchriformis			*Rhus natalensis*
Tetrapogon tenellus			*Salvadora persica*
Themeda triandra			*Solanum incanum*
Tragus berteronianus			*Vepris glomerata*
Tragus racemosus			*Ziziphus mucronata*
			Lantana rhodesiensis

Source: Abate et al. (2012).

This type of community composition is replicated almost everywhere where wooded grasslands are found, although with some modifications. In Lambwe valley in western Kenya, for example, the dominant tree species on the slopes is *Combretum molle*, but the genus *Acacia* becomes dominant in the valley bottom. The dominant grass here is *Hyparrhenia*. A similar grassland formation occurs in northern regions of Tanzania, around

Lake Chala where *Combretum* is predominant on the slopes but *Acacia* sp. takes over in the valley bottom.

Grassland

Here grasses form the dominant vegetation but trees and shrubs are also present at very low densities such that their canopy cover does not exceed 2 per cent (Pratt et al., 1966). Where there is no grazing, there is accumulation of explosive biomass which immediately catches fire in the slightest fire accident. Sub-types here are recognized based on the dominant grass species, vegetation height, dominance of annuals and degree of swampiness. The largest extents of pure grasslands occur in central and south-eastern Sudan, north-west Kenya and northern and western Tanzania. In Kenya, they occur in the plains, including the Athi, Laikipia, the Maasai Mara, Loita plains, etc.

Bushed Grassland

This formation is characterized by expansive grass dotted with scattered bushes whose total canopy cover does not reach 20 per cent. Fire is an important part of this structure. Like in the grasslands, the sub-types here are classified based on the dominant genera/species of the grass/shrub, the vegetation height or whether annual or perennial grass.

Wooded Grassland

This refers to a grassland type with scattered trees or tree formation, often with a canopy cover that does not exceed 20 per cent. Regular burning, often on an annual cycle, is observed. Classification of the subtypes is based on the dominant genera/species, water-logging frequency and vegetation height. Examples are the *Hyparrhenia-Combretum* wooded grassland; medium-height *Pennisetum mezianum-Acacia drepanolobium* seasonally waterlogged wooded grassland or short annual *Aristida-Acacia tortilis* wooded grassland.

Agronomic Use of Grasslands

Grasslands have accompanied human evolution and have been exploited throughout the human history. The evolution of the hominids is traced back to the east African grasslands (Stringer, 2003). Thus, over history, these grasslands have been inhabited by generations of *Homo sapiens* practicing different forms of agronomic activities, including hunting, gathering, pastoralism and crop cultivation as a way of livelihood.

Pastoralism, evolved as one of the earliest professions in which people traditionally moved their livestock from one place to other in search of forage and water. While the practice always required large areas of land for roaming, the migratory lifestyle allowed for the vegetation to periodically recover from heavy utilization by the livestock (Bonham, 1989). Livestock mobility also relieves areas of concentration and allows herds to exploit grazing resources that are unevenly distributed in time and space (Oba et al., 2000).

During the 19th century, colonial policies promoted cropping agriculture in the highland areas and livestock development in the lowland grasslands. In Tanzania and neighboring Kenya, for example, pastoralists were forcefully evicted from their traditional grazing lands into the more geographically isolated grasslands in order to pave way for the establishment of ranches, game reserves and national parks, first by the European settlers and then by the post-colonial regimes, completely ignoring the traditional communal tenure system (Reda, 2015; Hodgson, 2001; Fratkin et al., 2001). Unfortunately, these attitudes and misunderstandings about pastoralism were reinforced by the postcolonial governments in most countries of the region. They are still very evident in many countries: in land policies, in resettlements of pastoralists to make way for more 'commercial' investment and in allocations of development support and services (Nyberg et al., 2015). In Kenya, for example, the expansive pastoral grasslands were transformed from communal into group ranches in the 1960s (Graham, 1989). These are large parcels of land demarcated under the Land Adjudication Act of 1968 (Cap 284) and legally registered to one group duly constituted under the Land Act of 1968 (Cap 287). This in particular, forced the pastoralists into a settled lifestyle and significantly reduced the movement of their livestock. Conflicts among the ranchers, changing economic situations and shifting government policies have led to subdivision of the ranches into small parcels of land, further transforming the land use from extensive to intensive and continuous grazing (Burnsilver and Mwangi, 2007). In the process, grasslands have become home to violent conflicts, environmental degradation, drought and the related humanitarian crises and famines, which are the defining characteristics of the East Africa region (Catley et al., 2013; McDermott et al., 2010; Sumberg and Thomson, 2013; Thornton, 2010).

Tsetse fly and trypanosomiasis infestation have been a major challenge for livestock development in grasslands of the region. While it was viewed as a hindrance to economic exploitation of grasslands, it has played an important role in the conservation of some east African grasslands, especially in the greater Sudan, Ethiopia, Tanzania and Kenya, where known records exist. In Kenya for example, the grasslands in the Mara region and around Lake Victoria are known for their high densities of tsetse fly, which kept the human and livestock populations away from these wooded grasslands and savanna in the past. Eradication of tsetse in the 1970s however allowed for human settlement and increased livestock population, leading to degradation of natural grasslands. In the Lambwe valley of western Kenya, the presence of tsetse fly was vital for the preservation of this humid savanna until the late 1970s when concerted government efforts and interventions ensured its eradication. From 1975, there was massive influx of human and livestock population into the area and by the turn of 1980, the region had lost 42 per cent of its natural grassland to settlement, crop production and livestock grazing fields (Njoka et al., 2003). In Tanzania, degradation of grasslands is traced back to the colonial era policies associated with extensive vegetation clearing and burning aimed at eradicating the high population of tsetse flies and trypanosoma (Selemani et al., 2012).

A form of traditional conservation practice in grasslands was the exclusion of large areas of standing vegetation from grazing, from the beginning of rainy season and opening them during drought. This ensured that there were always centers of germ plasm dispersal, thus promoting biodiversity conservation. The promotion of settled lifestyle of the pastoral communities, however, has promoted grassland degradation. Rapid human population increase and the need for alternative livelihoods, have led to the conversion of grasslands into agricultural farms for crop production. Like in most developing countries, expanding human population in the grassland areas has been associated with increased livestock population. Equally, the area under cultivation increased by about 50 per cent in east Africa, 5 per cent of which was from converted grasslands between 1980 and 2000 (Gibbs et al., 2009). This, in combination with erratic rainfall, has promoted land degradation (Pye-Smith, 2010). In northern Tanzania, most of the areas under natural grasslands have been heavily cultivated, leaving a mosaic of crop lands, fallow lands and natural reserves that may limit the option for livestock migration (Selemani et al., 2012).

Current Practices of Grassland Management

As land diminishes and nomadic lifestyles become unattainable, a viable alternative to settled crop-livestock agriculture is to rely on sown pasture to support livestock production and avert pressure on grasslands and contribute to improved grassland quality. Sown pasture makes positive contribution to grasslands by improving vegetation cover through reseeding and soil quality through fertilization and manure addition. Sown pasture also reduces incidences of weeds, pests and diseases. The introduction of Rhodes grass (*Chloris gayana*) and elephant grass (*Pennisetum purpureum*) in the crop-livestock agriculture since the 20th century has seen their expansion in the higher potential areas, especially in the high rainfall areas of Ethiopia, Tanzania, Kenya and Uganda (FAO, 2005). In the highlands where land has become very scarce, most dairy farmers practice cut and carry zero grazing, a practice that has significantly promoted the cultivation of elephant grass (Staal et al., 1998). In the more arid areas, Rhodes grass has gained greater preference due to its affective rooting system that allows it to withstand water stress, with little reduction in productivity. It also plays an important role in soil erosion prevention, especially in dry areas, where heavy downpours occur after drought, washing off large volumes of topsoil due to lack of ground cover.

Cultivated forage has received little attention in the past in the east Africa region, despite an extensive history of grassland research. In Kenya, research on pasture grasses started as early as 1931 (Edwards and Bogdan, 1951). The first National Research Lab on pasture grasslands was set in Kitale in 1952, starting a vast screening programme for viable fodder cultivars for commercialization (Wandera, 1996). With increasing demand for forage, which now commands high prices, and a shift of dairy feeding from grazing to stall feeding, planted forages have grown in importance. In addition to the elephant grass, other forages, including fodder trees and shrubs and herbaceous legumes, are slowly being accepted and adopted on most smallholder farms. New fodder options, such as *Sesbania* spp. or *Calliandra* spp., and herbaceous legumes, such as *Desmodium* spp., have been coopted.

The development of dairy in most of the region has been, however, hindered by diseases and climate, since east Africa has major disease and parasite problems. Also, the selection for high quality cattle breeds is linked to water availability, given that water is scarce and that livestock breeds have varying demands for water. For example, in the highlands where temperatures are cool and water is abundant, exotic breeds, such as Jerseys, have been kept since 1992. In Uganda, despite its relatively cool temperatures and plenty of water available in most regions, raising of productive, exotic breeds was only possible after independence due to the 'tick resistance' policy of the colonial government. Over the years, dairying in the region has evolved from the colonial era through the state-owned farms of the 1960s, to the present smallholder farms. Most of the smallholder dairying systems were restricted to the sub-humid to semi humid agro-climatic zones (Pratt and Gwynne, 1977) and to a lesser extent in the humid and the transitional zones, which are of high agricultural potential. These are the areas with good potential for biomass production and markets for milk and that have significant potential for dairy development. About 65 per cent of dairy cattle in the region occur in the highlands under smallholder crop-dairy production systems. Dairying has become a significant source of income to smallholder forage growers and has played a role in sustaining soil fertility through nutrient cycling. Smallholder dairy farmers are now embracing intensive forage growing as an alternative source of income.

Table 13.2 Summary of annualized per-cow gross returns to dairy production in Uganda.

Cost Parameters		Dairy Production System		
		Extensive	Semi-intensive	Intensive
Total investment costs (a)	US$	4.18	37.18	139.17
	U Sh	4,426	39,335	1,47,237
Total operating costs (b)	US$	18.66	151.74	276.05
	U Sh	19,751	1,60,545	2,92,066
Total gross revenue (c)	US$	185.05	693.53	1,651.71
	U Sh	195.837	7,33,760	17,47,513
Gross margin, (c)-(b)-(a)	US$	162.20	504.61	1,236.45
	U Sh	171.660	5,33,880	13,08,210
Cost per litre of sold milk	US$	0.08	0.11	0.16
	U Sh	86	115	164
Cost per litre of total milk	US$	0.06	0.10	0.12
	U Sh	58	101	130

Note: Exchange rate in July 1996 was U Sh 1,058 = US$ 1.0.

Source: ILRI, 1996.

Smallholder dairy systems have evolved in east Africa, particularly around cities and towns that involve intensive and semi-intensive dairy practices. Intensive dairying includes intensive urban, peri-urban and rural dairy-manure production. The stocking rate is two to three cows on approximately a hectare of land which doubles for crop production. Most cattle are genetically heterogeneous *Bos taurus* (exotic) breeds or *Bos taurus* crosses with *Bos indicus* (zebu). Exotic dairy breeds include Friesian-Holstein, Ayrshire, Guernsey, Black and White Dane, their grades and crosses. The semi-intensive dairying includes rural and peri-urban dairy-meat-manure-draught production systems. Here, both the indigenous zebu and crosses of indigenous, exotic breeds and other ruminants, including goats, sheep and donkeys are kept. Cattle are paddocked, tethered or herded on communal land. Zebu milk is mainly consumed domestically, although farm-gate sales are common.

Table 13.3 Summary of annual per-cow gross returns to dairy production for the most important smallholder production systems in Tanzania.

Cost Parameters		Dairy Production System		
		Semi-intensive with Zebus	Intensive Rural with Exotic Cross	Intensive Urban with Exotic Crosses
Total variable costs	US$	126	260	747
	T Sh	75,500	1,56,234	4,48,348
Total revenue	US$	242	1,020	1,750
	T Sh	1,45,080	6,12,217	10,51,330
Gross margin	US$	116	760	1,000
	T Sh	69,580	4,55,983	6,02,982
Gross margin per litre	US$	0.26	0.38	0.52
	T Sh	159	230	315

Note: Exchange rate (1998, approximate) T Sh 600 = US$ 1.0.

Source: MOAC/SUA/ILRI, 1998.

Inadequate feed due to decreasing farm size and increasing competition for land between enterprises is, however, a challenge. In addition, feed quality fluctuates seasonally, since it is grain-fed. Inputs for intensified fodder production have become expensive and this factor can undermine the growth of dairy farming, since farmers only take up technologies that they consider profitable. Technical information and extension services are inadequate to transfer forage production information to smallholders. Generally, the growth of smallholder farming has been boosted by its capacity to integrate crops and dairy and the benefits from synergies between dairying, staple food and cash crops.

Table 13.4 Dairying in east Africa: cattle, milk production and per capita milk availability.

Parameter	Kenya	Tanzania	Uganda
Cattle ('000 head)			
Zebu	10,400	13,900	5,400
Dairy	3,045	250	150
Percentage dairy cattle	23	2	3
Annual milk production ('000 litres)	3,075	814	485
Annual per capita milk availability (litres LME)[1]	85	23	24

Notes: [1] LME = liquid milk equivalent.
Source: Muriuki and Thorpe, 2001.

Grasslands in east Africa have multiple uses. These include forage for livestock (especially herbage for grazers and browsers) which is usually seen as the principle product of rangelands. Grasslands also provide a variety of direct ecosystem goods and services of economic value. These include products, such as charcoal, gums and resin, honey and traditional plant uses (medicine, etc.) and water production (Friedel et al., 2000; Herlocker, 1999; Heady and Child, 1994).

Grasslands offer key ecosystem functions/services, which can be grouped as supportive (nutrient cycling, primary production, pollinator services, etc.), regulating (e.g., CO_2 sequestration, prevention of soil loss through erosion, soil fertility maintenance), provisioning (plant material and game) and cultural (eco-tourism, religious values, scenery). White et al. (2000) broadly grouped the services derived from most grasslands under four main heads—forage and livestock, food, biodiversity, carbon storage, and tourism and recreation. Additionally, grasslands play important provisioning services, including provision of drinking and irrigation water; as sources of genetic resources, they are habitats for human and wildlife, they are pollutant filters, they maintain the watershed functions, are a source of employment, supply oxygen, weather stabilization and contribute to aesthetic beauty.

Biodiversity

Nevertheless, despite local efforts at the grassroots level, loss of biodiversity has occurred on a massive scale in recent years due to persistent droughts, over exploitation, human encroachment and inappropriate government land use and land administration policies. Dispossession of pastoral land, population pressure and conflicts have resulted in a weakening of the customary institutions of resource management, thus exacerbating

Table 13.5 Reported monthly cash income by type of income and adoption status in Kenya.

Income Type	Households with No Cattle and No Planted Fodder	Households with Local Cattle but No Planted Fodder	Households with Grade or Cross-bred Cattle and Planted Fodder (Napier Grass)
Percentage of households with cash income from:			
Dairying	0	14	87
Poultry or eggs	9	8	14
Crops	75	80	73
Wages, salaries or off-farm activities	65	66	65
Remittances	39	35	22
Other income	13	25	18
Mean monthly cash income from dairying			
US$	0	5.18	109.82
K Sh	0[ac]	321[ab]	6,809[bc]
(SD)	–	(1,211)	(7,836)
Mean monthly cash income from poultry or eggs per month			
US$	10.50	4.31	31.68
K Sh	651	267	1,964
(SD)	(3,261)	(1,682)	(8,687)
Mean monthly cash income from crops			
US$	12.0	19.94	26.26
K Sh	744[c]	1,236	1,628[c]
(SD)	(991)	(2,242)	(2,967)
Mean monthly cash income from wages, salaries or off-farm activities			
US$	42.66	48.68	149.32
K Sh	2,645[c]	3,018[b]	9,258[bc]
(SD)	(5,585)	(4,169)	(17,079)
Mean monthly cash income from remittances			
US$	7.29	7.77	4.84
K Sh	452	482	300
(SD)	(965)	(902)	(852)
Mean monthly cash income from other sources			
US$	2.27	2.35	12.82
K Sh	141	146	795
(SD)	(823)	(350)	(4,115)
Mean total monthly cash income from all activities			
US$	75.27	87.78	337.29
K Sh	4,667[c]	5,439[b]	20.912[bc]
(SD)	(7,280)	(4,444)	(23,761)

SD = Standard deviation of the mean. The superscripts a, b and c indicates that the means for the two adoption categories [between adopters, i.e., households with at least one grade/cross bred (G/C) animal and planted fodder, and non-adopters, i.e., household owning no G/C animal and with no planted fodder] with the same letter are statistically different at the 95 per cent confidence level. Exchange rate (1999) K Sh 62 = US$ 1.

Source: Nicholson et al., 1999.

the problem of natural resource degradation in the drylands of east Africa. The current loss of biodiversity in the drylands should therefore be attributed to the aforementioned ecological, social and political factors rather than the inefficiencies of the pastoral system itself in managing the commons by east African pastoralists.

Carbon Storage

Tropical grasslands are some of the most productive ecosystems in the world, and as shown in Table 13.6, with biomass production exceeding many other ecosystems in the world (Grace et al., 2006).

With its 75 per cent coverage, grasslands in east Africa play an important role in ecosystem CO_2 exchange. It is becoming increasingly clear that these areas have a greater productivity (Santos et al., 2004; Grace et al., 2006), higher biodiversity and larger impact on global carbon cycles than previously realized (Scholes and Hall, 1996). Studies on CO_2 fluxes and storage in these grasslands are, however, still limited and it is not clear whether grasslands in east Africa are net C-sink or source. However, data obtained in the grasslands of Nairobi National Park over a 10 year period showed that east African grasslands are more productive than may have previously been reported (Kinyamario and Imbamba, 1992; Kinyamario, 1995). Previous estimates of tropical east African grassland productivity likely underestimated their net production, since they were based largely on the maximum-minimum method (shoots only) or standard IBP methods (Kinyamario and Imbamba, 1992; Long and Jones, 1992). Until these two reports, very few studies in this part of the world took into account either turnover of biomass or below-ground biomass and production. Conversion of savanna into croplands, however, has a net negative impact on C-sequestration and leads to a net C-release into the atmosphere (K'Otuto et al., 2013). In a moist grassland located in Ruma in the western part of Kenya, co-dominated by

Table 13.6 Primary production of some selected tropical grassland types from across the world.

Type of Grassland	Location	Net Production ($t\ C\ ha^{-1}\ year^{-1}$)	Author(s)
Closed bush island savanna	Calabozo, Venezuela	3.4	San Jose' (1995)
Open shrub savanna	Lamto, Cote d'Ivoire	6.4	Mordelet and Menaut (1995)
Dry grassland	Khirasara, India	1.7	Singh et al. (1985)
Mixed grass and forbs, tropical grassland	Kurukshetra, India	13.4	Singh et al. (1985)
Savanna woodland	Varanasi, upland, India	18	Singh et al. (1985)
Grass savanna	Nairobi National Park, Kenya	6.1	Kinyamario and Macharia (1992)
Grass savanna	Nairobi National Park, Kenya	6.2	Long et al. (1989)
Humid savanna woodland	Klong Hoi Khong, Thailand	9.5	Kamnalrut and Evenson (1992)
Tree savannas	Orinoco Lanos, Venezuela	7.0	San Jose' and Montes (2001)
Eucalyptus scrub	Northern Australia	11.0	Chen et al. (2003)

Source: Grace et al. (2006).

Hyparrhenia filipendula and *Themeda triandra*, peak net ecosystem CO_2 exchange (NEE) in the open grasslands, with scattered trees was around –10 µmol m^{-2} s^{-1} during the rainy season, but declined to near zero during drought (Otieno et al., 2010, 2015). In the same study, the amount of rainfall was the key factor determining grassland's Gross Primary Production (GPP), Net Ecosystem CO_2 Exchange (NEE) and Net Primary Production (NPP). The mean maximum GPP was 14.0 gC m^{-2} d^{-1}. In Sudan, Ardö et al., 2008 reported mean NEE rates of –14 µmol m^{-2} s^{-1} during the rainy seasons, but the rates declined to –2 µmol m^{-2} s^{-1} during drought. The general circulation model (GCM) predicts a strong future warming and general annual-mean rainfall increases for the Eastern Africa region. Given that rainfall is the main driver of CO_2 uptake and storage by the grassland ecosystems (K'Otuto et al., 2013; Otieno et al., 2010), general 18–36 per cent and 3–13 per cent increases in NPP and C-sequestration, respectively, are anticipated in the region's natural ecosystems between 2080 and 2099 (Doherty et al., 2010). A significant proportion will originate from grasslands, given its 75 per cent, coverage, assuming the influence of agricultural conversion.

Tourism and Recreation

East Africa is famous for its high diversity and number of grazing and browsing wildlife, which makes it a local and global tourist destination. Most of this rich wildlife population is found in the grasslands. Kenya, Tanzania, Uganda and Ruanda host the richest diversity and the largest number of grazing/browsing wildlife population in the east Africa region and the world (Reid et al., 1998). A large part of the remaining grasslands have been demarcated as national parks, game reserves or private ranches with a mix of native wildlife and livestock.

As a result of their rich biodiversity and heterogeneous landscapes, grasslands of east Africa provide valuable recreational and eco-tourist services for safaris, biking, photography, swimming, health baths, hiking, bird and game watching, cultural and spiritual needs, aesthetics, etc. Grassland tourism has become a major source of revenue for most east African countries, hence their continued decline is a major concern and questions arise as to for how long grasslands will continue to supply these services (World Resources, 2000–2001). In Tanzania, for example, the Serengeti National Park, one of the 12 national parks, with an area of 14,763 sq km supporting 4×10^6 animals, including zebras, wild beasts, gazelles, hippos, etc., is an economic boost to the country. Kenya, with its rich grasslands and large numbers of wildlife has been Africa's leading tourist destination since the 1960s. Tourism, mostly driven by Safari destinations in the grasslands is one of the leading foreign income earners in the country. Unfortunately, Kenyan tourism industry was negatively affected by post-election civil unrest between 2007–2008 and since then, recovery has been very slow. Kenya is making good progress with regard to eco-tourism development, support and promotion of local community shareholding of the wild resources and raising of environmental awareness among its population, which have contributed significantly to sustainable management of the grasslands.

Until the break of the civil war, Somalia, with its expansive grasslands was an important tourist destination, but this collapsed by 74 per cent after the war outbreak. Rwanda has made great strides in promoting tourism since the 1984 genocide, although most of its tourism is driven by forests, but the Akagera National Park in the north eastern part of the country is an expansive savanna, contributing significantly to Rwanda's tourism development (Mazimhaka, 2007). In Ethiopia, the success of tourism has been pegged on the Ethiopian grasslands being regarded as 'cradle of mankind'. Tourism in Ethiopia has grown steadily over the years (Frost and Shanka, 2002).

Threats to Grasslands

Grasslands have hosted humans for millions of years but currently they face a multitude of challenges, which threaten their existence. The main threats include fragmentation, agricultural expansion, invasive non-native species, lack of fire, urbanization or human settlement, desertification/climate change and overgrazing by livestock (Catley et al., 2013; McDermott et al., 2010; Sumberg and Thomson, 2013; Thornton, 2010). One of the greatest challenges to grassland existence is the rapid growth in human population during the last century, which has led to the original grasslands being invaded by human settlements. Most governments have also promoted a policy of sedentary lifestyle and establishment of settlements for nomads (FAO, 2005). To supplement their income and improve livelihoods, farmers have resorted to a mix of small-holder crop production and livestock farming—a departure from the traditional nomadic lifestyle. This has weakened or completely eliminated the traditional common property access and management regimes associated with transhumance, which developed over generations to exploit ecological heterogeneity and sustainably exploit the limited grassland resources. Grazing rotation and established grazing preserves are no longer possible. The remaining grassland area has therefore continued to degrade as the intensity of exploitation increases. In most of east African countries, original grasslands have been opened to invasion by non-nomadic communities for settlement purposes and in the last 40 years alone, a significant area originally under grasslands has been converted to either crop farms or settlement area (FAO, 2005).

In Kenya, for example, about 45,000 ha of grasslands around the Maasai Mara region was annexed for mechanized agriculture by 1975 and by 1980, about 60,000 ha was under intensive cultivation. This, however, was all abandoned by the year 2000 on collapse of wheat prices in the local market, low productivity due to uncertain weather and as a result of human-wildlife conflicts due to destruction of crops by wildlife. Similar situations occur elsewhere within the region where wild animals raise havoc in the surrounding farmlands. Elephant populations have been the main culprits, with marauding populations grazing and destroying several acres of croplands in Kenya and Tanzania (Walpole et al., 2003). They destroy a variety of crops including maize (*Zea mays*); millet (*Eleusine coracana*); sorghum(s); cassava (*Manihot esculenta*); banana (*Musa domestica*); sugarcane (*Saccharum officinarum*); tomato (*Lycoposicon esculentum*); kales (*Brassica* spp.); pumpkin (*Cucurbita maxima*); potatoes (*Ipomea batatas*); tobacco (*Nicotina tabacum*); beans (*Phaseolus vulgaris*); and Napier grass (*Pennisetum purpureum*). In Kenya, areas adjacent to wildlife conservation areas suffer from regular human-wildlife conflicts. For example, crop raiding has been recorded in areas of Marsabit farms. Ngene and Omondi (2009) reported that in a period of one year (2008–2009), Marsabit farmers lost over US$ 2,00,000 worth of crops and, considering these are subsistence farmers, such a loss is enormous.

In response to some of the human-wildlife conflicts, there has been an increase in the establishment of fences around crop farms, settlements and wildlife conservancies—a practice that appears to put the last nail on the existence of natural grassland ecosystems of eastern Africa (Løvschal et al., 2017). Between 1980s and 2000s, fences have led to more than 70 per cent decline in resident wildlife population in the Mara region of Kenya (Ogutu et al., 2011). Between 2014 and 2016, there has been a significant increase in fences in Greater Mara, expanding into the open grassland and threatening to lead to the collapse of unique grassland ecosystem within the next few years (Løvschal et al., 2017).Due to fences, pastoralism and nomadic lifestyle can no longer be sustained. A decline is seen in wildlife population due to collapse of migrating megafauna populations and an eminent collapse of the entire grassland Mara ecosystem (Ogutu et al., 2016). The situation is no better in Tanzania, where despite concerted management efforts to manage the Serengeti/Ngorongoro conservancies, the planned construction of the road that will cross the northern part of Serengeti National Park will greatly threaten the Mara-Serengeti migration (Dobson et al., 2010).

Human settlements have increased in areas that were originally extensive grasslands, leading to their fragmentation or complete loss. Around the Mara region of Kenya, human settlements have significantly increased during the last 50 years on areas that were originally grasslands (FAO, 2005). In Tanzania, human population around conserved grassland areas is increasing at an alarming rate. For example, population around Serengeti is increasing at a 4 per cent rate. An increase in livestock numbers has also been registered. Large hectares of land are being cultivated for crop production (WCMC, 2001; WWF, 2017). In the Ghibe Valley of Ethiopia, a wooded grassland, tsetse fly infestation kept the human population away and grazing intensity remained low, mainly by the wildlife, since cattle population was low (Bourn et al., 2001). Control of tsetse in the 1990s opened the area for human settlement and large areas of land were cleared for agriculture as cattle population and grazing intensity increased. These had strong impacts on the ecosystem (Reid et al., 2000).

Basically, a large part of the remaining grasslands are represented as protected grasslands, which are meant to host and preserve wildlife. In Tanzania, the Mkomazi Game Reserve straddles an area of 3,200 km^{-2} of savanna, with a mix of *Acacia-Commiphora* woodlands stretching from the Kenyan border down to the Pare and Usambara mountains. Kenya has patches of grasslands which stretch across the land as national parks or game reserves, otherwise, no protected area has been invaded by human settlements or other livelihood activities.

In most east African countries, especially Kenya, Tanzania, Uganda and Ethiopia pastoral populations inhabiting grazing areas are growing, especially as nomadism is replaced first by semi-nomadism and later by sedentarism, resulting in rising birth rates and falling death rates at each stage (Oxfam International, 2010). While the human birth rates have dropped in many agricultural areas of east Africa, this is not the case in drylands where grasslands abound. In many grasslands and grass-like ecosystems of east Africa, human population has remained higher than the national average due to higher birth rates and immigration. This increasing human population is highly dependent on the natural resource base for livelihood. This has led to decline in the area of grassland ecosystems resulting from expansion of farming operations, urbanization and mining activities (AGRA, 2014; Abdi et al., 2013), for example, in the pastoral areas of Narok and Kajiado in Kenya, grasslands and rangelands have been either converted to croplands to grow wheat and vegetables or transformed into game or wildlife conservancies (Campbell et al., 2005; Homewood et al., 2001; Reid et al., 2004; Tsegaye et al., 2010).

Changes in Management Types Threaten Grasslands

While their population is rapidly increasing, the pastoralists at the same time are experiencing a large-scale loss of rangeland due to expansion of farming operations, private ranches, energy development, mining operations and urbanization. Several reports (Campbell et al., 2005; Homewood et al., 2001; Reid et al., 2004; Tsegaye et al., 2010) mention loss of rangeland to agricultural encroachment as the single biggest threat to pastoralism. This phenomenon has been noted in Narok and Kajiado areas of Kenya where the Maasai pastoral communities reside. In these two regions, rangelands have been turned into farmlands, especially to wheat farming and horticultural activities. In the high agricultural potential areas of the fertile highlands of Kenya, the population is increasing and putting pressure on agricultural land, bringing about sub-division of holdings into small units. This has resulted in large populations moving from these areas to the lowlands (for example, the Laikipia Plateau) dominated by grasslands and other grass-associated vegetation types. Land fragmentation and accelerated land degradation is being experienced (Kiteme, 1998), resulting in loss of plant diversity.

Eutrophication

Eastern Africa does not have extensive flooded grasslands like those that occur in central Africa (Zambezi floodplains) and therefore, eutrophication is not a major problem. However, small areas around the Rift Valley lakes (for example, Lakes Naivasha and Baringo) are prone to the problem where grasslands are replaced with water sedges and aquatic weeds (Muthuri et al., 1989).

Invasive Species

In many arid and semi-arid grasslands of east Africa, overgrazing is a major problem that leads to land degradation. This is often followed by replacement of grass species with low-value forage invader species, such as *Ipomoea* in Kenya (Mganga et al., 2010; Mworia et al., 2008). Also, introduced species have the potential of taking over, as the case of *Prosopis* observed in some arid and semi-arid grasslands of Kenya. With seed from Brazil and Hawaii, *Prosopis juliflora* was first introduced in the coastal areas of Kenya in 1973 (Mwangi, 2005). In the early 1980s it was again introduced in the semi-arid districts of Baringo, Tana River and Turkana in order to promote energy self-sufficiency and reduce land degradation (Mwangi, 2005). Today, these three areas show large pockets of *P. juliflora* colonization and invasion. Recent literature (Mworia et al., 2011) shows that *P. juliflora* has already invaded many degraded grasslands in Kenya, including floodplains that are vital dry-season-grazing areas for herbivores. In addition, indigenous woody species diversity declined significantly as the density of *P. juliflora* increased. In many large areas where land has been degraded through overgrazing, such as the rangelands of Ethiopia and Kenya, woody species encroachment, for example by *Acacia* species, occur (Abate et al., 2012).

Climate Change and CO_2 Enrichment

Although not much research has been carried out on effects of climate change on east African grasslands, empirical models show that grasslands will be more affected than any other areas, leading to drier and longer dry seasons (Vizy and Cook, 2012). However, climate models by the Intergovernmental Panel on Climate Change for East Africa (Christensen et al., 2007) show that by 2080s there may be an increase of up to 2–4°C in temperature, while more intense rains are predicted to fall in the short rains (October–December) over much of Kenya, Uganda and northern Tanzania as soon as the 2020s, and becoming more pronounced in the following decades. This may be beneficial to wildlife and pastoralists as more rainfall could result in more dry-season pasture and longer access to wet-season pasture. With anticipated frequency of droughts becoming more common in these areas, it means that households will have no opportunity to rebuild their assets, including livestock and many will become sucked into a spiral of chronic food insecurity and poverty. Climate change will therefore affect large areas of east Africa, especially those arid and semi-arid areas. For example, more than 85 per cent of Kenya and Tanzania may be affected. This will affect large areas of east Africa and those arid and semi-arid areas, for example more than 85 per cent of Kenya's land surface consists of arid and semi-arid areas. In the end it will lead to loss of vegetation types, like grasslands, less available water for animals and human consumption and ultimately resulting in loss of livelihoods.

Management and Restoration of Grasslands

Emerging challenges in the management of grasslands and supply of resource needs has made it necessary for farmers to diversify and seek alternative approaches. In Kenya and Ethiopia, pastoral systems are evolving and farmers are seeking new approaches to avoid drought and related disasters and improve their incomes through diversification and risk management (Little et al., 2001). With increasing demand for food, land under crop production has expanded into grasslands, which are now becoming an integral part of the crop-livestock systems.

Conservation of Grasslands

In east Africa, high nature-value grasslands exist and make much of our national parks and nature reserves. These include the world famous Serengeti and Maasai Mara ecosystem known for its large wildlife migration and high wildlife biodiversity. They have remained pristine with little disturbances except the increased but controlled development of tourist industry, including construction of new hotel facilities and increased game drives as seen in Maasai Mara Game Reserve (Machogu, 2014). However, much of the areas outside game parks and reserves suffer from various degrees of land degradation due to overgrazing (Maitima et al., 2009). In 2010, it was estimated that about nearly 33 per cent of Ethiopian population (estimated at 80 million) lived in these dryland areas. Due to the continually increasing human population in these areas, as people tend to move from the highly degraded highlands where populations are rising to the fertile lowland drylands, the current carrying capacity of grasslands has increased exceeded, leading to land degradation (Maitima et al., 2009).

Legal Conservation Status (Fraction Covered in Protected Areas, Grasslands in the Conservation Legislation)

Many of the grasslands, except those found in national parks and game reserves that follow under the direct management of national and local governmental agencies like Kenya Wildlife Service, are under the local communities. The Wildlife Conservation and Management Act provides for protection, conservation and management of all wildlife and related matters in Kenya. The Act shall apply to all wildlife resources on public, community and private land, and Kenya territorial waters. Therefore, national and game reserves as regulated legally through laws and bylaws. However, community managed grassland are used as commons, where natural resources, including grazing areas and water, are shared among communities, leading to overgrazing. This has been witnessed in northern Kenya, southern Ethiopia, Tanzania and Sudan leading to severe land degradation (Sonneveld et al., 2010; Abdi et al., 2013).

Other Formal Instruments of Grassland Conservation (e.g., Payment Schemes to Farmers)

Several countries, including Kenya and Tanzania, have started projects of sharing the earnings from tourism with communities adjacent to national parks to encourage them to conserve wildlife and their habitats within and outside game parks. A lot of communities have seen these benefits and have also started their own game conservancies to attract tourists. For example, in the Ngorongoro Conservation Area (Melita and Mendlinger, 2013) and in the Serengeti Masai Mara ecosystem (Onchwati et al., 2010) such revenue sharing programmes have been started at community level.

Management and Restoration of High Nature Value Grasslands

Seasonal Periodic Burning

Literature has shown that human beings evolved in grasslands areas of east Africa (Cerling et al., 2011). Grasslands have evolved due to grazing by wildlife and livestock as part of their environment. Here, we find natural and semi-natural grasslands, whereby semi-natural grasslands may occur where woody vegetation was once cleared for agricultural purposes and have since been abandoned. In this case a return to the original vegetation is prevented by repeated burning or grazing. In wet tropical areas, like in western Uganda, such grasslands may be very dense and are dominated by elephant grass (*Pennisetum purpureum*). Fire, usually set by humans, is an important disturbance factor in the maintenance of grasslands by limiting woody species encroachment into grasslands. In wooded grasslands, particularly, periodic burning is necessary to keep down the rate of wood species encroachment (Menaut, 1983; Boonman, 1993; Gichohi et al., 1996). Prescribed burning has been widely recognized and accepted as an essential ecological management tool for removing moribund and/or unacceptable vegetation material, for creating or maintaining an optimum relationship between herbaceous and woody vegetation where necessary, and for encouraging animals to move to less preferred areas in order to minimize the overuse of preferred areas (Trollope et al., 2000; van Wilgen et al., 1990). Some animal species, including Thomson's and Grant's gazelles, impala and wildebeest have been seen to favor grazing on the green flush that emerges after burning especially in the Maasai Mara/Serengeti grassland ecosystem (Wilsey, 1996).

Periodic burning favors certain grass species over others. In grasslands of east Africa, *Themeda triandra* appears to dominate over other grass species, such as *Setaria phleoides* and *Pennisetum mezianum* (Deshmukh, 1986). In the absence of periodic fires, the vegetation tends to shift from herbaceous dominance towards a thick bush cover (Trollope, 1982; Knoop and Walker, 1985; Oba et al., 2000). Outside east Africa, studies demonstrate the role of fire on grassland ecosystems. For example, in Zimbabwe, Furley et al. (2008) found that after a long-term experiment running from 1953 to 1991, some grass and sedge species had flourished while others revealed greater susceptibility to fire. Fire-tolerant species predominated in the most frequently burned areas.

Wildlife Grazing

Grasslands on the African continent have co-evolved with herbivores and herbivory is an important disturbance factor that helps maintain these grasslands. Herbivores keep down the encroachment of woody species. For example, several studies show that when an herbivory on grasslands is removed, plant species composition and growth form changes relative to those under grazing (McNaughton et al., 1988) and that some more important grass species eventually disappear (Belsky, 1986). Kamau (2004) reported there was higher plant diversity in grazed areas than in non-grazed areas (Table 13.7).

Other studies have found that the grasses found in grazed patches are more productive than ungrazed patches (McNaughton, 1983), and that these patches are able to support larger concentrations of herbivores than the stocking rates would suggest (Hiernaux and Turner, 1996).

Emerging challenges in the management of grasslands and supply of resource needs have made it necessary that farmers diversify and seek alternative approaches. In Kenya and Ethiopia, pastoral systems are evolving and farmers are seeking new approaches to avoid drought the related disasters and to improve their incomes through diversification and risk management (Little et al., 2000). With increasing demand for food, land under crop production has extended into the grasslands, which are now becoming an integral part of crop-livestock systems.

Table 13.7 Comparisons of per cent plant cover of different growth forms in grazed and un-grazed sites. *Means significantly different at ($p < 0.05$), t-test. Source: (Kamau, 2004).

Growth Form	Grazed	Non-grazed	p-value
Forbs and grasses	72.5	17.4	0.04*
Shrubs	40	62.5	0.04*
Trees	25	62.5	0.12

Résumé and Future Prospects

Grasslands in the east African region host about 150 M people who are almost completely reliant on the ecosystem for livelihood (GRUMP, 2005). Grasslands here are sources of fiber and protein, medicine and the basis of significant income from tourism. The current Global Climate Models (GCMs) predictions for the east Africa region show general trends of increasing air temperature, which is above the global average for all seasons and increasing amount of precipitation (Christensen et al., 2006). Also, recent projections for six countries of the region, namely Ethiopia, Rwanda, Burundi, Uganda, Kenya and Tanzania, indicate a near-trebling in population between 2000 and 2050, to 498 million (UNPD, 2008). Thus, future climate-induced changes in land cover will occur concurrently with human-induced changes in land use, compounding their effects on grassland (Doherty et al., 2010). The future of grasslands in the region will, therefore, depend on how they will respond to future warmer and wetter or drier climates and on how much national governments in the region are able to curb human population increases and take mitigation measures against climate change.

References and FURTHER READINGS

Abate, T., E. Abule and L. Nigatu. 2012. Evaluation of rangeland in arid and semi-arid grazing land of South East Ethiopia. International Journal of Agricultural Sciences 2: 221–234.

Abdi, O.A., E.K. Glover and O. Luukkanen. 2013. Causes and impacts of land degradation and desertification: Case study of the sudan. International Journal of Agriculture and Forestry 3: 40–51. DOI: 10.5923/j.ijaf.20130302.03.

AGRA. 2014. Africa Agriculture Status Report: Climate Change and Smallholder Agriculture in Sub-Saharan Africa. Alliance for a Green Revolution in Africa (AGRA), Nairobi, Kenya, 218 pp.

Ardö, J., M. Mölder, B.A. El-Tahir and H.A.M. Elkhidir. 2008. Seasonal variation of carbon fluxes in a sparse savanna in semi arid Sudan. Carbon Balance Manag. 3: 7. DOI: 10.1186/17500680-3-7.

Beck, E., R. Scheibe and M. Senser. 1983. The vegetation of the Shira Plateau and the western slopes of Kibo (Mt. Kilimanjaro Tanzania). Phytocoenologia 11: 1–30.

Beck, E., H. Rehder, E.D. Schulze and J.O. Kokwaro. 1987. Alpine plant communities of Mt. Elgon—An altitudinal transect along Koitoboss Route. J.E. Afr. Nat. His. Soc. and Nat. Mus. 76: 1–12.

Belsky, A.J. 1986. Population and community processes in a mosaic grassland in the Serengeti. Journal of Ecology 74: 841–856.

Bob, R.B. and A.K. Behrensmeyer. 2004. The expansion of grassland ecosystems in Africa in relation to mammalian evolution and the origin of the genus. Homo Palaeogeography, Palaeoclimatology, Palaeoecology 207: 399–420.

Bonham, C.D. 1989. Measurements for Terrestrial Vegetation. John Wiley & Sons, NY.

Boonman, J.G. 1993. East Africa's Grasses and Fodders: Their Ecology and Husbandry. Springer Science, pp. 339.

Bonnefille, R. and T. Letouzey. 1976. Fruits fossils d'Antrocaryon dans la vallee de l'Omo (Ethiopie) 1976. AGRIS. http://www.nal.usda.gov/.

Bonnefille, R. and R. Dechamps. 1983. Data on fossil flora. Ann. Mus. Roy. Afr. Centr. Tervuren Sci. geol 85: 191–207.

Bonnefille, R. 1984. Cenozoic vegetation and environments of early hominids in East Africa grasslands. Science 247: 1325–1328.

Bonnefille, R. 1995. A reassessment of the Plio-Pleistocene pollen record of east Africa. pp. 299–310. *In*: E.S. Vrba, G.H. Denton, T.C. Partridge and L.H. Burckle (eds.). Paleoclimate and Evolution with emphasis on Human origins. Yale University Press, New Haven.

Burnsilver, S.B. and E.M. Mwangi. 2007. Beyond Group Ranch subdivision: collective action for livestock mobility, ecological viability, and livelihoods. Washington, DC, USA: CAPRi. Working Paper 66. 4

Campbell, D.J., D.P. Lusch, T.A. Smucker and E.E. Wangu. 2005. Multiple methods in the study of driving forces of land use and land cover change: A case study of SE Kajiado district Kenya. Human Ecology 33: 763–794.

Catley, A., J. Lind and I. Scoones. 2013. Pastoralism and Development in Africa: Dynamic Change at the Margins. Nomadic Peoples 17(1): 150(8).

Cerling, T.E. 1992. Development of grasslands and savannas in East Africa during the Neogene. Global Planetary Change 5: 241–247.

Cerling, T.E., J.R. Bowman and J.R. O'Neil. 1988. An isotopic study of a fluvial-lacustrine sequence: the Plio-Pleistocene Koobi For a sequence, East Africa. Paleaogeogr. Palaeoclimatol. Palaeocol. 63: 335–356.

Cerling, T.E., J. Queda, S.H. Ambrose and N.E. Sikes. 1991. Fossil Soils, grasses and carbon isotopes from Fort Ternan, Kenya. Grasslands or woodlands? Journal of Human Evolution 21: 295–306.

Cerling, T.E., J.G. Wynn, S.A. Andanje, M.I. Bird, D.K. Korir, N.E. Levin, W. Mace, A.N. Macharia, J. Quade and C.H. Remien. 2011. Woody cover and hominin environments in the past 6 million years. Nature 76: 51–56. DOI: 10.1038/nature10306.

Chen, X., L.B. Hutley and D. Eamus. 2003. Carbon balance of tropical savanna of northern Australia. Oecologia 137: 405–416.

Christensen, J.H., B. Hewitson, A. Busuioc et al. 2006. Regional climate projections. *In*: S. Solomon, D. Qin, M. Manning, Z. Chen, M. Marquis, K.B. Averyt, M. Tignor and H.L. Miller (eds.). Climate Change 2007: The Physical Science Basis. Contribution of Working Group I to the Fourth Assessment Report of the Intergovernmental Panel on Climate Change. Cambridge University Press, Cambridge, UK.UNPD, 2008

Christensen, J.H., B. Hewitson, A. Busuioc, A. Chen, X. Gao, I. Held, R. Jones, R.K. Kolli, W.-T. Kwon, R. Laprise and P. Whetton. 2007. 'Regional Climate Projections'. *In*: A. Solomon, D. Qin, M. Manning, Z. Chen, M. Marquis, K.B. Averyt, M. Tignor and H.L. Miller (eds.). Climate Change: The Physical Science Basis, contribution of Working Group I to the Fourth Assessment Report of the Intergovernmental Panel on Climate Change Cambridge: Cambridge University Press.

Deshmukh, I. 1986. Primary production of a grassland in Nairobi National Park, Kenya. Journal of Applied Ecology 23: 115–123.

Dixon, A.P., D. Faber-Langendoen, C. Josse, J. Morrison and C.J. Loucks. 2006. Distribution mapping of world grassland types. Journal of Biogeography 41: 2003–2019.

Dobson, A., M. Borner, A. Sinclair, P.J. Hudson and E. Wolanski. 2010. Road will ruin Serengeti. Nature 467: 272–273.

Doherty, R.M., S. Sitch, B. Smith, S.L. Lewis and P.K. Thornton. 2010. Implications of future climate and atmospheric CO_2 content for regional biogeochemistry, biogeography and ecosystem services across East Africa. Global Change Biology 16: 617–640. DOI: 10.1111/j.1365-486.2009.01997.x.

Edwards, D.C. and A.V. Bogdan. 1951. Important Grassland Plants of Kenya. Pitman and Sons: Nairobi.

Elias, E. and F. Abdi. 2010. Putting Pastoralists on the Policy Agenda: Land Alienation in Southern Ethiopia. International Institute for Environment and Development, London, UK.

FAO. 2000. The elimination of food insecurity in the Horn of Africa. A strategy for concerted government and UN agency action. Summary report of the Inter-agency Task Force on the UN Response to Long-term Food Security, Agricultural Development and Related Aspects in the Horn of Africa. FAO, Rome, Italy. 13 p. See: http://www.gm-unccd.org/FIELD/Multi/FAO/FAO9.pdf

FAO. 2005. Grasslands of the world. J.M. Suttie, S.G. Reynolds and C. Batello (eds.). Food and Agriculture Organization of the United Nations (FAO), Rome.

Fratkin, E. 2001. East African pastoralism in transition: Maasai, Boran, and Rendille cases. African Studies Review 44: 1–25.

Friedel, M.H., W.A. Laycock and G.N. Bastin. 2000. Assessing rangeland condition and trend. pp. 305–360. *In*: L. Mannetje and R.M. Jones (eds.). Field Laboratory Methods for Grassland and Animal Production Research, CABI International, Wallingford, UK.

Frost, F.A. and T. Shanka. 2002. Regionalism in tourism: The case for Kenya and Ethiopia. Journal of Travel & Tourism Marketing 11: 35–58. DOI: 10.1300/J073v11n01_03.

Furley, P.A., R.M. Rees, C.M. Ryan and G. Saiz. 2008. Savanna burning and the assessment of long-term fire experiments with particular reference to Zimbabwe. Progress in Physical Geography 32: 611–644.

Georgis, K., A. Dejene and M. Malo. 2010. Agricultural-based Livelihood Systems in Drylands in the Context of Climate Change: Inventory of Adaptation Practices and Technologies of Ethiopia. Environment and Natural Resources Management Working Paper, 38. FAO, Rome.

Gibbs, H.K., A.S. Ruesch, F. Achard, M.K. Clayton, P. Holmgren, N. Ramankutty and J.A. Foley. 2009. Tropical Forests were the Primary Sources of New Agricultural Land in the 1980s and 1990s. www.pnas.org/cgi/doi/10.1073/pnas.0910275107.

Gibson, D.J. 2009. Grasses and Grassland Ecology. https://books.google.co.ke/books, pp. 313.

Gichohi, H., C. Gakahu and E. Mwangi. 1996. Savanna ecosystems. pp. 273–298. *In*: T.R. McClanahan and T.P. Young (eds.). East African Ecosystems and their Conservation. Oxford University Press, Oxford.

Grace, J., J. San Jose, P. Meir, H.S. Miranda and R.A. Montes. 2006. Productivity and carbon fluxes of tropical savannas. Journal of Biogeography 33: 387–400.

Graham, O. 1988. Enclosure of the East African rangelands: recent trends and their impact. London, UK: Overseas Development Institute. Pastoral Development Network Paper 24a. 11 p.

GRUMP (The Global Rural-Urban Mapping Project). 2005. GRUMP Alpha version. Available at: http://www.ciesin. columbia.edu/download_data.html.

Heady, H.F. and D. Child. 1994. Rangeland Ecology and Management. Boulder (CO) Westview Press.

Herlocker, D. (ed). 1999. Rangeland ecology and resource development in Eastern Africa. GTZ, Nairobi, Kenya.

Hiernaux, P. and M.D. Turner. 1996. The effect of clipping on growth and nutrient uptake of Sahelian annual rangelands. Journal of Applied Ecology 33: 387–399.

Hodgson, D. 2001. Once Intrepid Warriors: Gender, Ethnicity and the Cultural Politics of Maasai Development, Bloomington, Indiana, University Press.

Homewood, K., E.F. Lambin, E. Coast, A. Kariuki, I. Kikulai, J. Kiveliai, M. Said, S. Serneels and M. Thompson. 2001. Long-term changes in Serengeti-Mara wildebeest and land cover: Pastoralism, population, or policies? Proceedings of National Academy of Sciences. Int. Livestock Res. Institute, Nairobi 98: 12544–12549.

ILRI [International Livestock Research Institute]. 1996. The Ugandan Dairy Sub-sector: A Rapid Appraisal, Carried Out by Ministry of Agriculture Animal Industry and Fisheries (MAAIF), Entebbe; Makerere University, Kampala; and Overseas Development Administration (ODA, UK), 68p.

Kamau, P. 2004. Forage Diversity and Impact of Grazing Management on Rangeland Ecosystems in Mbeere District, Kenya. LUCID Working paper No. 36. Int. Livestock Res. Institute. Nairobi.

Kamnalrut, A. and J.P. Evenson. 1992. Monsoon grassland in Thailand. pp. 100–126. *In*: S.P. Long, M.B. Jones and M.J. Roberts (eds.). Primary Productivity of Grass Ecosystems of the Tropics and Sub-tropics. Chapman and Hall, London.

Kennett, J.P. 1995. A review of polar climatic evolution during the Neogene, based on the Marine sediment record. pp. 49–64. *In*: E.S. Vrba, G.S. Denton, T.C. Partridge and L.H. Burckle. (eds.). Paleoclimate and Evolution, with emphasis on Human Origins. Yale Univ. Press. New Haven.

Kingston, J.D., A. Hill and B.D. Marino. 1994. Isotopic evidence for Neogene hominid paleo environments in the Kenya Rift Valley. Science 264: 955–959.

Kinyamario, J.I. and S.K. Imbamba. 1992. Savanna at Nairobi National Park, Kenya. pp. 25–69. *In*: S.P. Long, M.B. Jones and M.J. Roberts (eds.). Primary Productivity of Grass Ecosystems of the Tropics and Sub-tropics. Chapman and Hall, London, 267 pp.

Kinyamario, J.I. and J.N.M. Macharia. 1992. Above-ground standing-crop, protein content and dry matter digestibility of a tropical grassland range in the Nairobi National Park, Kenya. African Journal of Ecology 30: 33–41.

Kinyamario, J.I. 2015. NPP Grassland: Nairobi, Kenya. 1984–1994, R1. Data set. Available on-line [http://daac.ornl.gov] from Oak Ridge National Laboratory Distributed Active Archive Center, Oak Ridge, Tennessee, USA. http://dx.doi.org/10.3334/ORNLDAAC/151.

Kirui, O.K. and A. Mirzabaev. 2015. Costs of Land Degradation in Eastern Africa. 29th International Conference of Agricultural Economists, Milan, Italy.

Kiteme, B.P., U. Wiesmann, G. Kunzi and J.M. Mathuva. 1998. A highland-lowland system under transitional pressure: A spatio-temporal analysis. *In*: Towards Sustainable Regional Development in the Highland-Lowland System of Mount Kenya. Eastern and Southern Africa Geographical Journal 8: 45–53.

Knoop, W.T. and B.H. Walker. 1985. Interactions of woody and herbaceous vegetation in a southern Africa savanna. Journal of Ecology 73: 235–253.

K'Otuto, G.O., D.O. Otieno, B. Seo, H.O. Ogindo and J.C. Onyango. 2013. Carbon dioxide exchange and biomass productivity of the herbaceous layer of a managed tropical humid savanna ecosystem in western Kenya. J. Plant Ecology 6: 286–297. DOI: https://doi.org/10.1093/jpe/rts038.

Levin, N. 2002. Isotopic Evidence for Plio-Pleistocene environmental changeof Gona, Ethiopia. Msc thesis, University of Arizona, Tucson.

Little, P., J. McPeak, C.B. Barrett and P. Kristjanson. 1983. Challenging orthodoxies: Understanding poverty in pastoral areas of East Africa. Development and Change 39: 587–611. DOI: 10.1111/j.1467-7660.2008.00497.x.

Little, P.D., K. Smith, B.A. Cellarius, D.L. Coppock and C.B. Barrett. 2001. Avoiding disaster: Diversification and risk management among east African herders. Development and Change 32: 401–433.

Long, S.P., E. Garcia Moya, S.K. Imbamba, A. Kamnalrut, M.T.F. Piedade, J.M. Scurlock, Y.K. Shen and D.O. Hall. 1989. Primary productivity of natural grass ecosystems of the tropics, a reappraisal. Plant and Soil 115: 155–166.

Long, S.P. and M.B. Jones. 1992. Introduction, aims, goals, and general methods. pp. 1–24. *In*: Long, S.P., M.B. Jones and M.J. Roberts (eds.). Primary Productivity of Grass Ecosystems of the Tropics and Sub-tropics. Chapman and Hall, London, 267 pp.

Løvschal, M., P.K. Bøcher, J. Pilgaard, I. Amoke, A. Thuo, J.-C. Svenning and A. Odingo. 2017. Fencing bodes a rapid collapse of the unique Greater Mara ecosystem. Scientific Reports 7: 41450. DOI: 10.1038/srep 41450.

Machogu, J.O. 2014. Assessment of the Environmental Impacts of Wildlife based Tourism in Kenya's Protected Areas: A Case Study of Maasai Mara National Reserve. MA Project, University of Nairobi.

Magill, C.R., G.M. Ashley and K.H. Freeman. 2012. Ecosystem variability and early human habitats in eastern Africa. PNAS 110: 1167–1174.

Maitima, J.M., S.M. Mugatha, R.S. Reid, L.N. Gachimbi, A. Majule, H. Lyaruu, D. Pomery, S. Mathai and S. Mugisha. 2009. The linkages between land use change, land degradation and biodiversity across East Africa. African Journal of Environmental Science and Technology 3: 310–325.

Maslin, M.A., C.M. Brierley, A.M. Milner, S. Shultz, M.H. Trauth and K.E. Wilson. 2014. East African climate pulses and early human evolution. Quaternary Science Reviews 101: 1–17.

Mazimhaka, J. 2007. Diversifying Rwanda's tourism industry: A role for domestic tourism. Development Southern Africa 24: 491–504.

McDermott, J., S. Staal, H. Freeman, M. Herrero and J. van de Steeg. 2010. Sustaining intensification of smallholder livestock systems in the tropics. Livestock Science 130: 95–109.

McNaughton, S.J. 1983. Serengeti grassland ecology: The role of composite environmental factors and contingency in community organization. Ecological Monographs 53: 291–320.

McNaughton, S.J., R.W. Ruess and S.W. Seagle. 1988. Large mammals and process dynamics in African ecosystems. BioScience 38: 794–800.

Melita, A. and S. Mendlinger. 2013. The impact of tourism revenue on the local communities' livelihood: A case study of ngorongoro conservation area, Tanzania. Journal of Service Science and Management 6: 117–126. DOI: org/10.4236/jssm.2013.61012.

Menaut, J.C. 1983. The vegetation of African savannas. pp. 109–149. *In*: F. Bourlie`re (ed.). Ecosystems of the World: Tropical Savannas. Elsevier, Amsterdam.

Mganga, K.Z., N.K. Musimba, D.M. Nyariki, M.M. Nyangito, A.W. Mwang'ombe, W.N. Ekaya and W.M. Muiru. 2010. The challenges posed by *Ipomoea kituensis* and the grass-weed interaction in a reseeded semi-arid environment in Kenya. International Journal of Current Research 11: 001–005.

MOAC [Ministry of Agriculture and Co-operatives]/SUA [Sokoine University of Agriculture]/ILRI. 1998. The Tanzania Dairy Sub-sector: A Rapid Appraisal. Vol. 3. Main Report. Funded by the Swiss Agency for Development and Cooperation, Switzerland.

Mordelet, P. and J.C. Menaut. 1995. Influence of trees on above-ground production dynamics of grasses in a humid savanna. Journal of Vegetation Science 6: 223–228.

Muriuki, H.G. and W. Thorpe. 2001. Smallholder dairy production and marketing in Eastern and Southern Africa. Regional synthesis. *In*: D. Rangnekar and W. Thorpe (eds.). Smallholder Dairy Production and Marketing Opportunities and Constraints. Proceedings of the South - South workshop held at NDDB, Anand, India, 13–16 March 2001.

Muthuri, F.M., M.B. Jones and S.K. Imbamba. 1989. Primary productivity of papyrus (*Cyperus papyrus*) in a tropical swamp; Lake Naivasha, Kenya. Biomass 18: 1–14.

Mworia, J.K., J.I. Kinyamario and E.A. John. 2008. Impact of the invader *Ipomoea hildebrandtii* on grass biomass, nitrogen mineralization and determinants of its seedling establishment in Kajiado, Kenya. African Journal of Range and Forage Science 25: 11–16.

Mworia, J.K., J.I. Kinyamario, J.K. Omari and J.K. Wambua. 2011. Patterns of seed dispersal and establishment of the invader *Prosopis juliflora* in the upper floodplain of Tana River, Kenya. African Journal of Range and Forage Science 28: 35–41.

Ngene, S.M. and P.O.M. Omondi. 2009. The costs of living with elephants in the areas adjacent to Marsabit National Park and Reserve. Pachyderm 45: 77–87.

Nicholson, C.F., P.K. Thornton, L. Mohammed, R.W. Muinga, D.M. Mwamachi, E.H. Elbasha, S.J. Staal and W. Thorpe. 1999. Smallholder Dairy Technology in Coastal Kenya. An adoption and Impact study. ILRI Impact Assessment Series, No. 5. 59p.

Njoka, T.J., G.W. Muriuki, R.S. Reid and D.M. Nyariki. 2003. The use of sociological methods to assess land-use change: A case study of Lambwe Valley, Kenya. J. Soc. Sc. 7: 181–185.

Nyberg, G., P. Knutsson, M. Ostwald, I. Öborn, E. Wredle, D.J. Otieno, S. Mureithi, P. Mwangi, M.Y. Said, M. Jirström, J. Wernersson, S. Svanlund and J.N. Wairore. 2015. Enclosures in West Pokot, Kenya: Transforming land, livestock and livelihoods in drylands. Pastoralism: Research, Policy and Practice 5: 25. DOI: 10.1186/s13570-015-0044-7.

Oba, G. 1990. Effects of wildfire on a semi desert riparian woodland along the Turkwel river, Kenya, and management implications for Turkana pastoralists. Land Degradation and Rehabilitation 2: 247–259.

Oba, G., E. Post, P.O. Syvertsen and N.C. Stenseth. 2000. Bush cover and range condition assessments in relation to landscape and grazing in southern. Ethiopia 15: 535–546.

Obwodho, A.B. 2006. Intensive forage production for smallholder dairying in East Africa. *In:* Grasslands: Developments Opportunities Perspectives. Reynolds & Frame, 31 pp.

Ogutu, J.O., N. Owen-Smith, H.P. Piepho and M.Y. Said. 2011. Continuing wildlife population declines and range contraction in the Mara region of Kenya during 1977–2009. Journal of Zoology 285: 99–109.

Ogutu, J.O., H.P. Piepho, M.Y. Said, G.O. Ojwang, L.W. Njino, S.C. Kifugo and P.W. Wargute. 2016. Extreme wildlife declines and concurrent increase in livestock numbers in kenya: What are the causes? PLoS ONE 11: 10.1371/journal.pone.0163249.

Onchwati, J., H. Sommerville and N. Brockway. 2010. Sustainable tourism development in the Masai Mara National Reserve, Kenya, East Africa. WIT Transactions on Ecology and the Environment, Vol. 139. DOI: 10.2495/ST100281.

Opiyo, F.E.O., W.N. Ekaya, D.M. Nyariki and S.M. Mureithi. 2011. Seedbed preparation influence on morphometric characteristics of perennial grasses of a semi-arid rangeland in Kenya. African Journal of Plant Sciences 5: 460–468.

Otieno, D.O., G.O. K'Otuto, J.N. Maina, Y. Kuzyakov and J.C. Onyango. 2010. Responses of ecosystem carbon dioxide fluxes to soil moisture fluctuations in a moist Kenyan savanna. Journal of Tropical Ecology 26: 605–618.

Otieno, D., M. Ruidisch, B. Huwe, J. Ondier, B. Lee, S. Arnhold, A. Kolb, D. Okach and J. Onyango. 2015. Patterns of CO_2 exchange and productivity of the herbaceous vegetation and trees in a humid savanna in western Kenya 216: 1441–1456.

Oxfam. 2010. Pastoralism Demographics, Settlement and Service Provision in the Horn and East Africa: Transformation and Opportunities. Humanitarian Policy Group. Overseas Development Institute, London. Available at: http://www oxfamblogs.org/eastafrica/wp-content/uploads/2010/09/REGLAP-REPORTv2.

Phaulos, A., N. Kedir, B. Tekele Yohanis, K. Shambel, D. Tadel, B. Diriba, H. Bekele and T. Feyissa. 1999. Ethiopia agricultural research organization (Agro-ecological based Agricultural Production Constraints Identification Survey. SM1-1 sub agro ecology (Rayitu districts, Bale zone) Sinana Agriculture Research Center Oromiya Agricultural Bureau.

Pratt, D.J., P.J. Greenway and M.D. Gwynne. 1966. A classification of East African rangeland, with an Appendix on terminology. Journal of Applied Ecology 3: 369–382.

Pratt, D.J. and M.D. Gwynne. 1977. Range Management and Ecology in East Africa. London: Hodder & Stoughton Eds.

Pye-Smith, C. 2010. A Rural Revival in Tanzania: How agroforestry is helping farmers to restore the woodlands in Shinyanga Region. ICRAF Trees for Change no. 7. Nairobi: World Agroforestry Centre.

Reda, K.T. 2015. Natural resource degradation and conflict in the East African pastoral drylands. African Security Review 24: 270–278. DOI: 10.1080/10246029.2015.1059350.

Rehder, H., E. Beck, J.O. Kokwaro and R. Scheibe. 1981. Vegetation analysis of the upper Teleki Valley (Mt. Kenya) and adjacent areas. Journal of the East Africa Natural History Society 171: 1–8.

Reid, R.S., R.L. Kruska, C.J. Wilson and P.K. Thornton. 1998. Conservation Crises of the 21st Century: Tension Zones among Wildlife, People, and Livestock Across Africa in 2040. Paper Presented at the International Congress of Ecology, Florence, Italy, 19–26 July.

Reid, R.S., P.K. Thornton and R.L. Kruska. 2004. Loss and fragmentation of habitat for pastoral people and wildlife in East Africa: Concepts and issues. African Journal of Range and Forestry Science 21: 171–184.

Retallack, G.J., D.P. Dugas and E.A. Bestland. 1990. Fossil Soils and Grasses of a middle Miocene East. Ahyte (Ed), The evolution of the East African environment 2.Palaeobotany, Paleozoology, and Paleoanthropology. Univ. Hong Kong pp. 579–612.

Retallack, G.J. 1992. Middle Miocene fossil plants from Fort Ternan (Kenya) and evolution of African grasslands. Palaeobiology 18: 383–400.

San Jose', J.J. 1995. Environmentally sustainable use of the Orinoco savannas. Scientia Guaianae 5: 175–194.

San Jose', J.J. and R.A. Montes. 2001. Management effects on carbon stocks and fluxes across the Orinoco Llanos. Forest Ecology and Management 150: 293–311.

Santos, A.J., B. Quesada, C.a. Da Silva, G.t. Jair, F.m. Miranda, A.c. Miranda and J. Lloyd. 2004. High rates of net ecosystem carbon assimilation by Brachiara pasture in the Brazilian Cerrado. Global Change Biology 10: 877–885.

Schmidt, K. 1991. The Vegetation of the Aberdare National Park Kenya. Wagner. Innsbruck. 259.

Scholes, R.J. and D.O. Hall. 1996. The carbon budget of tropical savannas, woodlands and grasslands. *In*: A.I. Breymeyer, D.O. Hall, J.M. Melillo and G.I. Agren (eds.). Global Change: Effects on Coniferous Forests and Grasslands, SCOPE 56: 69–100.

Selemani, I.S., L.O. Eik, Ø. Holand, T. Ådnøy, E. Mtengeti and D. Mushi. 2012. The role of indigenous knowledge and perceptions of pastoral communities on traditional grazing management in north-western Tanzania. African Journal of Agricultural Research 7: 5537–5547.

Shackleton, N., J. Backman, H. Zimmermann, D.V. Kent, M.A. Hall, D.J. Roberts, D. Schnitker, J.G. Baldauf et al. 1984. Oxygen Isotope calibration of the onset of Ice rafting and history of glaciation in North Atlantic Region. Nature 307: 620–623.

Singh, J.S., Y. Hanxi and P.E. Sajise. 1985. Structural and functional aspects of Indian and southeast Asian savanna ecosystems. pp. 34–51. *In*: J.C. Tothill and J.C. Mott (eds.). International Savanna Symposium 1984. Australian Academy of Science, Canberra.

Sonneveld, B.G.J.S., S. Pande, K. Georgis, M.A. Keyzer, S.A. Ali and A. Takele. 2010. Land degradation and overgrazing in the afar region, Ethiopia: A spatial analysis. pp. 97–109. *In*: Land Degradation and Desertification: Assessment, Mitigation and Remediation, Springer Science. DOI: 10.1007/978-90-481-8667-0_8.

Staal, S.J., L. Chege, M. Kinyanjui, A. Kimani, B. Lukuyu, D. Njubi, M. Owango, J. Tanner, W. Thorpe and M. Wambugu. 1998. Characterization of dairy systems supplying the Nairobi milk market. A pilot survey in Kiambu District for the identification of target groups of producers. Smallholder Dairy (R&D) Project. KARI, ILRI and Livestock Production Department, Ministry of Agriculture.

Stringer, L.C. 2003. Applying the United Nations Convention to Combat Desertification in Africa: Local and Scientific Understandings of Environmental Degradation. PhD Thesis. University of Sheffield, Department of Geography, Sheffield.

Sumberg, J. and J. Thomson. 2013. STEPS Working Paper 52. Brighton: STEPS Centre. Revolution Reconsidered: Evolving Perspectives on Livestock Production and Consumption.

Thornton, P. 2010. Livestock production: Recent trends, future prospects. Philosophical Transactions of the Royal Society B 365: 2853–2867.

Trollope, W.S.W. 1982. Ecological effects of fire in South African savannas. pp. 292–306. *In*: B.J. Huntley and B.H. Walker (eds.). Ecology of Tropical Savannas. Springer, Berlin.

Trollope, W.S.W., C.J.H. Hines and L.A. Trollope. 2000. Simplified Techniques for Assessing Range Condition in the East Caprivi region of Namibia. Final Report. Windhoek, Namibia: Directorate of Forestry, Namibia-Finland Forestry Programme.

Tsegaye, D., S.R. Moe, P. Vedeld and E. Aynekulu. 2010. Land-use/cover dynamics in Northern Afar rangelands, Ethiopia. Agriculture, Ecosystems and Environment 139: 174–180.
UNDP. 2008. World population prospects: The 2008 Revision-UNDP www.un.org.
van Heist, M. 1994. Accompanying report with the Land unit map of Mount Elgon National Park. Mount Elgon Conservation and Development Project. Kampala. 83 pp.
van Wilgen, B.W., C.S. Everson and W.S.W. Trollope. 1990. Fire management in southern Africa: Some examples of current objectives, practices and problems. *In*: J.G. Goldammer (ed.). Fire in the Tropical Biota: Ecosystem Processes and Global Challenges Ecological Studies 84. Berlin: Springer-Verlag.
Vesey-Fitzgerald, D.F. 1963. Central African grasslands. Journal of Ecology 51: 243–274.
Vizy, E.K. and K.H. Cook. 2012. Mid-twenty-first-century changes in extreme events over northern and tropical Africa. Journal of Climate 25: 5748–5767. DOI: 10.1175/JCLI-D-11-00693.1.
Walpole, M., G. Karanja, N. Sitati and N. Leader-Williams. 2003. Wildlife and people: Conflict and conservation in Masai Mara, Kenya. *In*: IIED Wildlife & Development Series. London: International Institute for Environment and Development.
Wandera, J.L. 1996. Pasture and Cover Crops. Reports of the Kenya Agricultural Research Centre.
WCMC. 2001. Serengeti National Park, Tanzania. http://www.wcmc.org.uk/protected_areas/data/wh/serenget.html.
Wenchang, Y., R. Seager and M.A. Cane. 2014. The east African long rains in observations and models. Journal of Climate 27: 7185–7202. DOI: 10.1175/JCLI-D-13-00447.
Wesche, K. 2002. The high-altitude environment of Mt. Elgon (Uganda, Kenya): Climate, Vegetation and the impact of fire. Society for Tropical Ecology. Ecotropical Monographs 2: 253.
White, T.A., B.D. Campbell, P.D. Kemp and C.L. Hunt. 2001. Impacts of extreme climatic events on competition during grassland invasions. Global Change Biology 7: 1–13.
Wilsey, B.J. 1996. Variation in use of green flushes following burns among African ungulate species: The importance of body size. African Journal of Ecology 34: 32–38.
World Resources. 2000–2001. Peoples and Ecosystems. The Fraying Web of Life. Elsevier Science. Oxford, UK.
WWF. 2017. Eastern Africa: The Greater Serengeti Grassland Ecosystem in Northern Tanzania. http://www.worldwildlife.org/ecoregions/at0714.

14

Grasslands in South America

Nature and Extent in Relation to Provision of Ecosystem Services

Cesar Morales

What are Grasslands

Definition

There is a wide spectrum of definitions for grasslands. In 2010, the FAO (Food and Agriculture Organization of the United Nations), compiled the most well-known definitions (http://www.fao.org/agriculture/crops/thematic-sitemap/theme/spi/gcwg/definitions/en/). This profusion of definitions may be due to the difficulty in characterising the limits of grasslands, a less persistent canopy structure, more frequent disturbance regimes and their occurrence within a physiognomic continuum between forests and deserts.

The definition adopted by the World Wildlife Fund (WWF) will be used here, it says, Grassland is a type of habitat or biome which is dominated by grasses and other herbaceous (non-woody) flowering plants and a variety of scattered trees and bushes. Grasslands occur in areas with not enough regular rainfall to support the growth of a forest, but not so little as to form a desert. About one-quarter of the earth's surface is covered with grasslands World wildlife Fund.

There are a few trees in true grasslands either because the climate is too dry or the soil is too poor. In other areas, grasslands develop because grazing and browsing by wild animals, frequent fires, or both, prevent tree seedlings from growing (http://wwf.panda.org/about_our_earth/ecoregions/about/habitat_types/habitats/grasslands/).

Former Economic Affairs Officer of the Economic Commission of United Nations for Latin America and the Caribbean(ECLAC) International consultant.

Main Facts

There are many definitions about grasslands. The Global Land Cover Characteristics Database (GLCCD) of the US Geological Survey provides a global land area classification by ecosystem type as described by Loveland et al. (2000). It divides the earth's terrestrial area into a number of classifications (Table 14.1). Five of these (open or closed shrublands, woody and non-woody savannas and grasslands) are aggregated to form grasslands which are estimated to cover 5 billion hectares or 37 per cent of the earth's terrestrial area (excluding Greenland and Antarctica).

Based on their own statistics, FAO reported that permanent meadows and pastures cover 3.4 billion hectares or 69 per cent of the world's agricultural area (FAOSTAT accessed on November 2016). According the same source, grasslands contain about 20 per cent of the world's soil carbon stocks (FAO, 2009; Ramankutty et al., 2008; Schlesinger, 1977).

Grasslands have played a crucial role in food and forage production. It is estimated that around 20 per cent of the world's native grasslands have been converted to cultivated crops (Ramankutty et al., 2008) and production of world milk (27 per cent) and beef (23 per cent) occurs on grasslands managed solely for these purposes. The livestock industry, largely based on grasslands, provides livelihood to about one billion of the world's poorest people and one-third of global protein intake (Steinfeld et al., 2006).

The world population growth and changes in income are relevant factors to explain the increasing pressures on grasslands, mainly to produce more meat. That means more cattle grazing, particularly in Africa's rangelands which are highly vulnerable to climate change (Reid et al., 2004). As a result of past practices, 7.5 per cent of the world's grasslands have

Table 14.1 Global land area classification by ecosystem type.

Classification	km^2	(%)
Forest/Evergreen/Needleleaf	4,858,707	3.6
Forest/Evergreen/Broadleaf	13,479,749	10.0
Forest/Deciduous/Needleleaf	1,959,892	1.5
Forest/Deciduous/Broadleaf	2,229,308	1.7
Forest/Mixed	9,930,103	7.4
Shrublands/Open	2,636,901	2.0
Shrublands/Closed	20,706,263	15.4
Savannas/Woody	8,405,816	6.2
Savannas/Non-woody	7,607,497	5.6
Grasslands	10,541,721	7.8
Permanent wetlands	9,84,328	0.7
Croplands	15,206,323	11.3
Urban and built-up	2,56,332	0.2
Croplands/Natural vegetation mosaic	11,586,898	8.6
Snow or ice	2,621,872	1.9
Barren or sparsely vegetated	18,332,436	13.6
Water bodies	3,494,824	2.6
Total	**1,34,838,970**	**100.0**

Source: US Geological Survey.

been degraded due to overgrazing (Oldeman, 1994). Previous research has documented that improved grazing management could lead to greater forage production, more efficient use of land resources and enhanced profitability and rehabilitation of degraded lands (Oldeman, 1994).

Grasslands cover approximately 30 per cent of the earth's ice-free land surface and 70 per cent is agricultural land (Suttie et al., 2005; White et al., 2000). Drylands occupy 41 per cent of the land area and are home to more than two billion people (UNEP, 2006). Of the 3.4 billion ha of rangelands worldwide, an estimated 73 per cent is affected by soil degradation (WOCAT, 2009). Over one billion people depend on livestock, and 70 per cent of the 880 million rural poor living on less than USD 1.00 per day are partially dependent on livestock for their livelihoods (World Bank, 2007; Ashley et al., 1999). Livestock production can be found on two-thirds of the global drylands (Clay, 2004).

According O´Mara (2012), ruminants are efficient converters of grass into humanly edible energy and protein and grassland-based food production and have a comparable carbon footprint as mixed systems. As remarked previously, grasslands are a very important store of carbon and are continuing to sequester carbon with considerable potential to increase this further.

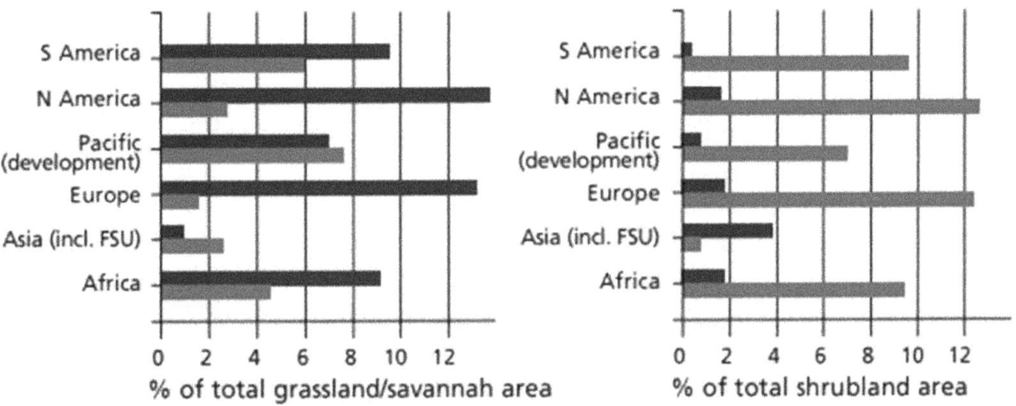

Fig. 14.1 Percentage of native grassland/savannah converted to cropland and pasture.
Source: Ramankutty et al., 2008.

Threats

Grasslands are exposed to pressures of different nature; from one side pressures related to the expansion of agriculture and its intensification, that means the replacement of grasslands by some crop, usually related with animal feed. The following graph shows the land use changes between grasslands and soy and corn in five states of the corn of belt of United States. As can be appreciated, the land use change is from grasslands to soy. The other important pressure is related to overgrazing of goats, sheep and cattle, usually in areas of recent expansion. Urban expansion also contributes to the diminishing of grassland surface area, as well as other productive activities, such as industrial development and mining. This one leads to destruction of soil due to pollution with some toxic minerals, such as mercury in the case of gold mining.

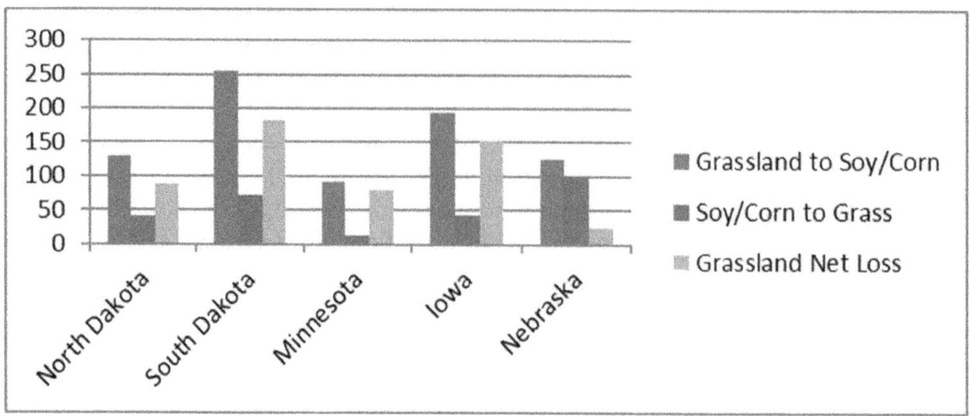

Fig. 14.2 Land use change affecting grasslands in Corn Belt States of USA (2006 to 2011).
Source: Recent land use change in the western Corn Belt threatens grasslands and wetlands.

Grasslands in South America

There are 605 million hectares of grasslands in South America, according to the WWF classification system for territorial ecoregions. This classification system defines four types of terrestrial ecoregions:

- Tropical and subtropical grasslands, savannas and shrublands
- Temperate grasslands, savannas and shrublands
- Flooded grasslands and savannas
- Montane grasslands and shrublands

The tropical and subtropical grasslands, savannas and shrublands have the most extended area, followed by the temperate grasslands. Both of them cover some of the best known grasslands, such as the Cerrados which is the largest savanna region in South America and biologically the richest savanna in the world, and the humid pampas, that are one of the most populated areas in Argentina (WWF, 2016). Table 14.2 presents the main characteristics of each ecoregion and its surface area.

Tropical and Subtropical Grasslands, Savannas and Shrublands

The climate is characterised by alternating wet and dry periods and high temperatures. The vegetation consists predominantly of grasses with different densities of trees. In general, the topography is flat to gently undulating. The principal crops are annuals, such as corn, soybean, sorghum, beans, cassava and cotton, grasses like *Brachiaria*, and legumes like *Stylosanthes*. In recent years, sugarcane and other biofuel crops have also been planted (Miyake et al., 2012).

Brazil, with the largest area of savannahs in South America (around 200 million ha), has 55 million ha of introduced pastures and 22 million ha under annual crops (Sousa, 2011). In general, the soils in this biome are low in organic matter, very infertile and acid throughout.

Erosion, compaction and nutrient imbalance are the main problems confronted here. Overgrazing, farm machinary intensive use, and insufficent fertilisers application together

Table 14.2 Grasslands in South America. Terrestrial ecoregions.

Terrestrial Ecoregions	Main Characteristics	Surface Area (Km²)
Tropical and subtropical grasslands, savannas and shrublands	Large expanses of land in the tropics do not receive enough rainfall to support extensive tree cover. The tropical and subtropical grasslands, savannas, and shrublands are characterised by rainfall levels between 90–150 cms per year. However, there may be great variability in soil moisture throughout the year. Grasses dominate the species composition of these ecoregions, although scattered trees may be common.	3,503,736
Flooded grasslands and savannas	These areas support numerous plants and animals adapted to the unique hydrologic regimes and soil conditions. Large congregations of migratory and resident water-birds may be found in these regions. However, the relative importance of these habitat types for these birds varies as the availability of water and productivity annually and seasonally shifts among complexes of smaller and larger wetlands throughout the region.	2,81,014
Temperate grasslands, savannas and shrublands	Temperate grasslands, savannas, and shrublands differ largely from tropical grasslands in the annual temperature regime as well as the type of species found here. Generally speaking, these regions are devoid of trees, except for riparian or gallery forests associated with streams and rivers.	1,454,797
Montane grasslands and shrublands	They are tropical, subtropical and temperate. The plants and animals of tropical montane paramos display striking adaptations to cool, wet conditions and intense sunlight. Around the world, characteristic plants of these habitats display features, such as rosette structures, waxy surfaces, and abundant pilosity. The paramos of the northern Andes are the most extensive examples of this major habitat type.	8,10,666
Total		**6,050,213**

Source: World Wild Life Fund

Table 14.3 Tropical and subtropical grasslands, savannas and shrublands.

Ecoregions/Countries	Area (Km²)	Status	Main Threats
Uruguayan savanna; Uruguay, Brazil and Argentina	448,845	Critical/ Endangered	Cattle ranching; overgrazing, land degradation
The Llanos; Colombia and Venezuela	389,016	Vulnerable	Agriculture and cattle—15 million head of cattle
Chaco Humedo; Brazil Argentina, Paraguay	334,885	Vulnerable	Overgrazing, clearing of trees, desertification
Savanna Northern Brazil, Guyana, Venezuela	104,377	Vulnerable	Degradation of forests into grasslands; fires, erosion
Montane savanna Córdoba, Argentina	58,275	Vulnerable	Overgrazing, erosion, habitat change
Cerrado Central Brazil, into Bolivia and Paraguay	1,916,850	Vulnerable	Intensive agribusiness and plants degradation
Campos Rupestres montane savannas Southeastern Brazil	26,418	Relatively Stable/Intact	Mining, extraction of native plants, cattle ranching, tourism, fires, agribusiness and urban expansion
Beni Savannas Northern Bolivia	126,132	Critical/ Endangered	Grazing cattle, seasonal burning for forage maintenance
Chaco Central Argentina	98,938	Vulnerable	Livestock grazing, desertification

Source: WWF, Grasslands.

with a intensification of the agricultural production are the most notorious causes behind these problems (Guimarães, 2013; Urquiaga et al., 2014).

With respect to impact, the increase in food production could be the main positive effect in this bioma. However, there are significant negative effects too—the carbon and nitrogen cycles affected by higher rate of organic matter decomposition, the loss of water quality due to erosion and sediment movement; together with biodiversity losses (Viglizzo and Frank, 2006; Viglizzo and Jobbagy, 2010).

This group includes nine ecoregions from which only one. Campos rupestres in the southern region of Brazil, is considered stable or intact. All the rest are in the category of vulnerable or critically endangered. Cattle and agriculture expansion are the trigger factors of the previously mentioned condition of critical/endangered and vulnerable. Another important factor is the use of fire to clear the forest to make space for agriculture and cattle, as well as intensification of the agribusiness in the case of the Cerrado, the biggest ecoregion of this group.

Temperate Grasslands, Savannas and Shrublands

This biome is predominantly located in the Argentinian Pampas. Its central plains are dominated by grasses on flat to gently-sloping lands with a temperate climate, and rains ranging from 1,500 mm in the northeast to 400 mm in the southwest. These areas have some of the most fertile soil in the world, although more than 13 million hectares with natural saline-sodic soil also appears in this biome. Despite the wide adoption of non-tillage, intensive annual cultivation (largely of soybean) and lack of rotation with other crops or pastures, soils are affected by degradation due to wind and water erosion, waterlogging, compaction, sealing/capping and soil fertility depletion (Satorre, 2005; Lavado and Taboada, 2009; Sainz Rozas et al., 2011; Viglizzo and Jobbagy, 2010). As has occurred in other places, clearing of forested lowlands to produce annual crops (soybean, cotton, etc.) has led to salination or sodification in areas where the groundwater table has risen (Paruelo et al., 2005; Viglizzo and Jobbagy, 2010).

Table 14.4 Temperate grasslands, savannas and shrublands.

Ecoregions/Countries	Area (Km^2)	Status	Main Threats
South-eastern Argentina	327,115	Critical/ Endangered	Change from natural habitat to agriculture; overgrazing and land degradation, burning and draining of lands
Patagonia. Argentina and south-eastern Chile	305,878	Critical/ Endangered	Overgrazing by sheep, erosion and desertification; regression of some species of fauna due to hunting
Eastern Argentina Humid Pampas	240,869	Critical/ Endangered	Conversion of natural habitats for agriculture; land degradation, over-grazing; burning and draining of land
Southern Argentina, stretching northward	408,959	Vulnerable	Increase of human activities; overgrazing (goats, sheep and cattle); clear cutting for fuel; and land clearance for agriculture, mining and oil exploration
Central Argentina	108,780	Critical/ Endangered	Intensification of agriculture and deforestation
Southern Argentina and Chile	63,196	Vulnerable	Desertification, grazing of cattle and introduction of herbivores leading to deterioration of the scarce vegetation cover

Source: WWF.

All the ecoregions of this group have the status of critically/endangered or vulnerable. This is because of expansion of human activities, intensification of agriculture, overgrazing and consequent erosion, land degradation and desertification. At the same time, the group includes ecoregions like the humid pampa with some of the richest soil in the world to be extremely overexploited. It is important to mention the Patagonian steppes which have been intensively overgrazed by sheep for many years. Another factor is hunting of the huanaco baby (chulengos) because of the high value of their skin.

Flooded Grasslands and Savannas

This biome is associated with tropical alluvial plains and is located mostly in Brazil, Bolivia and Paraguay. The Pantanal, which stretches across all the three countries, is the world's largest tropical wetland and is a highly productive environment. The most widespread use of this biome in South America is as pasture for cattle during the dry season. Their main ecosystem service is provision of food and fibre, which are dependent on water regulation.

Economic development in this Pantanal region, especially on the plateau of Rio Taquari Basin, has intensified the input of sediments in the Pantanal lowlands, causing serious social, economic and environmental impacts on the region (Galdino et al., 2006). Informal gold mines, using mercury, are contaminating severely the area, thus endangering provision of ecosystem services in this biome. This ecosystem is unique and rich in flora and fauna. Soil biodiversity is of prime importance and should be investigated and protected.

Table 14.5 Flooded grasslands and savannas.

Ecoregions/Countries	Area (Km^2)	Status	Main Threats
Mesopotamian savannas, north-eastern Argentina	77,700	Vulnerable	Overgrazing cattle ranching, land degradation
Flooded savanna, eastern Argentina	38,850	Critical/ Endangered	Large dykes, dams, waterways, roads, mining, petroleum, urban expansion, tourism
Pantanal southwestern Brazil, Bolivia and Paraguay	155,658	Critical/ Endangered	Pesticide runoff from agricultural lands within the 5,00,000 km^2 watershed of the Rio Paraguay and gold mining, mercury
Orinoco wetlands, northeastern Venezuela	5,957	Relatively stable/Intact	Oil extraction and exploration, dam construction, increase of human population
Guayaquil flooded grasslands, western Ecuador	2,849	Critical/ Endangered	Steady increase in human population and large agricultural irrigation programmes

Source: WWF, Grasslands.

Montane Grasslands and Shrublands

This biome, found at high altitudes throughout the Andes, mostly in Peru and Bolivia, is called Punas or Paramos. Occurring above 3,000 m or even higher, the region is dominated by grasses and small shrubs. Temperatures during the day may be quite high but frost occurs at night. In general, the area is quite dry but the biome is highly important for water production because it is located at the top of many watersheds and is a continuous source of pristine water from the thawing process.

Table 14.6 Montane grasslands and shrublands.

Ecoregions/Countries	Area (Km²)	Status	Main Threats
Western Argentina into Chile	178,191	Relatively Stable/Intact	Habitat conversion, increased ecotourism sports (littering, erosion, sewage disposal), burning of shrubs
Northern Colombia	1,295	Relatively Stable/Intact	Increase in cattle population; hunting pumas and condors to protect herds of cows and sheep
Ecuador into Colombia	30,044	Relatively Stable/Intact	
Western Venezuela	2,849	Relatively Stable/Intact	
Southern Ecuador and northern Peru	12,173	Relatively Stable/Intact	
Peru and Bolivia	117,326	Vulnerable	Deforestation, intensification of agriculture and firewood for heating and cooking; overgrazing and erosion
Central Andean Puna: Argentina, Bolivia and Peru	161,356	Vulnerable	Increasing mining activity, destruction of scarce plant cover and contamination of water and soil
Central Andean Dry Puna: Argentina, Bolivia, Chile	307,432	Relatively Stable/Intact	Destruction of Polylepis forests and land degradation

Source: WWF.

Most of the pastures are natural, but forage species have been introduced and adapted to the physical conditions. In some parts of northern South America, there are areas dedicated to intensive horticultural crops, including potatoes, carrots and quinoa. In these cases, conservation practices, irrigation and high inputs of organic and inorganic fertilisers are required throughout the year. Contamination of soil and water from excessive use of fertilisers and other agrochemicals is affecting the soil biodiversity.

This ecoregion along with central Andean Puna and Dry Puna is characterised by significant poverty in the population, most of whom are indigenous.

Grassland Ecosystem Services in South America

Introduction

Ecosystem services include those processes and conditions within which Nature flourishes and meets the needs, material and otherwise, of humankind. These include tangible goods, such as timber, fibre, fuel wood, foods and medicines, as well as an array of environmental services that support life on earth, such as water purification, carbon dioxide absorption, biogeochemical cycling and many others (Voeks and Rahmatian, 2008).

Grasslands produce an array of goods and services for humankind but only a few of them have market value. Meat, milk, wool and leather are the most important products currently produced in grasslands that have a market value. Simultaneously, grassland ecosystems confer on humans many other vital, and often under-recognised services, such as maintenance of the composition of the atmosphere, maintenance of the genetic library, amelioration of weather and conservation of soil. In many cases, the value of services provided by grasslands in terms of production inputs and sustenance of plant and animal life may be larger than the sum of the products with current market value (Sala and Paruelo, 1997). This paper examines two of those services from grasslands in South America: primary production of goods and carbon sequestration.

The list of ecosystem services usually includes climate regulation, protection and regeneration of soil fertility, pest and flood control, purification of water and air, and crop pollination.

Grasslands play a unique role as they link agriculture and environment and offer tangible solutions, ranging from their contribution to mitigation and adaptation to climate change, to improvement of land and ecosystem health and resilience, biological diversity and water cycles while serving as a basis of agricultural productivity and economic growth (FAO, 2010).

They are a major ecosystem and a form of land use giving us not only a range of useful products (meat, milk, hides, fur, etc.) but also 'ecosystem services'. The latter include the important role of grasslands in biodiversity, provision of clean water, flood prevention and, carbon (C) sequestration, the focus of this book desertification (FAO Rome, 2009).

Soil carbon sequestration in grasslands may mitigate rising levels of atmospheric carbon dioxide (CO_2). Carbon accumulation in grassland ecosystems occurs mainly below the ground where soil organic matter (SOM) is located in discrete pools (Jones, 2009).

Primary Production

Two of the world's most important productive activities depend on grasslands: agriculture and forest industry. A large part of the seeds, beef, dairy products, wool, leather and lumber consumed worldwide is produced on grasslands.

The Pampas and Campos, the biggest biomes of South America, form the basis of the commodity-exporting economy of Argentina, Uruguay and part of Brazil. Table 14.7 shows the estimated productivity of the most important grasslands in different countries of South America. Additionally Table 14.8 presents detailed information on the biggest grasslands biomes in South America, the Pampas and Campos. The Pampas of the Río de la Plata biome, covers 760,000 square km, in much of central Argentina, almost all of Uruguay and part of the southern Brazilian state of Rio Grande do Sul. Pampas biome and farmland form the basis of the economy of these three commodity-exporting nations. The Pampas and Campos biomes provide feed for 43 million heads of cattle and 14 million sheep. The biome is habitat of 4,000 native plant species, 300 species of birds, 29 species of mammals, 49 species of reptiles and 35 species of amphibians. Table 14.8, shows some indicators of productive 14 biomes of Argentina, Brazil, Bolivia, Chile, Colombia, Paraguay and Uruguay. The indicators are dry matter hectare^{-1} and year^{-1}, Animal Unit ha^{-1}, weaned rate and live weight gain/day.

As can be seen, grasslands of native pastures of Argentina, Brazil and Uruguay are more productive than the others if Animal Unit (AU) defined as a bovine of 350 kg live weight or equivalent is an indicator, which is consistent with the DM produced hectare/year.

Table 14.8 presents general information about the cattle production in biomes Pampa and Campos. As can be seen, most parts of the productive unit of growth cattle (cow and calf) and most parts of productive units belong to the producers.

As can be seen in the following figure, South America is a continent that produces the largest amount of cattle in the world. Africa occupies the first place in other ruminants, such as buffaloes and the second place in cows. Most parts of the grasslands are used for grazing

Table 14.7 Productivity of natural grasslands in important regions of South America.

Country/Region	Grassland Type	Rainfall mm	Dry Matter t ha^{-1} y^{-1}	(Au) ha^{-1}	Weaning Rate, %	LWG ha^{-1}
Argentina						
Temperate	Native pasture	700–1,000	1.0–6.0	0.2–0.8	60–70	50–130
Subtropical	Native pasture	1,000–1,500	2.0–5.0	0.3–0.8	45–50	30–50
	"Chaco"	400–600	0.8	0.07	–	–
Brazil						
South	Native pasture	1,200–1,800	2.0–6.0	0.3–0.8	50–60	30–50
SE	Savannas	900–3,000	0.8–2.0	0.4–0.8	50–60	30–50
"Cerrados"	Savannas	800–1,800	1.5–3.0	0.2–0.8	50–60	20–35
North	Tropical W/D	2,000–3,000	–	–	–	–
NE	"Caatinga"	300–500	1.0–4.0	–	–	20–40
Chile						
- Temperate	Native pasture	400–2,300	1.0–3.0	0.1–0.5	60–70	20–50
- Dryland	Native pasture	–	0.3–0.5	–	–	–
Colombia						
	Native pasture	-> 2,000	2.0–3.0	0.2–0.5	45–55	20–30
Paraguay						
	Savannas	800–1,000	0.8–4.0	0.03–0.5	50–60	30–50
Uruguay						
	Native pasture	1,000–1,300	2.5–5.0	0.7–1.0	70–75	80–100
Bolivia						
	Savannas W/D	1,800	0.8–2.0	0.07–0.2	–	–

Notes: LWG: Live weight gain/day; AU: Animal Unit; DM: Dry Matter.

Source: Production Potential of South America Grasslands; E. Maraschin Depto. de Pl. Forrageiras e Agrometeoroloiga-UFRG.

and to produce forage to feed cattle. Projections from different sources, conclude that the domestic and international demand for meat will continue to increase due to changes in diet and consumer patterns, as well as changes in the income, mainly in highly populated countries like China. This means new pressures on grasslands. The dynamic growth of South American production of cattle is in line with these trends. The rapid increase in the production of South American cattle during the period 2000 to 2014, surpasses the African continent and North America.

Within South America, ruminant production is concentrated mainly in Brazil, followed by Argentina, Colombia, Uruguay and Panama. Cattle production is most important in all these countries, though in some of them, it diminished in the period 2000 to 2014, as for example, from 4 million to three million heads in Chile. The internal consumption has been covered by imports and the production of pigs and chicken has increased. A significant

Table 14.8 Characteristics of the sub-regions of the Río de la Plata grasslands region ordered by biome and country.

Biome	Country	Area (million ha)	Subregion	% Area with Native Grasslands	Average Farm Size (ha)	Farms with Cattle (%)	Ownership (%)	Main Production System
Pampas	Argentina	9.3	Flooding	68	605	93	70	Cattle (cow-calf)
		8.3	Southern	29	697	81	66	Crops-cattle (finishing)
		12.9	Subhumid	17	526	72	66	Crops
		1.5	Semiarid	7	824	89	70	Crops
		7.4	Rolling	33	222	48	59	Crops
		3.2	Mesopotamic	50	388	84	67	Crops-cattle (finishing)
Campos	Uruguay	2.8	West sediment	54	357	78	59	Cattle (finishing), crops
		3.2	Basalt	39	728	87	67	Cattle (cow-calf), sheep
		1.2	Gondwanic sediment	79	362	65	66	Cattle (cow-calf)
		2	Eastern sierras	79	294	92	67	Cattle (cow-calf), sheep
		2.2	Graven Merin	71	393	74	63	Cattle (cow-calf), rice
		1	Graven Santa Lucía	41	57	74	64	Cattle (full cycle), dairy, horticulture
		2.6	Crystalline shield	69	395	93	59	Cattle (full cycle), dairy
	Brazil	0.7	Litoral	64	126	68	81	Horticulture, rice, cattle (cow-calf)
		3.9	SE Sierras	54	56	79	82	Cattle (cow-calf), sheep
		2.9	Missons	31	53	74	86	Crops
		3.3	Central depression	44	53	71	84	Rice, cattle (finishing)
		2.4	Campanha	70	204	82	83	Cattle (cow-calf)
	Argentina	2.7	Corrientes	85	928	82	80	Cattle (cow-calf)
Total		73.5						

Source: Land use change and ecosystem service provision in Pampas and Campos grasslands of southern South America; P. Modernel, W. Rossing, M. Corbeels, S. Dogliotti, V. Picasso and P. Tittonell.

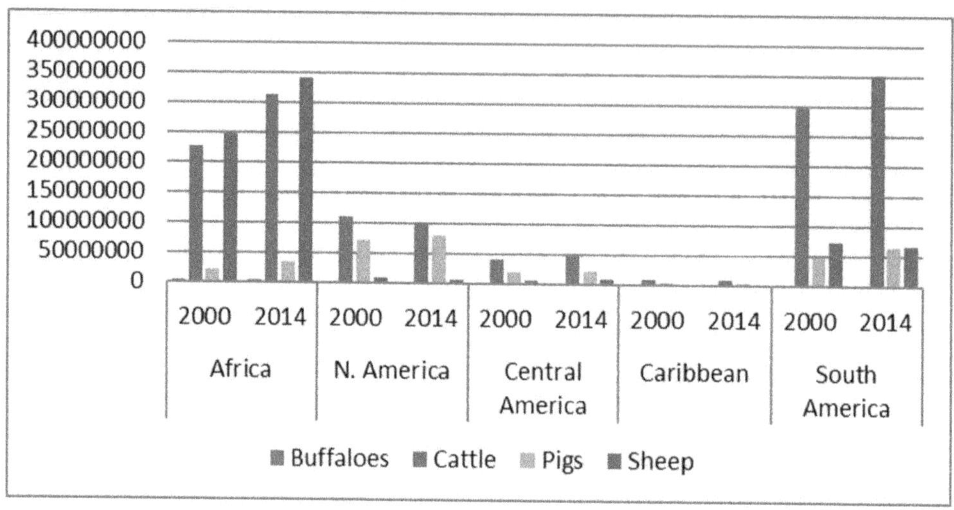

Fig. 14.3 Evolution of world main ruminant production (2000–2014).
Source: FAOSTAT, accessed in 2016.

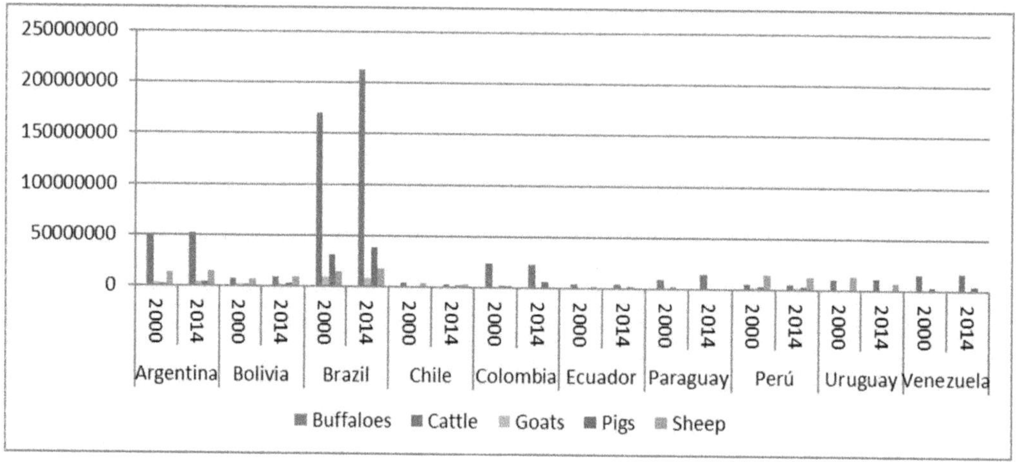

Fig. 14.4 South America: Evolution of production of ruminants during 2000 to 2014 countrywise.
Source: FAOSTAT, accessed in 2016.

part of cattle production is dependent on grasslands directly and/or through production of forage and other feeds for animals. The following figure illustrates the ruminant and cattle production countrywise for the period 2000 to 2014.

Grasslands Carbon Sequestration

Natural Grasslands Surface Area

Natural grasslands occur around the world and are characterised by using a number of methods. According to several authors, for a global characterisation, the methods can be

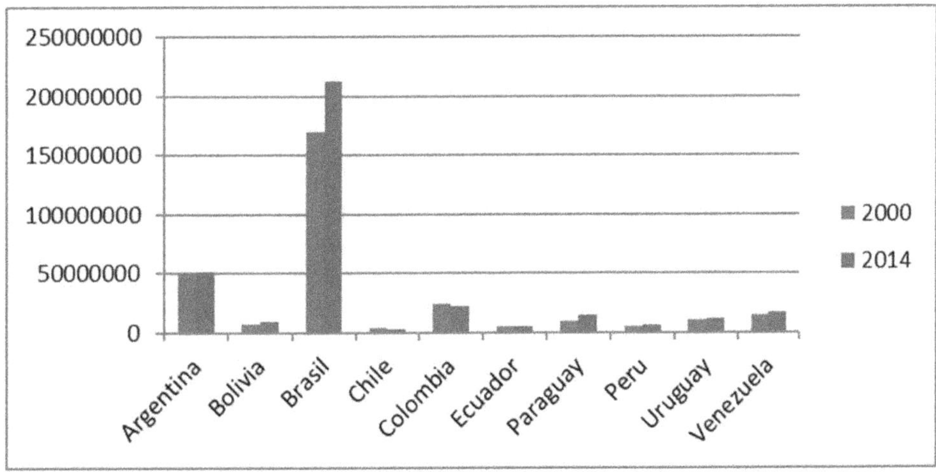

Fig. 14.5 South America: Evolution of cattle production during 2000 to 2014 countrywise. Source: FAOSTAT, accessed in 2016.

grouped into four types: vegetation composition; ecological and economic assessment; ecosystem mapping and remote sensing classification. The vegetation approach stresses the importance of species and growth forms as a primary expression of a terrestrial ecosystem and uses plant species assemblages to classify stands into plant community types (e.g., 'associations', 'alliances') and, combined with physiognomy, into broader vegetation types (e.g., classes, divisions, formations) (UNESCO, 1973; Ellenberg, 1988; Janssen and DiGregario, 1998; Faber-Langendoen et al., 2014).

The ecological and economic assessment approach characterises global grassland ecosystem health through an analysis of pressures exerted on the ecosystem and also reports on the connection to human well-being (Coupland, 1979; White et al., 2001; Suttie et al., 2005). The ecosystem-mapping approach emphasises the geographical or landscape delineation of ecosystem boundaries based on patterns present in biophysical factors, such as climate, landform and, sometimes, floral and faunal evidence (Holdridge, 1967; Uvardy, 1975; Walter and Box, 1976; Schultz, 1995; Bailey, 1996; Olson et al., 2001). The remote sensing method uses the vegetation approach in combination with satellite imagery to create global land cover datasets to describe the generalised spatial patterns in vegetation, that is abiotic and anthropogenic features on the earth's surface (Defries et al., 1995; Loveland and Belward, 1997; Bontemps et al., 2011).

The authors chose to develop a map of global grassland distribution using a combination of the vegetation approaches represented by the International Vegetation Classification (IVC), and a spatially explicit landscape-based approach, as manifested in the Terrestrial Ecoregions of the World (TEOW) framework. Both the systems offer a robust, hierarchical approach to describing global grassland biodiversity.

- International Vegetation Classification

  The International Vegetation Classification (IVC) (formerly called the International Classification of Ecological Communities or ICEC) is based on vegetation as it currently exists on the landscape. Landforms, soils and other features are not directly considered a part of the classification criteria, but ecological and biogeographical information help guide the structure of the classification. Because of conservation objectives, classification efforts focus on natural and semi-natural types of vegetation and *NatureServe Explorer* reports only these types. However, the system can be used to classify any vegetation as it currently exists across the landscape, including natural, modified, agricultural and urban vegetation types.

- Terrestrial Ecoregions of the World (TEOW)

  Here the biogeographic regionalisation of the Earth's terrestrial biodiversity is studied. The biogeographic units are ecoregions, which are defined as relatively large units of land or water containing a distinct assemblage of natural communities, sharing a large majority of species, dynamics, and environmental conditions. There are 867 terrestrial ecoregions, classified into 14 different biomes, such as forests, grasslands or deserts. Ecoregions represent the original distribution of distinct assemblages of species and communities.

Grassland Types in South America

The International Vegetation Classification (IVC), describes 46 global IVC grassland divisions that includes the Mediterranean Basin Dry Grassland IVC division, occurring in both Africa and Eurasia. Among these, almost 35 per cent belongs to South America.

Table 14.9 IVC divisions with dominant grassland types per continent.

Continent	Number of Dominant Grasslands Types	%
South America	16	34.8
Africa–Madagascar	12	26.1
Eurasia	9	19.6
Australia	5	10.9
North America	4	8.7
Total	46	100.0

Source: The International Vegetation Classification (IVC).

A more detailed approach is presented by Dixon, Faber-Langendoen, Josse, Morrison and Loucks 2014. They matched the TEOW framework with IVC divisions and identified 49 taxonomically and spatially distinct grassland types, creating a new global biogeographical representation of the Earth's grassland types. In these, 12 are located in South America. From these 12 grassland types, the total carbon sequestration and the rate of annual sequestration was estimated. The applied parameters for each grassland type were obtained mainly from the following sources:

- Good Practice Guidance for LULUCF, Chapter 3
- Grassland Afforestation in Southern South America: Carbon sequestration potential. E. Jobbágy, M. Nosetto, Universidad Nacional de San Luis, Argentina, K. Farley and R. Jackson, Dept Biology, Duke University, G. Piñeiro and J. Paruelo IFEVA, Universidad de Buenos Aires, Argentina
- Grasslands Carbon Sequestration; Management, Policies and economics, FAO, 2010.

Measuring Carbon Sequestration

IPCC has provided a framework for estimating and simulating emission reductions resulting from grassland management. Their approach makes it possible to estimate any change in SOC storage by assigning a reference C stock (total C stock in soil), which varies according to the climate, soil type and other factors, and then multiplying that value by factors representing the quantitative effect of changing grassland management on SOC storage. In order to develop such factors, IPCC analysed data from 49 studies that isolated the management effect (Ogle et al., 2004), discriminated the study sites by climate regions (temperate and boreal, tropical and tropical mountains) and derived coefficients for estimating changes in SOC stocks over a finite period following changes in management that impact SOC storage. In this study, data were compiled from the literature that furnished information on SOC stock rate change. These data are summarised by climate zone, management status and main grassland typology. Additionally, research made in specific places and biomes in South America produced carbon stock and carbon rates of sequestration that are applied in this paper. The obtained result reveals the enormous potential of grasslands in South America to contribute to store carbon.

Five of the most important biomes of South America, the Pampa with and without water, the Patagonia wet and the Patagonia dry, plus the Campos biome in Brazil were taken into consideration.

Table 14.10 Potential of regional carbon sequestration considering Pampa, Patagonia and Campos Biomes.

	Pampas + WG	**Pampas - WG**	**Patagonia Wet**	**Patagonia Dry**	**Campos**
N sites	4	5	12	5	Extrapolation
Mean Age	50	35	15.6	17.5	35
Δ Biomass (Mg Ha^{-1})	287	175	56	21	175
Δ Soil Organic matter (Mg Ha^{-1})	24	17	–13	2	17
Annual C (Mg Ha^{-1} $Year^{-1}$)	6.2	5.5	2.8	1.3	5.5
Total Area (Km^2)	1,58,000	3,15,000	20,000	31,000	40,700
Remnant Grassland (%)	85	5	95	100	40
Potential C Sequestration (Tg $Year^{-1}$)	83.5	8.7	5.3	4.1	89.6

Source: Grassland afforestation in Southern South America: Carbon sequestration potential.

As can be seen, the area of five biomes considered, reach 56.47 million hectares that can sequester 191.2 millions tons of Carbon each year at a rate of 3.39 tons per hectare per year. Considering a wider definition for grasslands, as this resulting from the matching of IVC and TEOW classification, which define a total of twelve biomes for South America, total carbon sequestration by year can reach 23.2 billion tons, as can be seen in the next table.

Table 14.11 Total area of five grasslands, their potential to C sequestration and the annual sequestration rate.

Total Area Grasslands (Hectares)	56,470,000
Total Potential C sequestration (Ton Year^{-1})	191,200,000
Annual Sequestration (Ton/Ha)	3,39

Source: Table 10.

Table 14.12 Carbon sequestration considering 12 grasslands defined by matching IVC and TEOW classifications.

IVC Division	Area (Ha)	C/ha (Stock)	C Total (Tons)
Brazilian-Parana lowland, shrubland, grassland and savanna	203,562,700	34	6,921,131,800
Pampean grassland and shrubland	75,105,100	44	3,304,624,400
Patagonian grassland and shrubland	55,480,400	95	5,270,638,000
Colombian-Venezuelan lowland shrubland, grassland and savanna	37,578,700	39	1,465,569,300
Patagonian cool semi-desert scrub and grassland	35,364,000	95	3,359,580,000
Pampean grassland and shrubland (semi-arid Pampa)	29,873,500	5	149,367,500
Tropical Andean cool semi-desert scrub and grassland	25,492,900	34	866,758,600
Tropical Andean shrubland and grassland	16,160,200	34	549,446,800
Guianan lowland shrubland, grassland and savanna	10,449,600	39	407,534,400
Amazonian shrubland and savanna	9,598,600	66	633,507,600
Guianan Montane shrubland and grassland	2,753,400	66	181,724,400
Brazilian-Parana Montane shrubland and grassland	2,624,700	34	89,239,800
Total	504,043,800		23,199,122,600

Source: Elaborated on the basis of the International Vegetation Classification (IVC).

Final Comments

Grasslands in South America are composed by a wide range of ecosystems and vegetation types that range from desert areas to steppes, sub-humid temperate, subtropical and tropical savannas and even portions of the tropical rain-forest environment.

Grasslands in South America cover one of the largest rangeland surfaces on earth—504 million hectares—and, in general, are developed ecosystems that have agronomic and ecological activities.

Applying the most conservative index of Animal Units per hectare, a total of 151 million cattle heads is the estimated annual population on 504 million hectares. In terms of carbon sequestration, it is estimated that grasslands can accommodate a total of 23.2 billion tons by year. The population growth, good diet and income increase, accompanied by an increase in the world demand for meat, are met partly by South American grasslands (Squires, 2018). South American grasslands are currently and will continue in the future to be a major source of feed for a wide range of grazing animals in arid, semiarid, sub-humid, humid, flat lands, valleys and mountains in remote areas for a long time. In spite of the changes in land use in favour of crops in many parts of the continent, the natural pasture will continue to be exploited to feed animals and one of the most important sources of economic returns.

Finally, grassland biomes in the world, and certainly in South America, are being exposed to intense pressures for land use change in favour of crops, mining activities, urban expansion, etc., endangering their sustainability and affecting severely their flora and fauna and, as a consequence, the environmental services are produced.

References

Ashley, S., S. Holden and P. Bazeley. 1999. Livestock in Development. U.K.: Outhouse Publishing Services.

Bailey, R. 1996. Ecosystem geography. New York: Springer.

Bontemps, S., P. Defourny, E. Van Boagert, O. Arino, V. Kalogirou and E. Perez. 2011. Retrieved from Globcover 2009 products description and validation report. European Space Agency & Université catholique de Louvain.: http://due.esrin.esa.int/globcover/LandCove.

Christopher, K. Wright and Michael, C. Wimberly. 2013. Recent land use change in the Western Corn Belt threatens grasslands and wetlands Geographic Information Science Center of Excellence, South Dakota State University.

Clay, J. 2004. World Agriculture and the Environment. A Commodity-by-Commodity Guide to Impacts and Practices. Islands press.

Conant, R.T., K. Paustian and E.T. Elliot. 2001. Grassland management and conversion into grassland: effects on soil carbon. Natural Resource Ecology Laboratory, Colorado State University, Fort Collins.

Coupland, R.T. (ed.). 1979. Grassland ecosystems of the world: IBP Vol. 18, Cambridge University Press, , 401pp.

Defries, R., M. Hansen and J. Townsend. 1995. Global discrimination of land cover types from metrics derived from AVHRR Pathfinder data. Remote Sensing of the Environment. Environment 54: 209–222.

Dixon Faber-Langendoen et al. 2014. EcoVeg: a new approach to vegetation description and classification. Ecological Monographs 84(4): 533–561.

Ellenberg, H. 1988. Vegetation ecology of Central Europe. Fourth edition, English Translation. Cambridge University Press, Great Britain.

Faber-Langendeon, D., T. Keeler-Wolf, D. Meidinger, D. Tart, B. Hoogland, C. Josse., . . . P. Comer. 2014. EcoVeg: a new approach to vegetation description and classification. Ecological Monographs, 533–561.

FAO. 2009. Land and Water Resources for Food and Agriculture Managing systems at risk. Rome: FAO.

FAO. 2009. Grassland carbon sequestration: management, policy and economics. Proceedings of the Workshop on the role of grassland carbon sequestration in the mitigation of climate change Integrated Crop Management (pp. Vol. 11–2010). Rome: FAO.

Galdino, S., L. Vieira and L. Pellegrin. 2006. Impactos Ambientais e Socioeconômicos na Bacia do Rio Taquari – Pantanal. Corumbá,. Embrapa Pantanal. Corumbá,, Mato Grosso Sul, Brasil: EMBRAPA.

Gibbs, H.K. and J.M. Salmon. 2015. Mapping the world's degraded lands. Applied Geography 57: 12–21.

Guimarães, L. 2013, October 3 October. Nutrient management challenges in Brazil and Latin America. The second global conference on Land-Ocean Connections (GLOC-2). Montego Bay, Jamaica.

Holdridge, L. 1967. Life zone ecology. Tropical . San Jose, Costa Rica: Science Center.

Janssen, L.J.M. and A. DiGregario. 1998. Land Cover Classification System (LCCS): classification concepts and user manual. Environment and Natural Resources Service, GCP/RAF/287/ITA. Africover – East Africa Project and Soil Resources, Management and Conserv. Africa: FAO.

Jobbágy, E., M. Nosetto et al. 2007. Grassland afforestation in Southern South America: Carbon sequestration potential & soil/water costs.

Jones, M.B. 2009. Potential for carbon sequestration in temperate grassland soils. Chapter I. *In*: Grassland carbon sequestration: management, Grassland carbon sequestration: management, policy and economics Proceedings of the Workshop on the role of grassland carbon sequestration in the mitigation of climate change.

Lavado, R. and M. Taboada. 2009. The Argentinean Pampas: A key region with a negative nutrient balance and soil degradation needs better nutrient management and conservation programs to sustain its future viability as a world agroresource. Journal of Soil and Water Conservation 65: 150–153.

Loveland, T. and A. Belward. 1997. The IGBP-DIS global 1 km land cover data set, DISCover: first results. International Journal of Remote Sensing 18: 3289–3295.

Loveland, T., B. Reed, J. Brown, D. Ohlen, Z. Zhu, L. Yang and J.W. Merchant. 2000. Development of a global land cover characteristics database and IGBP Discover from 1 km AVHRR data. International Journal of Remote Sensing 21(6-7): 1303–1330.

Maraschin, G.E. 2001. Production potential of South America grasslands. Depto. de Plantas Forrageiras e Agrometeoroloiga, Universidad Federal do Rio Grande do Sul, Porto Alegre, Brasil.

Miyake, S., M. Renouf, A. Peterson, C. Mcalpine and C. Smith. 2012. Land-use and environmental pressures resulting from current and future bioenergy crop expansion: A review. Journal of Rural Studies. 28: 650–658.

Modernel, P. et al. 2016. Land use change and ecosystem service provision in Pampas and Campos grasslands of southern South America. Environmental Research Letters 11(11).

Ogle, S., R. Conant and K. Paustian. 2004. Deriving grassland management factors for a carbon accounting method developed by the Intergovernmental Panel on Climate Change. Environ. Manage. 33(4): 474–484.

Oldeman, L.R. 1994. The global extent of soil degradation. in CAB International. pp. Pages 99–118. *In:* D.J.D.J. Greenland. Soil Resilience and Sustainable Land Use. Wallingford, UK.

Olson, D., E. Dinerstein, E. Wikramanayake, N. Burgess, G. Powell, E. Underwood, . . . K. Kassem. 2001. Terrestrial ecoregions of the world: a new map of life on Earth. BioScience 51(11): 933–938.

O'Mara, F.P. 2012. The role of grasslands in food security and climate change. The Irish Agriculture and Food Development Authority, Research Directorate, Head Office, Oak Park, Carlow, Ireland.

O´Mara, F.P. 2012. The role of grasslands in food security and climate change. Annals of Botany Nov 110(6): 1263–1270.

Paruelo, J., J. Guerschman and S. Verón. 2005. Expansión agrícola y cambios en el uso del suelo. Revista Ciencia Hoy 15(87): 14–23.

Ramankutty, N., A.T. Evan, C. Monfreda and J.A. Foley. 2008. Farming the planet: 1. Geographic distribution of global agricultural lands in the year 2000. Global Biogeochemical.

Raymond, S. Bradley, Mathias Vuille, Henry F. Diaz and Walter Vergara. 2006. Threats to water supplies in the tropical andes. Science Vol. 312, 23 June 2006.

Reid, R.T., P. Thornton, G. McCrabb, R. Kruska, F. Atieno and P. Jones. 2004. Is it possible to mitigate greenhouse gas emissions in pastoral ecosystems of the tropics? Environment, Development and Sustainability 6: 91–109.

Richard, T. Conant. 2006. Power Point. Grassland degradation—a global perspective. The Case of Greenhouse Gas Mitigation.

Sainz Rozas, H., H. Echeverría and H. Angelini. 2011. Organic carbon and pH levels in agricultural soils of the pampa and extra-pampean regions of Argentina. Ciencia del Suelo 29: 29–37.

Sala, O.E. and J.M. Paruelo. 1997. Ecosystem services in grasslands. Ecosystem services. Washington, DC, Island Press.

Schultz, J. 2005. The ecozones of the world. New York.: Springer-Verlag.

Satorre, E. 2005. Cambios tecnológicos en la agricultura Argentina actual. Revista Ciencia Hoy 15(87): 24–31.

Schlesinger, W.H. 1977. Carbon balance in terrestrial detritus. Annual Review of Ecology and Systematics 8: 51–81 (Volume publication date November 1977).

Sousa, D. 2011. Recurso fundamental para o desenvolvimento da Agricultura na Savana tropical. EMBRAPA-Cerrados. XIX Congreso Venezolano de la Ciencia del Suelo.Calabozo,. Calabozo, Venezuela.

Squires, V.R. et al. 2010. Interactions: Food, Agriculture and Environment—Volume II, Encyclopedia of Life Support Systems.

Squires, V.R. 2015. Rangeland Ecology, Management and Conservation Benefits. NOVA Science Publishers, N.Y.

Squires, V.R. Commentary on beef production systems in Brazil, Uruguay and Argentina. *In*: Squires, V.R. and Bryden, Wayne L. (eds). Livestock: Production, Management strategies and Challenges. NOVA Science Publishers, New York, USA (In Press).

Steinfeld et al. 2006. FAO, 2006. Livestock´s long shadow. Environmental Issues and Options.

Suttie, J., S. Reynolds and C. Botello. 2005. Grasslands of the World. FAO Plant Production and Protection Series No. 34, 2005. Rome: FAO.

Suttie, J.M., S.G. Reynolds and C. Batello (eds.). 2005. Food and agriculture organization of United nations. Grasslands of the Worlds. Plant Production and Protection Series No. 34 June 2006, 117(1): 109–134.

UNEP. 2006. Deserts and desertification. Don't desert drylands! World Environment Day . Nairobi, Kenya: UNEP.

UNESCO, United Nations Educational, Scientific and Cultural Organization. 1973. International classification and mapping of vegetation, UNESCO. UNESCO Ecology and Conservation Series 6., pp. 1–93.

Urquiaga, S.A., B. Alves, C. Antalia, M. Martins and R. Boddey. 2014. A cultura de milho e seu impacto nas emissões de GEE no Brasil. . pp. 1 edition. . (pp. 61–71.). Sete Lagoas, MG, ABMS.

Uvardy, M. 1975. A classification of the biogeographical provinces of the world. IUCN Occasional Paper no. 18. IUCN. Morges, Switzerland.: International Union for Conservation of Nature, IUCN.

Vera, R. 2005. The future for savanna and tropical grasslands: A Latin American perspective Article January 2005. Pontifical Catholic University of Chile.
Viglizzo, E. and F. Frank. 2006. Land-use options for Del Plata Basin in South America: Tradeoffs analysis based on ecosystem service provision. Economic Ecology 57: 140–151.
Viglizzo, E.F. and F. Frank. 2006. A rapid method for assessing the environmental performance of commercial farms in the Pampas of Argentina. Environmental Monitoring and Assessment.
Viglizzo, E.F. and E. Jobbágy. 2006. Expansión de la Frontera Agropecuaria en Argentina y su Impacto Ecológico-Ambiental Editores. INTA, Argentina.
Viglizzo, E.J. and E. Jobbágy. 2010. Expansión de la frontera agropecuaria en Argentina y su impacto ecológico-ambiental. Buenos Aires: Ediciones INTA.
Voeks, R.A. and M. Rahmatian. 2004. The providence of nature: Valuing ecosystem services, California State University.
Walter, H. and E. Box. 1976. Global classification of natural terrestrial ecosystems. Vegatatio 32: 75–81.
White, R., S. Murray and M. Rohweder. 2000. World Resources Institute, WRI. PILOT ANALYSIS OF GLOBAL ECOSYSTEMS. Grassland Ecosystems. WRI.
WOCAT. 2009. Assessing impacts of different grassland systems on land degradation and conservation.
World Bank. 2007. World Development Indicators. Washington DC. World Bank.
World Wild Life. Southern South America: Eastern Argentina. En https://www.worldwildlife.org/ecoregions/nt0803.
Wright, C.K. 2012. Recent land use change in the Western Corn Belt threatens grasslands and wetlands. Proceedings of the National Academy of Science of United States 110(10).
WWF, W.W. 2016. Living Planet Report 2016; Risk and resilience in a new era. WWF.

15

Grasslands in Australia

An Overview of Current Status and Future Prospects

Zhongnan Nie[a,*] and *David Chapman*[b]

Introduction

Australia is one of the hottest and driest continents in the world in terms of duration/intensity of heat and annual rainfall. Extreme air temperatures (> 45°C) have been recorded at almost all stations more than 150 km from the coast and in many areas on the northwest and southern coasts (NATMAP, 1986). Half of Australia's total land area has an average annual rainfall of < 300 mm (Fig. 15.1); consequently, a large part of central Australia is desert or semi-arid. The northern part of the continent is largely tropical rainforests, tropical and sub-tropical grasslands and desert. The more intensive farming systems are applied in the coastal regions and the south-east and south-west corners of Australia with a temperate climate (Nie and Norton, 2009; Nie et al., 2016).

Grasslands cover approximately 70 per cent of Australia and support the native game hunted and native plant foods harvested by the original Aboriginal inhabitants and the domestic grazing animals introduced by European colonists[1] (McIvor, 2005). The term 'grasslands' is defined as all herbaceous communities that are used for livestock (Moore, 1970), including both native pastures and pastures composed of mainly introduced plant species, either sown or volunteer. Grasslands are the dominant ecosystems and play an important role in agricultural production. The livestock industry accounts for more than 45 per cent of Australia's total value of agricultural production, within which grazing by ruminants to produce meat, fiber and dairy is the largest sector, both in dollar value and

[a] Department of Economic Development, Jobs, Transport & Resources, Private Bag 105, Hamilton, Victoria 3300, Australia.

[b] Dairy NZ, PO Box 160, Lincoln University, 7647, New Zealand.

* Corresponding author: Zhongnan.Nie@ecodev.vic.gov.au

[1] In some areas, this occurred for the first time as late as the 1880s.

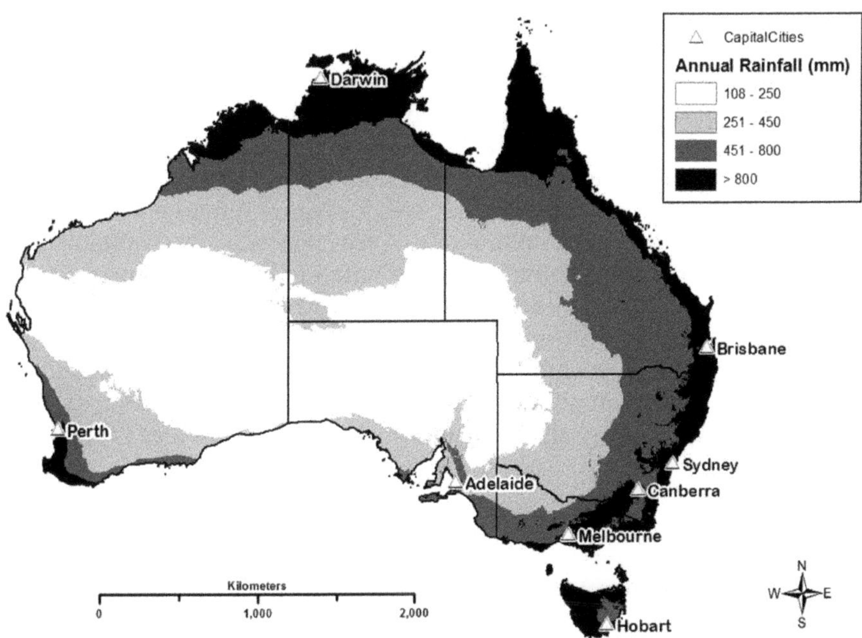

Fig. 15.1 Banded annual rainfall distribution in Australia (adapted from Nie et al., 2016).

land area (Nossal et al., 2008). Over 60 per cent of Australia's total land area is grazed by sheep and cattle (Hutchinson, 1992). Historically, the most important feature of Australian agriculture has been the evolution and expansion of sown pasture technologies, which commenced in the 1920s and peaked nationally in the early 1970s at 28 million ha (Donald, 1965; ABS).

There are currently many challenges/problems faced by the grazing industries and research, development and extension personnel attempting to address these issues in Australia. These have been largely associated with the aims to increase the grassland productivity, improve sustainability and environmental conservation and achieve an ecosystem balance between the two. Since the early 1980s, the economic and environmental pressures have brought about considerable changes across the grassland production systems (Bell et al., 2013). Much effort has been put into tackling ecological issues, such as dryland salinity, loss of biodiversity, poor ground cover, soil erosion and water and nutrient losses from agricultural land. Over the past decade, adaptation responses to climate change, mitigation strategies for agriculture's greenhouse gas emissions and carbon sequestration in grassland and cropping systems have been important topics in grassland research. Advances in information technology, precision and remote management, and genetic techniques, etc., have provided new opportunities for grassland research and management. However, the availability of grassland scientists and sources of funding for grassland research have been declining (Robson, 2013), which could be one of the biggest challenges for the grazing industry as well as ecological services in the future.

This chapter briefly reviews the current status of Australian grasslands including the types/zones of pastures and the problems and challenges associated with the pastures. Based on these analyses, the opportunities and strategies for future research and development (R&D) are proposed.

Grassland Production Systems in Australia

The grassland production systems in Australia cover a diverse range of environments from native rangelands with very low stock-carrying capacity in the arid interior to rainfed or irrigated improved pastures with high stock-carrying capacity in the south-eastern and south-western regions near the coast. The adoption of these systems in a region has been affected by many factors, such as the climate, soil, topography, markets, labour availability and access to information and technology. Conventionally, the systems have been considered in three geographic zones associated with rainfall by the Australian Bureau of Agricultural and Resources Economics (ABARE), which include the pastoral zone (PZ), the wheat (*Triticum aestivum*)-sheep (*Ovis aries*) zone (WSZ) and the high rainfall zone (HRZ) (McIvor, 2005). More recently, Bell et al. (2013) added a fourth production system—the intensive dairy system, in addition to the three zones through analysis of 10 broad agro-ecological regions in Australia (Williams et al., 2002). These four systems are briefly described below.

Pastoral Zone

The pastoral zone includes the arid (severe lack of available water) and semi-arid (precipitation below potential evapotranspiration but above arid/desert climate) regions and most of the northern tropical areas (Fig. 15.1). It covers 242 million ha of land and supports 63.3 million dry sheep equivalents (DSE), which is about 30 per cent of Australia's total grazing livestock (Bell et al., 2013). It is an extensive production system based on native vegetation with mostly wool production in southern Australia and beef production in central and northern Australia. The farm size is large, ranging from 5,000 ha to over 1,00,000 ha; however, the stocking rates are low (2–40 ha per sheep) (McIvor, 2005).

Native grassland communities in this zone include primarily Mitchell grass (*Astrebla* spp.) tussock grasslands, Spinifex (*Triodia* spp.) hummock grasslands, Eucalypt woodlands with wire grass (*Aristida* spp.) and blue grass (*Dichanthium* spp. and *Bothriochloa* spp.), and tall grass savannas with black spear grass (*Heteropogon contorus*), ribbon grass (*Chrysopogon* spp.) and native sorghum (*Sorghum* spp.) (Tothill and Gillies, 1992). Sown pastures are of very restricted and minor importance in most areas of the zone although they do play an important role in eastern and central Queensland and limited areas in the 'top end' of northern territory. Due to low rainfall and frequent droughts, the growing season of pastures is unreliable, leading to fluctuation in production and financial returns between years.

Over the decades ending 2010, improvements in pasture management (fencing, increased water points and better grazing management), supplementary feeding and the structure of production enterprises have led to increase in livestock densities and turn-off rates by 40 per cent (Bell et al., 2013). In the southern rangelands of the zone, there has been a strong move from the traditional Merino wool sheep to beef cattle and meat sheep breeders due to the declining profitability of wool. In regions such as western NSW and south-west Queensland, goats are also replacing or supplementing traditional enterprises. However, these changes have also brought about concerns on environmental sustainability. Since there has been little cropping hitherto in most areas, the zone is generally considered 'undisturbed' with valuable ecological value. The increased grazing intensity and changes in enterprise structure put pressures to integrate both production and ecological goals (Fitzhardinge, 2012). In response, the National Rangelands Natural Resource Management (NRM) Alliance was formed to improve and coordinate delivery of NRM goals across the zone (Forrest et al., 2010). In the last 20 years ending 2012, the areas used by grazing enterprises across PZ have been reduced partly due to increase of areas used for conservation, tourism and by Aboriginal (indigenous) communities (Bell et al., 2013).

Wheat-sheep Zone

The wheat-sheep zone is also named 'crop-livestock zone' given that mixed farming between crops and livestock is the dominant production system in this zone. The zone extends from south-eastern Queensland through the slopes and plains of NSW, northern Victoria and South Australia, to southern-western Australia. The annual rainfall ranges from 250 mm in part of Western Australia and South Australia to 750 mm in the subhumid subtropical slopes and plains of south-eastern Queensland with summer dominant rainfall (McIvor, 2005). The crops grown in this zone mainly include wheat, barley (*Hordeum vulgare*), oats (*Avena sativa*), grain legumes, pulses, oil seeds and sorghum (*Sorghum bicolor*; predominantly grown in the northern areas), and major animals include wool sheep, prime lamb and beef cattle (*Bos Taurus*). The self-regenerating Subterranean clover (*Trifolium subterranean*) is widely used on more acidic soils and annual medics (*Medicago* spp.) on more alkaline soils. The pasture phase is used between the sowing of crops. The farms with integrated crop-livestock (ICL) systems are mostly family owned with more stable income due to more reliable rainfall/seasonal conditions than in the PZ and the flexibility of multiple income sources.

There are various forms/types of ICL systems in WSZ and many names and terminology have been used in Australia to describe these systems. Nie et al. (2016) analyzed these systems and their terms commonly used in Australia and classified them into four major categories—crop-pasture rotation, crop-pasture intercropping, dual purpose crops and alley farming. Crop-pasture rotation is a most common type of ICL system, describing alternating phases of crop and pasture. The terms 'ley farming' and 'phase farming' are commonly used when describing crop-pasture rotation systems in Australia. Crop-pasture intercropping involves growing crop and pasture species together on the same paddock and at the same time. Other terms that are used in Australia and belong to this category include over-cropping, companion cropping, pasture cropping and native pasture cropping. Dual-purpose crops include temperate cereals and winter rape that are often grown to provide autumn/winter grazing for livestock at the vegetative stage of the plants. The animals are then removed to allow the crops to reach physiological maturity for grain harvest. While the dual-purpose crops provide valuable feed for animals in winter when pasture growth is slow, grazing dual-purpose crops may (e.g., Kelman and Dove, 2009) or may not (Virgona et al., 2006) have a negative impact on grain production. Dual-purpose crops research has become a hot topic in recent years. Alley farming involves planting annual crops or pasture between rows of trees/shrubs that can be grazed/used for forage. This system aims to use the different growth pattern of crops and pastures to achieve complementary outcomes and is essentially a type of agro-forestry.

There has been a shift towards increased cropping and farm size in the WSZ since 1990s, largely due to the low profitability of sheep enterprises. The average farm size increased from 1,700 ha in 1,990 to 2,300 ha in 2011 in the zone, and the proportion of cropped area has increased by 50 per cent in the southern and western Australia part of the zone (Bell et al., 2013). There has been less change in cropped area in the Queensland and NSW regions of the WSZ; however, sheep have been replaced by cattle in these regions.

High Rainfall Zone

The high rainfall zone by its name receives high annual rainfall with much longer growing season than the WSZ and PZ. It covers a large part of the coastal belt and adjacent tablelands of south-eastern Australian mainland, Tasmania, and small areas of south-eastern South Australia and south-western Western Australia with winter-dominant rainfall (annual rainfall > 600 mm) and the coastal regions and tablelands of northern New South Wales and Queensland with summer-dominant rainfall (annual rainfall > 750 mm) (Ewing and Flugge, 2004; Bell et al., 2013). Pastures commonly grown in this zone are diverse and include many improved temperate perennial [e.g., phalaris (*Phalaris aquatic*), tall fescue (*Lolium arundinaceum*), cocksfoot (*Dactylis glomerata*) and lucerne (*Medicago sativa*)] and annual (e.g., annual clovers and fodder crops) species in the south and tropical/sub-tropical species (e.g., *Brachiaria decumbens*, *Digitaria decumbens* and *Panicum maximum*) in the north. These improved pastures sown on 6 per cent of Australia's grazing lands primarily in the HRZ and the intensive dairy system (see next section) support 41 per cent of Australian domestic livestock (Nie and Norton, 2009). Native pastures are mostly native grasses, such as wallaby grass (*Rytidosperma* spp.), spear grass (*Austrostipa* spp.), weeping grass (*Microlaena stipoides*), kangaroo grass (*Themeda triandra*) and red-leg grass (*Bothriochloa macra*), which are often seen in low fertility and marginal land classes (Nie et al., 2015).

Over the past couple of decades, land use patterns on the farms of the zone have changed, largely due to changes in the prices of animal and crop products. In the temperate regions of this zone, there has been a move from traditional wool production to prime lamb and beef cattle production. Increasing grain and declining wool prices and development of new crop varieties and management strategies have also led to an expansion of cropping in the HRZ of southern Australia (Nie et al., 2016). This trend will continue in the foreseeable future.

Intensive Dairy System

The intensive dairy system is limited to areas with high rainfall or medium rainfall with irrigation over a range of environments—from the wet tropical tablelands in northern Queensland to the temperate tablelands and coastal regions of New South Wales, Victoria, Tasmania, South Australia and Western Australia. Victoria is the major dairy-producing state with 6,390 million liters of fresh milk produced in 2014–15, over 65 per cent of the national total (http://www.dairyaustralia.com.au/Markets-and-statistics/Production-and-sales/Milk.aspx—Dairy Australia statistics). Perennial pastures and annual forage crops [e.g., perennial ryegrass (*Lolium perenne*), tall fescue, white clover (*Trifolium repens*) and fodder crops] are the major source of feed for dairying with perennial ryegrass pasture supplying 60–70 per cent of the diet for lactating cows in southern Australia (Chapman et al., 2008; Tharmaraj et al., 2008). National dairy production rapidly grew from late 1980s (5,400 million liters) to early 2000s (> 11,200 million liters), and since then declined and remained between 9,000 and 10,000 million liters annually (Dairy Australia statistics). The decline in dairy production was at least in part attributed to the significant restrictions in irrigation water in the irrigated dairy regions of southern Australia from mid 2000s (Chapman et al., 2012).

Current Grassland Issues and Challenges

Over the past 30 years, economic and environmental pressures have brought about considerable changes in the diversity of Australia's grassland systems and the focuses for grassland research and development.

Balance Between Grassland Productivity and Sustainability

For a couple of decades before late 2000s, grassland management was focused on the ecological and environmental issues, such as dryland salinity, biodiversity, loss of groundcover and water/nutrient runoff. For instance, the Cooperative Research Centre (CRC) for Plant-based Management of Dryland Salinity was a seven-year national programme established in early 2000s to tackle the issues/problems associated with dryland salinity affecting 5.7 million hectares of land in southern Australia (Dear and Ewing, 2008; Hughes et al., 2008). The CRC was a partnership of State Department of Agriculture in NSW, Vic., SA and WA, CSIRO and many universities, aiming to control dryland salinity using deep-rooted perennial pastures over broad geographic scales of the landscape. Through extensive research and collaboration between the partners, the CRC collected a large quantity of data and information on salinity and its related problems and developed new perennial pasture plants and farming systems to address these issues (Dear and Ewing, 2008). The CRC was renewed at the end of its term with a changed name (Future Farm Systems CRC) until 2014 and refocused on productivity in farming systems. Over the past few years, grassland R&D projects in Australia have given priority both to productivity and environmental sustainability (e.g., mitigation of greenhouse gas emission and carbon sequestration) projects.

Indeed, grassland productivity and sustainability do not always complement each other, and sometimes increase in productivity is at the cost of environmental degradation and vice versa. Grassland management to increase pasture production such as fertilizer application, use of herbicide and pesticide and grazing management can, if misused, lead to negative impacts on the ecosystems and environment. Among a range of management practices, removal of nutrient deficiency, appropriate grazing management and control of weed and pests are the key factors that contribute to stress tolerance, plant survival and pasture production given widespread nutrient deficiency and unreliable rainfall and extreme weather conditions in Australia (Kemp, 1994). Soil fertility is the most important factor in agriculture in the sense that it is easiest to manipulate. Correction of any nutrient deficiency through fertilizer application is crucial to achieve optimum botanical composition, production and longevity of pastures in grazing systems (McCaskill and Quigley, 2006). Nitrogen (N) deficiency and declining use of phosphorus (P) are the key limitations to grassland productivity in Australia, despite the extensive use of legumes supported by P application and direct application of N fertilizer in more intensive systems, such as dairying (Simpson et al., 2011; Bell et al., 2013). However, when nutrient deficiency is minimal in the soil, further fertilizer application may significantly increase pasture yield (McKenzie et al., 2002; Nie et al., 2004, 2015), but lead to nutrient losses to rivers and groundwater (Gourley and Weaver, 2012).

Grazing management is one of the main management tools to increase pasture yield and utilization and maintain desirable composition and density (Hodgson, 1989; Chapman et al., 2003). While studies (Beattie, 1994; Nie and Zollinger, 2012) revealed that inappropriate grazing management, such as overgrazing, can reduce the proportion of desirable pasture species and production, many studies also demonstrated the benefits of adopting grazing management strategies to suit various pasture systems. Saul et al. (1998) reported that rotational grazing generally had positive effect on the production and persistence of perennials with most notable benefits with phalaris. Morley et al. (1969) showed that the percentage of phalaris increased and annual grasses decreased with increasing intensity of rotation and stocking rate. A rest from grazing of pasture over the summer/autumn period promoted the plant population density and hence the yield of perennial grass-based pastures (Waller et al., 1999; Nie et al., 2005; Nie and Zollinger, 2012).

Weeds, pests and diseases have become increasingly difficult to control in pasture and cropping systems due to frequent use of herbicides/pesticides and the consequent resistance of these plants and pests to chemicals (Trenbath, 1993; Ewing and Flugge, 2004). For instance, the evolved resistance to glyphosate, the most commonly used none-selective herbicide, was found in 58 populations of annual ryegrass (*Lolium rigidum*) in Australia (Powles et al., 1998). In Western Australia, annual ryegrass was found to have developed resistance to eight other major herbicide groups (Peltzer et al., 2009). Misuse of costly chemicals can also cause detrimental effects on sown pasture species and to beneficial insects and organisms, which make the problems worse.

Land Use Change and its Impact on Grasslands

Australia has traditionally been recognised as a country riding on the sheep's back and wool/sheep industry contributed significantly to the nation's economy. However, sheep numbers have fallen dramatically and the proportion of agricultural land cropped markedly increased over the past 50 years (Robson, 2013). Table 15.1 shows the changes in areas cropped and grazed from 1995 to 2015 in Australia. The cropped area was 17 million ha in 1995 and was increased by 57 per cent to 27 million ha in 2005 and by 84 per cent to 31 million ha in 2015. On the contrary, the grazed area was 416 million ha in 1995, which declined by 8 per cent to 382 million ha in 2005 and by 24 per cent to 317 million ha in 2015, largely due to a decline in demand for wool. On the other hand, growth in population and income in developing countries of Asia has led to increased demand for animal protein (meat and milk). This will greatly improve the profitability of Australian animal industry and place greater emphasis on grassland research.

There have been increasing differences in the relative profitability between cropping and animal enterprises in the past decades. Consequently, integrated crop and livestock systems have expanded particularly in southern Australia. This is also largely attributed to the superphosphate application to legume-based pastures that increase N fixation and pasture/crop yield. The introduction, selection and breeding of new and well-adapted pasture legumes and their associated rhizobia was a major factor in the success of legume ley agriculture in the WSZ of southern Australia (Nichols et al., 2012; Nie et al., 2016). More recently, cropping has expanded to the HRZ and R&D on integrated crop and livestock systems is facing new challenges as well as opportunities (Nie et al., 2016). In northern Australia, where there is little mixed farming, introduction and breeding of new pasture

Table 15.1 Change in area ('000 ha) cropped and grazed from 1995 to 2015 (ABS various dates).

Land Use	1995	2005	2015
Crops	17,040	26,700	31,397
Grazing	4,16,300	3,82,300	3,16,722

species contribute greatly to the grazing industry (Clements and Henzel, 2010). It was estimated that about 25 per cent of the northern beef cattle and almost the entire dairy-herd grazed sown pastures or forage crops at least for part of their lives (Clements, 1995).

Changes in Funding/Resources for Pasture R&D

The relative value of research funding for grasslands and agriculture has declined over the past decades. The growth in public agricultural R&D expenditure was 7 per cent per annum on an average from early 1950s to late 1970s, but only 0.6 per cent per annum from the late 1970s to late 2000s (Sheng et al., 2011). As a result, current and future skill shortages in agriculture and agricultural education sector have been widely discussed (Pratley and Leigh, 2008), and the situation in pasture science is more dire (Robson, 2013). The major organization involved in pasture research in Australia are CSIRO, the state departments, universities, private sectors (e.g., seed and fertilizer companies) and the CRCs. Using CSIRO, as an example, there were over 40 full-time pasture scientists (a similar number in the state departments and universities) in tropical and subtropical Australia from 1960s to 1980s. Since then, however, the number has dropped sharply and by 2011, there were only three pasture scientists among which two retired in 2012, leaving one scientist focused on pasture research in northern production systems (Bell et al., 2013). A similar decline, both in tropical and temperate pasture research capacity, has been occurring in other organizations. Many research roles, such as pasture breeding and evaluation, are expected to be taken by private sectors.

More importantly, the number of students enrolled in pasture and agricultural science has been declining. Teaching in agricultural science decreased by 18 per cent for commencing students and by 31 per cent for continuing undergraduate students between 2002 and 2010 (Robson, 2013). More recently, higher degree (e.g., Masters and Ph.D.) enrolments have also declined with fewer publications in indexed journals from Australian universities. This imposes huge risks for future R&D in grassland and pasture science.

Climate Change Related Pasture Research

Given the variable rainfall and temperature conditions and the large proportion of Australia's land surface being covered by grasslands, the role that grassland systems play in addressing many issues related to climate change cannot be underestimated. The evolving climate of Australia and the probable trend of further changes reinforce the need to develop adapted pasture plants, management strategies and technologies to cope with the changing climatic conditions (Howden et al., 2008). Temperature has risen by 0.7–1.1°C since 1910 and this trend is predicted to continue at higher rates in the next 50 to 70 years (Alexander et al., 2007; Hennessy et al., 2007). Rainfall change trends have shown that southern and eastern Australia have become drier since 1950, whereas some of the northern coastal regions have become wetter (Smith, 2004). Overall, the predicted reduction of rainfall in most areas of Australia together with the increase in temperature and evaporation could result in up to 20 per cent more droughts by 2030 (Mpelasoka et al., 2007).

Livestock production systems emit 37 per cent of the anthropogenic methane (Martin et al., 2010), 65 per cent of anthropogenic nitrous oxide (FAO, 2006) and 9 per cent of anthropogenic CO_2, most of which are from enteric fermentation by ruminants. The grazing (beef, sheep and dairy) industries account for 76 per cent of Australia's agricultural greenhouse emissions or 16 per cent of the total national emissions (DCCEE,

2011). Emissions from grasslands and livestock are also costly to the grazing industries themselves due to losses in energy and nutrients. For instance, methane emissions from grazing animals cost the grazing industries equivalent to 33–60 days' grazing per year for beef steers (Eckard et al., 2010).

Current strategies to address climate change in Australia are focused on two areas: adaptation and mitigation. 'Adaptation' is to develop new pasture plants and grazing management systems to adapt to the changing climate so that productivity and profitability can be maintained and improved. A number of studies have been conducted in this area in recent years, e.g., development of more persistent and productive pasture species for drier environments (e.g., Clark et al., 2016; Culvenor et al., 2016) and more efficient grazing and animal production systems (e.g., Raeside et al., 2016a,b). 'Mitigation' is to reduce greenhouse gas emissions through improved pasture/grazing management, increased carbon sequestration in the soil, modified livestock diets and breeding of animals with more efficient feed conversion. Grasslands provide one of the largest pool for carbon storage and the global soil organic carbon sequestration potential is estimated to be 0.01–0.03 Gt C/year on 3.7 billion ha of permanent pastures (Lal, 2004). Robertson et al. (2016) measured soil organic carbon at 615 sites across eight regions, five soil orders and four management classes in Victoria, Australia and found an extremely wide range of soil organic carbon content (2 to 239 t C/ha in 0–30 cm soil). This variation was largely attributed to annual rainfall and humidity, and to a much lesser degree to texture-related soil properties. Differences in soil organic carbon between management classes (continuous cropping, crop-pasture rotation, sheep or beef pasture and dairy pasture) were small and often in significant. Methods to reduce enteric emissions, such as genetic selection of animals and modification and manipulation of rumen biota that do not intervene in grassland management, have been well reviewed (McAllister and Newbold, 2008; Cottle et al., 2011).

Shift from Conventional to Molecular Breeding

Over the past couple of decades, there has been a major shift from conventional pasture breeding to using molecular technologies for plant improvement. Public funding ceased for conventional pasture breeding, which has now been primarily funded by private sectors. R&D, using molecular mark-based and transgenic technologies, has become increasingly important to more rapidly incorporate novel traits or species into systems and increase the rate of genetic gain in breeding programs (Smith and Spangenberg, 2014). In 2000s, molecular breeding programmes were focused on the development of genetic solutions for forage quality limitations, pest and disease resistance, nutrient acquisition efficiency, tolerance to abiotic stress and modification of plant growth and development (Smith et al., 2007). More recently, breeding and managing symbiosis, biotic stresses and disease tolerance, and breeding for new environments and production systems, have also been hot research topics. The application of molecular breeding technologies in pasture breeding has begun to deliver the promises in novel genetic variation (e.g., AMV-immune *Trifolium repens*) and better understanding of the nature of genetic variation underlying key phenotypic effects (Smith et al., 2007). The 4th International Symposium of Forage Breeding held at AgriBio, the Centre for AgriBioscience, Melbourne, Australia featured some of the key findings and progress made in molecular forage breeding in Australia and worldwide (Smith and Spangenberg, 2014). There is a need to increase the adoption of new pasture/forage cultivars and technologies developed from molecular breeding programs by grazing industries.

Opportunities and Strategies

New Technologies and Information Systems for Grassland R&D

Advances in new technologies and information systems have become an important part of our society over the past few decades, and grassland scientists, farmers and educators/consultants have been increasingly faced with emerging technologies that have been developed by engineering sciences. Schellberg and Verbruggen (2013) reviewed the recent developments in new technologies that have the potential to be used in grassland R&D, emphasizing four major areas—precision agriculture, remote sensing, geographic localization and biotechnology.

Precision agriculture in a broad sense is described as the application of information technology in agriculture (Cox, 2002). Initially it was used to adapt management to the special variability within fields in order to achieve optimum input with improved efficiency of resources. Nowadays, precision agriculture has a wide range of applications that improve the control of agricultural activities and the decisions of processes for production on farms. When principles and technology are used for animal management, it is also referred as precision livestock farming. This includes a range of activities, such as assessment of animal growth, grazing management, pasture/soil management, animal behavior monitoring and disease control (Wathes et al., 2008; Sun et al., 2013).

Remote sensing is defined as the art and science of acquiring information about an object without being in direct contact with the object (Jensen, 2007). It has been used to monitor the canopy characteristics, sward height, botanical composition and biomass of grasslands (Zha et al., 2003; Feilhauer et al., 2013; Maselli et al., 2013). The techniques used for grasslands and the technology itself, such as higher resolution images from new satellites, are still improving and the spatial and temporal variations of grasslands is expected to be assessed more accurately in the near future (Schellberg and Verbruggen, 2013).

Geographic location refers to a position on the Earth and is a useful tool to assist in precision agriculture. Global positioning system (GPS) has revolutionized positioning concept and has now become an international utility. Having precise location information through GPS allows plant, soil and water measurements to be mapped and users to return to specific locations to sample or treat those areas. This is particularly useful, given grassland diversity and complexity in its functional traits.

Biotechnology in agriculture is a collection of scientific techniques used to improve plants, animals and microorganisms, based on an understanding of DNA. It has been used to speed up breeding programmes (see section above) and extend the range of traits that can be addressed. Biotechnology is more than genetic engineering, which can also be used in disease diagnostics and for the production of vaccines against animal diseases.

Biophysical and Bio-economic Modeling

Large-scale field experiments, which are often required for grassland systems research, can be costly and have limitations given the complex nature of the systems and the decline in funding and resources. Biophysical and bio-economic modeling have been used to capture the integration of multiple elements of farming systems, which provide valuable information of the 'big' picture for management decision making of the systems (Salmon et al., 2004; Bell et al., 2008; Chapman et al., 2012; Cullen et al., 2014). The modeling could generate useful information in understanding the individual elements and their interactions of the grassland systems and the temporal and spatial variability of these elements for predicting the trend of the systems. Pre-experimental modeling can also help to narrow down the priority areas for further research, which may speed up the process of R&D. The challenge is to have valid data/information for particular modeling and the correct use of models since the modeling procedure depends on quality data to generate predictions for various scenarios. Without these data, or if the data needed are incomplete, the modeling cannot be run or the outcomes from the modeling can be misleading.

Allocation of Resources for Education and Cutting-edge Research

Grasslands and pastures have been declining in relative importance in the Australian economy. However, they will remain important in many areas since they provide the only means of producing valuable products and ecosystem services where there are no viable alternatives. With the expansion of integrated crop-livestock systems in the WSZ and HZ, for instance, grasslands will continue to play a role in nitrogen supply, disease breaks and weed control (Nie et al., 2016). Currently, lack of funding and resources for grassland research and education imposes a huge risk for the development of the grazing industries in the future. Research programmes need to be re-prioritized so that the limited resources can be used to address the most needed solutions to maintain industries on an internationally competitive edge. The current trend in skill shortages needs to be reversed to capitalise on and service the opportunities of grassland R&D in the future.

Collaboration in Grassland R&D in Australia and Internationally

Grassland research in Australia has long been world-class in quality. Australia projects an image of being clean and green for the nation's livestock products. This has attracted a large proportion of overseas students studying pasture science in Australian universities as well as many researchers coming to Australia for training and collaborative research. Over the decades ending 2010, the CRCs have played a critical role to ensure that extensive collaborations are made between scientists within Australia to address issues with national interests. Although there have been international projects through various sources of investment, such as Australia's aid programs, it is important for Australian grassland scientists to have more extensive and in-depth collaborations with scientists worldwide in the future. This will not only ensure information exchange on new technologies developed internationally, but also lift Australia's profile in grassland and livestock R&D at international stages.

Conclusion

Grassland is a very important natural resource in Australia given its area and value in the country's economy and ecosystem services. There are four major grassland zones in Australia, each of which represents different climate, soil and livestock production systems. There are many challenges faced by the grassland industries, such as how to balance between productivity and sustainability, change in land use patterns, climate change and decline in funding and resources. There are also opportunities to address these challenging issues. Strategies to address the shortage of funding and resources for grassland education and research must be developed and applied if Australia is to maintain a competitive edge for its grazing industries.

Acknowledgements

We thank Reto Zollinger for his helpful comments on the manuscript.

References and Further Reading

ABS. Various dates. Australian Bureau of Statistics, Canberra.

Alexander, L., P. Hope, D. Collins, B. Trewin, A. Lynch and N. Nicholls. 2007. Trends in Australia's climate means and extremes, a global context. Australian Meteorological Magazine 56: 1–18.

Beattie, A.S. 1994. Grazing for pasture management in the high-rainfall, perennial pasture zone of Australia. pp. 62–70. *In*: D.R. Kemp and D.L. Michalk (eds.). Pasture Management—Technology for the 21st Century. CSIRO Australia.

Bell, L.W., M.J. Robertson, D.K. Revell, J.M. Lilley and A.D. Moore. 2008. Approaches for assessing some attributes of feed-base systems in mixed farming enterprises. Australian Journal of Experimental Agriculture 48: 789–798.

Bell, L.W., R.C. Hayes, K.G. Pembleton and C.M. Waters. 2013. Diversity, trends, opportunities and challenges in Australian grasslands—meeting the sustainability and productivity imperatives of the future? Proceedings of the 22nd International Grassland Congress, Sydney, pp. 28–43.

Chapman, D.F., M.R. McCaskill, P.E. Quigley, A.N. Thompson, J.F. Graham, D. Borg, J. Lamb, G. Kearney, G.R. Saul and S.G. Clark. 2003. Effects of grazing method and fertiliser inputs on the productivity and sustainability of phalaris-based pastures in Western Victoria. Australian Journal of Experimental Agriculture 43: 785–798.

Chapman, D.F., S.N. Kenny, D. Beca and I.R. Johnson. 2008. Pasture and forage crop systems for non-irrigated dairy farms in southern Australia. 1. Physical production and economic performance. Agricultural Systems 97: 108–125.

Chapman, D.F., K. Dassanayake, J.O. Hill, B.R. Cullen and N. Lane. 2012. Forage-based dairying in a water-limited future: Use of models to investigate farming system adaptation in southern Australia. Journal of Dairy Science 95: 4153–4175.

Clark, S.G., Z. Nie, R.A. Culvenor, C.A. Harris, R.C. Hayes, G. Li, M.R. Norton and D.L. Partington. 2016. Field evaluation of cocksfoot, tall fescue and phalaris for dry marginal environments of south-eastern Australia. 1. Establishment and herbage production. Journal of Agronomy and Crop Science 202: 96–114.

Clements, R.J. 1995. Pastures for prosperity. 3. The future for new tropical pasture plants. Tropical Grasslands 30: 31–46.

Clements, R.J. and E.F. Henzell. 2010. Pasture research and development in northern Australia: an ongoing scientific adventure. Tropical Grasslands 44: 221–230.

Cottle, D.J., J.V. Nolan and S.G. Wiedemann. 2011. Ruminant enteric methane mitigation: a review. Animal Production Science 51: 491–514.

Cox, S. 2002. Information technology: The global key to precision agriculture and sustainability. Computers and Electronics in Agriculture 36: 93–111.

Cullen, B.R., R.P. Rawnsley, R.J. Eckard, K.M. Christie and M.J. Bell. 2014. Use of modelling to identify perennial ryegrass plant traits for future warmer and drier climates. Crop and Pasture Science 65: 758–766.

Culvenor, R.A., S.G. Clark, C.A. Harris, R.C. Hayes, G. Li, Z. Nie, M.R. Norton and D.L. Partington. 2016. Field evaluation of cocksfoot, tall fescue and phalaris for dry marginal environments of south-eastern Australia. 2. Persistence. Journal of Agronomy and Crop Science 202: 355–371.

DCCEE. 2011. Australia Greenhouse Emissions Information System. Department of Climate Change and Energy Efficience Canberra. Available at: http://ageis.climatechange.gov.au/NGGI.aspx.

Dear, B.S. and M.A. Ewing. 2008. The search for new pasture plants to achieve more sustainable production systems in southern Australia. Australian Journal of Experimental Agriculture 48: 387–396.

Donald, C.M. 1965. The progress of Australian agriculture and the role of pastures in environmental change. Australian Journal of Science 27: 187–198.

Eckard, R.J., C. Grainger and C.A.M. de Klein. 2010. Options for the abatement of methane and nitrous oxide from ruminant production: A review. Livestock Science 130: 47–56.

Ewing, M.A. and F. Flugge. 2004. The benefits and challenges of crop-livestock integration in Australian agriculture. Proceedings of the 4th International Crop Science Congress, Brisbane, Australia.

FAO. 2006. Livestock's Long Shadows: Environmental Issues and Options. Food and Agriculture Organisation of the United Nations. Rome, Italy.

Feilhauer, H., F. Thonfeld, U. Faude, K.S. He, D. Rocchini and S. Schmidtlein. 2013. Assessing floristic composition with multispectral sensors—A comparison based on mono-temporal and multiseasonal field spectra. International Journal of Applied Earth Observation and Geoinformation 21: 218–229.

Fitzhardinge, G. 2012. Australia's rangelands: A future vision. The Rangeland Journal 34: 33–45.

Forrest, K., J. Gavin, D. Green, M. Chuk and B. Warren. 2010. Putting our heads together. *In*: D.J. Eldridge and C. Waters (eds.). Proceedings of the 16th Biennial Conference of the Australian Rangeland Society. Bourke, NSW, Australia.

Gourley, C.J.P. and D.M. Weaver. 2012. Nutrient surpluses in Australian grazing systems: management practices, policy approaches, and difficult choices to improve water quality. Crop and Pasture Science 63: 805–818.

Hennessy, K., B. Fitzharris, B.C. Bates et al. 2007. Australia and New Zealand. pp. 507–540. *In*: M.L. Parry et al. (eds.). Climate Change 2007: Impacts, Adaptation and Vulnerability. Contribution of Working Group II to the Fourth Assessment Report of the Intergovernmental Panel on Climate Change. Cambridge University Press, Cambridge, UK.

Hodgson, J. 1989. Management of grazing systems. Proceedings of the New Zealand Grassland Association 50: 117–122.

Howden, S.M., S.J. Crimp and C.J. Stokes. 2008. Climate change and Australia livestock systems: Impacts, research, and policy issues. Australian Journal of Experimental Agriculture 28: 780–788.

Hughes, S.J., R. Snowball, K.F.M. Reed, B. Cohen, K. Gajda, A.R. Williams and S.L. Groeneweg. 2008. The system collection and characterization of herbaceous forage species for recharge and discharge environments in southern Australia. Australia. Australian Journal of Experimental Agriculture 48: 397–408.

Hutchinson, K.J. 1992. The grazing resource. Proceedings 6th Australian Society of Agronomy Conference, Armidale, pp. 54–60.

Jensen, J.R. 2007. Remote sensing of the environment: An earth resource perspective. Pearson Prentice Hall, 592.

Kelman, W.M. and H. Dove. 2009. Growth and phenology of winter wheat and oats in a dual-purpose management system. Crop and Pasture Science 60: 921–932.

Kemp, D.R. 1994. Pasture management research priorities. pp. 149–154. *In*: J.L. Wheeler et al. (eds.). Temperate Pastures—Their Production, Use and Management. Australian Wool Corp./CSIRO.

Lal, R. 2004. Soil carbon sequestration impacts on global climate change and food security. Science 304: 1623–1627.

Martin, C., D.P. Morgavi and M. Doreau. 2010. Methene mitigation in ruminants: From microbe to the farm scale. Animal 4: 351–365.

Maselli, F., G. Argenti, M. Chiesi, L. Angeli and D. Papale. 2013. Simulation of grassland productivity by the combination of ground and satellite data. Agriculture, Ecosystems and Environment 165: 163–172.

McAllister, T.A. and C.J. Newbold. 2008. Redirecting rumen fermentation to reduce methanogenesis. Australian Journal of Experimental Agriculture 48: 7–13.

McCaskill, M. and P. Quigley. 2006. Fertilising pastures (Chapter 5). pp. 43–52. *In*: Z.N. Nie and G. Saul (eds.). Greener Pastures for South-west Victoria. Victorian Department of Primary Industries, Hamilton.

McIvor, J.G. 2005. Australian grasslands. pp. 343–380. *In*: J.M. Suttie, S.G. Reynolds and C. Battello (eds.). Grasslands of the World (Chapter 9). Plant Production and Protection Series, FAO, Rome.

McKenzie, F.R., J.L. Jacobs and G. Kearney. 2002. The long-term impact of nitrogen fertiliser on perennial ryegrass tiller and white clover growing point densities in grazed dairy pastures in south-western Victoria. Australian Journal of Agricultural Research 53: 1203–1209.

Moore, R.M. (ed.). 1970. Australian Grasslands. Canberra ACT, Australia: Australian National University Press.

Morley, F.H.W., D. Bennet and G.T. McKinney. 1969. The effect of intensity of rotational grazing with breeding ewes on phalaris-subterranean clover pastures. Australian Journal of Experimental Agriculture and Animal Husbandry 9: 74–84.

Mpelasoka, F., K. Hennessy, R. Jones and B. Bates. 2007. Comparison of suitable drought indices for climate change impacts assessment over Australia: Towards resource management. International Journal of Climatology 27: 1673–1690.

NATMAP. 1986. Atlas of Australian Resources. Vol. 4. Climate. Division of National Mapping, Canberra, Australia.

Nichols, P.G.H., C.K. Revell, A.W. Humphries, J.H. Howie, E.J. Hall, G.A. Sandral, K. Ghamkhar and C.A. Harris. 2012. Temperate pasture legumes in Australia—Their history, current use, and future prospects. Crop & Pasture Science 63: 691–725.

Nie, Z.N., D.F. Chapman, J. Tharmaraj and R. Clements. 2004. Effects of pasture species mixture, management and environment on the productivity and persistence of dairy pastures in south west Victoria. 1. Herbage accumulation and seasonal growth pattern. Australian Journal of Agricultural Research 55: 625–636.

Nie, Z.N., P.E. Quigley and R. Zollinger. 2005. Impacts of strategic grazing on density and ground cover of naturalised hill pasture. Proceedings of the XX International Grassland Congress, Dublin, Ireland, p. 518.

Nie, Z.N. and M. Norton. 2009. Stress tolerance and persistence of perennial grasses—The role of the summer dormancy trait in temperate Australia. Crop Science 49: 2405–2411.

Nie, Z.N. and R.P. Zollinger. 2012. Impact of deferred grazing and fertiliser on plant population density, ground cover and soil moisture of native pastures in steep hill country of southern Australia. Grass and Forage Science 67: 231–242.

Nie, Z.N., R.P. Zollinger and R. Behrendt. 2015. Impact of deferred grazing and fertilizer on herbage production, soil seed reserve and nutritive value of native pastures in steep hill country of southern Australia. Grass and Forage Science 70: 394–405.

Nie, Z.N., T. McLean, A. Clough, J. Tocker, B. Christy, R. Harris, P. Riffkin, S. Clark and M. McCaskill. 2016. Benefits, challenges and opportunities of integrated crop-livestock systems and their potential application in the high rainfall zone of southern Australia: A review. Agriculture, Ecosystems and Environment 235: 17–31.

Nossal, K., Y. Sheng and S. Zhao. 2008. Productivity in the beef cattle and slaughter lamb industries, ABARE Research Report 08.13 for Meat and Livestock Australia. Commonwealth of Australia 2008, Canberra, 24 p.

Peltzer, S.C., A. Hashem, V.A. Osten, M.L. Gupta, A.J. Diggle, G.P. Riethmuller, A. Douglas, J.M. Moore and E.A. Koetz. 2009. Weed management in wide-row cropping systems: A review of current practices and risks for Australian farming systems. Crop and Pasture Science 60: 395–406.

Powles, S.B., D.F. Lorraine-Colwill and C. Preston. 1998. Evolved resistance to glyphosate in rigid ryegrass (*Lolium rigidum*) in Australia. Weed Science 46: 604–607.

Pratley, J.E. and R. Leigh. 2008. Agriculture in decline at Australian Universities. In Global Issues. Paddock Action. Proceedings of 14th Australian Agronomy Conference. Adelaide, South Australia, 21–25 September 2008.

Raeside, M.C., M. Robertson, Z.N. Nie, D.L. Partington, J.L. Jacobs and R. Behrendt. 2017a. Dietary choice and grazing behavior of sheep on spatially arranged pasture systems. 1. Herbage mass, nutritive characteristics and diet selection. Animal Production Science 57: 697–709.

Raeside, M.C., M. Robertson, Z.N. Nie, D.L. Partington, J.L. Jacobs and R. Behrendt. 2017b. Dietary choice and grazing behavior of sheep on spatially arranged pasture systems. 2. Wether lamb growth and carcass weight at slaughter. Animal Production Science 57: 710–718.

Robertson, F., D. Crawford, D. Partington, I. Oliver, D. Rees, C. Aumann, R. Armstrong, R. Perris, M. Davey, M. Moodie and J. Baldock. 2016. Soil organic carbon in cropping and pasture systems of Victoria, Australia. Soil Research 54: 64–77.

Robson, A. 2013. Australian grasslands research at the crossroads. Proceedings of the 22nd International Grassland Congress, Sydney, pp. 101–104.

Salmon, L., J.R. Donnelly, A.D. Moore, M. Freer and R.J. Simpson. 2004. Evaluation of options for production of large lean lambs in south-eastern Australia. Animal Feed Science and Technology 112: 195–209.

Saul, G., R. Waller, L. Warn and R. Hill. 1998. Grazing management for sheep pastures in Victoria—Update on current research results. Proceedings of 39th Annual Conference of Grassland Society of Victoria, pp. 87–95.

Schellberg, J. and E. Verbruggen. 2013. New frontiers and perspectives in grassland technology. Proceedings of the 22nd International Grassland Congress, Sydney, pp. 44–55.

Sheng, Y., E.M. Gray, J.D. Mullen and A. Davidson. 2011. Public investment in agricultural R&D and extension: An analysis of the static and dynamic effects on Australian broadacre productivity. ABARES Research Report 11.7.

Simpson, R.J., A. Oberson, R.A. Culvenor, M.H. Ryan, E.J. Veneklaas, H. Lambers, J.P. Lynch, P.R. Ryan, E. Delhaize, F.A. Smith, S.E. Smith, P.R. Harvey and A.E. Richardson. 2011. Strategies and agronomic interventions to improve the phosphorus-use efficiency of farming systems. Plant and Soil 349: 89–120.

Smith, I.N. 2004. Trends in Australian rainfall—Are they unusual? Australian Meteorological Magazine 53: 163–173.

Smith, K.F., J.W. Forster and G.C. Spangenberg. 2007. Converting genomic discoveries into genetic solutions for dairy pastures—An overview. Australian Journal of Experimental Agriculture 47: 1032–1038.

Smith, K.F. and G. Spangenberg. 2014. Forage breeding for changing environments and production systems: An overview. Crop and Pasture Science 65: i–ii.

Sun, Y., Q. Cheng, J. Lin, J. Schellberg and P. Schulze Lammers. 2013. Investigating soil physical properties and yield response in a grassland field using a dual-sensor penetrometer and EM38. Journal of Plant Nutrition and Soil Science 176: 209–216.

Tharmaraj, J., D.F. Chapman, Z.N. Nie and A.P. Lane. 2008. Herbage accumulation, botanical composition and nutritive value of five pasture types for dairy production in southern Australia. Australian Journal of Agricultural Research 59: 127–138.

Tothill, J.C. and C.C. Gillies. 1992. The Pasture Lands of Northern Australia: Their Condition, Productivity and Sustainability. Tropical Grassland Society of Australia: Brisbane.

Trenbath, B.R. 1993. Intercropping for the management of pests and diseases. Field Crops Research 34: 381–405.

Virgona, J.M., F.A.J. Gummer and J.F. Angus. 2006. Effects of grazing on wheat growth, yield, development, water use and nitrogen use. Australian Journal of Agricultural Research 57: 1307–1319.

Waller, R.A., P.E. Quigley, G.R. Saul, P.W.G. Sale and G.A. Kearney. 1999. Tactical versus continuous stocking for persistence of perennial ryegrass (*Lolium perenne* L.) in pastures grazed by sheep in south-western Victoria. Australian Journal of Experimental Agriculture 39: 265–274.

Wathes, C.M., H.H. Kristensen, J.M. Aerts and D. Berckmans. 2008. Is precision livestock farming an engineer's daydream or nightmare, an animal's friend or foe, and a farmer's panacea or pitfall? Computer and Electronics in Agriculture 64: 2–10.

Williams, J., R. Hook and A. Hamblin. 2002. Agro-ecological Regions of Australia: Methodologies for their Derivation and Key Issues in Resource Management. CSIRO Land and Water, Canberra.

Zha., Y., J. Gao, S. Ni, Y. Liu, J. Jiang and Y. Wei. 2003. A spectral reflectance-based approach to quantification of grassland cover from Landsat TM imagery. Remote Sensing of Environment 87: 371–375.

PART 4
Concluding Remarks and Summing Up

16

Climatic Change on Grassland Regions and Its Impact on Grassland-based Livelihoods in China

Limin Hua, Yujie Niu* and *Victor Squires*

Introduction

Climate change is a topic of interest to scientists, government officials, farmers, businessmen, etc., because it has vast and inescapable impacts on society as a whole. Agriculture and other industries are sensitive to climate change and the negative consequences flow through the economy. Climate change in Intergovernmental Panel on Climate Change (IPCC) usage refers to change in the state of the climate that can be identified by changes in the long-term mean and/or the variability of its properties and that persists for an extended period, typically decades or longer. Climate change can be seen at global, regional or even more local scales (Intergovernmental Panel on Climate Change, 2007).

China's landscape is vast and diverse, ranging from forest, steppes, gobi and deserts in the arid north to subtropical forests in the wetter south. The climate in China differs from region to region because of the country's highly complex topography and latitudinal position (Manfred and Peng, 2011). China's climate is mainly dominated by dry seasons and wet monsoons, which lead to pronounced temperature differences between winter and summer. In south-western China, the monsoon from the Pacific influences the climate, which is warm and moist in summer. However, a series of mountains in central China stop the northward movement of the monsoon, which causes dry and hot weather and the precipitation rate is relatively low and the daily range of temperature is relatively high. Generally, the majority of land in China is influenced by the typical temperate arid continental climate.

China has 400 million hectares (Mha) of grassland, which is the third largest area of grassland in the world. The grasslands cover approximately 41.7 per cent of the country and are mostly inhabited by 3.3 million households of various ethnic minorities (Squires

College of Grassland Science, Gansu Agricultural University, Lanzhou, China.
* Corresponding author: hualm@gsau.edui.cn

et al., 2010). The grassland ecosystems of China play important roles in servicing the ecological environment and regional socio-economic development and in maintaining biodiversity of plants and animals. China's grasslands are mainly situated in the drier and higher regions of north and north-west China, including Mongolia Plateau, Loess Plateau and Qinghai-Tibet Plateau (QTP). As a consequence, the grasslands are sensitive to climate change and human disturbance.

In 2016, the IPCC reported that the average temperature at global scale has risen 1.5°C above pre-industrial levels. Climate change has already produced visible adverse effects on China's agriculture and livestock-raising sectors, manifested by increased instability in agricultural production, severe damage to crops and livestock breeding caused by drought and high temperature in some parts of the country. Moreover, due to climate change, herders have to conduct appropriate approaches to adopt the changes in terms of their production pattern as well as livelihood style. Therefore, understanding the effect of climate change on grassland production and response of herders to climate change is vital to improve grassland management in China.

Distribution and Climate Zone of Grasslands in China

Grasslands are landscape where the vegetation is dominated by grass, sedge and shrubs. Grasslands are the biggest terrestrial ecosystem in China because the grassland area is twice that of forest area and three times that of cropping land. Grasslands have an important role on regulating climate, water and soil conservation, carbon sequestration, as well as supporting millions of livestock on which pastoral production depends. China's grasslands are distributed in the northern temperate region, south-western alpine region in Qinghai-Tibet Plateau and southern tropical and sub-tropical region (Fig. 16.1). Among the grasslands, the temperate steppe, desert, alpine grassland occupy almost two-thirds of the total grassland area.

North Temperate Grasslands

The grasslands are located in the regions bounded by the 400 mm isohyet. The geographical distribution of the grassland is from eastern Inner Mongolia, Losses Plateau, to west part of Xinjiang region. The area of the grassland covers around 41 per cent of total national grassland area and it was one of main livestock production areas in China. The horizontal zonality of grassland from east coast to west Tianshan mountain includes temperate meadow, typical steppe, desert steppe and desert. The dominant plants in these grasslands are *Leymus chinensis, Stipa baicalensis, Stipa grandis, Stipa krylovii, Artemisia frigida* and various halophytes, respectively.

Alpine Grassland in Tibet Plateau

The grasslands in this region involve alpine meadow and alpine steppe. The area accounts for 38 per cent of total national grassland area. The grasslands have vertical zonality from low altitude to high altitude. The horizontal zonality of the grassland from south-east Qinghai-Tibet plateau to north-west region includes alpine meadow, alpine steppe, alpine desert steppe and alpine desert. The dominant plants on these grasslands are *Kobresia humilis, Kobresia littledalei, Festuca rubra,* alpine shrub and various cold-salt tolerant herbs, subshrubs, and shrubs.

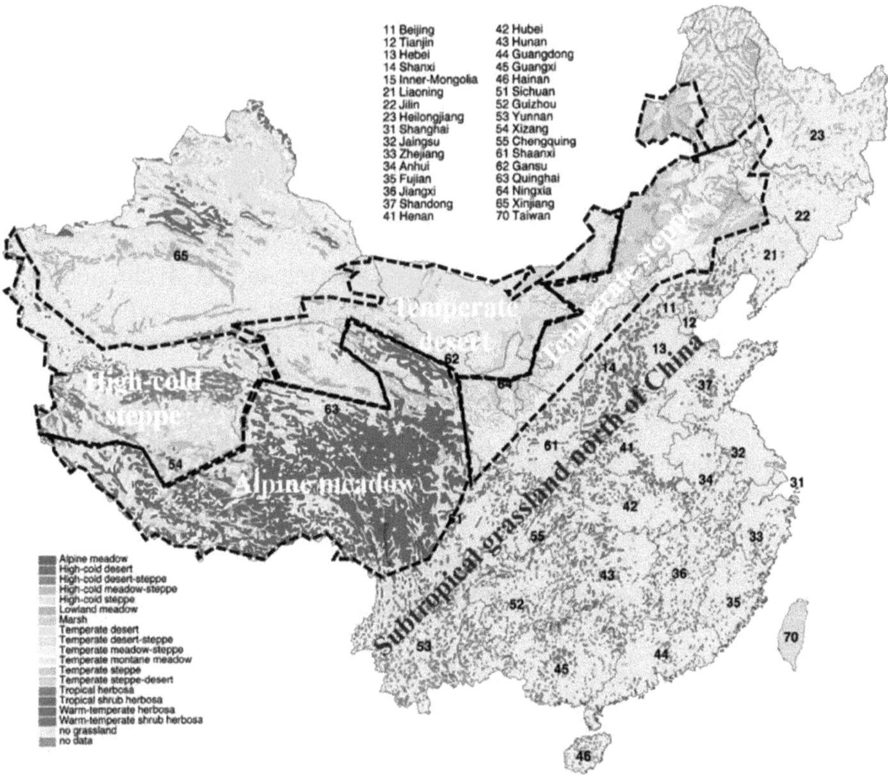

Fig. 16.1 Grassland types in China (Primary Data Sources: Commission for Integrated Survey of National Resources (CISNR) (1995): Atlas of Grassland Resources of China (1:1 million). Beijing, Chinese Academy of Science. China Map Press).

South Secondary Grasslands

The grasslands are mainly formed after deforestation and cover 12 per cent of total grassland area. The grasslands involve small areas of tropical grassland in Hainan province and south Guangdong province and a large area of sub-tropical grassland in south-eastern China. The dominant forages are *Pennisetum alopecuroides, Pennisetum sinense, Pennisetum purpureum*, but numerous other plant species, mainly forbs occur (Flora of China, http://flora.huh.harvard.edu/china/mss/intindex.htm).

Characteristic of Climate Change in Grassland Regions of China

According to third National Climate Assessment of China published in 2015, the temperature in China increased (on average) 0.9–1.5°C from 1909 to 2011. In the 15 years ending 2012, the trend of rising temperature slowed down; however, the average temperature in 2011 was the highest in the preceding hundred years. Although, the mean annual precipitation in 60 years prior to 2012 and over the preceding hundred years has not shown any obvious change, the change of rainfall distribution has regional characteristics, which the arid and semi-arid in north-west China becoming wetter because of more precipitation in the 30-year period ending 2011.

The northern steppe is located the temperate zone, which has sensitivity to climate change. IPCC reported that the steppe productivity had declined in the decades since the 1970s in arid and semi-arid regions, but not all can be attributed to climate change. Under the situation of the temperature increasing 2–4°C and rainfall decreasing, the steppe productivity of temperate grasslands will decrease 40–90 per cent. The Qinghai-Tibet Plateau has headwaters of seven major rivers (including transboundary rivers, such as the Ganges, Brahmaputra, and Mekong). The plateau has a typical highland climate and climate change there would precipitate a major impact on China (even extending to the Northern Hemisphere as a whole). Due to the high altitude and harsh climatic conditions, the alpine meadow and alpine grassland in this region are very fragile and sensitive to climate change (Shang et al., 2012). The plants of alpine meadow and alpine grassland are psychrophytes, which have short life, cold resistance and low biomass. Climate change, in particular warming, has influenced the plant productivity, reproductive success as well as dispersal of disseminules.

The southern tropical and subtropical grasslands in south China have a climate of hot summers and temperate winters, as well as higher precipitation. There is no specific dry season. The grassland productivity in the region is little impacted by the climate change because of higher rainfall, but species composition and other changes to the ecology may emerge.

Temperature Change in China's Grassland Regions

From 1951 to 2009, the ground surface temperature increased on an average by 1.38°C in China, the warming rates is 0.23°C/10a, which is similar to the level of global warming. The temperature in the north, north-west and south-west China has obviously increased over the 50 years ending 2012. The warming rate is 0.22°C/10a, 0.37°C/10a and 0.24°C/10a in north, north-west and south-west regions, respectively (Liang, 2014). Temperatures rose in the mid-and-late 1980s, in particular since 1978. The warming in north China and on the Qinghai-Tibet Plateau was greater than in other regions, especially in winter and spring. Most regions of QTP, including Xizang (Tibet) and Qinghai province, have a greater warming rate compared to other regions. The climate in west and north QTP has been changed from warmer-wetter to warmer-dryer, however, the climate in the east and south of the QTP has changed from warmer-dryer to warmer-wetter (Yu and Xu, 2009). In the three-river headwaters region of Qinghai, from 1961 to 2010, the annual and seasonal average temperature underwent several cold and warm fluctuations, but the average temperature had a significant rising trend at statistical significance level, especially since 2001. Spring, summer, autumn and annual average temperatures tended to be significantly warmer since the 1990s and the winter average temperature exhibited a significant increasing trend since 2001 (Yi et al., 2011). In Inner Mongolia, the temperature in meadow-steppe, steppe and desert-steppe rose between 1980 and 2014 (Ye et al., 2014). The magnitude of the warming in desert-steppe was greater than that in meadow-steppe and steppe. On desert-steppe, steppe and meadow-steppe, the warming rate is 0.46°C/10a, 0.30°C/10a and 0.36°C/10a (Ye et al., 2014). In the pastoral-agro regions, the warming rate was 0.45°C/10a, which was a higher rate than that of whole nation and globally. After 2008, the annual mean air temperature slowly decreased, which was mainly contributed by a higher rate of air temperature decrease in winter (Lin et al., 2016). In southwest China, including Sichuan, Chongqing, Yunnan and Guizhou from 1951 to 2000, the mean temperature, the daily maximum temperature and the daily minimum temperature annually increased after the mid1980s and the peak temperature occurred in 1998. During the 50 years ending 2012, the daily maximum temperature showed a decreasing trend. The annual variation of daily maximum temperature was larger than that of the mean temperature and also the daily minimum temperature. The decrease of daily maximum temperature had an obvious influence on the mean temperature. The daily minimum temperature showed an increasing trend and the increasing trend of temperature was much more obvious in winter than in summer. The warm period of summer and winter occurred respectively in 1950s and in 1980s (Bao et al., 2006).

Precipitation Change in China's Grassland Regions

In the 50 years from 1951 to 2012, the changes of precipitation had larger temporal-spatial characteristics in China. Gao et al. (2015) studied the daily precipitation data of 308 observational stations in the 15 provinces of the northern region of China during 1961–2010 and found that as the rainfall regime changed, the semi-arid zone expanded significantly since 1981 but the semi-humid zone in north-east China and the arid zone in the south-east of north-west China decreased. During the 50 years ending 2012, the annual rainfall increased in the arid zone, while the number of rainy days decreased in semi-arid and semi-humid zone. All precipitation indices in the second period (after 1981) were higher than those in the first period, especially at light and moderate rainfall in the north. The most obvious seasonal variation in the arid zone happened in winter. The summer and autumn precipitation in the second period was less than that in the first period in semi-arid and semi-humid zone. It is noteworthy that precipitation in the semi-arid zone significantly decreased in summer, while in autumn the seasonal precipitation variation was similar to that in the semi-humid zone. On the QTP, the precipitation rose in the 30 years ending 2012 (12.4 mm/10a) (Liang et al., 2014), in particular in the north of the QTP, which is close to south Xinjiang. The general decrease of precipitation extends from the south-eastern to north-western QTP. However, the interannual change of precipitation was not significantly different from 1981 to 2010. The rainfall in spring in the east and north-east of the QTP was obviously increasing; however, in summer, the higher rainfall occurred in the west of the QTP. The depth of snow and the number of days with snow cover on the QTP increased from 1961 to 1990 and decreased from 1991 to 2005 (Yao et al., 2013). In north Xinjiang, the maximum depth of snow cover was significantly positively correlated with winter precipitation; the maximum depth increased at an average annual growth rate of 0.8 per cent in 46 years from 1961 to 2009, which was smaller than that of winter precipitation. The days of snow cover and steady snow cover were also significantly positively correlated with days of temperature below 0°C, but the increasing trend but decreased after the 1990s (Wang et al., 2009). In Inner Mongolia, in past 50 years from 1961 of snow cover days was not significant; the days of snow cover increased from the 1960s to the 1980s, to 2005, the precipitation in meadow-steppe rose year by year, while the rainfall decreased in steppe with no obvious change in the desert-steppe (Han et al., 2010).

Characteristics of Extreme Climate in Grassland Regions

Extreme weather includes unexpected, unusual, unpredictable severe or unseasonal weather—weather at the extremes of the historical distribution—the range that has been seen in the past. Often, extreme events are based on a location's recorded weather history and defined as occurring in the top 10 per cent of unusual events (IPCC, 2007).

IPCC in 2014 reported that the occurrence of extreme events rose due to climate change (Cramer et al., 2014). In China's pastoral land, extreme drought and snow disaster are the main calamitous weather events. In 50 years from 1950 to 2000, the area and frequency of drought on grasslands rose and most drought occurred in west China (Li et al., 2003). Since 1990s, the effect and loss of grassland production caused by drought was higher than that previously reported. The occurrence of severe drought[1] accounted for 50 per cent of extreme events from 1989 to 1998 and it reached 60 per cent from 1999 to 2005. The annual increasing disaster area of drought in Qinghai and Xinjiang was 8,35,000 ha in 1990s. On QTP, the area of drought-affected land over the moderate level was 10,32,600 ha, which accounted for 40 per cent of the total plateau land (Li et al., 2016).

The snow disaster has been increasing in recent years. So far, the area of snow disaster on grasslands in China was 0.4 billion ha, which accounted for 10 per cent of the total grassland area. Most of snow disaster occurred in east Inner Mongolia, north Xinjiang as well as the Tibet Plateau. In the past 50 years, the occurrence of snow disaster in November and March–April in Xilingguole League (Inner Mongolia) was over 50 per cent and 40 per cent, respectively. In Xilingguole League and nearby Wulanchabu League, the snow disaster occurred every two or three years (Li et al., 2016). The frequency of snow disaster was 61.2 per cent of years in north Xinjiang and 38.8 per cent in south Xinjiang from 1960 to 2010. The Altay region in northern Xinjiang suffered 11 times, moderate four times and severe six times of mild snow disaster, most occurring in the front part of the mountains, hills and valley (Ma et al., 2014). In the 60 years ending 2011, the snow disaster occurred 30 times in Qinghai province, among which severe snow disaster was 12 times and extreme snow disaster was six times. From 1949 to 2002, the snow disaster occurred 80 times in Xizang (Tibet) and snow disasters that caused livestock mortalities in excess of 2 million head have occurred seven times (Li et al., 2016).

Effect of Climate Change on Grassland Ecosystem

The effect of climate change on grassland ecosystem has multiple facets. However, the process and direction of ecosystem change depends on one or several key factors of climate change. Therefore, the key point of studying climate change on grassland ecosystem is to find the principal factors that apply on each site. Climate change in grassland ecosystems involves the effect of temperature rise, precipitation change, CO_2 increase and extreme climate events on soil, plant, livestock, as well as the response of various components of grassland ecosystem to climate change (Fig. 16.2).

Effect of Climate Change on Grassland Soil

The soil is an important component of terrestrial ecosystems and plays a significant role in exchange of substances (mineral elements, water) and energy. Meanwhile, soil is a product of interaction between organisms (some microscopic) and the environment. Climate change has various impacts on grassland ecosystems through the soil. Rising temperature influences evaporation, resulting in changing soil moisture regimes. For example, in the steppe of Inner Mongolia, there is a positive correlation between temperature and

[1] As defined in Squires, 2017.

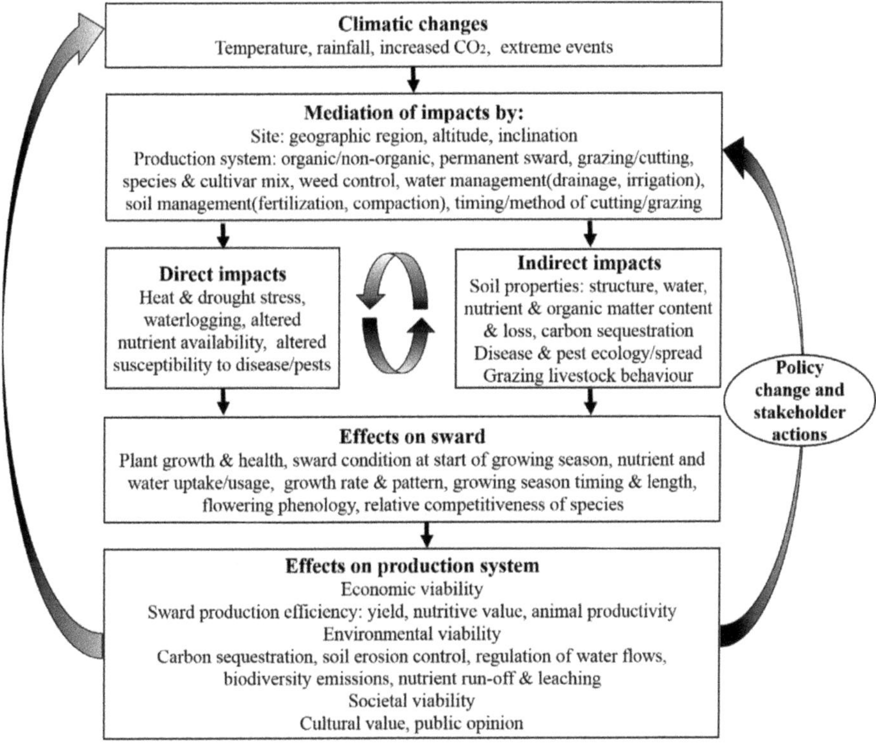

Fig. 16.2 Map of impacts of climate change on grassland systems, including feedbacks (Kipling et al., 2016).

evaporation, and a negative correlation between evaporation and soil moisture (Li et al., 2003). Soil moisture is one of the limiting factors of plant growth in steppe, which consists of mesophyte and xerophyte plants. Climate change influences distribution of soil moisture in time and space. Besides soil moisture, the content and distribution of soil organic carbon was related to productivity and scale of ecosystem. Climate change, grazing and land conversion to cropland influence the carbon and nitrogen storage in grassland ecosystem. The interaction and synergistic effect between climate change and carbon and nitrogen cycle in grassland ecosystem are the hot topic of current study. Carbon storage in grassland ecosystem accounted for 16.7 per cent of total terrestrial ecosystem in China (Fan et al., 2003). Research indicates that the grassland ecosystem as a whole in China is a carbon sink rather than carbon source, but grassland degradation has reduced carbon sequestration potential of many grasslands (Zhang et al., 2013). Grassland degradation causes carbon loss and disrupts the carbon cycle, resulting in weakening capacity of carbon sink. Under climate change, the organic carbon in surface soil is influenced by temperature and precipitation. The content of soil organic carbon decreases as air temperatures rise (Liang et al., 2014).

Effect of Temperature Change on Grassland Vegetation

Phenological characteristics of plants are direct and sensitive indicators influenced by environment changes within seasons and between years. Temperature is the main factor that influences plant phenophases under climate change. Liang et al. (2014) found that under warmer climate, the growing season was obviously prolonged in steppe, followed by desert steppe and meadow steppe in the period from 1983 to 2009. The climate on Stipa steppe in Inner Mongolia was warmer and drier, which resulted in prolonging of green-up stage of plants. From 1985 to 2002, the green-up stage and withering stage of Chinese wild rye advanced 2.4 days and 3.7 days when the temperature increased 1°C in the steppe in Inner Mongolia (Liang et al., 2014). On the QTP, some researchers found that simulated warming on alpine meadows with *Kobresia humilis* benefited the extension of plant growing season. Qi et al. (2006) found that the green-up stage of most forage plants in Qinghai advanced and the growing period by 2003 extended when compared with the situation in 1991.

Temperature change has different influence on net primary production (NPP) of grassland in different regions. Zhao et al. (2011) found that climate warming had a negative impact on NPP in China while the precipitation was less important. Li et al. (2003) found that the above-ground biomass of *Leymus chinensis* decreased significantly since 1993, especially during the period from 1982 to 1998. Increased temperature in winter aggravated the drought in the spring in the steppe, resulting in decreased grassland productivity. However, moderate temperature increase benefits the above-ground biomass production. In the past decades, some researchers studied the effect of simulating warming on plants by infra-red heater in order to evaluate the impact of climate change on ecosystem. Bai et al. (2010) found that the annual root biomass decreased 10.3 per cent under a simulated warming regime Liu et al. (2010) found that the above-ground biomass, average plant height and foliage cover of alpine meadow increased along a temperature gradient in open top chamber (OTC). Li et al. (2011), working in north-west Sichuan province, found that the above-ground biomass of grass and sedge increased and forbs decreased along an increasing temperature gradient. Qian et al. (2013) found daily NPP was simulated by the AVIM-GRASS (Atmosphere-Vegetation interaction model) from 1961 to 2007. Accumulated annual NPP was estimated and AHP (available herbage production) was derived (Fig. 16.3). Total annual AHP estimates showed a decrease from 1961 to 2007 across the northern grasslands and part of the western grasslands. The decrease in annual AHP in the middle and eastern parts of Inner Mongolia, south Ningxia, south-east Gansu and eastern Qinghai was 100–1,000 kg DM ha^{-1} in 10 years (where DM = dry matter). An annual decrease of 10–100 kg DM ha^{-1} in 10 years was recorded in north and east Xinjiang, east and south Qinghai, west Xizang and in the middle of Gansu. The annual AHP trend was not obvious or only increased by 20–100 kg DM ha^{-1} in 10 years in middle and south-west Xinjiang, most of the Qinghai-Tibet Plateau, west Gansu and west Inner Mongolia. The maximum rate of increase (100–800 kg DM ha^{-1} in 10 years) was in south-west Xinjiang, and middle and east Xizang. Overall these trends in AHP were driven more by the changes in moisture than by temperatures (Qian et al., 2012).

Effect of Precipitation Change on Grassland Vegetation

The effect of precipitation on phenological characters of plants shows regional differences. The increasing rainfall in August and September during 1985 to 2002 obviously prolonged the growing period of plants in Stipa grassland in Inner Mongolia (Liang et al., 2014). Under snow augmentation, the flowering date in phenological period of dominant plants was earlier in spring on alpine meadow in the Tibet Plateau. Mou et al. (2013) found that extreme drought at the beginning of a growing season significantly advanced the mid-flowering date of alpine meadow plant by two to three days, the extreme drought in the

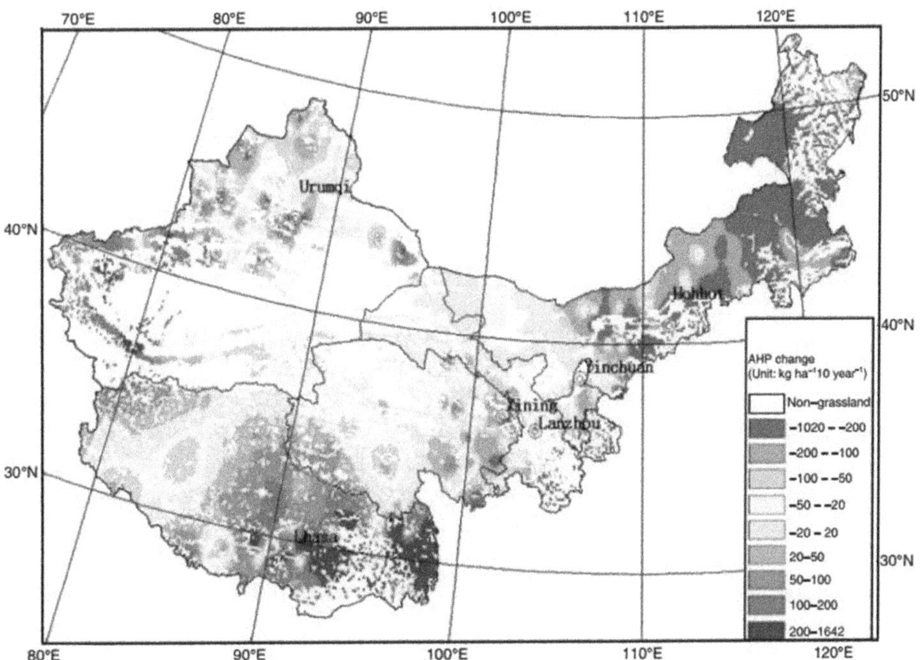

Fig. 16.3 Trends for annual available herbage production from 1961 to 2007 in the main grasslands of China.

peak growing season significantly compressed the flower duration by 2–3 days, the flower phenology responses of species were related to the seasonality of their reproduction and the forbs was more sensitive to the extreme drought than the sedges and grasses.

Precipitation is a dominant factor of influencing plant growth. Zhao (2007) used the meteorological data from 1961 to 2005 in the steppe of Inner Mongolia and found the annual precipitation dropped 40 mm and 27 mm in recent 20 and 45 years, respectively. Precipitation was the dominant factor constraining grassland production in this region. As a result, average forage productivity decreased 200.2 kg/ha over the 45 year period 199x to 20xx. Han et al. (2010) found that precipitation in 1961 to 2007 was the key factor, affecting forage potential climatic productivity (FPCP) in desert steppe of Inner Mongolia. Fang et al. (2005) found that NDVI of grassland and deciduous broad-leaf forest increased with rise in precipitation and more frequent precipitation significantly increased growth of grassland. Wang et al. (2003) found that excessive precipitation suppressed growth of grasses on alpine meadow, the above-ground biomass of forbs was sensitive to decreased rainfall and the above-ground biomass of sedges was stable.

Effect of Climate Change on Herder's Livelihoods

Herder's livelihood involves various aspects that affect productivity of plants and the herder's livestock on which he depends to make a living. The sustainable strategies are key approach of herders to reduce livelihood vulnerability, improve their resilience against weather disaster and improve the quality of life. Climate change influences the herder's livelihood through its effect on forage yield and livestock production. In the face of climate change, herders have to adjust their production pattern and living style to adapt to the new conditions. In China, research of climate change on family revenue has just started and most research depends on qualitative assessments. Yin et al. (2014) found that in Three Rivers source region (Qinghai), the rising temperature had negative correlation with household income while the precipitation had positive correlation with the income because the rising temperature benefits the grass growth in alpine rangeland and the continuous rising rainfall (more cloudy skies) causes temperature decline, which is not good for plant growth in alpine meadow.

Increase the Planting Area of Sown Grassland on Tibet Plateau

Climate change, in particular wetter and warmer regimes, provide more precipitation and temperature for plant growth on the QTP. The beneficial climate condition is especially good for sown grassland development. In Tianzhu county, in north-east QTP, each household increased the area planted to oats on plowed grassland. In the 1980s, there was an average 0.5 ha of oat planted by each household, but gradually, the planted area increased to 1 or 2 ha per household. Oat hay is fed to livestock in winter. Sown grassland is needed to fill the feed gap caused by more serious grassland degradation and concomitant decrease in forage. Climate change has made it possible to extend the area planted. Therefore, the local herders changed grazing on natural rangeland to a combination of grazing and indoor pen feeding with fodder supplement. Besides the extension of planting area of sown grassland, the altitudinal band that is suitable for forage crops, has become higher and higher due to climate warming. In Tianzhu county, the average elevation of oat planting was from 2,200 to 2,800 m in 1980s, but by 2014, the elevation had increased 100–200 m. Although, more hay could solve the shortage of fodder in winter, plowing destroys the natural rangeland, resulting in soil erosion and other ecological functional loss. In addition, the *Grassland Law of China* bans plowing of rangeland to sow artificial grassland (improved pastures). How to balance the grazing pressure on natural grassland (livestock numbers) and constraints on extension of the area of sown grassland to fill the feed gap is still a challenge in pastoral land.

Change of Rangeland Use Pattern

In the alpine mountain on the QTP, the local herders conduct traditional seasonal grazing in higher mountains in summer and in low valleys in winter. However, the climate warming has increased the growing period of plants in highland, in particular delaying the withering of forage plants at summer's end, which results in extension of the grazing period in summer. In Maqu county, in east QTP, the herders grazed their livestock in summer for two months in summer pasture in 1980s, but now they can graze the livestock for three months. In addition, the rangeland in higher mountains is the habitat for wildlife. Overgrazing by livestock just prior to the first snow of winter will reduce the forage available for wildlife and, ultimately, destroy the biodiversity in the rangeland ecosystem.

Effect of Climate Change on Livestock

Climate warming, especially temperature increase in winter, leads to less snowfall and therefore, fewer snow disasters. The warm temperature benefits the livestock against cold winter, in particular calves and lambs. On the one hand, the improvement of livestock survival rate increases the livestock number on rangeland, which contributes to the problem of overgrazing. On the other, the warm winter increases pathogenic microorganism in soil and plant, which threaten the health of livestock. Therefore, expenditure on livestock disease control also increases.

Increase in Household Expenditure

In west Inner Mongolia and Ningxia province, the climate is becoming drier and drier on desert steppe. The local herders had to dig more wells for livestock watering and for irrigating sown pastures and fodder crops. Digging wells means extra cost but a better outcome for local herders because they used to water their animals in some small lake or stream. Regretably, those water resources have been disappearing because of climate warming. On the steppe in Xilingguole Prefecture, Inner Mongolia, the herders face increasing cost of hay for their livestock in winter because the steppe productivity declines due to less rainfall in this region. Besides, the cost of renting pasture increases in pastoral land because expanding herd/flock size cannot be supported on low productivity pastures that were assigned to each household under the Grassland Household Responsibility Contract System (GHCRS, see below). Many herders have to rent other pastures for their livestock to maintain their income to support their families.

Herder's Response to Climate Change

Adjust Livestock Breed to Adapt to Climate Change

Ye (2014) studied the effects of climate change and adaptation measures of herdsmen to the climate change in Inner Mongolia. She found that most herders believed that the climate change is the first reason influencing livestock production. Grassland of lower productivity could not support the livestock's intake because of more serious grassland degradation caused by climate change. The herders had the option to reduce their livestock numbers, but, facing the challenge of maintaining their livelihood, the herders chose to adjust livestock species instead of reducing livestock number on their pasture. Under the policy of Grassland Household-contract Responsibility System (GHRS), the herders have to use rights to their own pastures within a fixed area (Hua et al., 2015; Squires and Hua, 2015). They can't increase the grazing area through traditional seasonal grazing. In Inner Mongolia, the herders reduced the number of goat or sheep, which are a grazing type livestock that are adapted to grazing rangelands, and switched over to raise small-tail Han sheep. The Han sheep is not good at grazing, but, gives multiple births if fed well. The herders produced or bought more hay to feed the Han sheep indoors instead of grazing on rangeland. Certainly, the cost of forage production also increased. On the QTP, the pastoralists also reduced the number of Tibet sheep because of lower grassland productivity and higher labor cost for small ruminants. In the alpine pastoral land located in north-west Sichuan and south Gansu, the local herders raise more yak than Tibet sheep because management of yak grazing needs less labor and yak can survive in the harsh conditions imposed by colder weather and poor nutrition of forage.

Purchase More Forage and New Type Fodder for Livestock Production

In the 1980s, herders in the main pastoral land, such as Inner Mongolia, Tibet and Xinjiang seldom purchased forage for their livestock. They grazed their livestock on nature pasture according to the varied conditions in terms of climate and grassland. But, with grassland degradation caused by climate change, in particular less rainfall in spring, the herders in Siziwang Banner in Inner Mongolia have had to purchase more hay for their livestock since 2012. Besides purchasing more forage, the herders also buy new style fodder, such as alfalfa pellet feed, mineral lick block, etc. These fodders provide high nutrition to the livestock and make the animals more resilient to weather disasters. However, the herders believe that the increasing cost of forage has led to a decrease in household income (Liang et al., 2014).

Renting Pasture or Allied Household Grazing to Respond the Climate Change

The Householder Contract Responsibility System (HCRS) is the fundamental tool for rangeland management in China. A key part of the policy was to assign user rights to individual households from land that was formerly communally used. The area of household's contracted pasture depends on the family size and livestock number. Under the policy, each household only uses its own pasture. With the grassland degradation caused by climate change, small households have to rent pastures to graze their livestock because their own allocated grazing area of pasture cannot support the increased number of livestock. In the steppe of Inner Mongolia, each household rented 46.25 ha of rangeland besides renting seasonal pasture for three to five months (Zhang et al., 2010). Besides renting pasture, some herders, in particular in the agro-pastoral region, embraced the allied household contract system in which small individual household pasture was combined to form a bigger pasture. The rain-fed cropping system is particularly sensitive to climate change because of variable rainfall. Hua and Squires (2015) studied the allied household contract system in agro-pastoral region of west China and found that larger scales of operation gave more flexibility and allow rest rotation grazing and range improvement measures to be practiced.

Off-farm Work and Diversified Livelihood Against Climate Change

Under climate change, some herders gave up their livestock production because of grassland degradation and changed to other livelihood activities. Herders with a small area of pasture and a few livestock, cannot make a profit from their livestock production, especially as the subsistence lifestyle is no longer viable in the market (cash) economy. Therefore, off-farm work is a priority for them. Herders are moving to urban areas as migrant workers after accepting training programmes started by local governments. Some herders changed from livestock production to operate tourism-related enterprises involving agri-tourism, eco-tourism and cultural tourism. In Sunan and Tianzhu county located in Qilian mountain, some Tibetan herders and Yugu Monitory herders run family tourism, including meals, horse riding, etc., for the tourists. The new countryside tourism earns more profit for herders and the local governments make the policy to subside them as part of the national programme to lift millions out of poverty.

References and Further Reading

Bai, W., S. Wan, S. Niu et al. 2010. Increased temperature and precipitation interact to affect root production, mortality and turnover in a temperate steppe: Implications for ecosystem C cycling. Global Change Biology 16(4): 1306–1316 (in Chinese).

Bao, Q., Y.M. Liu, J.C. Shi and G.X. Wu. 2010. Comparisons of soil moisture datasets over the Tibetan Plateau and application to the simulation of Asian summer monsoon onset. Advances in Atmospheric Sciences 27: 303–314.

Cramer, W., G.W. Yohe, M. Auffhammer, C. Huggel, U. Molau, M.A.F. da Silva Dias, A. Solow, D.A. Stone and L. Tibig. 2014. Detection and attribution of observed impacts. pp. 979–1037. *In*: C.B. Field et al. (eds.). Climate Change 2014: Impacts, Adaptation, and Vulnerability. Part A: Global and Sectoral Aspects. Contribution of Working Group II to the Fifth Assessment Report of the Intergovernmental Panel on Climate Change. Cambridge University Press, Cambridge, United Kingdom and New York, NY, USA.

Fan, J.W., H.P. Zhong, B. Liang, P.L. Shi and G.R. Yu. 2003. Carbon stock in grassland ecosystem and its affecting factors. Grassland of China 25(6): 51–58 (in Chinese).

Fang, J., S. Piao, L. Zhou et al. 2005. Precipitation patterns alter growth of temperate vegetation. Geophysical Research Letters 322(21): 365–370.

Gao, J.Q., X.G. Yang, C.Y. Dong and K.N. Li. 2015. Precipitation resource changed characteristics in arid and humid regions in Northern China with climate changes. Transactions of the Chinese Society of Agricultural Engineering 31(12): 99–110 (in Chinese).

Han, F., J.M. Niu and P.T. Liu. 2010. Impact of climate change on forage potential climatic productivity in desert steppe in Inner Mongolia. Chinese Journal of Grassland 32(5): 57–65 (in Chinese).

Hua, L.M., S.W. Yang, V. Squires and G.Z. Wang. 2015. An alternative rangeland management strategy in an agro-pastoral area in western China. Rangeland Ecology & Management 68: 109–118.

IPCC. 2207. Climate Change 2007: Impacts, Adaptation and Vulnerability. A Report of the Intergovernmental Panel on Climate Change. Cambridge University Press, Cambridge U.K.

Kipling, R.P., P. Virkajärvi, L. Breitsameter et al. 2016. Key challenges and priorities for modeling European grasslands under climate change. Science of the Total Environments 566-567: 851–864.

Li, L. and F.J. Hou. 2016. The main natural disasters affecting grassland productivity in China. Pratacultural Science 33(5): 981–989 (in Chinese).

Li, M.S. and Y.H. Li. 2003. Studies on drought in the past 50 years in China. Agricultural Meteorology 24(1): 6–10 (in Chinese).

Li, N., G. Wang, Y. Yan et al. 2011. Short-term effects of temperature enhancement on community structure and biomass of alpine meadow in the Qinghai-Tibet Plateau[J]. Acta Ecologica Sinica 31(4): 895–905 (in Chinese).

Li, Z.Q., Z.G. Liu, Z.Z. Chen et al. 2003. The effects of climate changes on the productivity in the Inner Mongolia steppe of China. Acta Pratacultural Science 12(1): 4–10 (in Chinese).

Liang, Y. Ganjurjav, W.N. Zhang et al. 2014. A review on effect of climate change on grassland ecosystem in China. Journal of Agricultural Science & Technology 16(2): 1–8 (in Chinese).

Lin, C.C., K. Wang and Y.M. Sun. 2016. Study on the change of temperature time series in medium section of agro-pastoral ecotone of Northern China during the last 60 years. Acta Agrestia Sinica 24(4): 747–753 (in Chinese).

Liu, W., C.T. Wang, J.Z. Zhao et al. 2010 Responses of quantity characteristics of plant community to simulating warming in alpine Kobresia humilis meadow ecosystem. Acta Botanica Boreali-Occidentalia Sinica 30(5): 995–1003 (in Chinese).

Ma, Y., Y. Wang and X. Wang. 2014. Classification of the snow disasters and circulation features of the blizzard in Altay region, China. Journal of Arid Land Resources and Environment 28(8): 21 (in Chinese).

Manfred, D. and G.B. Peng. 2011. The Climate of China. Springer. Berlin, pp. 20–29.

Mou, C., G. Sun, P. Luo et al. 2013. Flowering responses of alpine meadow plant in the Qinghai-Tibetan plateau to extreme drought imposed in different periods. Chinese Journal of Applied & Environmental Biology 19(2): 272–279 (in Chinese).

Qi, R.Y., Q.L. Wang and H.Y. Shen. 2006. Analysis of phenological phase variation of herbage plants over Qinghai and impact of meteorological conditions. Meteorological Technology 34(3): 306–310 (in Chinese).

Qian, S., L.Y. Wang and X.F. Gong. 2013. Climate change and its effects on grassland productivity and carrying capacity of livestock in the main grasslands of China. Rangeland Journal 34(4): 341–347.

Shang, Z.H., M.J. Gibb and R.J. Long. 2012. Effect of snow disasters on livestock farming in some rangeland regions of China and mitigation strategies a review. The Rangeland Journal 34: 89–01.

Squires, V.R. and L.M. Hua. 2015. On the failure to control overgrazing and land degradation in China's pastoral lands: Implications for policy and for the research agenda. pp. 19–42. *In*: Victor R. Squires (ed.). Rangeland Ecology, Management and Conservation Benefits. NOVA Science Publishers, N.Y.

Squires, V.R. 2017. Desertification and drought. *In*: Saeid Eslamian (ed.). Handbook of Drought and Water Scarcity, Vol. 1. Principles of Drought and Water Scarcity. CRC Press, Boca Raton.

Squires, V., L.M. Hua, D. Zhang and G. Li. 2010. Towards Sustainable use of Rangelands in north-west China. Springer, Dordrecht.

Wang, C.T., Q.J. Wang, Z.X. Shen et al. 2003. Response of biodiversity and productivity to simulated rainfall on an alpine *Kobresia humilis* meadow. Acta Botanica Boreali-occidentalia Sinica 23(10): 1713–1718 (in Chinese).

Yao, T.D., D.H. Qin, Y.P. Shen, L. Zhao and N.L. Wang. 2013. Cryospheric changes and their impacts on regional water cycle and ecological conditions in the Qinghai Tibetan Plateau. Chinese Journal of Nature 35: 179–186 (in Chinese).

Ye, H.R. 2014. Study on the Effects of Climate Change and Adaptation Measures of Herdsmen to the Climate Change Inner Mongolia. Inner Mongolia University (in Chinese).

Yi, X.S., Y.Y. Yin, G.S. Li and J.T. Peng. 2011. Temperature variation in recent 50 years in the Three-River Headwaters Region of Qinghai Province. Acta Geographica Sinica 66(11): 1451–1465.

Yin, F., X. Deng, Q. Jin, Y. Yuan and C. Zhao. 2014. The impacts of climate change and human activities on grassland productivity in Qinghai Province, China Front. Earth Sci. 8(1): 93–103.

Yu, H.Y. and J.C. Xu. 2009. Effects of climate change on vegetation on Qinghai-Tibet plateau: A review. Chinese Journal of Ecology 28(4): 747–754 (in Chinese).

Zhang, Y.D., H.J. Meng and N. Ta. 2010. The transfer of the right to grassland contractual management in pastoral area and its impact on pastoralist's livelihood. Pratacultural Science 27: 130–135 (in Chinese).

Zhang, Y.J., G.W. Yang, N. Liu, S.J. Chang and X.Y. Wang. 2013. Review of grassland management practices for carbon sequestration. Acta Prataculturae Sinica 22(2): 290–299 (in Chinese).

Zhao, D.S., S.H. Wu, Y.H. Yin et al. 2011.Variation trends of natural vegetation net primary productivity in China under climate change scenario. Chinese Journal of Applied Ecology 22(4): 897–904 (in Chinese).

Zhao, H.Y. 2007. Impacts of climate change on forage potential climate productivity in typical grassland. Chinese Journal of Agrometeorology 28(3): 281–284 (in Chinese).

17

Future Prospects of Grasslands and Ranching on the North American Scene

Paul F. Starrs[1,*] and *Lynn Huntsinger*[2]

Introduction

In this day and age, it makes little sense to discuss a biome or environment, such as grasslands, without taking into account the effects of human uses and abuses, conscious or otherwise, on habitat and ecological systems. Grazed grasslands are, in fact, almost invariably rangelands, until they are plowed or paved or put to another use, and we are their most aggressive agents of transformation. This volume would be remiss to ignore some of the complicated yet predictable ways that humans alter grassland environments. Grazing, plowing, harvesting, afforesting, urbanizing, crop-sowing or accidentally seeding invasive species are but a few of the ways in which the prospects of grasslands are changed. The dominant use of grasslands in North America remains grazing and, therefore, ranching can be either a force for stability, or an economy on the verge of substantial alteration, transformation, or even obliteration (Kolbert, 2014).

The examples are legion and telling, and one case is perhaps sufficiently illustrative—every state in the US (and many provinces of Canada and Mexico) now has a form of viticulture. The wine industry in California, a state that remains the jewel in the crown of American wine-making with 91 per cent of wine grape production, in 2010 created 3,09,000 jobs and attracted 20 million visitors, generating something like $125 billion for US tourism and wine-quaffing economy (Starrs and Goin, 2010; 235). But the kicker is that vast acreages of former grassland and rangeland, of late much of it upslope, have seen oaks and native forests torn out and grasslands disked and sprayed to provide space for trellised grapes and the pseudo-chateaus dear to the tourist trade (Fig. 17.1). This development is case by case—a broader pattern of land-use conversion to plowed and irrigated grain cropping is long afoot in the remaining tallgrass prairie, and especially, in the shortgrass prairie reaches of the Great Plains, which extend from west-central Canada to central Texas. These changes are not benign, since user changes in grasslands are often irreversible.

[1] Department of Geography, University of Nevada, Reno.
[2] Department of Environmental Science, Policy, and Management, University of California, Berkeley.
* Corresponding author: starrs@unr.edu

Fig. 17.1 Grasslands and woodlands in California, as in so many areas of North America, are threatened with conversion to other uses, as in this vineyard and winery in Napa, California, where grape vines now extend into the high slopes. (Photograph by Paul F. Starrs.)

Compelled change is something of a constant on grazed grasslands, even though livestock ranch-owners tend toward a philosophical conservatism that seeks to protect past practice and tradition. Because throughout North America extensive grazing often requires negotiated access to grassland and to upland territory in a variety of different ownerships, compromise and planning ahead are essential. Market forces, and in particular the effects of innovation on product demand, lead to changes in how livestock are raised, in the management and preservation of grassland environments, and to novelties in the marketing of grassland products (Starrs, 1998). So too do changing tastes influence matters of land management, especially where public lands are involved. Accepted wisdom for one generation may become anathema to the next, and often the 'victims' are grassland owners and managers, or federal or state agency employees whose role is stewardship of the public land. The machinations are indeed complex.

Some of the novelties involve new means of gaining financial profit from land; others actually employ grasslands to produce significant non-livestock products. These uses can involve recreation, watershed protection, agrotourism, carbon markets, mining via fracking, favoring wildlife over domesticated livestock, even building pipelines and forging other rights-of-way across land in ways that displace pastoralism and threaten a once relatively-stable grassland environment. Ecosystem services attainable with tweaks to grassland management are many and still evolving, and as in the European Union, may actually bring in additional and previously unattainable income from government, nonprofit, conservation, amenity-seekers and pollution-abatement sources. What geographer John Holmes has called in the Australian context a 'post-productivist transition' includes recognition of user-rights of indigenous First Peoples, policy initiatives, shifting societal views of what constitute appropriate uses of grasslands and rangelands and his writing raises the issue of who should reign over land-use decision-making (Holmes, 2002). Traditional uses are put to the question and a contest for power and authority over land is not just an urban versus rural issue; it bears directly on the state-derived authority of government. While these issues show a somewhat different face between Canada, the United States and Mexico, they do apply to each and merit mention in large measure because in an urbanizing and radicalizing world of the 21st century, there are signs aplenty of an accelerating confrontation over the most appropriate disposition and use of grasslands.

Adaptation to change is a necessity hundreds of years old for landowners and livestock graziers, and therefore no one should be entirely surprised by the resulting upsets. But nearly two decades along into the 21st century, there are novelties deserving a discussion.

Rewilding Grasslands: Buffalo Commons, International Efforts, Wildlife Triumphalism

Rewilding is conservation as a large-scale endeavor, now embraced as an action plan by a subset of the applied conservation biology scientific community. The origins of rewilding are rooted in the late 1980s, when environmental advocate Dave Foreman began migrating toward conservation activism, after nearly a decade of work with the direct action group, Earth First, which he helped found. A key concept in the rewilding vision is improving ecosystem function and mitigating lost biodiversity, emphasising the confidence of its supporters in 'cores, corridors and carnivores' (Soulé and Terborgh, 1999; Foreman, 2004). Alliteration aside, that description of rewilding emphasizes the role of ecological restoration, in part by aiming toward a mix of plant communities and wildlife species that more closely matched the function of biomes and ecotones from past eras, now undone by human intervention (Kolbert, 2012). Yet 'rewilding' is itself an active plan, not an argument for a passive removal that would abandon an environment to 'let nature take its course'. In the world of conservation, biology, and adaptive management, considerable attention focuses on wildlife and especially carnivorous mammals, which are seen as keystone species and whose return or population decline is taken to be an indication of broader environmental health. The circumstances in which such attention turned to endangered plant species or larger plant community ecological health are rare but not unknown: vernal pools (ephemeral spring-season ponds) in the eastern San Joaquin Valley of California is one case in point and, relatively speaking, a preservation success story; the plowing for cropland agriculture of the Nebraska Sandhills and drawing down of the Ogallala Aquifer constitute a thirty-year and larger-scale crisis (Wise, 2015; Parker and Wilson, 2016).

There are considerable costs and perhaps insuperable complications involved in returning grassland to an earlier state, and also controversies, evident in a 2005 proposal described in *Nature*, which argued that reserves in parts of the Great Plains should be re-wilded with an introduction of exotic carnivores and large mammals (Donlan et al., 2005). Working from a history of land abandonment and social justice, Frank and Deborah Popper argued in the late 1980s that areas of the north-western United States Great Plains—many of which hit a peak population in the 1870s and have ever since lost population—might be returned to native Americans and repopulated with bison (the American buffalo, *Bison bison*) (1987). A proposal for such a park, they note, was first voiced in 1842 (Popper and Popper, 2006). That initiative gained force with a Nature Conservancy effort to acquire properties appropriate for bison. In 2013, on 13 reserves totaling some 50,600 ha, including one reserve in Chihuahua province, Mexico, there grazed 5,600 bison. Polemics about presumptive 'wilderness' aside, there is an increasing emphasis on rewilding in various parts of the world, including the United States, Canada, and Mexico. International endeavors to re-wild by bringing back carnivores that once freely traveled grasslands and forested areas include U.S. Mexico collaboration on the Northern Jaguar project and the Mexican Gray Wolf project, with effort shared between the United States and Mexico, including some reserves purchased by non-profit organizations. Of course, attempts to bring back high-level predators is likely to be disrupted if an impermeable 'wall' is built along the US–Mexico border.

Government-Landowner Disputes: Feral/Wild Horses, Center for Biological Diversity, the Bundy Clan, Starving Governance Agencies

The question of grassland ownership complicates the available management choices. The divide between overwhelmingly private and substantially public grazing land affects grasslands as a matter of proprietorship. But public lands contain much of the grassland and rangeland west of the 100th meridian, a traditional boundary marking 'West' from the eastern half of the United States (Fig. 17.2) simply because large areas of those lands remained after the disposition of western territory, and Native American claims went largely unhonored in the 19th century. Protected areas in Mexico are not as common as in the United States; public domain lands in North America do have uses besides livestock grazing on leases taken out from federal or regional governments. The crucial issue is which the 'public' has the greatest claim to the use of public lands: adjoining locals with interests in livestock or mining or water, or, as an increasingly vocal alternative, distant populations that may consider public lands to be just as much theirs as the neighbors, who see nearby public lands as theirs for use by right of nearness and tradition. These parties often end up at odds. The 'goods' being fought over are the public lands themselves and the questions are about user rights: local or distant users? Use for financial profit, or ecological end? Managed by government agency, by local stewards or by more remote, but still entitled, interests?

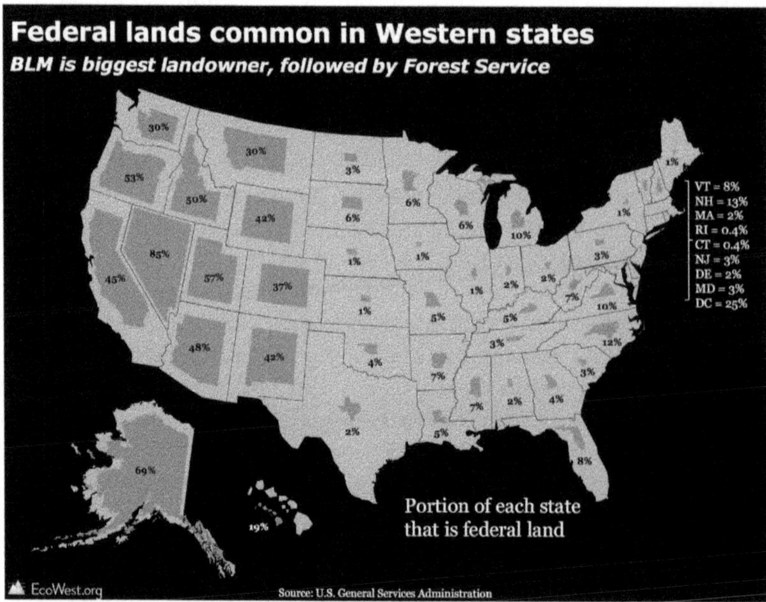

Fig. 17.2 While significant portions of the United States are public lands, largely in the American West, there is a striking absence of public land in the grasslands of the Midwest and Great Plains, in the center of the country. (Modified from EcoWest.org map redrawn from public data Ecowest, 2013).)

No shortage exists of appropriate studies illustrative of dilemmas in grassland management. One case, compelling because it involves ideology, invasive species, hard-edged single-issue NGOs, and resource-user anger, turns on issues associated with feral horses and burros, so-called wild horses and mustangs. These animals have a several hundred-year history of presence and use in grassland environments, and in the west of the United States, resoundingly caught the public's interest beginning in the post-World War II years (NRC, 2013).

There was in the United States a tradition dating back to the Civil War of finding horses or mules appropriate for artillery-hauling or mounted cavalry use, but a formal program needed remaking. During World War I, motorized cavalry was only in its infancy, so horses and mules remained an important part of a quartermaster's responsibilities for moving material and men. Western ranchers were paid more than reasonable fees for raising quality mounts and pack animals for the military, and across the American West, it was a common practice for mares to be turned out onto the range, once they were no longer prime animals for draft or riding use on a ranch. Although an informal programme in 1918, just three years later, in 1921, the United States Army Remount Service—a division within the Quartermaster Corps—was up and running. It continued through the next World War until 1948. The horses would periodically be gathered, the colts separated and kept, and every few years the studs would be removed from the herd to reduce consanguination or inbreeding; sometimes the Q-Corps would provide studs to be turned out with the now-loosed mares. The next couple of decades were problematic—demand was much reduced, except for horseflesh that could be sent to Mexico for ranch use, or to slaughterhouses, where demand abroad was met by harvesting meat from the mustang herds.

At issue, however, is as much public sentiment as the carrying capacity of grassland and rangeland. While these are feral horses by virtue of origin, their supporters—often a fervent lot—describe them pointedly as wild horses, and at times even argue that due to the Pleistocene presence of horses in North America, they should be treated as a native species. There is US Congressional precedent for the use of 'wild', in the 'Wild and Free-roaming Horse and Burro Act' of 1972. In the 45 years since that Act was passed, increases in herd size from Wyoming to south-eastern Oregon and across Nevada and parts of Utah are notable—yet also difficult to quantify with precision, according to a 2013 National Academy of Sciences report (Fig. 17.3; NRC, 2013). Horse and burro populations on the range certainly increased from 2005 onward and the number of animals in short-term and long-term holding facilities likewise increased (NRC, 2013). The costs of rounding up animals, completing veterinary inspection and care, and maintaining troughs and water supplies to sustain the horses and burros, are sizable. In fact, feeding animals kept in holding facilities have come to consume much of the BLM and US Forest Service budget originally allocated to developing a management plan and caring for herds in the wild (Garrott and Oli, 2013). Populations are estimated to grow 15 to 20 per cent per year, which suggests that there will be no letup of competition for public-land grazing-resources among the three contending groups: protected horses and burros, wildlife including deer, antelope, bighorn sheep and carnivores, and, finally, domesticated livestock. Wolves and mountain lions are not common in the areas where wild horses and burros reside, so there is little or no natural control of animal numbers, especially since current management practices keep herds to a size where animal numbers are not affected by limits to their food supply.

Costs involved in wild horse and burro management in grassland and rangeland environments are anything but incidental; across North America, with the arrival in the 1980s of neoliberalism, conviction grew among elements of the public—and assuredly among some legislators—that federal (Central) government involvement in land

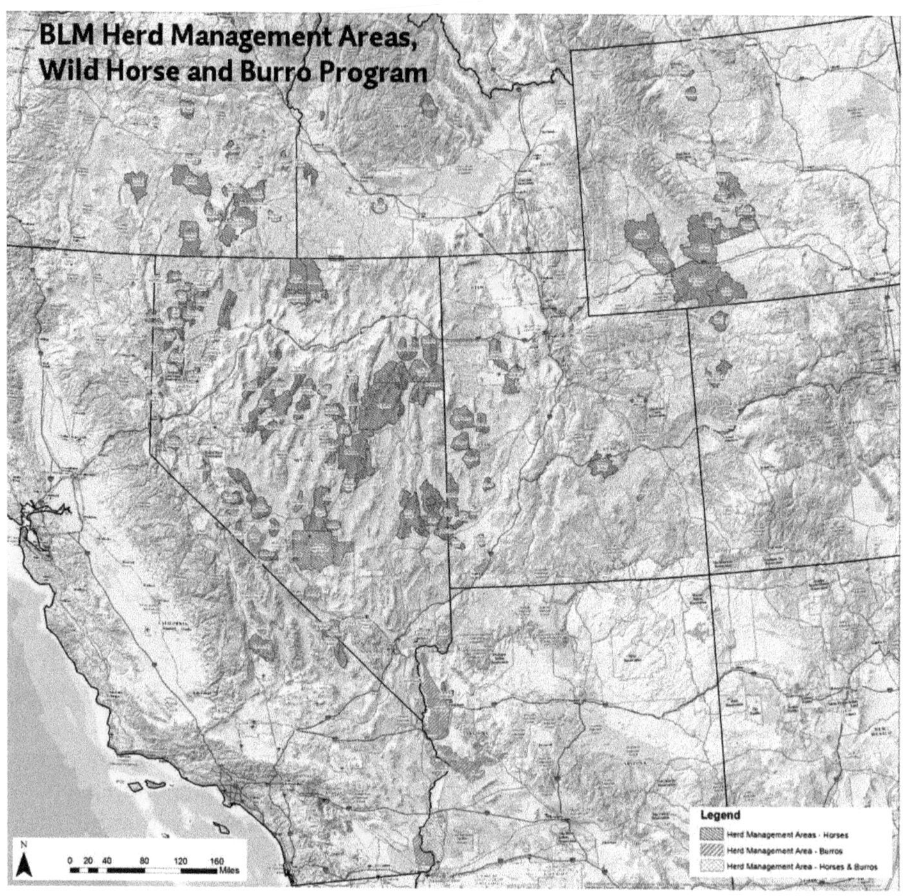

Fig. 17.3 Herd management areas for wild horses and burros on BLM land. (Map modified from original BLM product.)

management questions posed a problem. Starting with the presidency of Ronald Reagan in the United States, increasing effort was turned toward budget reduction for federal management agencies, which put fewer managers and land stewards on the ground. That has meant ever larger areas to be cared for but with ever fewer personnel, equipment, and budgetary resources: starvation, of a sort, by attrition and financial short-changing. Since wild (feral) horses are perhaps the quintessence of charismatic megafauna, in the public eye (Notzke, 2016), polemics about the most effective, humane, and needed management techniques will continue eddying as components of hard-fought opinion, and the limits of possibility are ultimately political and philosophical, rather than what pragmatically could be done (NRC, 2013; Boyd et al., 2017).

Nor do other prominent rangeland-grassland use questions fall from view. A second case in point, taken from circumstances in the western United States and Canada, involves both historical and current population distributions of the sage grouse (Fig. 17.4), which is referred to as an 'umbrella' species (Roberge and Angelstam, 2004). That term, along with keystone or flagship species, designates a species considered an important indicator of the success or failure of protection for habitats and communities of other animals (or plants). The status of an 'umbrella' species is examined especially when the topic at hand

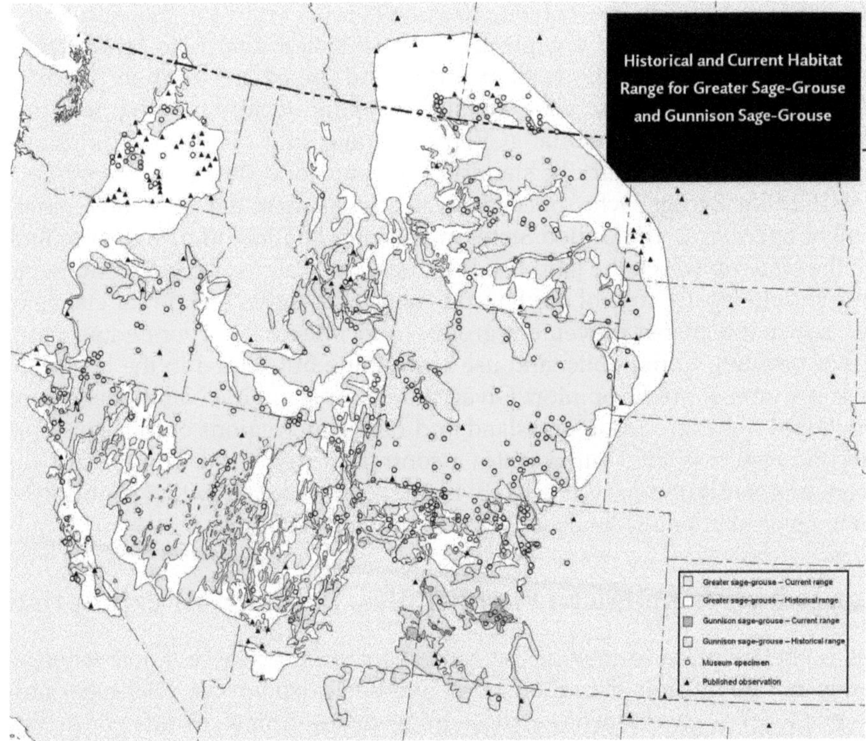

Fig. 17.4 The range of two species of Sage grouse in the western United States (light colors are historical range; darker shades show current distribution). (Modified from WDF&W, 2017; Schroeder et al., 2004.)

is identification of a conservation reserve or area and the protection of broader ecosystems is sometimes keyed to the well-being and population rise or fall of the umbrella species.

In California's Sierra Nevada, the fisher (an aggressive small predator, *Pekania pennanti*) is considered a keystone species. In the Great Basin and the Rocky mountains, extending into Canada, a similarly important umbrella species is believed to be the sage grouse (*Centrocercus minimus* and -*urophasianus*), a stout but handsome bird species that does best in relatively undisturbed habitat (France et al., 2008). Western states in the United States, and most prominently Nevada and Utah, have sought to come to management agreements that will modify grazing livestock regimes in a way that will protect the populations but will also prevent listing of the sage-grouse as a threatened or endangered species (Boyd and Svejcar, 2009). However, as of 2017, amid turnover in the federal administration, it seems that agreements reached as a result of long negotiations and extensive study may be summarily reexamined, overturned and stricken from the books (Flatt, 2017). Matters of interpretation of federal edicts and regulations are subject to hugely divergent interpretations, depending on the source of information and the politics of a report's author (USDI, 2017; D'Angelo, 2017).

The issues generated by the wild horse and burro debates and arguments associated with sage grouse and their habitat are typical enough, worldwide, of grassland and rangeland controversies, especially where the properties used are public lands and the use of those lands is subject to dispute. In other land-use and access arguments across the United States, and to a lesser degree in Canada and Mexico, strongly partisan opinions exist, as between the Tucson, Arizona-based Center for Biological Diversity, a potent nonprofit membership organization whose stated goal is to 'systematically and ambitiously use biological data, legal expertise and the citizen petition provision of the powerful Endangered Species Act to obtain sweeping, legally binding new protections for animals, plants and their habitat' (CBD, 2017), and such fair-use partisans as the 2014 Bundy Gold Butte and Bunkerville standoff protestors and the occupiers of the Malheur National Wildlife Refuge, who pitched an argument that the BLM and other federal government agencies in the United States are required by law to turn over federal public lands to the states in which the land resides. Legal trials are continuing, but the argument over who should legally control public lands and what rights remain for claims by states, counties, sovereign citizen movement groups, and potential corporate and family users desirous of profiting from public land use is anything but settled in the law of the courts and the courtroom of public opinion. Given the example of First Nation people in Canada and traditional rights groups in grassland and rangeland regions of Mexico, and with the 2016 election results in the United States a contrarian rightward turn toward support of central government supremacy, it is likely to the point of certainty that continued disputes over governance and public lands will be a fixture for years ahead.

Cession to Native American First Peoples, Border Wall Expropriation

There is no better place to look at the contest for property access and management of grasslands and rangelands than the United States-Mexico border. The issues at hand are not just one level of government versus another (federal, state, county) or private-public landowner. They extend far beyond, to questions of one sovereign state (Native American, as with the Tohono O'odham Nation) versus another state—the US federal government, for example. Who rules the roost? The answer is anything but simple, given the land-use controversies and history along the borderland. And given the post-2016 election results and a neo-conservative push to 'build the wall, a beautiful wall' between the United States and Mexico (Schmidt, 2016). The Tohono O'odham Nation, with some 25,000 members and a 75-mile-long swath of traditional land in tribal management totaling more than 1.1 m ha (2.8 million acres) straddling the United States–Mexican border, takes a collectively dim view of the federal government's 2017 proposal that would install a wall-blocking traditional Native American access to Mexico or, conversely, to the United States (Fig. 17.5). Traditions and tolerance dating back hundreds of years allow Tohono O'odham members ready passage across the United States-Mexico border with grazing animals, for health care and social visits, and to observe religious holidays, including the Feast of St. Francis of Assisi, who is a revered figure for some tribal members. An impermeable wall would interfere not only with these traditional rights of passage, but would disrupt the local economy, usurp land rights and challenge tribal sovereignty, divide wildlife populations and prevent migrations, and create a massive scar on a much-respected landscape.

The Tohono O'odham Nation's skepticism about the usefulness—and legality—of any such proposed 'wall' makes it a profoundly contested space. Recognized as a Native American sovereign nation within the United States, Tohono O'odham Nation tribal members have for centuries readily crossed to and from what the United States and Mexico claim are an international boundary. It is for the Tohono O'odham simply traditional territory, with a pesky series of bollards erected across the line, as a kind of prophylactic border barrier. But the Nation sees the issue of 'passage' as otherwise, with members in their view retaining ready anytime passage as a tribal right. Who, then, controls the rangeland,

Fig. 17.5 United States Border Patrol (ICE) officers at United States-Mexico border crossing on Tohono O'odham Nation lands, Arizona; photograph taken from T-O territory in the Mexican State of Sonora, on the south side of the boundary. (Photograph by Paul F. Starrs, 2014.)

livestock, and decision-making along this northern border section of Mexico? The grazing and other agricultural resources have been well known and much used through time, but the question of who controls access to those resources, and who makes binding decisions about profitable use of those lands has for some decades—and arguably for centuries—been vested in the Tohono O'odham Nation, a tenure arrangement which suggests that the 'border' is a beast of altogether another color. Ownership and governance among First Nation peoples, wherever on the North American continent, makes for complicated jurisdictions and rights adjudications (Dear, 2013).

In 2018, attempts by the administration to extract US budget advances for a putatively impenetrable wall defining and sealing the US-Mexico border, along a traverse of nearly 2,000 never easily traveled miles (3,200 km), are of concern to those living along the border (Berenstein and Santos, 2017). An addition to a continuing budget resolution in March 2017 implied a scoping budget of USD $3 billion for design and testing of a small section of the wall, but the total cost is anticipated to total $36–44 billion, depending on the design and how much land must be purchased outright or claimed, via legal action, through eminent domain proceedings—an eventuality with which the Tohono O'odham Nation, for one, is most unlikely to cooperate. These diplomatic negotiations taking place between three and often more than three sovereign powers are anything but clear cut. But the most damaging effects of the wall would certainly be on wildlife transit, disruption of property rights, religious pilgrimage observance and movement of livestock—none of them trivial. Attempts by federal and state wildlife officials to create transit paths across First Nation/Native American traditional lands, many of them long used for grazing and including sacred space, establish significant points of controversy that from time to time escalate to violence.

Disputes between nation-state sovereign powers and First Nation/Native American peoples in Canada, the United States of America and Mexico are unlikely to dissipate, especially when autocratic government edicts appear involved. In each of these three nation-states, the management of grassland-rangeland territories, frequently utilized by traditional peoples as grazing resources, is subject to periodic controversies and contentious outside claims of user-rights. And indigenous populations with a recognized claim to sovereignty can break from the rulings of a nation-state government, as has happened prominently in Canada and the United States, and may yet happen in Mexico. One example, taking place entirely on grasslands and rangelands, is a reintroduction by the Confederated Tribes of the Colville Reservation, in eastern Washington state. There, tribal wildlife biologists and reservation residents have brought in outside populations of pronghorn antelope to reintroduce on tribal land an animal once native there, but hunted to near disappearance (Tobias, 2017). Such disputes actually pull together many of the themes of this chapter, including rewilding, government/landowner disputes, land use and management needs of traditional/Native peoples, borderland access and transit rights, and significant ecosystem services and nonmonetary uses of land—in effect, a witches brew of complex political-economic-policy issues.

Mining and Pipelines and Rights-of-way

Grasslands and rangelands are not only public and private properties and economic resources, they are, in addition, venues. In, on, and under grasslands occur a number of activities that, if not necessarily directly destructive, can nonetheless have ancillary effects on grassland health and impede some of the more effective management techniques. These involve, most of all, the extraction or movement-across of resources. Think of it as a transhumance of economically-significant products, which are either obtained from underneath or above grasslands (as with the fracking of oil-rich shale or wind energy, collected from towers installed in grassland environment), or the movement of oil, accumulated oil-shale residue, or the wheeling of electricity above and across grasslands or rangeland (Fig. 17.6, Fig. 17.7). The Great Plains, with their vast extent from Canada, nearly into Mexico, are relatively unaffected by urbanization. They offer a great surface, relatively level and minimally modified, across which above-ground pipelines, high-voltage transmission lines, unit-train rail lines, or subsurface conveyances can be established to move goods across long distances. These generally run north-to-south, but sometimes the reverse, gridding the North American continent. Changes in technology that have made fracking possible—as one case in point—alter not only regional ecological dynamics, they reshape economic, transportation, and international treaties and political relationships. These are not trifling matters; they indicate novel ways that the three countries on the continent are being compelled to deal with one another.

The legal predicates are anything but simple. Laws associated with land use and land ownership in North America sometimes grant tenure in fee-simple, with ownership of surface, subsurface, and air rights vested in the title holder. But it is quite common in Canada, the United States, and Mexico, for land rights to be separated, for tenure to involve a 'package' or a bundle of rights and goods that can be severed from the property title. This has made conservation easements possible; for example, since rights to develop a property can be split off, donated, or sold and held by another party than the person who owns and occupies a terrain. In the case of conservation easements that draw on surface rights, a property may be retasked from a feature that is likely to be converted to ranchettes or smaller parcels into a site saved for grazing, cropping, other agriculture uses, or preserved as a cultural or otherwise protected feature. But the bundle of rights can extend to other features of a property. Air rights, for example, are critical for wind energy production, or

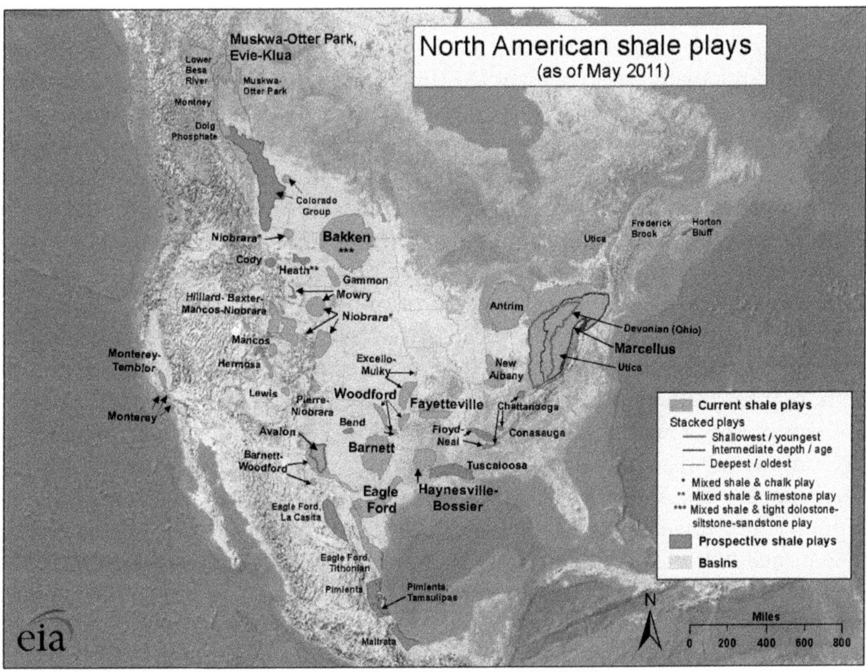

Fig. 17.6 'North American Shale Plays', undertaken to remove oil and natural gas products from shale deposits, take place on grassland and rangeland areas from Canada and Alaska through the rest of the United States and extend into Mexico. (Map by EIA, 2011.)

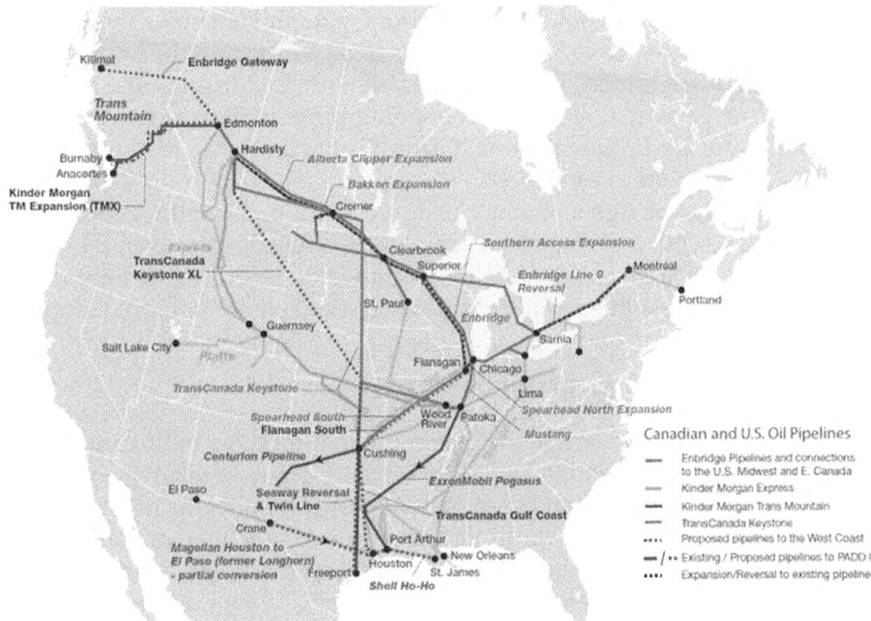

Fig. 17.7 Generalized map showing Canadian and US Oil Pipelines in 2012; includes in-progress (2017) TransCanada Keystone XL. (Map by NOAA-ORR, 2012.)

for the establishment of electrical transmissions lines. But while they may seem ephemeral or even illusory to the casual observer, they are not in terms of the law of property and persuasion. And the installation of sizable numbers of wind energy turbines or solar panel industrial-scale fields, which have served to make all the prairie-plains states substantial renewable energy producers also requires high-voltage electrical transmission lines, since some of the locations where the aeolian power is produced, like western Nebraska and the Dakotas, have in fact been losing residents, overall, for something on the order of 150 years. Energy sources, whether produced renewably on-site, or whether transported as a raw source to distant power plants by way of pipelines or other transit systems, must move their products, which requires a means of transportation.

A proliferation of exploitive opportunities that help 'develop' grassland and rangeland environments take advantage of what it perceived as open space, but it requires a degree of protection from transit routes across an otherwise unimpeded surface. Wind energy turbines, prominent from Canada to Texas, generate not just electricity from downslope Rocky Mountain winds, they also produce controversy, from landowners who may not fancy the high towers and scarred landscapes produced by installation. Nor are high-tension powerlines always seen as an undiluted good; they present obstacles of various sorts: visual, aesthetic, in rights-of-way, to wildlife (aerial, subterranean, surficial), and to the transit of other denizens of the terrain. The same can be said, but even more so, of movement systems for extractive resources, including coal, natural gas, and oil shale, which require substantial disturbance of a grassland environment. Public dissent, once relatively complacent, is rising as the profits and the prevalence of such uses become more prominent—and accidental pipeline spills and derailment damage terrain.

While recognizably useful for moving energy from one location to another, pipelines have on occasion produced sizable controversy, as in the prominent cases of the TransCanada Keystone XL pipeline and the Dakota Access Pipeline. The first, after much discussion, was blocked by the Obama Administration—until the new administration decided in 2017 that the Keystone XL pipeline could indeed be built and implemented. The Dakota Access Pipeline, designed to move up to 5,70,000 barrels of oil a day from the Bakken oil fields in North Dakota to western Illinois, will cross beneath both the Missouri and the Mississippi rivers. The pipeline generated sizable and ongoing protests from elders of the Standing Rock Sioux Tribe, but US President Donald Trump signed an executive order to advance construction of the pipeline in early 2017.

Pipelines, mining, and rights of way constitute at a minimum three separate features that can disrupt grassland and rangeland ecosystems. They do, however, make it possible for important natural resources to be tapped, exhumed or generated, and transferred from one location to another—at the cost of considerable social and ecological disruption.

Savory and Alternative Grazing Systems

Grazing systems, or ways to manage the grazing of livestock, are built around controlling factors that influence the amount and selectivity of plant defoliation by the grazing animal. By choosing particular kinds of animals of particular phenological states (growing, lactating, maintenance, diet, etc.), controlling the time of year they graze, how long they graze, and what forage they have access to, different patterns of defoliation and hoof traffic can be created. Selectivity is one of the more interesting factors driving animal grazing. Unlike fire or mowing, animals select the plants they eat, sometimes preferring younger or older plants, or one species over another, for example. At the landscape scale, grazing animals are selective, preferring to go downhill, as with cattle, or uphill, as with bighorn sheep, or near water, as with cattle, or far from water, as is sometimes the case with horses and so on. Knowing the way grazing animals behave and respond to their environment is the key to managing grazing.

The selectivity of animals can be easily manipulated. The same number of animals, when confined to a smaller area, will be less selective, hence the idea of rotational or mob grazing. Even if the total amount of forage is same and the total area and number of livestock is same, if you confine the animals to smaller subsections of the total area for shorter periods of time, they will not be able to be as selective as if they were allowed to roam the full area freely through the entire grazing period. Of course, the total amount of forage versus animals remains important in both cases. But in general, if the goal is to obtain a more even use, or to make sure animals eat even the things they really don't want to eat, creating effects similar to those from mowing, then reducing selectivity is a good strategy. On the other hand, if you are working with animal preferences, for example if the goal is to reduce grasses and leave flowers for pollinators, or short plants around vernal pools (ephemeral spring ponds), allowing animals like cattle, that prefer grass, to fully express their selectivity is ideal.

A variety of strategies for creating 'grazing systems' using these factors have been promulgated, most notably, Holistic Resource Management (HRM), which is a package of practices, goals, and ideas that is marketed to livestock producers and includes practices for working with one's family and setting personal goals. While most rangeland scientists would prefer to draw on the large menu of available choices in grazing management and develop a grazing system from scratch based on the specific goals of the manager and the characteristics of the ecosystem, HRM has proven popular with some livestock producers. Livestock grazing in the United States is blamed wholesale for a variety of ills, and research shows that society's negative attitudes toward grazing frustrate rangeland livestock producers (Liffman et al., 2000). HRM puts a positive spin on grazing, but to a degree and in ways that some find unjustified by scientific results (Bartolome, 1989; Briske et al., 2013; Ketcham, 2017). Nevertheless, some conservation organizations and individuals find it a satisfactory way to reassure their constituencies and perhaps themselves that they are doing a good thing. In fact, the HRM emphasis on human relationships, including family goal setting, and holistic economic planning that recognizes intangible values, has historically been much too absent from classic extension approaches (Ruyle, 2000).

Grassland Conservation and Ecosystem Services

Threats to grasslands occur at multiple scales (Huntsinger and Oviedo, 2014). At the landscape scale, conversion of grassland to cropland is a threat to rangeland ecosystems and to grazing, wildlife habitat and numerous ecosystem services that they provide. For example, Canada for over a hundred years attempted to slow and stop aggressive homesteading in the western reaches where arriving land claimants sought to convert large areas of southern Manitoba, Alberta, Saskatchewan and British Columbia to cropland. Instead, provincial governments, with permission from Central government in Ottawa, came around to provide long-term grazing leases throughout what Canadians refer to as the Prairie region. While nearly 4 million head of cattle graze there, wildlife populations are not only important in regional history, they are increasingly significant in their influence on plants, soils, hydrology and ecosystem services. Combined systems producing ecosystem services and livestock products on grasslands have come to be known as working landscapes (Huntsinger and Sayre, 2007). Working landscapes have become an important justification for grassland or rangeland conservation with grazing, because such grasslands provide multiple public and private goods, as well as food (Barry, 2015).

Beyond conversion to cropland, there are still ample threats at the landscape scale to low-intensity grazing use, which complicate the preservation of a grassland-rangeland-prairie ecosystem. These include land-use conversion into housing, aggressive use of some formerly public lands for recreation and alternative crops, the installation of turbines for the production of wind energy and sizable projects that create farms of solar panels for regional electricity generation. Further, there are ongoing forest harvests, invasive exotic range plant species, powerline rights of way, and oil and natural gas extraction that increasingly rely on the use of fracking techniques, each representing land use changes that disrupt habitat, corridors and wildlife populations. Wildland fires as a result of long-enforced fire suppression regimes, climate change and poor vegetation management have for nearly 50 years created an ongoing enhanced likelihood of devastation by flame.

At the plant community level, too often the impacts of livestock grazing are conflated with the impacts of poor grazing management. Grazing is not a black box. Livestock grazing is a form of agricultural production and as with other forms of agricultural production, it changes ecosystems. There are complex tradeoffs involved in these changes linked to practices and the way it is managed. The governance, economic conditions, cultures and social values of producers influence grazing management as much as crop management, sometimes, particularly in poverty-stricken areas, leading to abuses as people eke out a living. Where distant investors drive agricultural development, production may lack sensitivity to local human needs and environmental conditions, also resulting in abuses. In fact, numerous techniques can be used to create grazing impacts that minimize impacts on species of concern, maximize impacts on undesirable species, or that actively improve habitat for wildlife. Grazing can be used to reduce fuel loadings, changing wildfire characteristics and impacts, potentially protecting forests and habitations. In the United States and Canada, livestock grazing is seldom subsistence oriented, and is often a local enterprise, leaving considerable flexibility in grazing management. In Mexico, a mix of subsistence ranchers and capitalist ventures, as well as a diversity of cultures and lifeways, create a more complex portfolio of grazing drivers, practices, and outcomes.

At the landscape scale, in the United States, conservation easements are an increasingly popular market for ecosystem services. Conservation easements are created when a rancher sells or donates the development rights to private property to a conservation organization in perpetuity. However, at present in the United States, the major market for ecosystem services from rangeland is largely a private one (Oviedo et al., 2017). To address poor management, poverty alleviation, regulation, education and incentives are used (Huntsinger and Oviedo, 2014).

Carbon Sequestration on Grasslands

While the per-hectare carbon capture potential of grasslands is less than either cropland or forests, because grasslands are so extensive, existing research suggests modest changes in carbon storage on rangelands can potentially alter the global carbon cycle (Ritten et al., 2012). Rangelands harbor around 30 per cent of the terrestrial plant carbon on the earth (Booker et al., 2013). Grasslands store a large proportion of their carbon underground in roots, where it is relatively safe from the fires that plague forests and shrublands in arid areas. On the arid and semi-arid sites, typical of rangelands, annual fluxes are small and

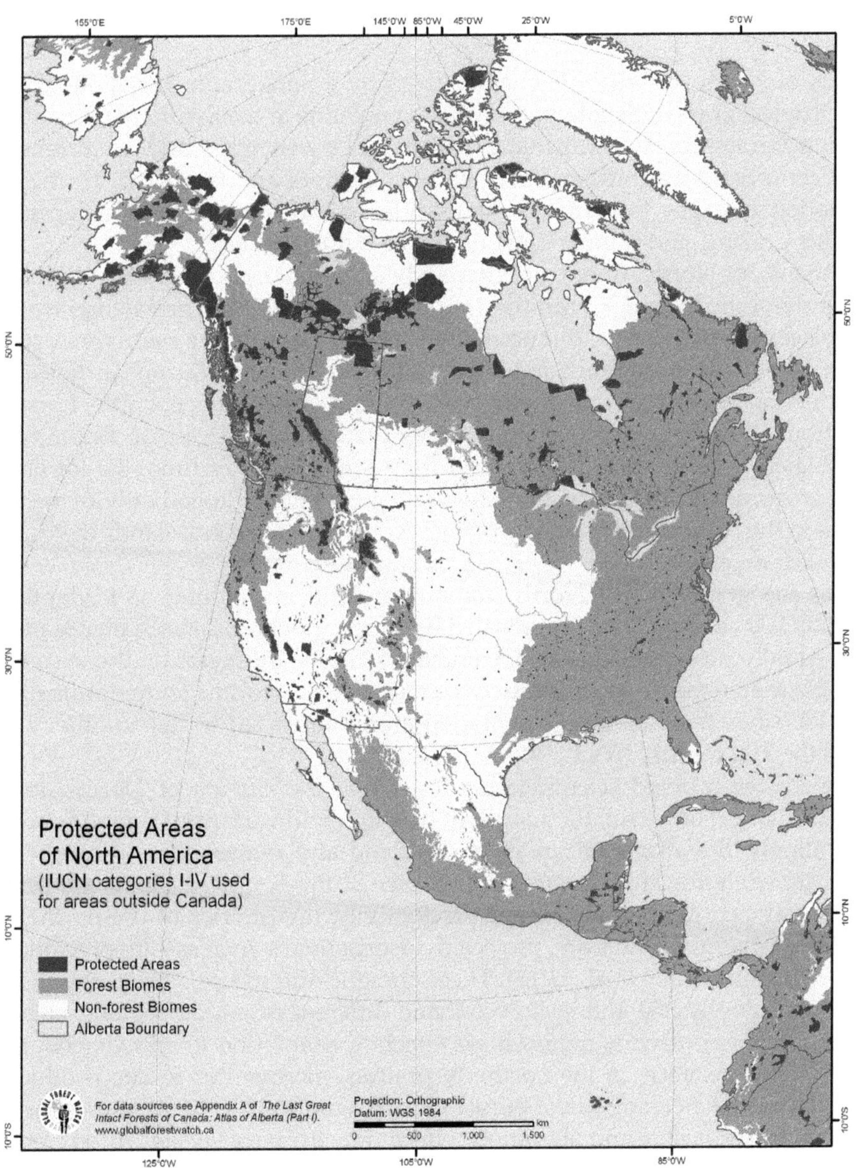

Fig. 17.8 Protected areas of North America. (Sourced from Global Forest Watch Canada, 2009.)

unpredictable over time and space, varying primarily with precipitation, but also with soils and vegetation. Unfortunately, revenue streams from carbon markets alone are modest for rangelands, but if bundled with wildlife, watershed, and recreational benefits, could augment the values paid under the ecosystem services scheme, or in certain markets, as with those for conservation easements in the United States (Fig. 17.8). Policy and management initiatives should seek long-term protection of rangelands and rangeland soils to conserve carbon and a broader range of environmental and social benefits (Booker et al., 2013).

Conclusion

Grasslands of North America are characterized by a remarkable variation across space. From subtropical grasslands interspersed with wetlands in the Southern Coastal Plains, to the arid bunchgrass-thorny shrub matrix of the Hot Deserts, grassland environments of the continent comprise a magnificent diversity of soils, flora and fauna (Ch 11, this volume). This variation emerges from the complex interaction of geo-topo-edaphic properties, climate, fire, grazing, and human inhabitation on a huge canvas.

The history of North American grasslands is one of a grassland not exposed to the grazing of domestic livestock until the 16th century. Before that, grasslands were grazed, in some places intensively, for thousands of years by large native herbivores, such as the American bison, elk, pronghorn antelope and deer, but with European livestock came the introduction of seed and managed herd animals (goats, sheep, cattle, horses, swine, poultry) from all over the world. As with other places where the plant communities were largely isolated from the rest of the world for a long period of time, the introduction of livestock, crops, and European technologies had great impact. Together with the plow, and alterations in the pattern and types of grazing, North American grasslands have undergone a profound change.

Just as the grasslands of North America are still responding and adapting to the introduction 500 years ago of domesticated livestock, global climate change is projected to wield an equally profound—but unfortunately swifter—influence on the system. Due to anthropogenic increases in greenhouse gas emissions, the world's temperature has already risen by 1°C since the Industrial Revolution and is projected to rise another 1°C by the middle of the 21st century (IPCC, 2014).

There is now a broad scientific consensus that the climate of North America will continue to change; however, the nature of that flux will vary from region to region and, correspondingly, the effects on areas of grassland and rangeland will not be uniform. For instance, while the Hot Deserts and prairies of the Southern Plains are projected to experience reduced rainfall and forage productivity, the prairies of the northern United States and southern Canada are projected to experience increase in precipitation and forage production (Polley et al., 2013). These climatic differences among regions, coupled with inherent biophysical and socio-economic differences among regions, suggests that a region-specific approach is required for effective adaptation to climate change (Briske et al., 2015). For instance, in the northerly prairies, increases in forage production may allow for increased stocking rates, but producers will need to be prepared for potential reduction in forage *quality* and alterations effected by invasive species (Briske et al., 2015). To the south, from the central Great Plains and the American southwest through central Mexico, livestock production will likely require novel strategies including using smaller ruminants, such as sheep and goats, or cattle highly adapted to arid conditions (Anderson et al., 2015), or shifting operations from cow-calf to stocker cattle, opportunistically as conditions change (Torell et al., 2010; Briske et al., 2015). Overall, while all producers stand to benefit from enhancing their climate contingency planning, opportunities and constraints are likely take on a highly regional nature.

The term 'anthropocene' has come on to the scene with considerable vigor—Google Scholar counts 16,400 published uses of the term since 2016, yet grasslands around the world have long been shaped and manipulated by human influence. In the case of North American grasslands and rangelands, which are the remit of this chapter, it is no exaggeration to say that the pace of change certainly has picked up in the last century or two, thanks largely to technological innovations (the steel plow, wire-and-steel-post fencing, improved pumping and irrigation technology, novelties in transportation and the wheeling of energy through power grids). But the most fundamental changes on the North American grassland scene involve freer global markets, agricultural specialization, global climate change and perhaps, most of all urbanization, which thanks to growth in the human population imposes different requirements on lands that once were the domain largely of undomesticated native herbivores and their human predators (Kolbert, 2014; Ceballos et al., 2017). Grasslands had a central role in human evolution. The fossil record lends weight to the idea that humans developed key traits, such as flexible diets, large brains, complex social structures and the ability to walk and run on two legs while adapting to the spread of open grasslands. Humans had to descend from the arboreal homes of our primate ancestors, develop the ability to move rapidly over open ground, and evolve the social skills needed for survival (Uno et al., 2016). Today we live in an era of human dominance—but that dominance needs to make room for large and contiguous grasslands that connect us to our past and provide us and our animal and biotic neighbors sustenance, biodiversity, habitat, carbon sequestration, watershed and vast beauty.

1 *The Central Coast Rangelands Coalition: Veronica Yovobich*

We exist to catalyze awareness of rangeland stewardship, including a cooperative, experimental approach to sustain biological diversity and natural resources and to enhance the wisdom and economic viability of rangeland stewards in the Central Coast area of California.

—Central Coast Rangeland Coalition, Holistic Goal

Over the past century, the traditional ranching operations found in the oak savannas, grasslands and coastal prairies of California's central coast have been subdivided and converted to a vast sea of human development. It has become increasingly important to manage the remaining rangelands with an eye towards preserving both ecological and economic sustainability.

In recognition of the unique challenges faced by rangeland managers and livestock operators in the greater San Francisco Bay Area, the Central Coast Rangeland Coalition (CCRC) formed as a group committed to using current science to promote sound rangeland stewardship. To meet ecosystem and civic goals, a collaborative partnership was formed between ranchers, conservation groups, agency personnel, researchers, students, land managers, policy makers, consultants, and others. Together, this group acts synergistically to exchange the resources, ideas, and expertise necessary to succeed in addressing conservation goals across the landscape. As a team, the group fosters constructive dialogue, strives to provide benefits across the community, and serves as a model for cooperative conservation and economic problem-solving beyond the Central Coast. One of the hallmarks of a CCRC meeting is the use of breakout groups after each significant segment of a workshop, to allow everyone to think and discuss reactions and ideas about material just presented.

The CCRC's diverse members represent some 52,000 ha of private rangeland, and many more thousands of acres of public rangeland across California's Central Coast, including Sonoma, Napa, Marin, Contra Costa, Alameda, San Francisco, Santa Clara, San Mateo, Santa Cruz, San Benito, Monterey, and San Luis Obispo counties. Established in 2002, the CCRC was born from the understanding that this ecoregion has a distinctive combination of climate, plant and animal communities, human presence, and rangeland use opportunities.

The group's primary focus is on collating and disseminating information on contemporary advances in relevant tools and scientific research. They agreed early on that they would not advocate for any specific management style and that anything presented should be validated by scientific research or expert opinion. Topics of workshops have included forage production, soil stability, watershed function, nutrient cycling, biological diversity, habitat quality, livestock safety, water quality, native and invasive species, vegetation structure, rancher economics, and grazing systems.

The CCRC holds meetings twice each year where experts present and discuss their work and share their experiences; members participate in on-site, field-based excursions to see these practices in action; and members participate in small group activities to facilitate information sharing. In addition, the CCRC is in the sixth year of the Rancher, Manager, and Scientist Forum on Rangeland Conservation, a funding opportunity that supports graduate student literature reviews on priority topics and maintains an online archive of past workshop presentations and related information.

Ecological processes do not begin and end at property boundaries; appropriate management requires the kind of holistic and collaborative approach that groups like the CCRC provide.

—by Veronica Yovobich, PhD cand., University of California, Santa Cruz

2 *Janos Biosphere Reserve, Chihuahua, Mexico: Rodrigo Sierra-Corona*

The Janos Biosphere Reserve (526,482 ha) is located in the north western corner of the Mexican state of Chihuahua. It borders the Mexican state of Sonora to the west and New Mexico, USA, to the north. The Reserve was established in 2009 by the Mexican federal government principally to protect vast expanses of native Chihuahuan Desert grasslands, one of the most threatened ecosystems in Mexico. As in many of the grasslands on our planet, a history of poor cattle management and ongoing expansion of industrial agriculture is menacing the area's ecological integrity and function (List et al., 2010; Sierra-Corona et al., 2015; Ponce-Guevara et al., 2016; Hruska et al., 2017).

The Reserve's grasslands comprise the southern extent of distribution for species like black-tailed prairie dogs (*Cynomys ludovicianus*), American bison (*Bison bison*), and northern porcupine (*Erithizon dorsatum*), and the northern extent for jaguar (*Panthera onca*) and the thick billed parrot (*Rhyncopsyta pachirrinca*). The Reserve is also home to species of conservation interest, like the golden eagle (*Aquila chrysaetus*), black bear (*Ursus americanus*), pronghorn (*Antilocapra americana*), and Mexican wolf (*Canis lupus baileyii*) present in the region (Ceballos et al., 2010; List et al., 2010).

Unlike in the United States, black-tailed prairie dogs are protected by Mexican federal law and its presence within this grassland mosaic is a main conservation justification (Ceballos et al., 2010; Davidson et al., 2010; Hruska et al., 2017). The Janos-Casas Grandes black-tailed prairie dog complex was in 1988 the biggest in North America, at 45,000 ha (Ceballos et al., 2010). But after the poisoning efforts carried out by the U.S. government during the late 1800s and early 1900s, black-tailed populations were wiped out from Arizona, southern New Mexico, and more than 90 per cent of its historical range (Miller et al., 2007). While Mexican ranch owners independently expend time and resources trying to eradicate this species, the lack of government support has at least contributed to this colony complex's ability to persist (Ceballos et al., 2010).

Over the last 25 years, the Janos-Cases Grandes black-tailed population has declined dramatically, with no more than 5,000 hectares of active colonies. This decline is still poorly understood and appears to have many causes, of which only the most obvious is the expansion of irrigated agriculture. Heavy grazing and changing annual rainfall patterns are likely contributors. Climate change, which is already felt in more frequent severe droughts, hard freezes, and short winters, is expected to worsen with time.

Currently, a consortium of NGOs, universities and the Mexican federal government (CONANP) are jointly engaged in efforts to halt the decline of the Janos grasslands. Projects include wildlife reintroductions (e.g., black-footed ferret and Mexican wolves), extensive monitoring programmes and outreach to cattle producers—both large ranchers and smallholders. Grazing workshops, demonstration projects, and informal partnerships aim to convince cattle producers that changing management practices can generate greater economic returns while providing better wildlife habitat. While the Janos Biosphere Reserve management plan's prohibition on agricultural expansion has yet to be seriously enforced, the protected area designation has succeeded in garnering much greater attention and funding for the area. Ongoing research on both ecological and agricultural dynamics, paired with landowner collaboration, appear to be making slow progress in the long-term conservation of this remarkable region.

—by Rodrigo Sierra-Corona, Laboratorio de Ecologia y Conservacíon de Fauna Silvestre Instituto de Ecologia, UNAM, Mexico

References

Anderson, D.M., R.E. Estell, A.L. Gonzalez, A.F. Cibils and L.A. Torell. 2015. Criollo Cattle: Heritage genetics for arid landscapes. Rangelands 37: 62–67.

Barry, S.B. 2015. Understanding working landscapes: The benefits of livestock grazing California's annual grasslands. ANR Publication 8517. http://anrcatalog.ucanr.edu/pdf/8517.pdf <accessed June 2017>.

Bartolome, J.W. 1989. Book review: Holistic resource management. J. Soil Water Cons. 44: 591–592.

Berenstein, E. and F. Santos. 2017. How cowboys make deals at the border. *New York Times*, 14 July. https://www.nytimes.com/video/us/100000005173759/how-cowboys-make-deals-at-the-border.html <accessed July 2017>.

Booker, K., L. Huntsinger, J.W. Bartolome, N. Sayre and W. Stewart. 2013. What can ecological science tell us about opportunities for carbon sequestration on rangelands? J. Global Envi. Change 23: 240–251. http://dx.doi.org/10.1016/j.gloenvcha.2012.10.001 <accessed June 2017>.

Boyd, C.S. and T.J. Svejcar. 2009. Managing complex problems in rangeland ecosystems. Rangeland Ecol. & Mgmt. 62: 491–499.

Boyd, C.S., K.W. Davies and G.H. Collins. 2017. Impacts of feral horse use on herbaceous riparian vegetation within a sagebrush steppe ecosystem. Rangeland Ecol. & Mgmt. 70: 411–417.

Briske, D.D. et al. 2013. The Savory Method can not green deserts or reverse climate change. Rangelands 35(5): 72–74.

Briske, D.D., L.A. Joyce, H.W. Polley, J.R. Brown, K. Wolter, J.A. Morgan, V.A. McCarl and D.W. Bailey. 2015. Climate-change adaptation on rangelands: Linking regional exposure with diverse adaptive capacity. Front Ecol. and the Envi. 13: 249–256.

CBD [Center for Biological Diversity]. 2017. Center for Biological Diversity: Our Story. http://www.biologicaldiversity.org/about/story/index.html <accessed 04 June 2017>.

Ceballos, G., A. Davidson, R. List, J. Pacheco, P. Manzano-Fischer, G. Santos-Barrera and J. Cruzado. 2010. Rapid decline of a grassland system and its ecological and conservation implications. PLoS One 5. DOI: 10.1371/journal.pone.0008562.

Ceballos, G., P.R. Ehrlich and R. Dirzo. 2017. Biological annihilation via the ongoing sixth mass extinction signaled by vertebrate population losses and declines PNAS 2017; published ahead of print July 10, 2017. DOI: 10.1073/pnas.1704949114.

D'Angelo, C. 2017. Interior Secretary takes aim at Obama-era sage-grouse protections: Ryan Zinke's oil-friendly order 'might just have landed the decisive blow' against the imperiled bird, a conservationist said. Huff Post, 07 June, http://www.huffingtonpost.com/entry/ryan-zinke-sage-grouse_us_593847a0e4b0c5a35c9b5791 <accessed 11 June 2017>.

Davidson, A.D., E. Ponce, D.C. Lightfoot, E.L. Fredrickson, J.H. Brown, J. Cruzado, S.L. Brantley, R. Sierra-Corona, R. List, D. Toledo and G. Ceballos. 2010. Rapid response of a grassland ecosystem to an experimental manipulation of a keystone rodent and domestic livestock. Ecology 91: 3189–3200.

Dear, M. 2013. Why Walls won't Work: Repairing the US-Mexico Divide. New York: Oxford University Press, 288 p.

Donlan, J., H.W. Greene, J. Berger et al. 2005. Re-wilding North America. Nature (18 August) 436: 913–914.

Ecowest. 2013. Portion of each state that is federal land. Slide 4. https://www.slideshare.net/ecowest/land-ownership-in-the-american-west <accessed July 2017>.

EIA [US Department of Energy, Energy Information Administration]. 2011. Natural Gas Explained. https://www.eia.gov/energyexplained/index.cfm?page=natural_gas_where <accessed July 2017>.

Flatt, C. 2017. Interior Head orders reconsideration of sage grouse protections. *OPB News*, 07 June. http://www.opb.org/news/article/zinke-orders-sage-grouse-plan-review/#.WTmxP02zqQ0.twitter <accessed 21 June 2017>.

Foreman, D. 2004. Rewilding North America: A Vision for Conservation in the 21st Century. Washington DC: Island Press, 219 p.

France, K.A., D.C. Ganskopp and C.S. Boyd. 2008. Interspace/undercanopy foraging patterns of beef cattle in sagebrush habitats. Rangel. Ecol. & Mgmt. 61: 389–393. https://doi.org/10.2111/06-072.1 <accessed July 2017>.

Garrott, R.A. and M.K. Oli. 2013. A Critical crossroad for BLM's wild horse program. Science 23 August 341(6148): 847–848. DOI: 10.1126/science.1240280 <accessed July 2017>.

Global Forest Watch Canada. 2009. Protected areas of North America, in maps—The Last great intact forest landscapes of Canada: Atlas of Alberta: Part I; Map Section A. http://globalforestwatch.ca/publications/20090402A_MapsI <accessed 14 June 2017>.

Holmes, J. 2002. Diversity and change in Australia's rangelands: A post-productivist transition with a difference? Trans. Instit. of Brit. Geog. 27: 362–384.

Hruska, T., D. Toledo, R. Sierra-Corona and V. Solis-Gracia. 2017. Social-ecological dynamics of change and restoration attempts in the Chihuahuan Desert grasslands of Janos Biosphere Reserve, Mexico. Plant Ecol. 218: 67–80.

Huntsinger, L. 2016. The tragedy of the common narrative: Re-telling degradation in the American West. pp. 293–326. *In*: R.H. Behnke and M. Mortimore (eds.). The End of Desertification? Disputing Environmental Change in the Drylands. Berlin: Springer Earth Systems Sciences, Springer-Verlag, 560 p. DOI: 10.1007/978-3-642-16014-1 <accessed July 2017>.

Huntsinger, L. and N. Sayre. 2007. Introduction: The working landscapes special issue. Rangelands 23: 9–13.

Huntsinger, L. and J.L. Oviedo. 2014. Ecosystem services may be better termed social ecological services in a traditional pastoral system: The case in California Mediterranean rangelands at multiple scales. Ecol. and Soc. 19: 14 pp. http://dx.doi.org/10.5751/ES-06143-190108 <accessed July 2017>.

IPCC [Intergovernmental Panel on Climate Change]. 2014. Climate Change 2013: The Physical Science Basis: Working Group I Contribution to the Fifth Assessment Report of the Intergovernmental Panel on Climate Change. http://www.ipcc.ch/report/ar5/wg1/ <accessed 28 Jun 2017>.

Ketcham, J. 2017. Allan Savory's holistic management theory falls short on science: A critical look at the holistic management and planned grazing theories of Allan Savory. *Sierra Magazine* (March-April). http://www.sierraclub.org/sierra/2017-2-march-april/feature/allan-savory-says-more-cows-land-will-reverse-climate-change <accessed June 2017>.

Kolbert, E. 2012. Recall of the wild: The Quest to engineer a world before humans. The New Yorker 24 & 31 December, pp. 50–60. http://www.newyorker.com/magazine/2012/12/24/recall-of-the-wild <accessed July 2017>.

Kolbert, E. 2014. The Sixth Extinction: An Unnatural History. New York: Henry Holt & Company, 319 p.

Liffman, R., L. Huntsinger and L. Forero. 2000. To ranch or not to ranch: Home on the urban range? J. Range. Mgmt. 53: 362–370.

List, R., J. Pacheco, E. Ponce, R. Sierra-Corona and G. Ceballos. 2010. The janos biosphere reserve, northern Mexico. Intl. J. Wilderness 16: 35–41.

Miller, B.J., R.P. Reading, D.E. Biggins, J.K. Detling, S.C. Forrest, J.L. Hoogland, J. Javersak, S.D. Miller, J. Proctor, J. Truett and D.W. Uresk. 2007. Prairie dogs: An ecological review and current biopolitics. J. Wildlife Mgmt. 71: 2801–2810.

NOAA-ORR [National Oceanic and Atmospheric Administration, Office of Response and Restoration]. 2012. What are the increased risks from transporting tar sands oil? Blog entry, 13 December. http://response.restoration.noaa.gov/about/media/what-are-increased-risks-transporting-tar-sands-oil.html <accessed 14 June 2017>.

Notzke, C. 2016. Wild horse-based tourism as wildlife tourism: The Wild horse as the other. Curr. Iss. in Tourism 19: 1235–1259. DOI: 10.1080/13683500.2014.897688 <accessed July 2017>.

NRC [Committee to Review the Bureau of Land Management Wild Horse and Burro Management Program, National Research Council, National Academy of Sciences]. 2013. Using Science to Improve the BLM Wild Horse and Burro Program: A Way Forward. Washington DC: The National Academies Press, 450 p. DOI: 10.17226/13511 http://www.nap.edu/catalog.php?record_id=13511 <accessed July 2017>.

Oviedo, J.L., L. Huntsinger and P. Campos. 2017. The contribution of amenities to landowner income: The case of Spanish and Californian hardwood rangelands. J. Rangel. Ecol. Mgmt. 70: 518–528.

Parker, L. and R. Wilson. 2016. What happens to the U.S. Midwest when the water's gone? National Geographic (August). http://www.nationalgeographic.com/magazine/2016/08/vanishing-midwest-ogallala-aquifer-drought/ <accessed 26 June 2017>.

Polley, H.W., D.D. Briske, J.A. Morgan, K. Wolter, D.W. Bailey and J.R. Brown. 2013. Climate change and North American rangelands: Trends, projections and implications. Rangel. Ecol. & Mgmt. 66: 493–511.

Ponce-Guevara, E., A. Davidson, R. Sierra-Corona and G. Ceballos. 2016. Interactive effects of black-tailed prairie dogs and cattle on shrub encroachment in a desert grassland ecosystem. PLoS One 11: e0154748.

Popper, F.J. and D.E. Popper. 2006. The Buffalo Commons: Its antecedents and their implications online. J. of Rural Res. & Policy 1: https://dx.doi.org/10.4148/ojrrp.v1i6.34 <accessed July 2017>.

Ritten, J.P., C.T. Bastian and B.S. Rashford. 2012. Profitability of carbon sequestration in western rangelands of the United States. J. Rangel. Ecol. Mgmt. 65: 340–350.

Roberge, J.M. and P. Angelstam. 2004. Usefulness of the umbrella species concept as a conservation tool. Conserv. Biol. 18: 76–85. http://onlinelibrary.wiley.com/doi/10.1111/j.1523-1739.2004.00450.x/abstract <accessed July 2017>.

Ruyle, G.B. 2000. Book review: Holistic management: A New framework for decision making. Ecol. Engineering 15: 161–162.

Schmidt, S. 2016. A 75-mile-wide gap in Trump's wall? A tribe says it won't let it divide its land. Washington Post, 15 November 2015. https://www.washingtonpost.com/news/morning-mix/wp/2016/11/15/a-75-mile-wide-gap-in-trumps-wall-a-tribe-says-it-wont-let-the-wall-divide-its-land/?utm_term=.cab79b04a2b8 <accessed 04 June 2017>.

Schroeder, M.A., C.L. Albridge, A.D. Apa, J.R. Bohne et al. 2004. Distribution of sage-grouse in North America. The Condor 106: 363–376.

Sierra-Corona, R., A. Davidson, E.L. Fredrickson, H. Luna-Soria, H. Suzan-Azpiri, E. Ponce-Guevara and G. Ceballos. 2015. Black-tailed prairie dogs, cattle and the conservation of North America's arid grasslands. PLoS One 10. DOI: 10.1371/journal.pone.0118602 <accessed July 2017>.

Soulé, M.E. and M. Terborgh (eds.). 1999. Continental Conservation: Scientific Foundations of Regional Reserve Networks. Washington DC: Island Press, 238 p.

Starrs, P.F. 1998. Let the Cowboy Ride: Cattle Ranching in the American West. Baltimore, Maryland: Johns Hopkins University Press, 396 p.

Starrs, P.F. and P. Goin. 2010. A Field Guide to California Agriculture. Berkeley and Los Angeles: University of California Press, 503 pp.

Tobias, J. 2017. This native tribe is reintroducing a disappeared species on its own land, and the federal government can't do much about it. Pacific Standard 31 May, https://psmag.com/magazine/native-tribe-reintroducing-disappeared-species-on-own-land <accessed July 2017>.

Torell, L.A., S. Murugan and O.A. Ramirez. 2010. Economics of flexible versus conservative stocking strategies to manage climate variability risk. Rangel. Ecol. & Mgmt. 63: 415–425.

Uno, K.T., P.J. Polissar, K.E. Jackson and P.B. deMenocal. 2016. Neogene biomarker record of vegetation change in eastern Africa. PNAS 113: 6355–6363. DOI: 10.1073/pnas.1521267113. http://www.pnas.org/content/113/23/6355 <accessed July 2017>.

USDI [United States Department of the Interior]. 2017. Secretary Zinke signs order to improve sage-grouse conservation, strengthen communication and collaboration between states and feds [Press Release]. 08 June. https://www.doi.gov/pressreleases/secretary-zinke-signs-order-improve-sage-grouse-conservation-strengthen-communication <accessed 21 June 2017>.

WDF&W [Washington Department of Fish & Wildlife: Conservation]. 2017. Species & ecosystem science, greater sage-grouse ecology. http://wdfw.wa.gov/conservation/research/projects/grouse/greater_sage-grouse/ <accessed 09 June 2017>.

Wise, L. 2015. A drying shame: With the Ogallala Aquifer in peril, the days of irrigation for western Kansas seem numbered. Kansas City [Kansas, USA] Star, 24 July. http://www.kansascity.com/news/state/kansas/article28640722.html <accessed July 2017>.

18

What Future for the World's Grasslands Under Global (not only Climate) Change?

Victor R. Squires,[1,*] *Haying Feng*[1] and *Limin Hua*[2]

Grasslands had a central role in human evolution. The fossil record lends weight to the idea that humans developed key traits, such as flexible diets, large brains, complex social structures and the ability to walk and run on two legs while adapting to the spread of open grasslands. Humans had to descend from the arboreal homes of our primate ancestors, develop the ability to move rapidly over open ground and evolve the social skills needed for survival. Today we live in an era of human dominance—but that dominance needs to make room for large and contiguous grasslands that connect us to our past and provide us sustenance, biodiversity, habitat, carbon sequestration, watershed and vast beauty.

—Starrs, Huntsinger and Spiegal, 2018

Preamble

Global change is upon us. It is affecting the economics (cash economy, trading relations), sociology (breakdown of traditional cultures, urbanization) and the ecology all of which are under great pressure from the rising human population and the increasing pollution load that humans impose. Add to this the inevitable consequences of changing climates and the grasslands will suffer more from land use change and land degradation.

As was stated in the editors' Preface, the aim of this book is to provide readers with an up-to-date summary of the situation across the globe. To this end we have commissioned writing teams to explore and report on the current status of grasslands across the globe and to analyse the problems faced and prospects for situation betterment.

[1] Qinzhou University, Guangxi, China.
[2] Gansu Agricultural University, Lanzhou, China.
* Corresponding author

This concluding chapter examines a set of themes arising from the chapters that make up the bulk of this book. The following provide a focus for the text that follows:

- Recent history of grassland biomes—brief recap of current thinking and recent trends with special reference to dry grasslands in the Palearctic regions
- The current status of grasslands and germplasm resources (biodiversity)—an overview
- Management systems that ensure sustainability
- How to recover degraded grasslands?
- Socio-economic issues and considerations in grassland management
- The impact of environmental problems on grasslands, such as future climate change and intensification
- The problems/prospects facing pastoralists and other grassland-based livestock producers.

Areal Extent and Geographical Distribution of Grassland Biomes

According to the current estimate, about 40 per cent of the global land surface is grassland if we exclude Greenland and Antarctica (White et al., 2000). This seemingly high estimate for extant grasslands, however, results from including not only 'non-woody grasslands' but also savannas, woodlands, shrublands and tundra in the definition of grasslands. As pointed out by Anderson (2006) this estimate of existing grassland is potentially misleading, because most ecologists would not include woodlands, shrublands and tundra in a definition of grasslands, although de Haan et al. (1997) said that approximately 60 per cent of the world's pasture land (just less than half the world's usable land surface) is classified as grazing land (not all of which is grassland).

In addition, while it is recognized that temperate grasslands have experienced heavy conversion to agriculture, White et al., 2000 state that at least 5 per cent of grasslands world-wide are 'strongly to extremely denuded'. This relatively low percentage possibly results from the inclusion of non-traditional landscapes in the definition of 'grasslands', giving rise to a biased result. Applying a more widely used definition of grasslands, many of the world's temperate grassland ecosystems have been essentially destroyed by human activities. In many cases the 'destruction' has been wrought by large-scale conversion of grassland *senso lato* to cropland. For example, see Otieno and Kinyamario (this volume) and Morales (this volume) and other authors who document the changes in places as diverse as Kenya, Chile, Japan, northern and eastern Europe and the Mediterranean region. Intensive production that has become a feature of grassland management in parts of western Europe (Dengler and Tischew, this volume), North America (Starrs, Huntsinger and Spiegal, this volume), South America (Morales, this volume), East Africa (Otieno and Kinyamario), Japan (Ushimaru, this volume) and elsewhere must not be allowed to lead to further environmental degradation and threats to sustainability. These are serious challenges, easily forgotten in many areas of Europe where the immediate problems are of balancing livestock production with local demand, consistent with maintaining rural economies and ecosystems (Dengler and Tischew this volume for northern Europe; Ambarli et al., this volume) for Mediterranean Europe and the Middle East.

Threats to Grasslands Worldwide

Land use patterns are changing everywhere in response to rising human population pressure. Land use change is closely related to socio-economic development and environmental changes; therefore, land use change has become a major area for research. Examples of this are given in several chapters in this volume (e.g., Morales, this volume; Starrs, Huntsinger and Spiegal, this volume). Land use is undergoing dramatic alterations not only in the Mediterranean (e.g., Symeonakis et al., 2007; Kilic, 2006; Ambarli et al., this volume) but in eastern, central and northern Europe as well (Török et al., this volume; Dengler and Tischew, this volume), usually at the expense of grasslands (Hodgson et al., 2005; Chuvardas and Vrahnakis, 2009). Russia is not exempt from such changes (Reinecke et al., this volume). In Europe, one of the major driving forces for these alterations may be found in the European Community's (EC's) system of subsidies under the Common Agricultural Policy (CAP) that favors intensification and extensification of arable lands (Hodgson et al., 2005; Chuvardas and Vrahnakis, 2009). The total grassland area in the EU decreased by 13 per cent from 1990 to 2003 (FAO, 2006) and the trend continues. Persistent threats to grasslands in Turkey are conversion to croplands and unsustainable grazing (Ambarli et al., this volume). Novel threats acting for the past half century are uncontrolled urbanization, afforestation, road construction, climate change, mining and excessive use of underground water. The 'recent' large-scale destruction and degradation of grasslands took place between 1950–1980 with subvention (subsidization) of agricultural mechanization, allocation of governmental land for agriculture, opening of marginal lands (Bragina et al., this volume) and rangelands for crop production (see Otieno and Kinyamario [about Kenya], this volume; Morales [about South America], this volume).

Land Abandonment

On the other hand, land abandonment is a serious problem in areas as different as the converted savanna in Kenya's Massai Mara where many thousands of ha of rangelands were plowed to plant wheat but later abandoned (Otieno and Kinymario, this volume) or the vast areas in Eastern Europe (Török et al., this volume) and Mediterranean Europe (Peco et al., 2005). Some of the area is abandoned cropland (see earlier) but others are grazing lands that are too far away from the now sedentary livestock owners' homes. Many upland or remote areas lack access because of damaged infrastructure (broken bridges, damaged roads, lack of water for livestock and shepherds). This a problem in Middle Asia (Bragina et al., this volume; Strong and Squires, 2012). Shrub and woodland encroachment is a feature of some abandoned land and is either a cause of abandonment of former cropland as in parts of Argentina (Yanneli et al., 2014) and Chile (Morales, this volume) or a consequence of removal of grazing, fire or regular mowing that was once part of the management regime as in Japan (Ushimaru, this volume) or in parts of Europe (Dengler and Tischew, this volume; Török et al., this volume; Peco et al., 2005). The area of Mediterranean mountain pastures (mostly pseudoalpine and alpine natural and semi-natural grasslands) have undergone rapid decline mostly due to human population decline that led to land abandonment, cessation of traditional practices, livestock pressure reduction, scrub and shrub encroachment and increase of catastrophic wildfires (Zomeni et al., 2008; Lasanta et al., 2015). With the increase in rural population and mechanization of agriculture, surplus laborers and non-agrarian workers started to migrate internally in 1950s to West European countries since 1960 which initiated large-scale land abandonment (Ambarli et al., this volume).

Land Fragmentation—Consolidation

Fragmentation of land is a process that dissects natural systems into spatially isolated parts (Hobbs et al., 2008). In fact, development of human society has necessitated fragmentation of land use. From northern America, eastern Asia to southern Africa, grasslands, woodlands and wetlands have been converted for social, economic and political purposes. The survival of human society has depended on it as forests were cleared and land was put under cultivation while other areas were subsumed by urban and infrastructural developments. In China, Australia and Africa, herder and farmer livelihoods and financial viability are affected because grassland fragmentation results in land degradation, poverty, biodiversity loss and so forth (Flintan, 2011; Squires and Hua, 2017). The fragmentation of land, in particular on grasslands, has become a focus of attention for researchers and policy makers. There are two dimensions: preserving ecological integrity (Woodley and Kay, 1993) and ensuring viable agricultural/pastoral production systems. The priority in rangelands the world over is to avoid further fragmentation and rehabilitate already degraded rangeland ecosystems (Squires et al., 2009). Grassland clearing for cultivation, woody and invasive plant encroachment and indigenous woody plant regeneration are the main causal factors behind such observed grassland losses. The remaining grassland patches are more fragmented, the average patch size is smaller and the total number of grassland patches have increased and the remaining grassland patches are more isolated. The largest, least fragmented grassland patches are often isolated. However, land-cover change is only the physical expression of the complex interactions between socio-economic factors. To create effective and sustainable conservation plan for the grassland biome, with an aim to reducing habitat loss, requires an action plan to address these factors as the ultimate drivers of land cover change.

Developmental planning and policy making must take into consideration the predictions of species-rich grassland elimination (and the decrease of the area of open shrublands) in any future landscapes, given that the ecosystems they support are important for sustaining the ecological integrity and social relevance of the landscape.

Restoring Degraded Grassland

Restoring and repairing damaged or destroyed ecosystems is a high priority in a great diversity of grassland ecosystems (Squires, 2016). Many of these ecosystems have been degraded or altered by biological invasions and/or years of natural resource extraction or by such practices as fire exclusion in abandoned grasslands lands. Restoration is becoming an increasingly important management response to use-associated habitat degradation. Managers of grasslands seek to achieve several objectives, some of which may be difficult to reconcile.

Natural grasslands, are often extensive areas that are highly variable over space and time (Squires, 2015). This spatio-temporal variability means that the environment, at this scale, is also more than a collection of individual phenomena. The emergent ecological properties of landscapes at the scale of grassland communities (including the humans who live and work there) are spatial pattern and diversity (Robbins et al., 2002). Conservation of biodiversity and the restoration of biological communities and landscapes, especially in the world's remaining intact grasslands, poses huge challenges for managers of natural resources. Restoration is becoming an increasingly important management response to use-associated habitat degradation (Bonet, 2004). The recovery of damaged ecological

Fig. 18.1 Invertebrates are important ecosystem components in grasslands. Some (A) compete with grazers for forage while others (B) help in nutrient cycling.

systems, through conserving and maintaining ecological systems enriches the lives of their inhabitants (including wildlife and invertebrates [Fig. 18.1]).

According to Whisenant (2002)

> Repairing damaged areas requires realistic objectives that consider the extent of damage, ecological potential, land use goals, and socio-economic constraints. Since grassland ecosystems are dynamic and constantly changing, rather than static and predictable, it is unrealistic to set out pre-defined species groups as goals. Instead redirecting essential ecosystem processes toward preferred trajectories should repair damaged ecosystems.

Ecological principles and management practices involved in restoring and rehabilitating grassland ecosystems after disturbance have been enunciated (Walker and Jansenn, 2002; Whisenant, 2002; du Toit et al., 2014; Squires, 2016). Proven practices to return damaged ecosystems to a productive and stable state have site- or situation-specificity (Stafford Smith et al., 2007). This makes it difficult to promote replication and scaling up (Squires, 2013). Relative heterogeneity of patterns is an important factor in vulnerability to degradation and underpins the success, or otherwise, of attempts to restore damaged or degraded grasslands, especially those in rangelands (Whisenant, 2002; du Toit et al., 2012). Degradation, especially in the more arid areas, occurs as a result of a suite of processes in the coupled human–environment system. Each process calls for different understanding, measurement techniques and management responses.

Mining in Mongolia, for example, is a major cause of land degradation. Desert areas have expanded and now cover approximately 11 per cent of Mongolia (Aralova et al., 2018). Mining areas are an important factor in creation of bare land. In other instances, they cause soil erosion (Fig. 18.2).

In total, 38.5 million ha or 24.6 per cent of the whole country in Mongolia is covered by mining leases. Large land area is destroyed by mining operations and/or by mining surveys and prospecting (Pfeiffer et al., this volume; Batkhishig, 2013). Oil and gas exploration and exploitation cause grassland degradation (and fragmentation) on a large scale in some countries. Off road vehicle tracks criss-cross the desert grasslands in roadless regions of Mongolia, Africa and Australia.

Restoring degraded land is a challenge for those in places, especially in Middle Asia (Bragina et al., this volume) where, due to economic situation, there is a heavy reliance on extracting natural resources. For example, oil and gas in Kazakhstan, Turkmenistan and Uzbekistan and mining for gold and other minerals in Kyrgyzstan. These industries and the associated infrastructure (mines, pipelines, roads, railways, water diversions, mine waste disposal, etc.) are an additional potential danger and threat to the fragile grassland ecosystems.

Fig. 18.2 Mining on grasslands can be quite destructive.

Recovering Functional Integrity in Grasslands

In the last decades, due to climate changes, soil deterioration and Land Use/Land Cover Changes (LULCCs), land degradation risk has become one of the most important ecological issues at the global level. Land degradation involves two interlocking systems—the natural ecosystem and the socio-economic system. The complexity of land degradation processes should be addressed using a multidisciplinary approach. The principal aim of grassland ecosystem management is to help develop and promote strategies to protect functional integrity and balance social/economic and biodiversity values (Squires, 2016).

Grassland ecosystems have a major role in maintaining livelihoods of millions of people. Some of those people rely on commercial-scale production systems where large capital outlays are involved, while others are small-scale and are classified as 'subsistence level' production units. Many involve livestock in the production system. Although livestock grazing contributes only a small percentage to the world's GDP (circa 1.5 per cent) maintaining the long-term stability of these industries is vital because of the high social and environmental consequences of a collapse. Up to one billion of the world's poor people are dependent on livestock grazing for food and income from these industries that occupy about 25 per cent of the world's land base (Steinfeld et al., 2006). Grassland-based meat and milk production systems account for a large proportion of both the global land area and of the production of meat and milk. In addition, grasslands play an important part in many mixed agricultural systems, providing a break in rotational cropping systems (Nie and Chapman, this volume) and providing feed to livestock in systems in which the diet comprises both crop residues or by-products and grazed or conserved grass. The increasing global demand for meat and milk, environmental concerns about the sustainability of intensive production systems and, in the Western world at least, issues of food quality, safety and animal welfare are increasingly having an impact on the buying patterns of consumers. These factors, coupled with the fact that grass is a relatively cheap source of feed for ruminants and that grassland-based systems are seen as being environmentally sustainable and 'welfare friendly', suggest that there is considerable scope for continuing the production of meat and milk from grass-based systems. However, there is a real challenge to develop production systems that maximize the utilization of grass per unit of animal product, that commensurate with a sound environmental practice. This necessitates developing grazing management systems that maximize the use of grass and forage in the diet of animals.

The long-term sustainability of livestock grazing is also crucial for the environment. In fact, five of the most important and serious environmental problems are global warming, land degradation, air and water pollution, and the loss of biodiversity due to livestock (Steinfeld et al., 2006). For these reasons, finding more effective approaches that guide the sustainable management of grassland ecosystems, especially grazed ones, is urgently needed. Sustainability is an elusive goal. The notorious vagueness of the term and its scope for varied and seemingly legitimate interpretation by different parties, appear to make it all but useless as an operational guide (O'Riordan, 1988). One has only to consider simple questions—Sustain what? How? For whom? Over what time period? Measured by what criteria? To appreciate that, sustainability can never be precisely defined. Regardless of the ambiguity of the term, however, there appears to be a general consensus that achieving sustainability will place new demands on individuals, society and science. The challenge facing science is how best to structure and undertake research to meet the diverse, and often apparently conflicting, needs of society, local communities and individual natural resource users (Squires, 2015).

Getting serious about sustainability means acknowledging the intricate interdependency of environmental, economic and social issues on a finite planet. Much of the apparent conflict surrounding many resource management issues relates to the fact that different interest groups fail to appreciate the perspectives and values inherent in the actions of others (Starrs et al., this volume). If these groups can be encouraged to share their experiences and viewpoints, there will be a greater understanding of why these differences exist. Equally important, the involvement of different groups may well provide useful ideas and strategies that lie outside the normal perspective of those with the primary responsibility of managing any particular resource.

Management Systems that Ensure Sustainability in European Grasslands

Most grasslands in Europe were created and/or their biodiversity is maintained by an extensive form of management (Fischer and Wipf, 2002). This means in most cases, grazing or mowing management as discussed by Dengler and Tischew (this volume) and by Török et al. (this volume). To conserve grassland biodiversity, it is crucial to maintain extensive management regimes (best represented by a traditional agricultural regime) to avoid both abandonment and too high land-use intensity. In the case of already degraded grassland stands, the change of management intensity is also suggested, but in the case of completely destroyed grasslands, the recovery by spontaneous succession or technical reclamation methods is recommended (Török et al., 2011). The positive effects of re-introduction of traditional management by mowing or grazing have been demonstrated. For pastures, low intensity grazing (i.e., < 0.5 animal units per hectare) is recommended with a strong preference for traditional herding of local cattle breeds or free grazing by wild horses and cattle. In USA too, there is enthusiasm by some for rewilding. A key concept in the rewilding vision is improving ecosystem function and mitigating lost biodiversity (Starrs et al., this volume). As the re-introduction of traditional management practices is often not feasible or economically sustainable, conservation authorities are seeking alternative management practices like prescribed burning during the dormant season. Valkó et al. (2013) suggest that prescribed burning with long fire-return periods (i.e., at least three consecutive years without burning) might be a cost-effective and appropriate tool in eliminating accumulated litter and sustaining grassland biodiversity. The importance of small-scale, low-intensity farming in the conservation of European biodiversity and the maintenance of cultural landscapes has been recognized for decades and led to the development of the High Nature Value (HNV) concept in the 1990s (see below).

Climate Change

Another emerging potential threat for grasslands is climate change, especially for habitats that are supported by dry grassland vegetation (Rodwell et al., 2007). Grasslands, whether natural or managed, are the result of interactions between climate, soil, interspecies competition and natural or human disturbance. They are vital to human sustenance but vulnerable to human mismanagement and to climatic stress. Large areas that are presently classified as grasslands are the result of human intervention. In many instances, this was first achieved by hunter-gatherers through the use of fire to clear woodlands so as to allow more grazing by animals (sometimes to attract game animals). Control of weeds and feral animals has enabled increased harvesting of organic matter (forage, fodder, firewood, building materials, etc.). Large-scale mechanical clearing of forests and woodlands, irrigation, cultivation and fertilization have all been used to add to the area of grassland (including sown pastures) and to the opportunities for harvesting even more biomass.

Clearly climate is relevant to the existence, location and nature of natural or climax grasslands, and to at least some aspects of grasslands created or modified by human activities, notably their continued viability, productivity and economics. Expected climatic changes include not only increasing temperatures, but also changes in total rainfall and its seasonality, and increases in rainfall intensity. Higher temperatures mean greater losses of water due to evapotranspiration that could increase by 5–15 per cent. If this were to occur in regions where summer rainfall did not increase much or even declined (e.g., southern Europe and western USA) increased aridity and fire potential, which would favor grasslands over woodlands, would prevail. Longer and more frequent dry spells could be accompanied by a greater frequency of lightning strikes. Lightning strikes are a major initiator of wildfires that periodically destroy woody plants and help tip the balance toward grasslands. Shrubland encroachment on grasslands in south-western USA over the last century or so has been variously ascribed to overgrazing by livestock, climatic change, fire suppression and rising CO_2 concentration (Starrs, Huntsinger and Spiegal, this volume; Squires, 2018).

Along with the direct effects of increased carbon dioxide (CO_2), concentrations are also of importance. There has been wide discussion about the potential effect of increasing CO_2 concentrations on the competition between C_4 and C_3 plants; C_4 plants evolved largely in response to low CO_2 concentrations (Ehleringer et al., 1991) with the mechanism that concentrates CO_2 within the cells so that ambient CO_2 concentrations do not limit net assimilation rates. It is, therefore, thought they will benefit relatively little from increases in CO_2 concentration in the atmosphere, compared to C_3 plants, in which photosynthesis is usually CO_2 limited. However, C_4 grasses are clearly adapted to higher temperatures than are C_3 species and present-day distributions of C_3 and C_4 grasses on several continents suggest that temperature is the key climatic variable that determines their relative distribution. Other factors that come into play are water-and nitrogen-use efficiencies, and the high light requirements of C_4 plants. The competition between C_4 grasses and C_3 woody shrubs is another story. Many woody C_3 plants are well adapted to the climatic range of C_4 grasses, but the structural form of the vegetation in many tropical areas is savanna rather than the 'more canopy' woodlands. This savanna structural formation is more suited to the long dry season as it allows the C_4 grasses access to the high light levels needed for their survival. Higher CO_2 levels may favor the spread of closed canopy C_3 shrublands, especially if monsoon rainfall increases.

Natural disturbance (fire, pest and disease outbreaks) may occur due to extreme weather events. Climatic fluctuations cause major impacts on grasslands and savannas, often in conjunction with overstocking and other human-induced stress (Gaur and Squires, 2018). The empirical evidence from a number of grassland biomes suggests that if climate change were to lead to a long-term drying trend, this would be of international importance. The grasslands of the world are in the frontline as vital suppliers of goods and services vital to human sustenance, and are areas highly vulnerable to stress. While it is generally classified as a 'slow onset phenomenon', climate change impacts are already evident as climate change gathers pace.

Projections averaged across a suite of climate and vegetation modeling show a progressively increasing drought risk across much of the dry grasslands. For example, it is estimated that over the last century, the mean annual temperature has increased by 0.8 per cent in Europe, whilst annual precipitation has increased by 10–40 per cent in northern Europe and decreased up to 20 per cent in parts of southern Europe (Parry, 2000). Similar

Table 18.1 Physical climate related factors.

Temperature	Diurnal and annual maximum, minimum, variability, extremes
Rainfall	Total, seasonality, variability, extremes
Evaporation	
Soil moisture	
Water table	
Salinity	
Soil erosion/loss	
Photosynthetically available radiation	
Fire frequency and intensity	
OTHER PHYSICAL FACTORS	
Ultraviolet radiation	
Local air pollution	
Cloud condensation nuclei	
Sulfate aerosols	
Carbon dioxide concentrations	
BIOLOGICAL FACTORS	
Inter-specific competition	
Pest animals and herbivory	
	Generation time, dispersal
Plant diseases and fungi	
	Generation time, dispersal
Soil carbon and nutrient status	
Plant response to carbon dioxide (C_3,C_4 plants)	
	Water use efficiency
	Plant morphology
	Relative genotype/species advantage
	C/N ratio
HUMAN FACTORS	
Greenhouse gas emissions	
Management strategies	
	Stocking rates
	Irrigation
	Fire ignition and suppression
	Fertilizer use
	Pest and disease control
Economics and finance	
Population growth, immigration, emigration	

increases have occurred elsewhere and in some places the temperature increase has been even higher (Hua et al., this volume). According to Morgan et al. (2011) global warming is predicted to induce desiccation in many of the world's grassland biomes through increases in evaporative demand. Rising CO_2 may counter that trend by improving plant water-use efficiency. However, it is not clear how important this CO_2-enhanced water use efficiency might be in offsetting warming-induced desiccation because higher CO_2 also leads to higher plant biomass and therefore, greater transpirational surface. Furthermore, although warming is predicted to favor warm season, C_4 grasses, rising CO_2 should favor C_3, or cool-season plants. In semi-arid grasslands, elevated CO_2 may completely reverse the desiccating effects of moderate warming. In a warmer, CO_2-enriched world, both soil-water content and productivity in semi-arid grasslands may be higher than previously expected.

The structure and the floristic elements of the dry grassland habitats are supported by drought-resistant species and the prospect of a drier climate is expected to result in changes of current biodiversity status, land use and land cover change (Poyatas et al., 2003). Although the extent of global change impacts on natural and human systems depends largely on adaptation capacities, not every societal approach to adaptation is mitigating or focused on sustainability especially in many areas of Europe where the immediate problems are of balancing livestock production with local demand, consistent with maintaining rural economies and ecosystems. Pressures to adopt unsustainable practices as yields drop in response to a changing climate, may increase grassland degradation. This fact should further motivate support of policies and programmes that encourage the implementation of sustainable grassland management practices. A key challenge is the large number of smallholders and pastoralists who may be among the hardest hit by climate change (FAO, 2008). Their challenge is often exacerbated because uncertain land tenure discourages investments that pay dividends in the long term. Thus, efforts to spread knowledge on sustainable grassland management practices are essential for ensuring their successful implementation and must address tenure-related motivations to implement sustainable practices. Not all categories of producers have the same potential for implementing sustainable land-management practices and some producers will benefit more and sooner than others. Development-mitigation-adaptation strategies must be evaluated within the framework of local environmental conditions, institutions and capacities. Priority should be given to investments in sustainable land management practices that:

- show strong evidence of enhancing near- and longer-term productivity and profitability for farmers and pastoralists
- offer opportunities to enhance production, mitigate GHG emissions and enable adaptation to climate change; develop incentives that foster sustainability of existing resources—soil, water, air, labour, etc.
- rehabilitate lands that can be improved at modest cost and adopt low-tech changes in management practices
- support research and education as best practices for maintaining fertility and production
- align with existing investment programmes.

High Nature Value Grasslands—A European Union (EU) Initiative

Grasslands with high levels of biodiversity are important for Europe (in particular) as they support healthy soils and large amounts of wildlife. Extremely valuable and diverse areas are termed high-nature-value grasslands and become part of EU agricultural policy, often maintained under long-term management agreements. To prevent degradation of high nature value (hereafter HNV) grasslands, common practices include mowing the land or grazing with livestock. Land managers are also usually restricted from applying fertilizers. High biodiversity-farmed landscapes dominated by semi-natural pastures and meadows (HNV landscapes) deliver a whole range of ecosystem services—biodiversity conservation, recreational landscapes, clean water, resistance to fire and flood, mitigation of climate change, high quality food produced at low carbon cost and viable rural communities. European Mediterranean countries have the highest percentage of HNV farming systems in Europe (Veen et al., 2009). Mediterranean pastures fulfill many conditions necessary for designation as HNV—small-scale low-input farming, extensive grazing, different types and levels of disturbance creating a diverse mosaic of habitats, high quality products and multifunctional services and goods (Escabano et al., 2015). Dehesas, an agroforestry system typical of the southwestern quadrant of the Iberian peninsula (Gaspar et al., 2009), are good examples of HNV (Diaz et al., 1997). They form small-scale landscape mosaics, composed of cereal cropping and livestock grazing under open canopy of *Quercus* spp. They have several economic and social benefits, such as an increase in farm income through multiple use of natural resources, employment stability, increase security against price change and combat depopulation. Among ecosystem services, that they provide are included combating desertification, resilience against climate change, decreasing fire risk and maintaining ecological corridors and refuges for large mammals and birds of prey. Another HNV example is seen in pseudo-steppes with grasses and annuals of the Thero-Brachypodietea. An extensive grazing system takes place in these semi-natural grasslands in which cereal production takes place on ancient croplands in the habitat mosaic (Ambarli et al., this volume).

Climate Change and Mountain Grasslands

Mountain regions encompass nearly 24 per cent of the total land surface of the earth and are home to approximately 12 per cent of the world's population (Huber et al., 2005). Their ecosystems play a critical role in sustaining human life, both in the highlands and the lowlands. During recent years, resource use in high mountain areas has changed mainly in response to globalization of the economy and increased world population. As a result, mountain regions are undergoing rapid environmental change, exploitation and depletion of natural resources, leading to ecological imbalances and economic unsustainability (Beever and Belant, 2011; Kreutzmann, 2012; Grover et al., 2015). Moreover, the changing climatic conditions have stressed mountain ecosystems through higher mean annual temperatures with melting of glaciers and snow (Hagg, 2018). Altered precipitation patterns have also had an impact (Hua et al., this volume). A number of critical issues have emerged as reflected in the downward spiral of resource degradation, increasing rural poverty and food and livelihood insecurity in mountain regions. New and comprehensive approaches to mountain development are needed to identify sustainable resource development practices (SLM) to strengthen local institutions and knowledge systems, and increase the resilience of both mountain environments and their inhabitants.

Biodiversity Matters in Production Landscapes

Ecological evidence has accumulated over the decades to show that an increased number of species within a grassland community can produce beneficial side-effects to ecological functions, such as production of biomass and nutrient cycling (Loreau et al., 2001). There is agreement in ecological science that biodiversity matters intrinsically, but also for other values, although how and why it matters remains contested (Naeem and Wright, 2003). The benefits of diversity are not limited to the ecosystem functions, such as biomass production and nutrient cycling but also increased resilience or stability, defined as the ability to recover after disturbance (Walker and Salt, 2012). This idea is based on the insurance hypothesis of ecosystem stability (McCann, 2000): an ecosystem with more species, contains diverse set of traits and therefore a higher likelihood of recovering from disturbance (Suding et al., 2008). Because of the role biodiversity plays in maintaining key functions and building resilience to change, there is increasing recognition that encouraging diversity in grazed grasslands is not antithetical to production. Instead, focusing management activities on both biodiversity and productivity could, in the long run, ensure the sustainability of production and environmental integrity (Firn, 2007). The issues of sustainable production and resilience are highly topical considering predictions for an increase in extreme rainfall variability as the effects of global climate change take hold (Gaur and Squires, 2018). What matters most about biodiversity of plants in grasslands is not just the number of species in the assemblage but the quality of biodiversity. In particular, the collective traits of the species present and how they respond to perturbations and, in turn, how these responses affect functioning (Suding et al., 2008). An increase in 'quality' species within a plant community may contribute considerably to function and stability. It is clear that with increased species diversity, there is a higher likelihood that the 'right' species will be present to improve and/or maintain functioning.

Current Status and Future Prospects for the World's Grassland Biomes

One of the most important conservation issues in ecology is the imperiled state of grassland ecosystems worldwide due to land conversion, desertification and the loss of indigenous, especially endemic, populations and species (Ceballos et al., 2010). These losses are rapidly reducing Earth's life support systems and the services that nature provides, such as the clean air and water on which we all depend. They are also impacting our food supplies (Flora, 2010).

There are three key drivers that are likely to affect the production of meat and milk from grass-based systems in the next 20 years. These are: (i) a general increase in global demand for milk and meat; (ii) concern about the environmental impact of agricultural practices; and (iii) increasing concern by consumers about food quality, including food safety and animal welfare, especially in the developed world (Squires et al., 2015). The global demand for animal products is projected to increase considerably, driven mainly by population growth, economic growth and urbanization in China and South-east Asia (Flora, 2010; Wang et al., 2017). The environmental impacts of grazing systems vary considerably, depending on the climatic conditions, but are likely to be an increasingly important factor. Food quality, food safety and animal welfare will also probably continue to be an important aspect in the market for milk and meat in developed countries as they would provide the opportunity for producers of meat and milk from grass-based systems.

Grassland-based meat and milk production systems account for a large proportion of the production of meat and milk. Approximately 60 per cent of the world's pasture land (just less than half the world's usable surface) is classified as grazing land (de Haan et al., 1997). In addition, grassland plays an important role in many mixed agricultural systems, providing a break in rotational cropping systems and providing feed to livestock in systems in which the diet comprises both crop residues or by-products and grazed or conserved grass.

The annual production of meat and milk products from grazing systems and mixed farming systems is considerable. Only about 9 per cent of global meat production comes from grazing systems (defined by de Haan et al. (1997) as systems based almost exclusively on pasture with little or no integration with crops, mainly based on native pasture). Mixed farming systems account for 54 per cent of meat production and 90 per cent of milk production. A considerable proportion of the feed consumed by livestock in these mixed systems will be from grazed and stall-fed pasture.

However, these global figures hide a huge range in the importance of grassland-based systems in different regions and countries. In temperate regions, grass-based systems of milk production predominate. For example, in New Zealand, virtually all dairy production is based on grassland, with over 90 per cent of the total nutrient requirements coming from grazing (Hodgson, 1990).

In the European Union, over 95 per cent of milk production is based on grassland often managed relatively intensively. Even though the level of supplementary feeding may be relatively high in dairy systems, grazed or conserved grass still accounts for 50–70 per cent of the nutrient requirements of the cows, and 75–100 per cent of requirements in most beef and sheep systems can be met from grazed or conserved grass. Compared with other sources of fodder for meat and milk production of grass, especially grazed grass, can prove a cheap source of feed. The costs of grazed grass are generally about half of those of conserved hay or silage. In addition, the support-energy requirements for harvesting and storing conserved grass, especially as silage, are often considerable (Wilkins, 2000; Gornall et al., 2010).

Many of the world's arid and semi-arid rangelands are grazed by domestic livestock, and although the stocking density may be relatively low, animals provide livelihoods for pastoralists and are often central to their cultural heritage. As climates change and the impact of the market economy sets in, the likely outcome is that the pattern of all agricultural production will have to change to fit whatever climatic and environmental conditions prevail. The main issues relate to who will produce what and where. There is no doubt that ruminants will have to move to wetter, hillier and more range-type conditions and the ruminant population is likely to decrease in size as a consequence. The areas of drier arable land are likely to increase and more food and feed-grains are likely to be produced on them.

References and Further Reading

Allen, W.J., O.J.H. Bosch, R.G. Gibson and A.J. Jopp. 1998. Co-learning our way to sustainability: An integrated and community-based research approach to support natural resource management decision-making. pp. 51–59. *In*: S.A. El-Swaify and D.S. Yakowitz (eds.). Multiple Objective Decision Making for Land, Water and Environmental Management. Lewis Publishers, Boston. Ch. 4.

Ambarli, D., M. Vrahnakis, S. Burrascano, A. Naqinezhad and M. Pulido. 2018. Grasslands of the Mediterranean and the Middle East: Problems and prospects. pp. 89–112 (this volume).

Anderson, R.C. 2006. Evolution and origin of the Central grassland of North America: Climate, fire and mammalian grazers. J. Torrey Botanical Society 133(4): 626–247, Applied Ecology 6(2): 107–118.

Aralova, D., J. Kariyeva, L. Menzel, T. Khujanazarov, K. Toderich, U. Halik and D. Gufurov. 2018. Assessment of land degradation process and identification of long-term trends in vegetation dynamics in the drylands of Greater Central Asia. pp. 133–153. *In*: V.R. Squires and Lu Qi (eds.). Sustainable Land Management in Greater Central Asia: An Integrated and Regional Perspective. Routledge, London and New York.

Bakker, J.P. and F. Berendse. 1999. Constraints in the restoration of ecological diversity in grassland and heathland communities. *In*: Trends in Ecology and Evolution 14(2): 63–68.

Bajocco, S., A. De Angelis, L. Perini, A. Ferrara and L. Salvati. 2012. The impact of land use/land cover changes on land degradation dynamics: A Mediterranean case study. Environmental Management 49: 980–989.

Batkhishig, O. 2013. Human impact and land degradation in Mongolia. pp. 265–280. *In*: J. Chen, S. Wan, G. Henebry, J. Qi, G. Gutman, S. Ge and M. Kappas (eds.). Dryland East Asia: Land Dynamics Amid Social and Climatic Change, Higher Education Press, Beijing.

Beever, E.A. and J.L. Belant. 2011. Ecological Consequences of Climate Change: Mechanisms, Conservation and Management. CRC Press, Boca Raton, Florida, USA, 314 pp.

Bonet, A. 2004. Secondary succession of semi-arid mediterranean old-fields in south-eastern Spain: Insights for conservation and restoration of degraded lands. Journal of Arid Environments 56(2): 213–233.

Bookshire, E.N.J. and T. Weaver. 2015. Long-term decline in grassland productivity driven by increasing dryness. Nature Communications 6: 7148. DOI: 10.1038/ncomms8148.

Bragina, T.M., A. Nowak, K.A. Vanselow and V. Wagner. 2018. Grasslands of Kazakhstan and Middle Asia: The ecology and land use of a vast area of global importance. pp. 139–167 (this volume).

Britaňák, N., L. Hanzes and I. Ilavská. 2009. Alternative use of different grassland types: I. Influence of mowing on botanical composition and dry matter production. *In*: Grassland Science in Europe 14: 68–71.

Butaye, J., D. Adriaens and O. Honnay. 2005. Conservation and restoration of calcareous grasslands: A concise review of the effects of fragmentation and management on plant species. *In*: Biotechnol. Agron. Soc. Envir. 9(2). Available on: http://popups.ulg.ac.be/Base/document.php?id=1516.

Ceballos, G., A. Davidson, R. List, J. Pacheco, P. Manzano-Fischer, G. Santos-Barrera et al. 2010. Rapid decline of a grassland system and its ecological and conservation implications. PLoS ONE 5(1): e8562.

Chuvardas, D. and M. Vrahnakis. 2009. A semi-empirical model for the near future evolution of the lake Koronia landscape. Journal of Environmental Protection and Ecology 10(3): 867–876.

Crepet, W.L. and G.D. Feldman. 1991. The earliest remains of grasses in the fossil record. Am. J. Bot. 78: 1010–1014.

de Haan, C., H. Steinfeld and H. Blackburn. 1997. Livestock and the Environment: Finding a Balance. Wrenmedia, Suffolk, UK.

Dengler, J. and S. Tischew. 2018. Grasslands of Western and Northern Europe—Between intensification and abandonment. pp. 27–63 (this volume).

Devendra, M., Ajoy Kumar Roy and Kaushal, Pankaj. 2018. Grasslands/Rangelands of India: Current status and future prospects. pp. 221–238, this volume.

Diaz, M., P. Campos and F.G. Pulido. 1997. The Spanish dehesas: A diversity of land uses and wildlife. pp. 178–209. *In*: D. Pain and M. Penkowski (eds.). Farming and Birds in Europe: The Common Agricultural Policy and its Implications for Bird Conservation. Academic Press, London.

Du Toit, J., R. Kock and J. Deutsch. 2012. Wild Rangelands: Conserving Wildlife While Maintaining Livestock in Semi-arid Systems. John Wiley. N.Y., 448 p.

Du Toit, J., R. Kock and J. Deutsch. 2014. Wild Rangelands: Conserving Wildlife While Maintaining Livestock in Semi-Arid Ecosystems. John Wiley & Sons.

Ehleringer, J.R., R.F. Sage, L.B. Flanagan and R.W. Pearcy. 1991. Climate change and the evolution of C_4 synthesis. Tree 6: 95–99.

Escabano, A.J., P. Gaspar, F.J. Mesias, M. Escribno and F. Pulido. 2015. Comparative sustainability of extensive beef cattle farms in a high nature value agroforestry system. pp. 65–86. *In*: V.R. Squires (ed.). Rangeland Ecology, Management and Conservation Benefits. Nova Science Publishers, N.Y.

FAO. 2006. Statistical Yearbook 2006, Rome.

FAO. 2006b. Livestock's Long Shadow: Environmental Issues and Options, Rome.

FAO. 2008. Investing in Sustainable Agricultural Intensification: The Role of Conservation Agriculture; A Framework for Action. Food and Agriculture Organization of the United Nations, Rome, Italy.

Fischer, M. and S. Wipf. 2002. Effect of low-intensity grazing on the species-rich vegetation of traditionally mown subalpine meadows. Biol. Conserv. 104: 1–11.

Firn, J. 2007. Developing strategies and methods for rehabilitating degraded pastures using native grasses. Ecological Management & Restoration 8(3): 182–186.

Flintan, F. 2011. Broken Lands. Broken Lives? Causes, Processes and Impacts of Land Fragmentation in the Rangelands of Ethiopia, Kenya and Uganda. Report for REGLAP, Nairobi.

Flora, C. 2010. Food security in the context of energy and resource depletion: Sustainable agriculture in developing countries. Renewable Agriculture and Food Systems 25(2): 118–128. DOI: 10.1017/S1742170510000177.

Gaspar, P., F.J. Mesías, M. Escribano and F. Pulido. 2009. Sustainability in Spanish extensive farms (Dehesas): An economic and management indicator-based evaluation. Rangeland Ecology & Management 62(2): 153–162.

Gaur, M.K. and V.R. Squires. 2018. Climate Variability Impacts on Land Use and Livelihoods. Springer, N.Y., 348 p.

Gornall, J., R. Betts, E. Burke, R. Clark, J. Camp, K. Willett and A. Wiltshire. 2010. Implications of climate change for agricultural productivity in the early twenty-first century. Phil. Trans. R. Soc. B 365: 2973–2989. DOI: 10.1098/rstb.2010.0158.

Grover, V.I., A. Borsdorf, J. Breuste, P. Chandra, F. Tiwari and F. Witkowski. 2015. Impact of Global Changes on Mountains: Responses and Adaptation. CRC Press, Boca Raton, 517 p.

Hagg, W. 2018. Water from the mountains of Greater Central Asia: A resource under threat. pp. 237–248. *In*: V.R. Squires and Lu, Qi (eds.). Sustainable Land Management in Greater Central Asia. Routledge, Abingdon.

Hobbs, N.T., R.S. Reid, K.A. Galvin and J.E. Ellis. 2008. Fragmentation of arid and semi-arid ecosystems: Implications for people and animals. pp. 25–44. *In*: K.A. Galvin, R.S. Reid, R.H. Jr. Behnke and N.T. Hobbs (eds.). Fragmentation in Semi-Arid and Arid Landscapes: Consequences for Human and Natural Systems, Springer, Dordrecht.

Hodgson, J. 1990. Grazing Management: Science into Practice. Harlow, UK: Longman. https://www.ipcc.ch/publications_and_data/publications_.

Hodgson, J.G., J.P. Grime, P.J. Wilson, K. Thompson and S.R. Band. 2005. The impacts of agricultural change (1963–2003) on the grassland flora of Central England: Processes and prospects. Basic and Applied Ecology 6(2): 107–118.

Hua, L.M., Y. Niu and V.R. Squires. 2018. Climatic change on grassland regions and its impact on grassland-based livelihoods in China. pp. 355–368 (this volume).
Huber, U.I., H.K.M. Bugmann and M.A. Reasoner (eds.). 2005. Global change and mountain regions: An overview of current knowledge. Advances in Global Change Research, Vol. 23, Springer, Dordrecht, 652 pp.
Kreutzmann, H. 2012. Pastoral practices in High Asia: Agency of 'development' effected by modernization, resettlement and transformation. Springer Dordrecht.
Kilic, S., F. Evrendilek, S. Berberoglu and A.C. Demirkesen. 2006. Environmental monitoring of land-use and land-cover changes in a mediterranean region of Turkey. Environ. Monit. Assess. 114: 157–168.
Lasanta, T., E. Nadal-Romero and J. Arnaez. 2015. Managing abandoned farmland to control the impact of re-vegetation on the environment. The state of the art in Europe Environmental Science and Policy 52: 99–112.
Loreau, M., S. Naeem, P. Inchausti, J. Bengtsson, J.P. Grime, A. Hector, D.U. Hooper, M.A. Huston, D. Raffaelli, B. Schmid, D. Tilman and D.A. Wardle. 2001. Biodiversity and ecosystem functioning: Current knowledge and future challenges. Science 294: 804–808.
Maltitz, G.P. 2018. Southern African Grasslands in an Era of Global Change. pp. 273–292 (this volume).
McCann, K.S. 2000. The Diversity–Stability Debate Nature 405(6783): 228.
Millennium Assessment. 2005. Ecosystems and Human Well-being. World Resources Institute, Washington, DC. © 2005 10 G Street NE, Suite 800, Washington, DC 20002 millenniumassessment.org/documents/document.793.aspx.pdf, 100 p.
Morales, C. 2018. Grasslands in South America: Nature and Extent in Relation to Provision of Ecosystem Services. pp. 319–332 (this volume).
Morgan, J.A., D.R. LeCain, E. Pendall et al. 2011. C_4 grasses prosper as carbon dioxide eliminates desiccation in warmed semi-arid grassland. Nature 476(7359): 202–205.
Naeem, S. and J.P. Wright. 2003. Disentangling biodiversity effects on ecosystem functioning: Deriving solutions to a seemingly insurmountable problem. Ecology Letters 6(6): 567–579.
Nie, Z. and D. Chapman. 2018. Grasslands in Australia—An Overview of Current Status and Future Prospects. pp. 338–354 (this volume).
O'Riordan, T. 1988. The politics of sustainability. *In*: R.K. Turner (ed.). Sustainable Environmental Management: Principles and Practice. Westview Press, Boulder, Colorado.
Otieno, D. and J. Kinyamario. 2018. Grasslands of Eastern Africa: Problems and Prospects. pp. 293–318 (this volume).
Parry, M.L., H. Harasawa and S. Nishiok (eds.). 2000. Assessment Report. IPCC, Cambridge University Press, Cambridge.
Peco, B., I. de Pablos, J. Traba and C. Levassor. 2005. The effect of grazing abandonment on species composition and functional traits: The case of Dehesa grasslands. Basic and Applied Ecology 6(2): 175–183.
Pfeiffer, M., C. Dulamsuren, Y. Jäschke and K. Wesche. 2018. Grasslands of China and Mongolia: Spatial Extent, Land Use and Conservation. pp. 168–196 (this volume).
Poyatas, R., J. Latron and L. Pilar. 2003. Land use and land cover change after agricultural abandonment. Mountain Research and Development 23(4): 362–36.
Reinecke, J.S.F., I.E. Smelansky, E.I. Troeva, I.A. Trofimov and L.S. Trofimova. 2018. Land Use of Natural and Secondary Grasslands in Russia. pp. 113–138 (this volume).
Retallack, G.J. 2001. Cenozoic expansion of grasslands and climatic cooling. J. Geol. 109: 407–426.
Robbins, P.F., N. Abel, H. Jiang, M. Mortimer, M. Mulligan, O.S. Okin, M. Stafford Smith and B.L. II Turner. 2002. Desertification at the community scale: Sustaining dynamic human-environment systems. pp. 325–355. *In*: J.F. Reynolds and M. Stafford Smith (eds.). Global Desertification: Do Humans Cause Deserts? Dahlem Univ. Press, Dahlem.
Rodwell, J.S., V. Morgan, R.G. Jefferson and D. Moss. 2007. The European Context of British Lowland Grasslands. JNCC Report, No. 394.
Sluiter, R. and S.M. de Jong. 2007. Spatial patterns of mediterranean land abandonment and related land cover transitions. Landscape Ecology 22: 559–576.
Squires, V.R., X. Lu, Q. Lu, T. Wang and Y. Yang. 2009. Degradation and Recovery in China's Pastoral Lands. CABI, Wallingford.
Squires, V.R. 2013. Replication and scaling up. pp. 445–459. *In*: G.A. Heshmati and V.R. Squires (eds.). Combating Desertification in Asia, Africa and the Middle East: Proven Practices. Springer, Dordrecht.
Squires, V.R., L. Hua and G. Wang. 2015. Food security: A multi-faceted and multi-dimensional issue in China. Journal of Food, Agriculture & Environment 13(2): 24–31.
Squires, V.R. 2015a. Rangeland Ecology, Management and Conservation Benefits. Nova Science Publishers, N.Y., 201 p.

Squires, V.R. 2015b. Sustainable rangeland management: An ecological and economic imperative. pp. 3–18. *In*: Victor R. Squires (ed.). Rangeland Ecology, Management and Conservation Benefits, Nova Science Publishers, N.Y.

Squires, V.R. 2016. Ecological Restoration: Global Challenges, Social Aspects and Environmental Benefits . Nova Science Publishers, N.Y., 309 p.

Squires, V.R. and L.M. Hua. 2017. Land fragmentation: A scourge in China's pastoral. Livestock. Livestock Research for Rural Development 26/10.

Squires, V.R. 2018. Aridlands of North America—current status and future prospects. pp. 151–179. *In*: M.K. Gaur and V.R. Squires (eds.). Climate Variability Impacts on Land Use and Livelihoods. Springer, N.Y.

Stafford, Smith, D.M., G.M. Mckeon, I.W. Watson, B.K. Henry, G.S. Stone, W.B. Hall and S.M. Howden. 2007. Learning from episodes of degradation and recovery in variable australian rangelands. Proceedings of the National Academy of Sciences of the United States of America 104(52): 20690–20695.

Suding, K.N. 2011. Toward an era of restoration in ecology: successes, failures, and opportunities ahead. Annual Review of Ecology, Evolution, and Systematics 42: 465–487.

Starrs, P.F., L. Huntsinger and S. Spiegal. 2018. North American Grasslands & Biogeographic Regions. pp. 239–272 (this volume).

Starrs, P.F. and L. Huntsinger. 2018. Future Prospects of Grasslands and Ranching on the North American Scene. pp. 369–389 (this volume).

Steinfeld, H., P. Gerber, T. Wassenaar, V. Castel, M. Rosales and C. de Haan. 2006. Livestock's Long Shadow. FAO, Rome.

Stromberg, C.A.E. 2011. Evolution of grasses and grassland ecosystems. Annu. Rev. Earth Planet. Sci. 39: 517–44.

Strong, P.J.H. and V.R. Squires. 2012. Rangeland-based livestock: A vital subsector under threat in Tajikistan. pp. 213–27. *In*: V. Squires (ed.). Rangeland Stewardship in Central Asia: Balancing improved livelihoods, biodiversity conservation and land protection. Springer, Dordrecht.

Symeonakis, E., A. Calvo-Cases and E. Arnau-Rosalen. 2007. Land use change and land degradation in south-eastern Mediterranean Spain. Environmental Management 40(1): 80–94.

Török, P., E. Vida, B. Deák, S. Lengyel and B. Tóthmérész. 2011. Grassland restoration on former croplands in Europe: An assessment of applicability of techniques and costs. Biodivers. Conserv. 20: 2311–2332.

Török, P., M. Janišová, A. Kuzemko, S. Rūsiņa and Z. Dajić Stevanović. 2018. Grasslands, their Threats and Management in Eastern Europe. pp. 64–68 (this volume).

Ushimaru, A., K. Uchida and T. Suka. 2018. Grassland biodiversity in Japan: Threats, Management and Conservation. pp. 197–220, this volume.

Valkó, O., P. Török, G. Matus and B. Tóthmérész. 2012. Is regular mowing the most appropriate and cost-effective management maintaining diversity and biomass of target forbs in mountain hay meadows? Flora 207: 303–309.

Valkó, O., B. Deák, T. Magura, P. Török, A. Kelemen, K. Tóth, R. Horváth, D.D. Nagy, Z. Debnár, G. Zsigrai, I. Kapocsi and B. Tóthmérész. 2016. Supporting biodiversity by prescribed burning in grasslands—A multi-taxa approach. Science of the Total Environment 572: 1377–1384.

Veen, P., R. Jefferson, J. de Smidt and J. van der Straaten (eds.). 2009. Grasslands in Europe of High Nature Value. KNNV Uitgeverij.

Walker, B. and D. Salt. 2012. Resilience Practice: Engaging the Sources of Our Sustainability. Island Press, USA, 240 p.

Walker, B.H. and M.A. Janssen. 2002. Rangelands, pastoralists and governments: Interlinked systems of people and nature. Phil. Trans. Roy. Soc. Lond, Series B 357: 719–725.

Wang, G., L. Hua and V.R. Squires. 2017. Development impacts on beef and mutton production from the pastoral and agro-pastoral systems in China and the economic and cultural factors that influence it. Livestock Research for Rural Development 26/10.

Webb, N.P. and C.J. Stokes. 2012. Climate change scenarios to facilitate stakeholder engagement in agricultural adaptation. Mitigation and Adaptation Strategies for Global Change 17(8): 957–973.

Whisenant, S.G. 2002. Repairing damaged wildlands: A process-oriented landscape scale approach. Biological Conservation, Restoration and Sustainability. Cambridge University Press, N.Y. 3rd printing.

White, R.P., S. Murray and M. Rohweder. 2000. Pilot Analysis of Global Eosystems: Grassland Ecosystems. Washington, D.C.: World Resources Institute, 81 p.

Wilkins, R.J. 2000. Grassland in the Twentieth Century. IGER Innovations. Institute for Grassland and Environmental Research, Aberystwyth, United Kingdom.

Woodley, S. and J. Kay. 1993. Ecological Integrity and the Management of Ecosystems. CRC Press, Boca Raton.

Yannelli, F.A., S. Tabeni, L.E. Mastrantonio and N. Vezzani. 2014. Assessing degradation of abandoned farmlands for conservation of the Monte Desert Biome in Argentina. Environmental Management 53(1): 231–239.

Zomeni, M., J. Tzanopoulos and J.D. Pantis. 2008. Historical analysis of landscape change using remote sensing techniques: An explanatory tool for agricultural transformation in Greek rural areas. Landscape, Urban Plan 86: 38–46.

Index

Editors

Dr. Victor Roy Squires

Dr. Squires is an Australian—is a former Dean of the Faculty of Natural Resources, University of Adelaide and is an internationally recognized dryland management expert. He has a background in teaching and applied researching and holds a PhD degree in Natural Resources Management and Range Ecology, Masters Degree in Ecology and Botany, and Bachelor in Botany and Geography. He is former Dean of the Faculty of Natural Resources, University of Adelaide. As an educator he taught graduate and post graduate students in Australia, and conducted applied research and training programs for institutions and government agencies throughout the world. He has worked in many developing countries, e.g., China, Mongolia, East Africa and Central Asia.

He has authored/edited 15 books and more than 150 papers.

He has been recognized by State Council in China and awarded Gold Medals for International Science and Technology Cooperation (2008) and a Friendship (*YouYi*) award (2011).

Dr. Jürgen Dengler

Professor and head of the Vegetation Ecology Group at the Institute of Natural Resource Sciences (IUNR) of the Zurich University of Applied Science (ZHAW) in Switzerland. Dr. Dengler received his PhD in Ecology from the Kiel University and his Habilitation in Vegetation Ecology and Macroecology from the University of Hamburg (both Germany). His research focuses on vegetation ecology, experimental plant ecology, macroecology, conservation biology and ecoinformatics. He is founder and one of the chairs of the **Eurasian Dry Grassland Group (EDGG)**, an international organization dealing with research and conservation of Palaearctic grasslands, with more than 1,200 members from nearly 70 countries, and member of the Steering Committees of EVA and sPlot, the two largest vegetation-plot databases on Earth. Dr. Dengler is author/editor of more than 240 scientific publications, including about 20 Special Issues/Features in international journals dealing with grassland diversity, management and conservation and 10 books, notably ***European Red List of Habitats—Part 2*** (European Union, Luxembourg, 2016); ***Vegetation Databases for the 21st Century*** (University of Hamburg, 2012); ***Biodiversity in Southern Africa—Volume 1*** (Klaus Hess Publishers, Göttingen, 2010) and ***The Plant Communities of Mecklenburg Vorpommern and their Vulnerability*** (2 volumes, in German, Weissdorn, Jena, 2001 and 2004). Dr. Dengler is serving as Chief Editor of **Phytocoenologia**, the leading journal in vegetation classification.

Dr. Haiying Feng

Dr. Feng is a rural sociologist whose PhD is from China Agricultural University, Beijing, She is currently a Professor, Qinzhou University, Guangxi Autonomous Region, China. People's Republic of China. Her previous affiliation was the Qinghai Institute of Management and Administration, Xining. Her research interests are in the fields of rural development, ethnology, sociology and the influence of Coupled Human Natural Systems (CHANS) on climate change, land use and livelihoods among China's ethnic minority pastoralists in undeveloped areas of western China.

Recent publications are *Humans as agents of change in arid lands with special reference to Qinghai-Tibet Plateau (China)*, and *"Climatic variability and impact on livelihoods in the Cold Arid Qinghai Plateau"*.

In 2015 Dr. Feng received National-level sponsorship and spent one-year as Visiting Scholar at Charles Sturt University in Australia and she has accepted an invitation to be a Visiting Research Fellow in University of Adelaide, Australia in 2018.

Dr. Hua Limin

PhD student supervisor and Deputy Dean, of College of Grassland Science of Gansu Agricultural University (GAU) in Lanzhou, Gansu Province, China. The College of Grassland Science of Gansu Agricultural University has the longest history of grassland science research in China and developed many Outstanding talents, involving Academician Ren Jizhou and Nan Zhibiao. Professor Hua has a PhD in Grassland Ecology from GAU and is devoting himself to combating grassland degradation, in particular on grassland rodent ecological research and pest rodents control. He received several awards, including Provincial Prize for Progress in Science and Technology twice, National Grassland Science Technology Awards. Professor Hua is author/editor of 11 books notably *Grassland Improvement and Livestock Production Technology in Pastoral Land in China* (in Chinese). Beijing, China Agriculture Press, 2015; *Rangeland Ecology, Management and Conservation Benefits*, USA, NOVA, 2015; *Rangeland Stewardship in Central Asia*, Netherlands, Springer, 2013; *Towards sustainable use of rangelands in north west China*, Netherlands, Springer, 2010.